OILWELL DRILLING ENGINEERING

PRINCIPLES AND PRACTICE

The Drill Floor (courtesy of BP)

OILWELL DRILLING ENGINEERING

PRINCIPLES AND PRACTICE

—H. RABIA—

University of Newcastle upon Tyne

Published by

Graham & Trotman

First published in 1985 by

Graham & Trotman Limited
Sterling House
66 Wilton Road
London SW1V 1DE

Graham & Trotman Inc.
13 Park Avenue
Gaithersburg
MD 20877
USA

British Library Cataloguing in Publication Data

Rabia, H.
Oilwell drilling engineering: principles and practice.
1. Oil well drilling
I. Title
622'.3382 TN871.2

ISBN 0-86010-661-6
ISBN 0-86010-714-0 Pbk

Typeset in Great Britain by H. Charlesworth, Huddersfield, UK
Printed in Great Britain by The Alden Press, Oxford, UK

CONTENTS

Preface

The aim of this book is to provide a clear, up to date text covering the principles of oil well drilling engineering. It is the author's contention that it is necessary to start from first principles in order to give both student and practising engineer a comprehensive understanding of the subject. Hence, no prior knowledge of any topic is assumed, and all the necessary equations and calculations required in well design are presented. Full derivation of most equations is included, to enable both the engineer and the scientist to appreciate the limitations of the equations, and to allow the researcher to incorporate further refinements.

A number of friends from the oil industry and various universities have contributed to this book by reviewing and proof-reading the manuscript. I am particularly indebted to the following:

Dr H Kendall, Australian Petroleum Company
Mr E Roberts, National Supply Company (UK) Ltd
Mr C Frederick, Reed Rock Bits Company, Houston, USA
Mr A Aitken, Christensen Diamond Products (UK) Ltd
Mr S Mitchell, Dowell Schlumberger, Aberdeen, UK
Mr R Harmon, Imco Services Division, Aberdeen, UK
Mr M Bacon, International Drilling Services Ltd, Sheffield, UK
Mr P Dymond, BP, Aberdeen, UK
Mr D Smith, Eastman Whipstock (North Sea) Ltd, UK
Dr G Buksh, Shell, UK
Mr M Watson, Hydril, Aberdeen, UK
Professor J F Tunnicliffe, University of Newcastle upon Tyne, UK
Dr N Brook, University of Leeds, UK
Mr N Tomlin, University of Newcastle upon Tyne, UK

I am also grateful to a large number of manufacturing firms for allowing me free use of their literature. Each company is acknowledged against the appropriate illustration. In addition, thanks go to Mrs Edna Gannie and Mrs Beryl Leggart for typing the manuscript and my sincere thanks and appreciation are reserved for my wife for proof-reading the manuscript and for her support during the writing of the book. Finally, the author would welcome any offer of constructive criticism with a view to extending specific sections of the book.

Hussain Rabia
1985

Foreword

The publication of any new book on drilling engineering provides a welcome addition to the inventory of available information on the subject. This book contains a presentation of the fundamental principles of drilling engineering. The theory behind the operation of a rig and the design of an oilwell is well known to those directly involved but has rarely been recorded so comprehensively. This accumulation of basic information will both guide the training of new engineers and offer help to those working in the field.

Drilling engineering today has a truly international flavour. This has not always been the case, as a review of its recent history shows. The discipline of drilling engineering developed gradually over many years from limited origins in a few American universities, the major oil companies and some US independent operators. There are two main reasons for the slow emergence of drilling engineering during the 1950s and 1960s. At that time, oil was being produced from the least troublesome and most easily located fields, reducing the need for the development of more sophisticated drilling techniques. Secondly, well costs were less significant than the cost of the production equipment and refining and distribution networks that the companies were setting up, so little attention was paid to optimising the drilling process.

In the '60s, however, strenuous efforts to optimise the oilwell began to pay dividends in the USA, where competition between the many independent operators resulted in the essential need to drill cost-effectively and competitively. During the '70s, as oil prices rose and major North Sea oil fields were developed the biggest problems faced by European operators were the establishment and integrity of the drilling and production structures themselves. Methods of drilling, well tried and developed overseas, were used by staff who had experienced the immense variety of problems presented by fields in the Middle East and the USA. Many of the attitudes of the expatriate drillers were reflected in the early days of the North Sea. Drilling risks were minimised because the lack of suitable repair equipment would frequently prevent adequate remedial operations being undertaken. With this 'safety first' approach, little incentive continued to be given to engineers to pursue the new ideas that are now changing drilling engineering so rapidly today.

This situation persisted into the late '70s with the North Sea being treated similarly to other inaccessible overseas locations. In the last few years, however, technology has gained a firm foothold. The size and expense of the drilling operation in Europe has lead to the creation of a major support infrastructure network where all of the very latest British, Continental and US-based technologies have become available to the North Sea operators and their staff. At the same time, information technologies have been providing the engineers with well and rig performance data at their fingertips. Time analysis charts that once had to be laboriously plotted and cross-checked between rigs can now be compared on colourful histograms at the press of a button. Nearby, well histories can be called up and instantly reviewed during the preparation of a new drilling programme. Real-time rig data can be displayed in the operator's office where the company's specialist can advise not just on one, but on several rigs' operations. These developments, exciting as they are, bring with them increasing responsibilities to those involved with drilling today. Not only must the drilling engineer be an expert in the traditional aspects of well planning, he must also be familiar with computing, accountancy, risk analysis, metallurgy, chemistry and so on — the list is almost endless.

The multi-disciplinary nature of drilling engineering is well illustrated by the broad spectrum of topics covered in this book. Few texts exist which cover the full range of subjects collectively called 'drilling engineering'. The need has long existed for the various publications produced by the API, equipment manufacutrers and oil company specialists to be gathered together under one cover. Dr Rabia has also explained many of the principles involved with illustrations and worked examples presented from wells in the North Sea and Middle East. His thorough treatment of subjects, often referred to by drillers in words best described as 'jargon' provide a ready reference to those less familiar with the rig and the processes involved in drilling a well.

The importance to the trainee drilling engineer of understanding and being familiar with the contents of this book must not be underestimated. On many occasions the engineer will find himself faced by a problem described in this book. On the majority of those occasions, he will have no one to turn to for advice. The nature of drilling is one of self-sufficiency. Decisions have to be made quickly and effectively. The rig engineer in the North Sea only has back up office staff available for 8 hours/day, 5 days/week. The overseas engineer in a remote location will have a great deal less support, if any at all. His ability to react will depend on the extent of his training and the experience that he has gained in the few years before he has taken on the responsibilities of engineering of his own well.

While it is not practical to refer to a book for each and every problem, a text of this kind can serve three major functions. It can provide the engineer with the basic knowledge needed to enhance his understanding of field operations so that he can gain the greatest possible advantage from the onsite training he should receive early in his career. Secondly, it can fill in gaps that almost inevitably occur even in the most well planned, practical and theoretical training programmes — the variety of problems and situations are so extensive in the drilling business that the engineer is unlikely to face all of them, even within an entire career. Lastly, the book provides under one cover a range of necessary information previously only available from a number of different sources. This is a book that the rigsite engineer should keep available as a ready reference document whenever the need arises.

While there are many areas of new state-of-the-art technology not referred to in this book, the need to strengthen our knowledge of the fundamentals still exists. A statement made in 1960 by J.E. Brantley in the preface to his *Rotary Drilling Handbook* is still true today: "It may be said, safely, that tools and equipment are further advanced than the ability of the industry to use them to their maximum possibilities".

By thoroughly understanding the basics of drilling engineering, the engineer of the future should be able to use both the current and latest of technologies to best effect. Future challenges will continue to press the drilling industry and its engineer to the limit of their abilities. Oil and gas fields exist today which can only be made economic by the determination of those involved to push back the frontiers of technology. Only by responding successfully to these future challenges will stable hydrocarbon production be maintained well into the 21st Century.

I would like to thank the author, Dr Rabia, and those who have contributed to the preparation of this book. In doing so, they have created what now becomes a standard reference book in *Oilwell Drilling Engineering.*

D. C. G. Llewelyn
(Manager, Drilling & Completions Branch,
BP Research Centre,
Sunbury-on-Thames, UK)

Chapter 1

The Oil Well: A Brief Outline

In this chapter the drilling rig and its associated equipment will be briefly introduced, together with the stages required to drill an oil well. This will, it is hoped, set the pace for a more detailed discussion in the following 12 chapters. Experienced engineers can, therefore, skip this chapter. A quick review of the units used in this book will be presented.

The topics which will be discussed include:

Rig components
The oil well
- Drilling
- Making a connection
- Tripping operations
- Casing and cementing
- Logging, testing and completion

System of units

RIG COMPONENTS

Referring to Figure 1.1, the most important rig components include: Rig engines or prime movers; derrick and substructure; hoisting equipment; rotary equipment; mud pumps; and blowout preventers (BOPs).

Rig engines or prime movers (item N, Figure 1.1)

Most modern oil rigs use internal combustion engines as their main power plants. Diesel oil is the principal fuel used, largely because of its availability; however, some rigs still use natural gas. Both number and size of engines required depend on the size and rating of the rig. Rigs suitable for shallow drilling (<5000 ft) use two engines to develop a power of 500–1000 h.p. (373–746 kW)[2]. Deep drilling (deeper than 10 000 ft) is achieved by heavy-duty rigs utilising three to four engines which are capable of developing up to 3000 h.p. (2237 kW).

Transmission of the developed power to various parts of the rig is achieved either mechanically or electrically. In mechanical transmission (Figure 1.1) the power developed by each engine is gathered in a single arrangement, termed the compound (see Fig. 1.2). The compound delivers the engines' power to draw-works and rotary table through roller chains and sprockets. In mechanical transmission, rig pumps are powered by the use of large belts.

In electric transmission, diesel engines are mounted on the ground some distance away from the rig and are used to drive large electric generators. The generators produce electricity that is sent through cables to electric motors attached directly to draw-works, rotary table and mud pump.

The major advantage of the diesel-electric system is that it does away with the compound system, thereby eliminating the need for aligning the compound engine and draw-works. Also, in the diesel-electric system, the engine noise is kept away from the drilling crew.

Derrick and substructure (items 1, 2, 3, Figure 1.1)

A derrick is a four-sided structure of sufficient height and strength to allow the hoisting (lowering and raising) of equipment in and out of the well. It also provides a working place (a platform, item 2) for the derrick man during tripping operations.

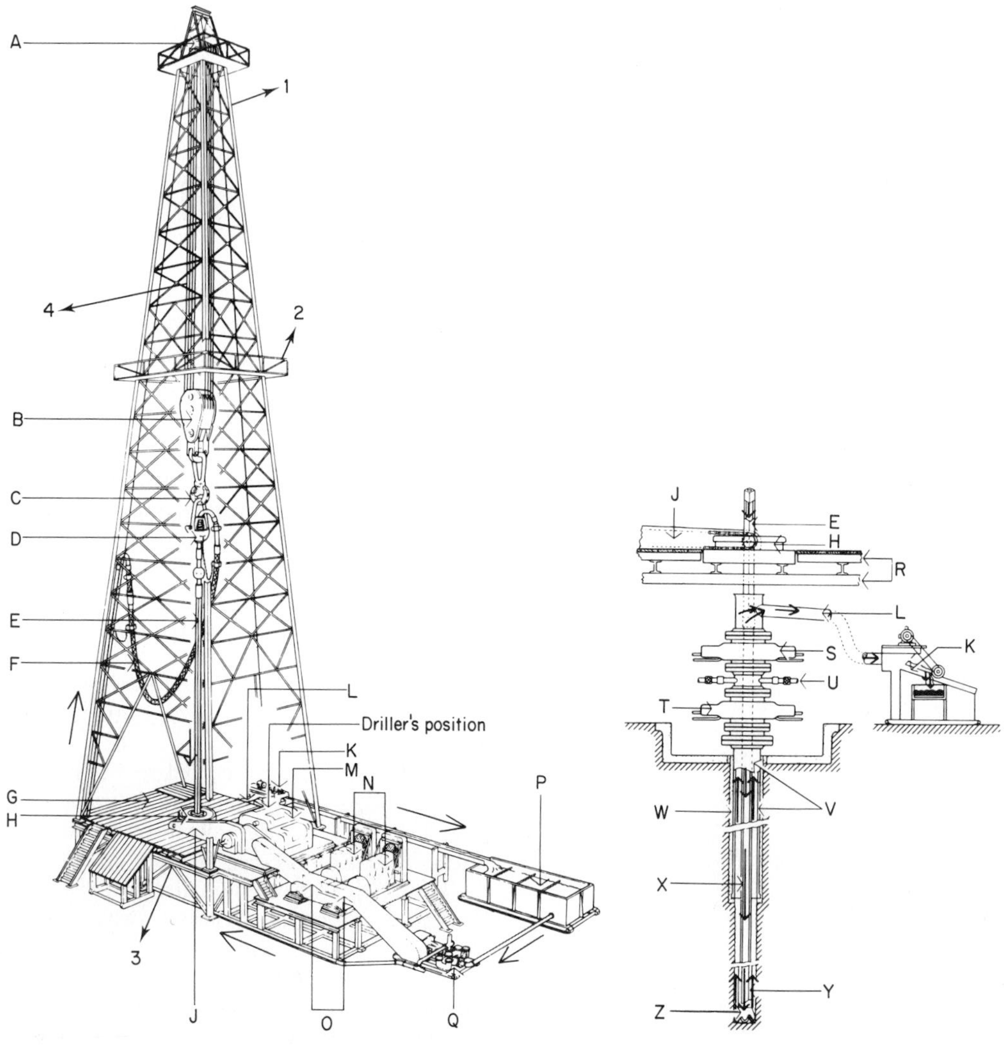

(a) (b)

Fig. 1.1. (a) Diagrammatic view of rotary drilling rig: 1, derrick; 2, platform; 3, substructure; 4, drilling line; A, five-sheave crown block; B, four-sheave travelling block; C, hook; D, swivel; E, kelly; F, standpipe and kelly hose; G, derrick floor; H, rotary table; J, rotary table drive from main transmission; K, shale shaker; L, drilling fluid return line; M, draw-works; N, main engines; O, main transmission (engines to draw-works and to pump); P, suction pit; Q, pump. (b) Diagrammatic section showing equipment below the derrick floor, the borehole and drill string: E, kelly; H, rotary table; J, rotary table drive; K, shale shaker; L, drilling fluid return line (flow line); R, cut-out section of drilling floor; S, T, hydraulically-operated blowout preventers; U, outlets provided with valves and chokes for drilling fluids when upper blowout preventer is closed; V, surface casing; W, cement bond between casing and borehole wall; X, drillpipe; Y, heavy, thick-walled pipe (drill collars) at bottom of drill string; Z, bit. Heavy arrows indicate flow of drilling fluid. (Courtesy of IMCO Services Division)

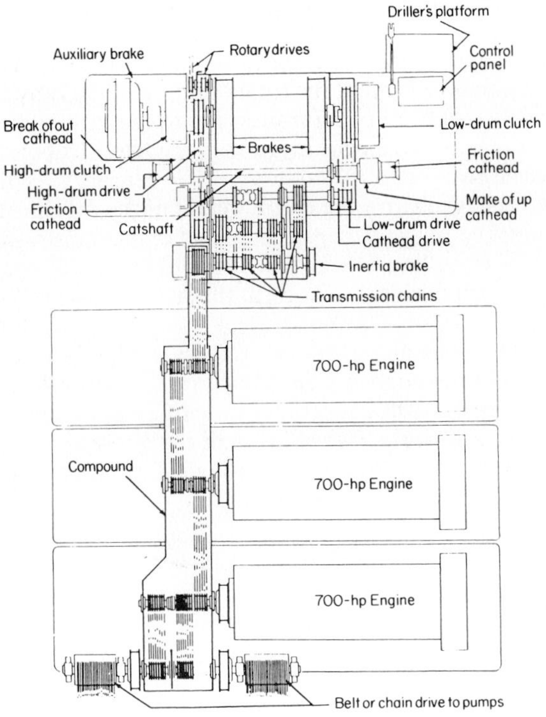

Fig. 1.2. Multi-engine and chain-drive transmission for a mechanical rig[2]. (Reproduced by kind permission of University of Texas, Petroleum Extension Service (PETEX) from *A Primer of Oilwell Drilling*)

The substructure provides support for the derrick, draw-works and drill string.

Hoisting equipment

Hoisting equipment includes: (a) draw-works; (b) hoisting tackle; and (c) drilling line. (See Chapters 2 and 3 for details.)

DRAW-WORKS

The draw-works is the hoisting mechanism on the drilling rig enabling heavy loads to be raised or lowered by means of wire rope wound on a drum. The draw-works also enable the driller, through the cat-heads, to make or break drillpipe, drill collars and other connections.

Hoisting tackle (items A and B, Figure 1.1)

There are two blocks, namely a crown block and a travelling block. The crown block is static and rests at the top of the derrick; the travelling block moves up and down the derrick during making/breaking of drill-pipe joints. Each block has a number of pulleys. The drilling line is wound a number of times on each block, with the end coming out of the crown block clamped to a deadline anchor underneath the derrick substructure. The other end is wound on the drum of the draw-works. The reason for having several lengths of drilling line is to enable the lines to carry heavier loads which a single line cannot do. After a number of trips the drilling line is unclamped and approximately 30 ft of new drilling line is fed out. This process ensures that the same portion of the drilling line does not remain at points of high stress. (See Chapter 3 for cut-off practices.)

Drilling line (item 4, Figure 1.1)

The drilling line is a heavy-duty wire rope used to raise or lower the thousands of pounds of drilling equipment during the course of drilling and completion of the well.

Rotary equipment (items H, E, F, D, X and Y, Figure 1.1)

Rotary drilling equipment includes: rotary table and master bushing; kelly and kelly bushing; swivel; and drill stem.

Rotary table and master bushing

The main function of the rotary table is to transfer rotary motion through a master bushing to the kelly, to drillpipe and, eventually, to the drill bit. Rotation of a drill bit is necessary for rock breakage and, in turn, for 'making hole'.

Figure 1.3 shows a rotary table and a master bushing. Besides transferring rotation to the kelly, the master bushing also acts as a seat for slips. Slips (Figure 1.4) are wedge-shaped devices, lined with tooth-like gripping elements. They are needed to hold the drill string suspended in the hole when adding or breaking joints of drillpipe or drill collars.

Power required for the rotation of the table is transmitted from the main rig engines through a chain drive from the compound, as in mechanical rigs (Figure 1.1), or the rotary table is independently powered through a cable connected to a motor attached to the rotary table (Figure 1.5).

Kelly and kelly bushing

The kelly has a hexagonal or a square shape, and its main function is to transfer motion to the drillpipe when the kelly bushing is engaged with the master bushing (see Chapter 2). The kelly serves also as a medium for transporting mud down to the drill pipes and to the bit. During tripping in or out, the kelly complex rests in a side hole, called a rathole, drilled especially for this purpose.

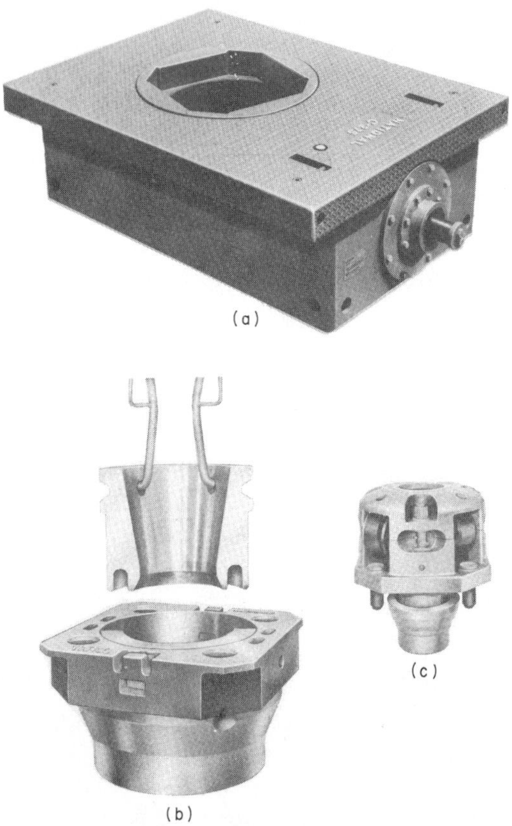

Fig. 1.3. Rotary table (a), master bushing (b) and kelly drive bushing (c).

Swivel

The swivel is installed above the kelly, and its main function is to prevent the rotary motion of the kelly (or drill string) from being transferred to the drilling line. This is achieved by the rotation of the lower half of the swivel on a set of heavy-duty roller bearings. As the swivel has to carry the entire weight of the drill string, it must be ruggedly constructed and of the same rating as the travelling block. The swivel (Figure 1.6) has a bail which fits inside a hook at the end of the travelling block.

The swivel also allows mud to be pumped through the kelly through a side attachment, described as a 'gooseneck'. A flexible rotary hose (item F, Figure 1.1) connects with the swivel through the gooseneck. The rotary hose is connected through a standpipe and surface lines to the mud pumps.

Drill stem

The drill stem consists of: drillpipe; drill collars; accessories; and drill bit.

(a) The drillpipe serves as a medium for the transmission of rotary motion to the bit and also acts as a passage for mud.

(b) Drill collars are heavy-duty pipes with large outside diameters. They are used primarily to put weight on the bit during drilling operations. Experience has shown that a maximum of 85% of the total weight of drill collars should be applied to the drill bit. The remainder is used to keep the drill pipes in tension, thereby avoiding buckling of the pipes.

(c) Accessories normally include heavy-walled drill pipes, stabilisers and shock subs. The heavy-walled drillpipe (HWDP) is used to ensure that the drillpipe is always kept in tension (see Chapter 2 for more details).

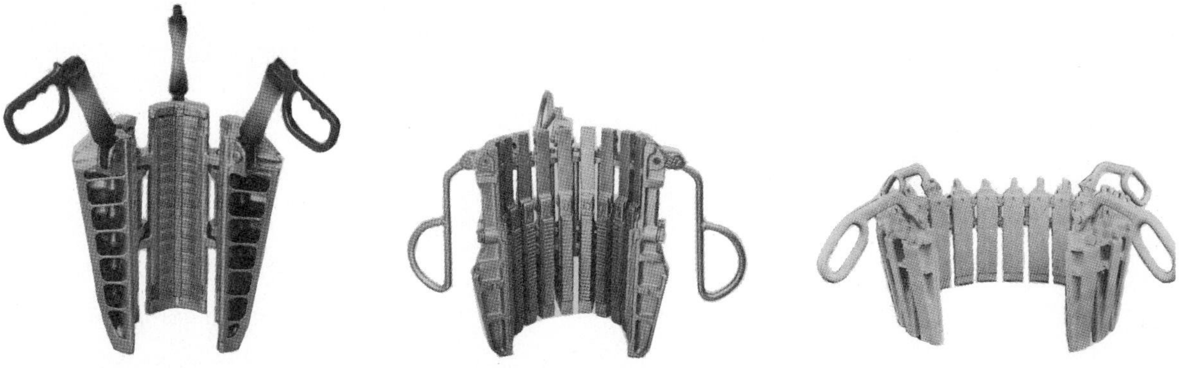

(a) (b) (c)

Fig. 1.4. Drillpipe slips (a), drill collar slip (b) and casing slips (c). (Courtesy of Joy Manufacturing Company)

(a)

(b)

Fig. 1.5. Independently driven rotary table[1]: (a) two-speed transmission, side-mounted with support feet; (b) complete, skid-mounted independent rotary drive unit. (Courtesy of National Supply Company)

A stabiliser is a special tool with an outside diameter (OD) close to the hole diameter. The main functions of a stabiliser are to prevent buckling/bending of drill collars and to control drill string direction. Stabilisers are run between drill collars and drill bit. A shock sub is included in the bottom hole assembly to absorb shocks when the bit bounces off hard formations, thereby protecting the drill string and surface equipment from damaging effects of bit vibrations.

(d) The drill bit is the main component of the drill string and is used to cut the rock for the purpose of making hole. Drill bits can have one cutting head, as in diamond and polycrystalline diamond compact bits, or can have two or three cutting heads, commonly known as cones, as in bicone or tricone bits. The latter is the most commonly used in the oil industry and will be fully discussed in Chapter 4.

Mud pumps

The oil well rig pump is basically a piston reciprocating inside a cylinder such that the reciprocation action creates pressure and volume. Mud pumps are used to circulate huge quantities of drilling mud (300–700 gpm) down many thousands of feet of drillpipe, through small nozzles at the drill bit and, finally, up the hole to arrive at the surface again. The pump must, therefore, produce pressure to overcome the frictional or drag forces in order to move the drilling mud.

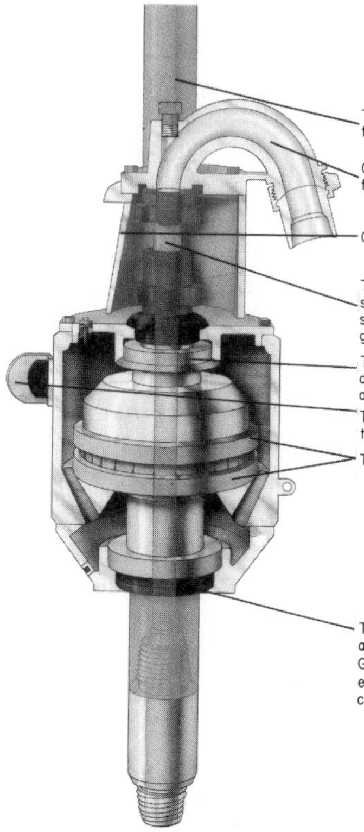

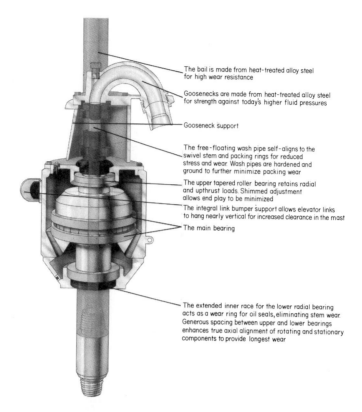

The bail is made from heat-treated alloy steel for high wear resistance
Goosenecks are made from heat-treated alloy steel for strength against today's higher fluid pressures
Gooseneck support
The free-floating wash pipe self-aligns to the swivel stem and packing rings for reduced stress and wear. Wash pipes are hardened and ground to further minimize packing wear
The upper tapered roller bearing retains radial and upthrust loads. Shimmed adjustment allows end play to be minimized
The integral link bumper support allows elevator links to hang nearly vertical for increased clearance in the mast
The main bearing
The extended inner race for the lower radial bearing acts as a wear ring for oil seals, eliminating stem wear. Generous spacing between upper and lower bearings enhances true axial alignment of rotating and stationary components to provide longest wear

Fig. 1.6. Swivel[1]. (Courtesy of National Supply Company)

Mud pumps are available in two types:

(1) Duplex pumps using two double-acting pistons. In this type the piston creates pressure on both the forward and backward strokes.
(2) Triplex pumps using three single-acting pistons. In this type the piston produces pressure on the forward stroke only.

Variations in volume and pressure can be obtained by changing the inside diameter of the cylinder through the use of liners or changing piston sizes.

Blowout preventers (BOPs)

A well kick is an unwanted flow of formation fluids into the well bore hole which may (if not controlled) develop into a blowout.

Basically, blowout preventers (see Figure 1.1b) are valves that can be closed any time a well kick is detected. Blowout preventers are of three types:

(1) Annular preventers are manufactured to close around any size or shape of pipe run in the hole and are normally closed when the well is threatened by a kick.
(2) Pipe rams. Two types of pipe ram are available: standard and variable. The standard pipe ram is sized for one particular dimension of drill pipe likely to be used during drilling operations. The variable pipe ram has the ability to pack off a range of pipe sizes.
(3) Blind and shear rams. Blind rams are designed to close the bore when no drill string or casing is present. A shear ram is a type of blind ram which has the ability to cut pipe and pack off an open hole.

A full discussion of BOPs and kick detection is given in Chapter 13.

THE OIL WELL

Drilling

Once it has been established that a potential oil-bearing structure exists, the only way of confirming the existence of oil is to drill a well. In practice, the frequency of striking oil in an unexplored area is one in nine[2].

In areas where there is a great deal of vegetation

and soft ground, a stove pipe (30–42 in OD) is driven by a pile driver to a depth of approximately 100 ft. This is necessary to protect surface formations from being eroded by the drilling mud, causing a large washout with eventual loss of the rig.

The oil well proper starts off with a hole size ranging from $17\frac{1}{2}$ to 36 in and is drilled to a depth of 200–300 ft. The bottom hole assembly required to drill a large hole to a shallow depth normally consists of drill collars and one stabiliser. For deep holes a more rigid bottom hole assembly using three stabilisers is required to keep the hole straight or to maintain existing hole deviation. The basic bottom hole assembly consists of bit, near bit stabiliser, two drill collars, stabiliser, two or three drill collars, stabiliser, drill collars, HWDP and drillpipe to surface.

The first string of casing ($13\frac{3}{8}$–30 in OD) is described as conductor pipe, and is run mainly to provide a conduit for the drilling mud.

Once the conductor pipe is cemented, a smaller bit, with a different bottom hole assembly, is run inside the conductor pipe and another hole (described as surface hole) is drilled to a prescribed depth. The depth is dictated by hole conditions and formation pressures (see Chapters 9, 10 and 12). Another casing string (described as surface casing) is run and cemented.

The process of drilling a hole and running casing continues until the oil or gas zone is reached. The last casing string is described as a production string (see Chapter 10). Typical hole/casing sizes for a development area (i.e. where oil is known to exist from previous exploration drilling) are given in Table 1.1.

It should be noted that combinations of hole/casing sizes other than those listed in Table 1.1 are also in use. However, the arrangement predominates in the Middle East, the North Sea and Brunei.

A full discussion of bottom hole assemblies will be given in Chapter 8.

Making a connection

Prior to the drilling of a hole, the bottom hole assembly (see Chapter 8) is made up on the derrick floor. The bit sub is first made up on the bit, and then stabilisers and drill collars are connected. The bottom hole assembly is then run in hole and held in the rotary table at the last joint.

Initially, the drillpipe joints are stacked singly on a pipe rack adjacent to the rig floor. A winch line, operated by a portable air hoist on the rig floor, is used to pick up the individual drillpipe joints. Each joint of drillpipe is first placed in a mousehole prior to being run in the well. The kelly and its drive bushing are kept in a special hole described as a 'rathole' drilled adjacent to the mousehole. The derrick substructure is equipped with openings for the rathole and mousehole. Both holes are lined with pipe.

The kelly and its drive bushing are lifted from the rathole and connected to the drillpipe joint in the mousehole. The whole assembly is then lifted and swung to the rotary table to be connected to the top of the bottom hole assembly.

The drillpipe is joined to the top joint of drill collars (or heavy-walled drillpipe) using a pipe spinner (Figure 1.7) and a special pipe wrench described as tongs (Figure 1.8). The pipe spinner is used for hand-tightening, while the tongs are used for final tightening. The drill string assembly is then lowered into the hole and the rotary table is activated to impart rotation to the drill string. The kelly is slowly lowered until the drill bit just touches the bottom of the hole. This will be observed at the surface as a decrease in the hanging weight of the drill string (or, as it is known, 'weight on bit'). This weight is read via a weight indicator on the driller's console connected by a hydraulic hose to a tension sensor attached to the deadline anchor (see Chapter 3 for details).

TABLE 1.1 Typical hole size/casing size arrangements

Hole size (*in*)	*Hole size* (*mm*)	*Casing size* (*in*)	*Casing size* (*mm*)	*Description*
36	914.4	30	762	Stove pipe
24/26	609.6/660.4	$18\frac{5}{8}$/20	473.1/508	Conductor pipe
17.5	444.5	$13\frac{3}{8}$	339.7	Surface casing
12.25	311.2	$9\frac{5}{8}$	244.5	Intermediate casing
8.5	215.9	7	177.8	Production casing
6	152.4	$4\frac{1}{2}$/5	114.3/127	Full production casing or a production liner

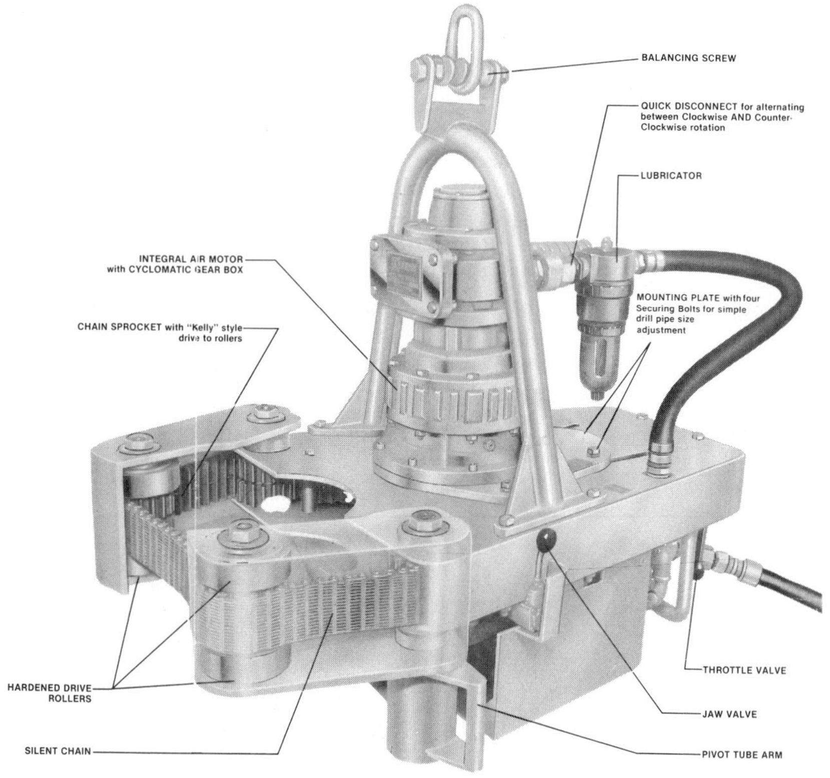

Fig. 1.7. Pipe spinner. (Courtesy of Klampon Corporation)

The driller adjusts the weight on bit according to drilling programme requirements as prepared by the Engineering Department. Each rock type will require a different combination of weight on bit and rotation to produce maximum penetration (see Chapter 4 for details). Hole is, therefore, made by the combined effect of compression (weight on bit) and rotation and by flushing the resulting cuttings.

Fig. 1.8. The tong. (Courtesy of Joy Manufacturing Company)

Most kellies are 40 ft long, which allows 40 ft of hole to be drilled when the top of the kelly reaches the rotary table. Further hole can be made by adding extra joints of drillpipe. Normally, one joint of drillpipe is added at a time by first lifting the entire kelly length above the rotating table. Slips are then placed around the top joint of drillpipe to hold it in the rotary table. The kelly is then broken, swung to the mousehole and stabbed into the box of an awaiting drillpipe single. A kelly spinner situated at the top of the kelly spins up the kelly to provide hand-tightening of the joint, and a power tong is then used to provide final tightening.

The kelly is then raised (by the draw-works) and the new joint is made up on the top of the drill string that is held in the rotary table. The new length of drill string is run in hole to start drilling and making hole. Figure 1.9 is a schematic drawing of the process of making a connection. This process of adding a new joint of drillpipe is repeated until the bit is worn out or when hole total depth (TD) is reached. Then the entire drill string is removed (or 'tripped') from the hole.

Tripping operations

The term 'tripping operations' refers to the process of lowering or raising the drill string into or out of the hole. The drill string is frequently pulled out of hole (POH) in order to change the drill bit or when the hole is drilled to its final TD before the casing is run. Tripping in or running in hole (RIH) is the process of lowering the entire drill string after a bit change or for reaming and circulation purposes.

Figure 1.10 is a schematic drawing of the sequences used in tripping out. The process starts by first raising the kelly above the rotary table, setting slips and then breaking the kelly, kelly bushing and swivel from the topmost joint and stacking them in the rathole, as shown in Figure 1.10.

The drillpipe is then removed by attaching pipe elevators to the drillpipe and using the drum hoist to raise the drillpipe above the rig floor. Pipe elevators for drillpipe, casing and tubing are shown in Figure 1.11. Basically, an elevator is a set of clamps that latches onto the pipe, which allows the drill string to be lifted out of the hole.

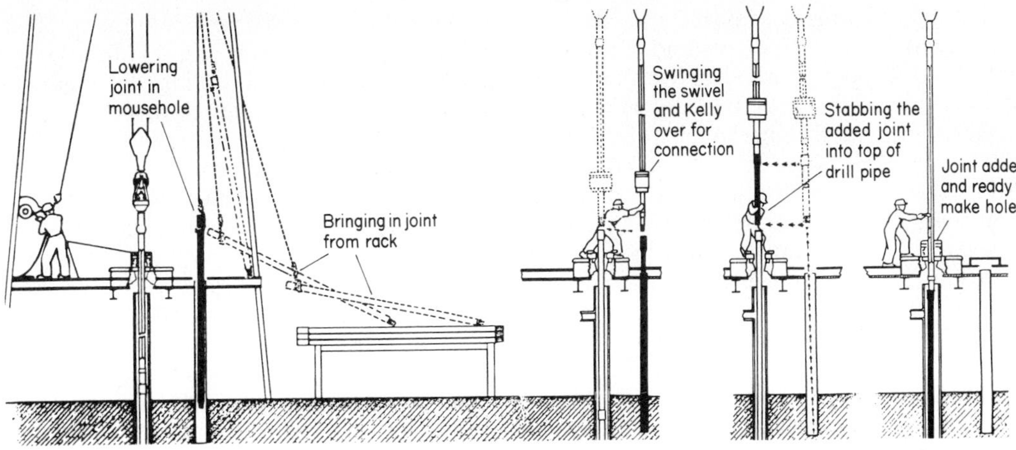

Fig. 1.9. The process of making a connection[2].

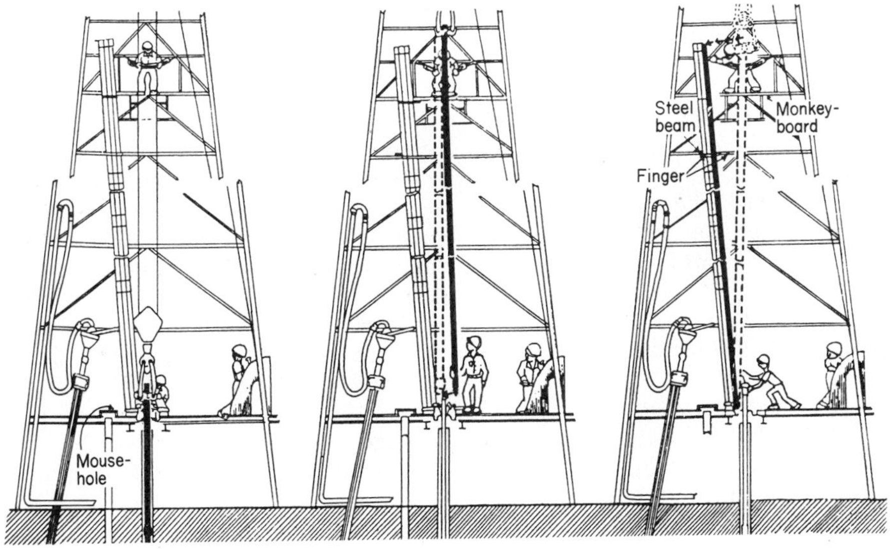

Fig. 1.10. Tripping-out operation[2]. (Reproduced by kind permission of University of Texas, Petroleum Extension Service (PETEX) from *Primer of Oilwell Drilling*)

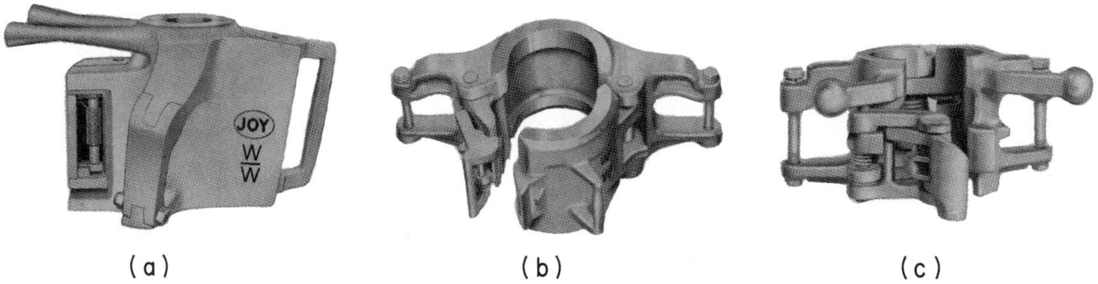

Fig. 1.11. Pipe elevators: (a) drillpipe elevators; (b) casing elevators; (c) tubing elevators. (Courtesy of Joy Manufacturing Company)

The drill string is normally removed in a stack of three joints, described as a 'stand'. A stand of drillpipe (approximately 93 ft long) is raised above the rotary table and then disconnected, or broken, with tongs and spun out of hole with the pipe spinner or by back-rotating the rotary table. The top of the stand is then picked up by a derrickman working on the platform (or monkey-board), from where he unlatches the pipe from the elevator. The top of the stand is then moved into a specially designed finger-board within the platform, just after the workmen (known as 'roughnecks') on the rig floor swing the lower part of the stand aside on the rig floor immediately below the monkey-board.

The empty elevators are then lowered and latched onto the top of the remaining drill string, the slips are removed from the rotary table and another stand is removed from the hole. This process is continued until the entire length of drill string is removed from the hole and stacked in the derrick.

Tripping-in operations use the same process as tripping out, but in reverse, i.e. pipe is picked up from the stack with the elevator. The bottom hole assembly, including drill bit and drill collars, is, of course, run first inside the hole.

It should be observed that, when the well is finally drilled, tested and completed, the drillpipe stands are broken into individual joints inside the mousehole prior to rig move.

Casing and cementing

Casing is run and cemented after each section of hole is drilled. Referring to Table 1.1, the conductor pipe is run and cemented to surface. Normally, this pipe does not carry any wellhead equipment (i.e. BOPs) and the drilling flowline is nippled directly to this string. (An exception to this is in offshore oil operations or areas where shallow gas may be encountered.)

Drilling of surface hole (e.g. $17\frac{1}{2}$ in) is carried out through the conductor pipe until the complete section is drilled. Surface casing (e.g. $13\frac{3}{8}$ in OD) is then run and cemented. In land drilling, the conductor pipe is cut so that it is level with cellar floor.

The $13\frac{3}{8}$ in casing will carry the blowout preventers which are necessary to provide control while the next hole is being drilled. The $13\frac{3}{8}$ in casing is provided with a connection described as casing head housing (CHH).

Cementing of the $13\frac{3}{8}$ in casing is facilitated by using a landing joint to which the cementing equipment will be attached. Later the landing joint is removed and the CHH is screwed onto the topmost joint of the casing. In some cases a landing joint cannot be used, owing to difficulty of spacing out, and the cementing equipment will have to be attached to the casing itself. In this case, after cementing, the casing is cut off and dressed, and the CHH is welded onto the top of the casing. Figure 1.12 shows the screw-type and slip and weld-on CHH. The BOPs equipment can now be attached to the top of the CHH to allow safe drilling of the next hole. The casing head housing is also used to suspend the next string of casing. The wellhead arrangement after cementing the surface casing is shown in Figure 1.13.

The intermediate hole is drilled inside the surface casing. The intermediate casing (e.g. $9\frac{5}{8}$ in OD) is run and cemented in the same manner as the surface casing. This casing is then landed by hanging it inside the CHH with a special casing hanger, as shown in Figure 1.14. The casing hanger employs gripping teeth which hold the casing tightly in place; the casing hanger is also provided with teeth on the outside, to engage with the CHH.

The top of the intermediate casing is then cut and dressed, and another spool, described as an 'intermediate casing spool', is attached to the CHH. The intermediate spool provides the following functions: (a) to pack off the top of the intermediate casing, thereby restricting well-bore fluid pressure to the

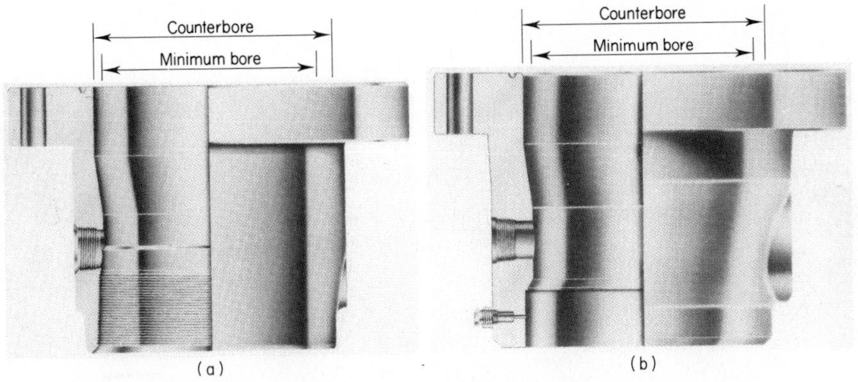

Fig. 1.12. Casing head housing: (a) with female casing threads; (b) with slip-on weld casing connection. (Courtesy of Cameron Iron Works)

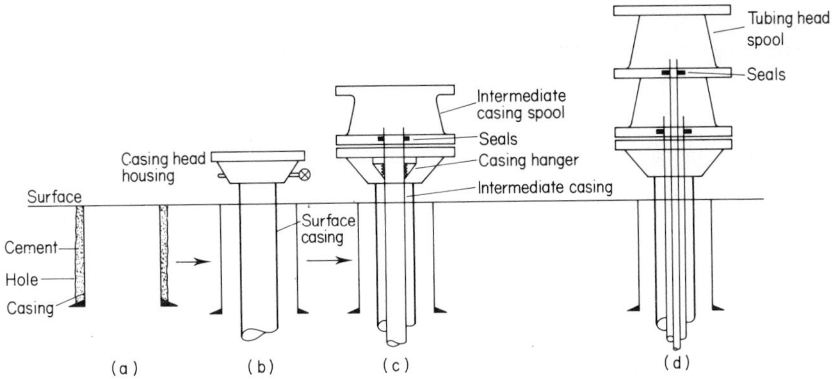

Fig. 1.13. Wellhead arrangement after setting: (a) conductor pipe; (b) surface casing; (c) intermediate casing; (d) production casing.

inside of the casing; (b) to carry the next stack of BOPs required to control hole during the drilling of the next hole; (c) to provide suspension for the next casing string.

The CHH and intermediate spool(s) have recesses cut inside their bodies to provide seating for the casing hangers.

The last hole is drilled through the intermediate casing, and the producing string is run and cemented as before. The production string is hung inside the intermediate spool with a casing hanger. The production casing ($4\frac{1}{2}$, 5 or 7 in OD) is cut and dressed, and a final spool, described as 'tubing head spool', is screwed on. The tubing head spool provides a pack-off to the top of the production string and also provides a seating for the tubing hanger which will hold the production tubing. Oil or gas is normally produced through tubing (rather than through casing) which is sealed in a production packer placed just above the producing zone(s). A section through a complete wellhead, together with the Christmas tree, is shown in Figure 1.15.

Logging, testing and completion

After the well is drilled to TD, the hole is normally logged (in open or cased hole) by running specialised equipment on wireline. The primary object of open hole logging is to determine porosity, water saturation and boundaries of producing zone(s). These parameters are necessary in the determination of the quantity of movable oil and, in turn, the producing life of the reservoir. Detailed discussion of well logging can be found in Reference 3.

In most development and exploration wells, routine testing of producing or potentially producing (in exploration) wells is carried out to determine the reservoir pressure, type and quality of hydrocarbon

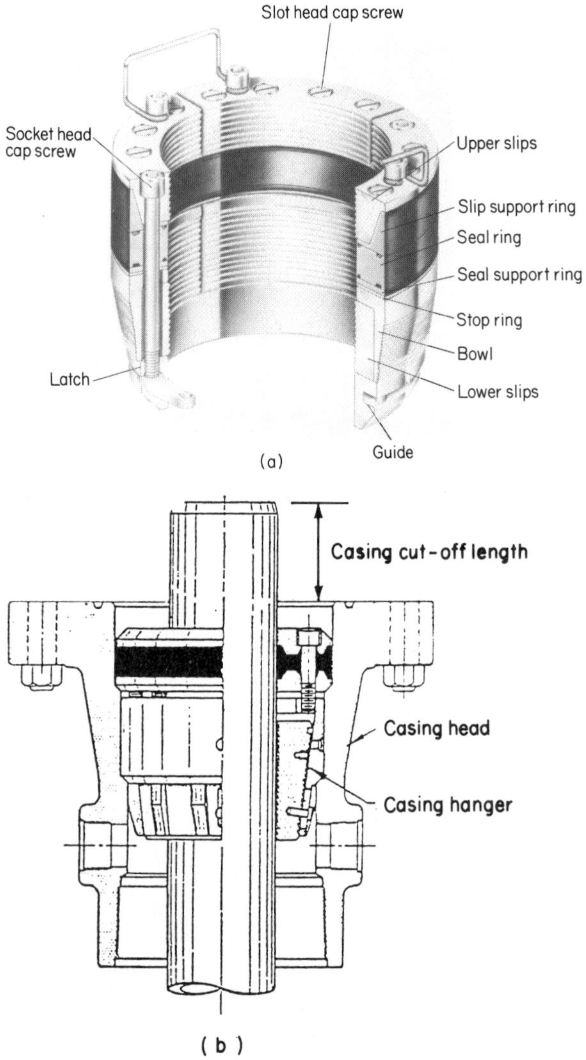

Fig. 1.14. Casing hanger (a) in set position (b).

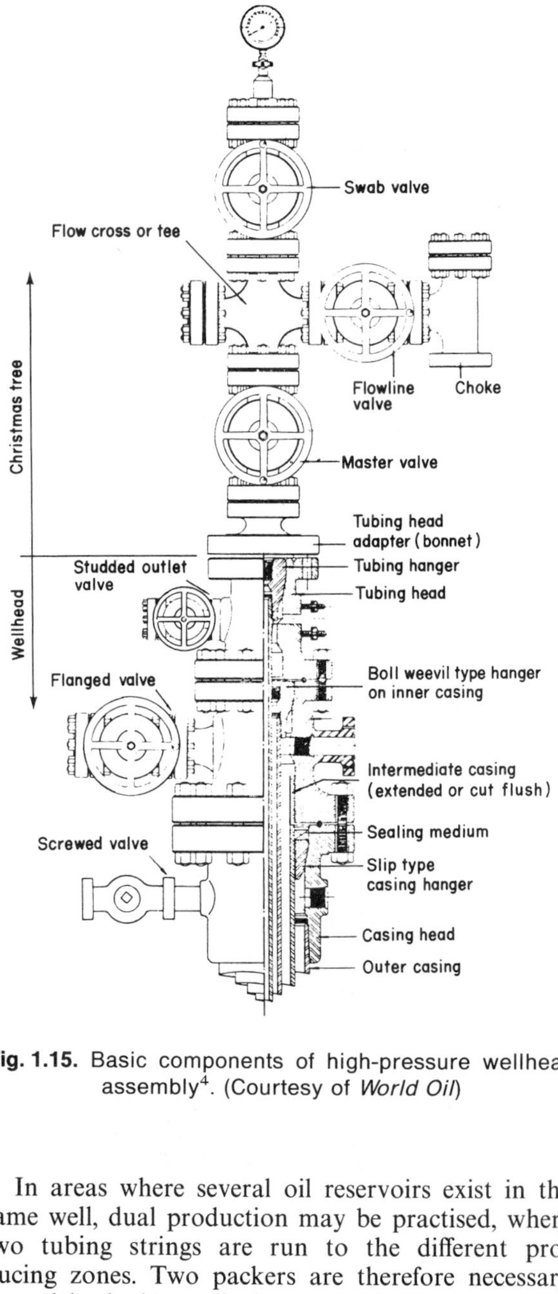

Fig. 1.15. Basic components of high-pressure wellhead assembly[4]. (Courtesy of *World Oil*)

fluids. Production testing is carried out to determine the productivity index of the well in terms of barrels of oil or million cubic feet of gas per day. Production and drill stem testing are carried out to monitor well deliverability characteristics, fluid types, and, through pressure response, several reservoir parameters.

Completion of an oil well involves setting a production packer, running the production tubing(s) and, finally, perforating the producing zone(s). A production packer is set, just above the producing zone, to seal the annulus from the reservoir pressure and confine fluids to the producing tubing. The tubing is screwed at the surface to a tubing hanger (Figure 1.15); the latter is landed in the tubing head spool.

In areas where several oil reservoirs exist in the same well, dual production may be practised, where two tubing strings are run to the different producing zones. Two packers are therefore necessary to seal both the producing zones from the annulus. Figure 1.16 is a schematic drawing of the various equipment required for dual production.

Finally, a Christmas tree is attached to the top flange of the tubing head spool. The Christmas tree is a solid piece of steel with a hollow passage connected to the top of the tubing. It has a number of valves to control the flow of hydrocarbon from the well. Figure 1.15 is a schematic drawing of a completed well, together with the Christmas tree.

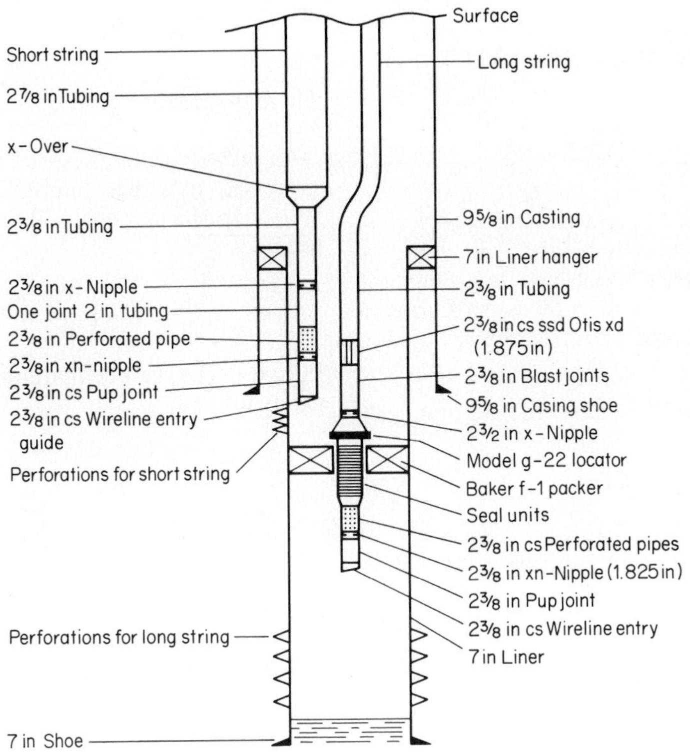

Fig. 1.16. Dual completion diagram: short string = $2\frac{3}{8}$ in and $2\frac{7}{8}$ in tubing; long string = $2\frac{3}{8}$ in tubing.

SYSTEM OF UNITS

Since the 1950s a great deal of research has been directed towards each aspect of the oil industry, including drilling, production, reservoir engineering, etc. The new accumulation of knowledge has led to an improvement in practices and safer operations.

Unfortunately, the use of new technology has not influenced the system of units used in engineering calculations. The use of the old Imperial system of units has continued to the present day, rather than the more elegant SI system. Some countries such as Canada have recently adopted a metric system, while some European countries such as France and West Germany have been using the SI system for some time. However, the major producing countries of the world, such as the USA, Middle East, UK, etc., are still using the Imperial system. Hence, this book will use the Imperial system as the main units, with translations into the SI system where needed. Eventually, the author hopes to present all calculations in the SI system, when the latter has been adopted by the major producing countries.

The following section is a brief introduction to the derivation of units in both systems. Further details can be found in most basic engineering textbooks. Also, throughout the book, derivation of units will be presented where necessary.

Imperial units

Fundamental units

The units of mass, length and time are described as fundamental units and form the basis for deriving the units of all other quantities.

Units of *mass* are expressed in pounds, abbreviated as ‘lbm’.

Units of *length* are expressed in feet, abbreviated as ‘ft’.

Units of *time* are expressed in seconds, abbreviated as ‘s’.

Derived units

Force Force is defined as the product of mass × acceleration. A 1 lb mass will exert a force, F, given by

$$F = 1 \times g$$

where

$$g = \text{acceleration due to gravity} = 32.17 \text{ ft/s}^2$$

Hence,

$$F = 32.17 \frac{\text{lbm ft}}{\text{s}^2}$$

In practice, the units of force in the Imperial system are expressed as pounds force and the value of acceleration due to gravity is not included. Thus, the force exerted by a mass of 1 lb in the earth's gravitational field is 1 pound force, normally abbreviated as 'lb'.

Pressure Pressure (P) is defined as force per unit area, i.e.

$$P = \frac{F}{A}$$

Units of force are lb and units of area are ft^2; thus, pressure is given by

$$P = \frac{F}{A} \frac{\text{lb}}{\text{ft}^2} \tag{1.1}$$

In drilling engineering, the use of the units lb/ft^2 produces large numbers and, instead, a different unit is adopted. The unit of pressure is lb/in^2 (abbreviated to psi). Also, the unit of area is measured in inches^2 rather than in the fundamental units of ft^2. Thus, even in the Imperial system, there are two types of unit — consistent and field.

Consistent units are defined as units containing the fundamental dimensions of mass (lbm), length (ft) and time (s). Thus, the consistent unit of pressure is lb/ft^2.

Field units are those which use a modification of the fundamental dimensions. Drill pipe diameter, for example, is measured in inches instead of feet. The units inches are described as 'field units'. Similarly, 'psi' (=lb/ft^2) is the field unit of pressure.

In practice, the pressure equation is conveniently expressed in terms of density, acceleration due to gravity and height (or hole depth). This expression can be easily obtained by simplifying the force term in Equation (1.1) as follows:

$$\begin{aligned} \text{force} &= \text{mass} \times \text{acceleration due to gravity } (g) \\ &= (\text{density} \times \text{volume}) \times g \\ &= (\text{density} \times \text{area} \times \text{height}) \times g \end{aligned} \tag{1.2}$$

The expression for pressure (P) is obtained by combining Equations (1.1) and (1.2), viz:

$$P = \frac{\text{density} \times \text{area} \times \text{height} \times g}{\text{area}}$$

or

$$P = \rho g h \tag{1.3}$$

where ρ = density; g = acceleration due to gravity; and h = height.

In oil well drilling, density in the Imperial System is expressed in either lbm/ft^3 (pcf) or lbm/gal (ppg). When density is expressed in lbm/ft^3, Equation (1.3) becomes

$$P = \rho \left(\frac{\text{lbm}}{\text{ft}^3}\right) \times g \times h \text{ (ft)} \tag{1.4}$$

Equation (1.4) can be manipulated to give pressure in psi as follows:

$$\begin{aligned} P &= \rho \left(\frac{\text{lbm} \times g}{\text{ft}^3}\right) \times h \text{ (ft)} \times \left(\frac{\text{ft}^2}{144 \text{ in}^2}\right) \\ &= \frac{\rho h}{144} \frac{\text{lb}}{\text{in}^2} \text{ or psi} \end{aligned} \tag{1.5}$$

where lbm $\times g$ = lb.

When density is expressed in lbm/gal, Equation (1.3) can be modified to give pressure in psi as follows:

$$\begin{aligned} P &= \rho \left(\frac{\text{lbm}}{\text{gal}} \times \frac{7.48 \text{ gal}}{\text{ft}^3}\right) \times g \times h \text{ (ft)} \\ &= \rho h \left(\frac{\text{lbm} \times g}{\text{ft}^2}\right) \times 7.48 \\ &= \rho h \left(\frac{\text{lb}}{\text{ft}^2}\right) \times \left(\frac{\text{ft}^2}{144 \text{ in}^2}\right) \times 7.48 \\ &= 0.052 \rho h \text{ psi} \end{aligned} \tag{1.6}$$

In this book, both units of density (pcf and ppg) will be used.

Power Power is defined as the product of force (F) $\times$ velocity (V).

Consistent units

$$\begin{aligned} \text{Power} &= F \text{ (lb)} \times V \text{ (ft/s)} \\ &= FV \frac{\text{lb ft}}{\text{s}} \end{aligned}$$

Field units

$$\begin{aligned} \text{Power} &= F \text{ (lb)} \times V \left(\frac{\text{ft}}{\text{min}}\right) \\ &= FV \frac{\text{lb ft}}{\text{min}} \end{aligned}$$

Normally, power is expressed in units of horsepower, where 1 horsepower = 33 000 lb ft/s. Thus,

$$\text{power} = FV\frac{\text{lb ft}}{\text{min}}\left(\frac{\text{min}}{60\text{ s}}\right) \times \frac{\text{horsepower}}{33\,000\left(\frac{\text{lb ft}}{\text{s}}\right)}$$

$$= \frac{FV}{33\,000}\text{horsepower}$$

Viscosity (μ) Viscosity is defined as

$$\mu = \frac{\text{shear stress}}{\text{shear strain}}$$

$$= \frac{\text{force/area}}{\text{change in velocity per unit length}}$$

$$= \frac{F/A}{\text{d}v/\text{d}L}$$

Consistent units

$$\mu = \frac{\text{lb/ft}^2}{\frac{\text{ft}}{\text{s}}\Big/\text{ft}} = \frac{\text{lb s}}{\text{ft}^2}$$

Field units

$$\mu = \frac{F/A}{\text{d}v/\text{d}L}$$

where $F = \text{lb}$; $A = \text{in}^2$; $V = \text{ft/min}$; and $L = \text{ft}$.

$$\mu = \frac{\text{lb/in}^2}{\frac{\text{ft}}{\text{min}}\Big/\text{ft}} = \frac{\text{lb} \times \text{min}}{\text{in}^2}$$

The above equations can be further simplified by replacing in^2 by ft^2 and min by s to finally give the units of viscosity as

$$\frac{\text{lb} \times \text{s}}{\text{ft}^2}$$

The field unit of viscosity is the centipoise and is derived from the metric system (see below).

SI system

Fundamental units

Mass is expressed in kilogrammes (kg).
Length is expressed in metres (m).
Time is expressed in seconds (s).

Once again, the oil industry does not use these fundamental units in all measurements and modified units are used for practical purposes. For example, the units of diameter are expressed in millimetres (1 mm = 0.001 m) instead of metres. The field version of the SI system is conveniently described as the 'metric system'.

Force

$$F = \text{m} \times g$$

$$F = \text{kg} \times \frac{\text{m}}{\text{s}^2}$$

The product kg m/s^2 is given the name newton after the English physicist, Sir Isaac Newton. The unit newton is normally abbreviated as 'N'.

Pressure

$$P = \frac{F}{A}$$

$$= \frac{\text{N}}{\text{m}^2}$$

The unit N/m^2 is very small and the field units (or metric units) are given in either: (a) kilonewtons/m^2, where $1\ \text{kN/m}^2 = 1000\ \text{N/m}^2$, or (b) bars, where $1\ \text{bar} = 10^5\ \text{N/m}^2$.

Sometimes the unit N/m^2 is abbreviated as Pa or pascal, after the French scientist B. Pascal. The units of pressure can be expressed in kPa ($= 1000\ \text{N/m}^2$) or megapascals, MPa ($= 10^6\ \text{N/m}^2$). Also, pressure can be expressed as

$$P = \rho g h$$

When ρ is in kg/m^3, g in m/s^2 and h in m, the unit of pressure is N/m^2.

Viscosity

$$\mu = \frac{F/A}{\text{d}v/\text{d}L} = \frac{\text{N/m}^2}{(\text{m/s})/\text{m}} = \frac{\text{NS}}{\text{m}^2}$$

Replacing the unit N by kg m/s^2, we obtain

$$\mu = \frac{\left(\frac{\text{kg m}}{\text{s}^2}\right) \times \text{s}}{\text{m}^2}$$

$$= \frac{\text{kg}}{\text{m s}} \tag{1.7}$$

The kg/m s is the consistent units of viscosity in the SI system. However, the metric system uses the units of centipoise (cP), where 1 poise = 100 cP and

$$1\ \text{poise} = \frac{\text{g}}{\text{cm s}} \tag{1.8}$$

(1 kg = 1000 grams (g), 1 m = 100 centimetres (cm)). Comparing Equations (1.7) and (1.8) shows that

$$1\ \text{cP} = 10^{-3}\ \frac{\text{kg}}{\text{m s}}$$

Power

$$\text{Power} = \text{force} \times \text{velocity}$$

$$= \text{N} \times \frac{\text{m}}{\text{s}}$$

The product N × m is given the units joules. The units of power are given in watts, where 1 watt = 1 joule/s. The watt is a very small unit and the kilowatt is used instead, where 1 kW = 1000 watts.

Using the basic derivations presented for force, pressure, viscosity and power, one can easily convert from one system to the other. Appendix A.2 summarises the conversions between Imperial and metric systems for the quantities of length, area, volume, mass, force, density, pressure, energy, power, viscosity and temperature.

References

1. National Supply Company (1981). *Manufacturing Catalogue.*
2. Baker, R. (1979). *A Primer of Oilwell Drilling.* Petroleum Extension Services, University of Texas at Austin.
3. Schlumberger (1972). *Log Interpretation, Volume I: Principles.* Schlumberger Publications.
4. Snyder, R. and Suman, G. (1979). *Handbook of High Pressure Well Completion. World Oil.*

Chapter 2

Drill String Design

The main components of a drill string include:

Kelly
Drillpipe
Drill collars
Accessories, including heavy-wall drillpipe, jars, stabilisers, reamers, shock subs and bit subs
Drill bit

In this chapter, only the first four components will be discussed. Discussion of the drill bit will be given in Chapter 4.

KELLY

Figure 2.1 shows the main components of the drill string, including the position of the kelly within the assembly.

A kelly is used to transmit rotation and weight to the drill bit via drillpipe and drill collars. Weight on bit and rotation are the principal factors in breaking the rock and making hole. The kelly also carries the total weight of the drill string and is, therefore, the most heavily loaded item. Kellys are manufactured either from bars with an as-forged drive section or from bars with fully machined drive sections.

Kellys are manufactured from high grades of chrome molybdenum steel and heat treated by one of two processes: (1) the full length quenched and tempered;[4] or (2) the drive section normalised and tempered and the ends quenched and tempered.

Kellys that are treated by process (1) show greater impact strength and are suitable for use on drill ships subjected to stresses from pitch and roll of the ship. Most kellys have a Brinell hardness ranging from 285 to 341.

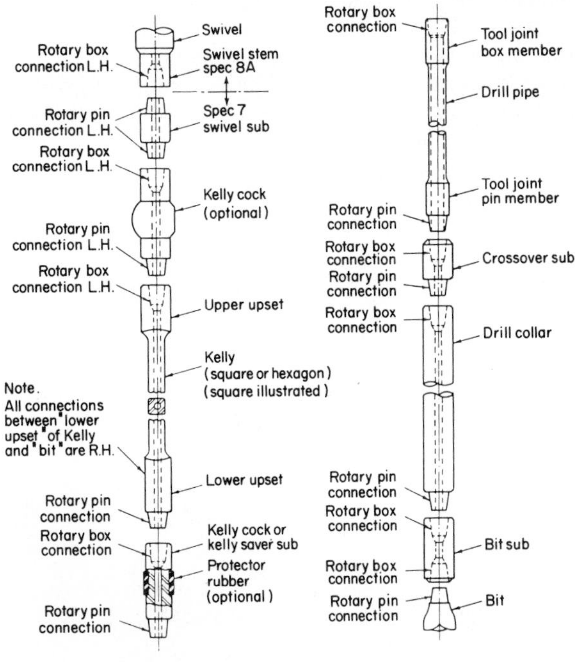

Fig. 2.1. Typical drill-stem assembly[3]. (Courtesy of API)

Two different shapes of kelly are also available: square and hexagonal. The drive section of a hexagonal kelly is stronger than the drive section of the square kelly when the appropriate kelly is selected for a given casing size[4].

Kellys are manufactured in two lengths: 40 ft (12.2 m) and 54 ft (16.5 m). Kellys are also manu-

TABLE 2.1 Outside diameters of kellys (inches)

Square	*Hexagonal*
$2\frac{1}{2}$	3
3	$3\frac{1}{2}$
$3\frac{1}{2}$	$4\frac{1}{4}$
$4\frac{1}{4}$	$5\frac{1}{4}$
$5\frac{1}{4}$	6
6	

factured in various sizes, as shown in Table 2.1. Further detail on kelly dimensions can be found in References 3 and 4.

Rotation of the kelly (and, in turn, the drill string) is derived from the rotary table by means of a kelly drive bushing and a master bushing. The master bushing (Figure 2.2) fits in a recess in the rotary table. It serves two purposes: (1) it provides engagement of the kelly drive bushing with the rotary table; and (2) it provides a tapered seating for the slips which hold drillpipe in the rotary table (see Chapter 1).

The kelly drive bushing (Figure 2.3) engages with the master bushing by either (a) drive pins fitted on the bottom of the kelly bushing which fit into holes bored into the master bushing or (b) a square section on the bottom of the kelly bushing which fits into a square recess in the master bushing. As the table turns, the kelly drive bushing turns with it to drive the kelly[2]. At the same time as the kelly works down, the rollers in the kelly bushing allow the kelly free movement and keep it centred in the rotary bore.

Kelly accessories include:

Kelly saver sub — a small sub connected to the bottom of the kelly to protect its threads from excessive wear as successive joints of drillpipe are made up and broken out during tripping and drilling operations.

Kelly cock — a small sub installed on top of the kelly, or below the kelly saver sub. When used above the kelly, it acts as a back-pressure valve, protecting equipment above the kelly (e.g. swivel and rotary hose) from the high surging pressures coming from below. When installed below the kelly, a kelly cock can be used to shut off drillpipe pressure, as in kick situations. Figure 2.4 shows typical positions where a kelly cock is normally installed and a section through a kelly cock.

DRILLPIPE

As shown in Figure 2.1, the main function of drillpipe is to transmit rotary motion and drilling mud under high pressure to the drill bit. The drillpipe is subjected to several types of loading, including axial loading due

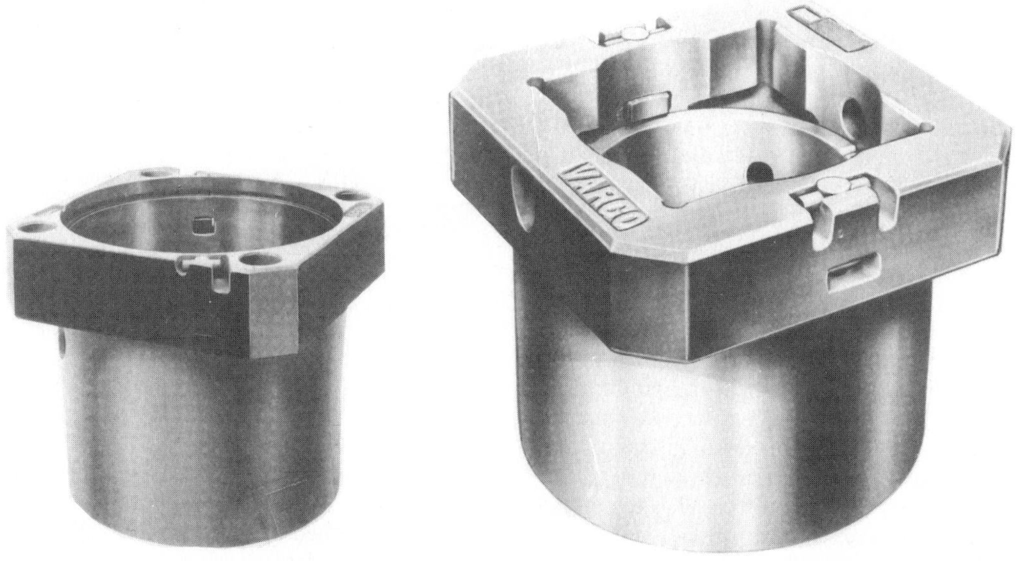

(a) (b)

Fig. 2.2. Types of master bushing[2]: (a) pin drive; (b) square drive. (Courtesy of Varco Oil Tools)

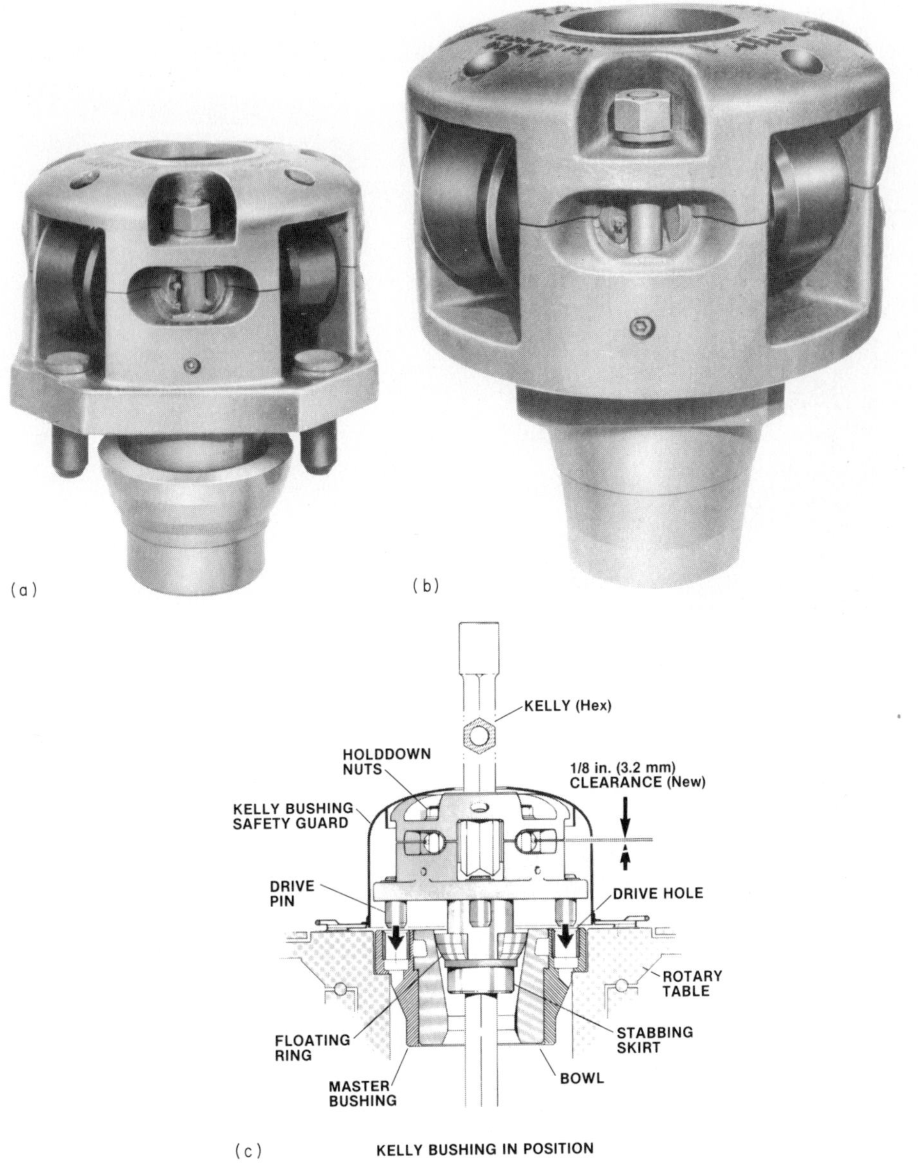

Fig. 2.3. Kelly drive bushing: (a) pin drive roller; (b) square drive roller; (c) bushing in position. (Courtesy of Varco Oil Tools)

to weight carried and its own weight, radial forces due to well-bore pressure, torque due to rotation, and cyclic stress reversals when the drillpipe is bent, as in dog-legged holes. The drillpipe must therefore be capable of withstanding all types of imposed loading and must have a reasonably long service life.

The drillpipe is manufactured as a seamless pipe, with an external, internal or external and internal upset. The term 'upset' refers to the manufacturing process, involving increasing the metal thickness near the pipe end where a coupling is attached in order to increase the pipe strength at that position. According to the type of upset, drillpipe can be described as internal upset (IU), external upset (EU) or internal–external upset (IEU).

Drillpipe is manufactured in three ranges: Range One, 18–22 ft; Range Two, 27–30 ft; and Range Three, 38–45 ft. Five grades of drill pipe are also available, to

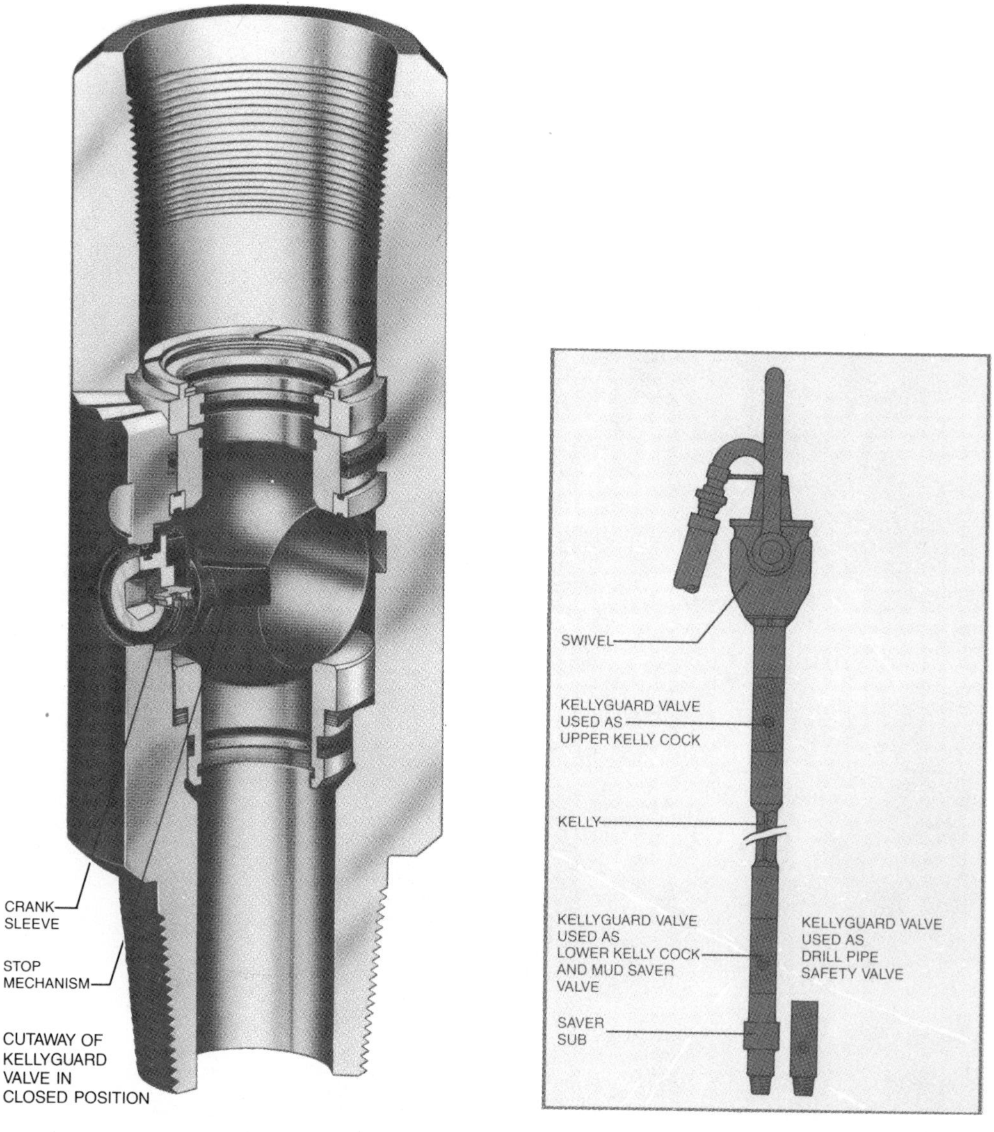

Fig. 2.4. Kelly cock. (Courtesy of Hydril)

suit different hole depths and loading requirements. API[3] specifies these grades as D, E, X95, G105 and S135. The five grades are also manufactured in different sizes, ranging from 2.375 in (60.3 mm) to 6.625 in (168.3 mm) OD. Each grade and size is specified by: (a) nominal weight per foot (or metre); (b) internal diameter; (c) collapse resistance; (d) internal yield pressure; and (e) pipe body yield strength. These factors will be discussed in some detail in Chapter 10.

Properties (a)–(e) of drillpipe change with time, owing to the severe stresses to which the drillpipe is subjected during drilling, coring, etc. These stresses (or forces) result from the combined weight of drillpipe and drill collars, bending stresses due to rotation, shock loading during the arresting of drillpipe in the slips and cyclic fatigue stresses. During drilling operations, these stresses result in pipe wear and, in turn, in a reduction in the strength properties of new drillpipe. For this reason, API[3] classifies drillpipe according to degree of wear, as follows:

Class One: New drillpipe
Premium: Pipe having a uniform wear and a minimum wall thickness of 80%

In practice, once drillpipe has been in hole, it is downgraded to premium.

Class Two: Pipe having a minimum wall thickness of 65% with all wear on one side provided that the cross-sectional area is the same as that of the premium class

Class Three: Pipe having a minimum wall thickness of 55% with all wear on one side

The minimum properties of the five grades and four classes of drillpipe are given in Tables 2.2–2.5. These properties will be used in drill string design (see page 37).

Tool joints

Tool joints, or couplings, are short, cylindrical pieces attached to each end of a drillpipe joint. The tool joints are attached to the ends of drillpipe by a flash weld or inertia weld process (Figure 2.5). Tool joints are threaded either externally or internally. The ex-

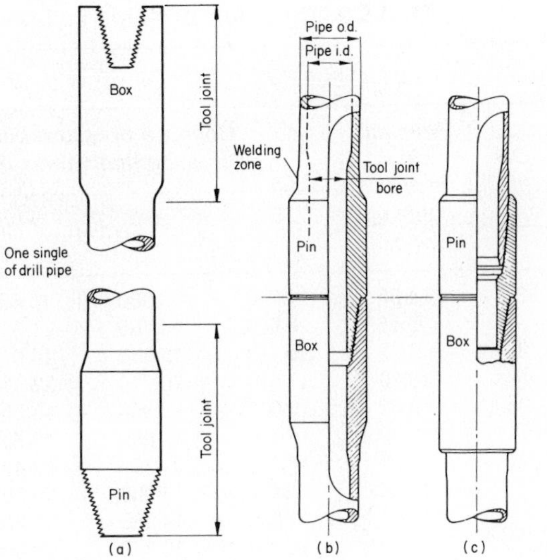

Fig. 2.5. Drillpipe and tool joints: (a) a drillpipe single; (b) welded drillpipe tool joint; (c) shrink-fit threaded drillpipe tool joint.

TABLE 2.2(a) New drillpipe torsional and tensile data. (Courtesy of API[4])

1	2	3	4	5	6	7	8	9	10
Size OD (in)	*Nominal weight thds and couplings (lb)*	*Torsional data torsional yield strength (ft-lb)**				*Tensile data based on minimum values load at the minimum yield strength (lb)*			
		E	*X95*	*G105*	*S135*	*E*	*X95*	*G105*	*S135*
$2\frac{3}{8}$	4.85	4760	6030	6670	8570	97 820	123 900	136 940	176 070
	6.65	6250	7920	8750	11 250	138 220	175 080	193 500	248 790
$2\frac{7}{8}$	6.85	8080	10 240	11 320	14 550	135 900	172 140	190 260	244 620
	10.40	11 550	14 640	16 180	20 800	214 340	271 500	300 080	385 820
$3\frac{1}{2}$	9.50	14 150	17 920	19 800	25 460	194 270	246 070	271 970	349 680
	13.30	18 550	23 500	25 970	33 390	271 570	343 990	380 190	488 820
	15.50	21 090	26 710	29 520	37 950	322 780	408 850	451 890	581 000
4	11.85	19 470	24 670	27 260	35 050	230 750	292 290	323 050	415 350
	14.00	23 290	29 500	32 600	41 920	285 360	361 460	399 500	513 650
	15.70	25 810	32 690	36 130	46 460	324 120	410 550	453 770	583 420
$4\frac{1}{2}$	13.75	25 910	32 820	36 270	46 630	270 030	342 040	378 040	486 050
	16.60	30 810	39 020	43 130	55 450	330 560	418 700	462 780	595 000
	20.00	36 900	46 740	51 660	66 420	412 360	522 320	577 300	742 240
	22.82	40 910	51 820	57 280	73 640	471 240	596 900	659 740	848 230
5	16.25	35 040	44 390	49 060	63 080	328 070	415 560	459 300	590 530
	19.50	41 170	52 140	57 600	74 100	395 600	501 090	553 830	712 070
	25.60	52 260	66 190	73 160	94 060	530 150	671 520	742 200	954 260
$5\frac{1}{2}$	19.20	44 070	55 830	61 700	79 330	372 180	471 430	521 050	669 920
	21.90	50 710	64 230	70 990	91 280	437 120	553 680	611 960	786 810
	24.70	56 570	71 660	79 200	101 830	497 220	629 810	696 110	895 000
$6\frac{5}{8}$	25.20	70 580	89 400	98 810	127 050	489 470	619 990	685 250	881 040

* Based on the shear strength equal to 57.7% of minimum yield strength and nominal wall thickness.

TABLE 2.2(b) New drillpipe collapse and internal pressure data. (Courtesy of API[4])

1	*2*	*3*	*4*	*5*	*6*	*7*	*8*	*9*	*10*
Size OD (in)	*Nominal weight thds and couplings (lb)*	*Collapse pressure based on minimum values (psi)*				*Internal pressure at minimum yield strength (psi)*			
		E	*X95*	*G105*	*S135*	*E*	*X95*	*G105*	*S135*
$2\frac{3}{8}$	4.85	11 040	13 980	15 460	19 070	10 500	13 300	14 700	18 900
	6.65	15 600	19 760	21 840	28 080	15 470	19 600	21 660	27 850
$2\frac{7}{8}$	6.85	10 470	12 930	14 010	17 060	9910	12 550	13 870	17 830
	10.40	16 510	20 910	23 110	29 720	16 530	20 930	23 140	29 750
$3\frac{1}{2}$	9.50	10 040	12 060	13 050	15 780	9520	12 070	13 340	17 150
	13.30	14 110	17 880	19 760	25 400	13 800	17 480	19 320	24 840
	15.50	16 770	21 250	23 480	30 190	16 840	21 330	23 570	30 310
4	11.85	8410	9960	10 700	12 650	8600	10 890	12 040	15 480
	14.00	11 350	14 380	15 900	20 170	10 830	13 720	15 160	19 490
	15.70	12 900	16 340	18 050	23 210	12 470	15 790	17 460	22 440
$4\frac{1}{2}$	13.75	7200	8400	8950	10 310	7900	10 010	11 070	14 230
	16.60	10 390	12 750	13 820	16 800	9830	12 450	13 760	17 690
	20.00	12 960	16 420	18 150	23 330	12 540	15 890	17 560	22 580
	22.82	14 810	18 770	20 740	26 670	14 580	18 470	20 420	26 250
5	16.25	6970	8090	8610	9860	7770	9840	10 880	13 990
	19.50	10 000	12 010	12 990	15 700	9500	12 040	13 300	17 110
	25.60	13 500	17 100	18 900	24 300	13 120	16 620	18 380	23 620
$5\frac{1}{2}$	19.20	6070	6930	7300	8120	7250	9190	10 160	13 060
	21.90	8440	10 000	10 740	12 710	8610	10 910	12 060	15 510
	24.70	10 460	12 920	14 000	17 050	9900	12 540	13 860	17 830
$6\frac{5}{8}$	25.20	4810	5310	5490	6040	6540	8280	9150	11 770

ternally threaded end of the drillpipe tool joint is described as the 'pin' and the internally threaded end is described as the 'box'. Individual drillpipes are joined by stabbing the pin of one joint into the box of another and torquing with power tongs.

API[3] specifies that all drill-stem tool joints should be of the weld-on type and have the following mechanical properties when new: minimum yield strength = 120 000 psi; minimum tensile strength = 140 000 psi; and minimum elongation percentage = 13.

Figure 2.6 gives a section through a tool joint connection, together with the dimensional description required to define a tool joint. Table 2.6 gives the dimensional data of tool joints suitable for use with Grades E, X, G and S drillpipe. Dimensions of tool joints are required for the calculation of the approximate weight of the drillpipe and tool joint assembly.

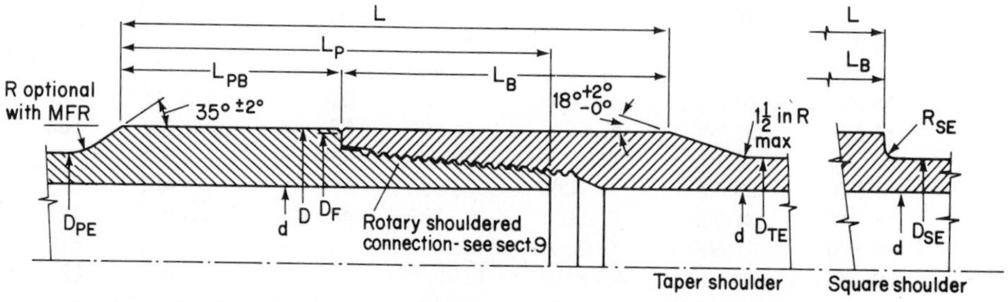

Fig. 2.6. Tool joint, taper shoulder and square shoulder[3] (see Table 2.6). (Courtesy of API)

Thread forms

The threads on the pins and boxes of drill string members may be distinguished by the following: width of thread crest, radius of thread root, angle between the flanks of adjacent threads and number of threads per inch. API[3] adopted the format V-numerical number to distinguish various oil field threads. The number following the V refers to either the width of the crest flat or the radius of the root. The letter R follows the number referring to root radius.

Figure 2.7 gives the current thread forms used on rotary shouldered connections of drill stem members.

Rotary shouldered connections

The term 'rotary shouldered connection' refers to the threads of the pin or box of drillpipe or drill collars. The threads of the pin of one joint engage with the threads of the box of another joint during make-up. The actual seal is provided by metal contact of the shoulders of the tool joints. The engaging threads are not made to provide a seal and open channels between the threads exist, even when the joint is torqued.

Besides thread form and number of threads per inch, a connection can also be distinguished by dimensional data relating to the small and large diameters of pin, box bore, length of pin and box, etc. Figure 2.8 gives the API[4] nomenclature for rotary shouldered connections.

API[3] suggests the use of the term 'number connection' (NC) to distinguish the various sizes and styles of rotary connections. The NC refers to the pitch diameter of the pin thread at gauge point (defined as C in Figure 2.8) when rounded to units and tenths of inches. Thus, if C is 1.063 in, the first two figures are used, i.e. 10, to provide a description of the connection as NC10. Table 2.7 gives dimensional data of all oil field rotary connections (refer to Figure 2.8 for symbols).

TABLE 2.3(a) Used drillpipe torsional and tensile data API Premium Class. (Courtesy of API[4])

1	2	3	4	5	6	7	8	9	10
Size OD (in)	*New wt. nom. wt./ thds and couplings (lb/ft)*	*Torsional yield strength based on uniform wear (ft-lb)*†*				*Tensile data based on uniform wear load at minimum yield strength (lb)†*			
		E	*X95*	*G105*	*S135*	*E*	*X95*	*G105*	*S135*
$2\frac{3}{8}$	4.85	3730	4720	5220	6710	76 880	97 380	107 640	138 380
	6.65	4810	6090	6730	8660	107 620	136 330	150 680	193 730
$2\frac{7}{8}$	6.85	6330	8020	8860	11 400	106 950	135 470	149 730	192 510
	10.40	8850	11 220	12 400	15 940	166 500	210 900	233 100	299 700
$3\frac{1}{2}$	9.50	11 090	14 050	15 530	19 970	153 000	193 800	214 200	275 400
	13.30	14 360	18 190	20 100	25 850	212 250	268 850	297 150	382 050
	15.50	16 140	20 450	22 600	29 060	250 500	317 300	350 700	450 900
4	11.85	15 310	19 390	21 430	27 560	182 020	230 560	254 840	327 640
	14.00	18 200	23 050	25 470	32 750	224 180	283 960	313 850	403 520
	15.70	20 070	25 420	28 090	36 120	253 880	321 580	355 430	456 980
$4\frac{1}{2}$	13.75	20 400	25 840	28 560	36 730	213 220	270 080	298 510	383 800
	16.60	24 130	30 570	33 790	43 450	260 100	329 460	364 140	468 180
	20.00	28 680	36 330	40 150	51 630	322 950	409 070	452 130	581 310
	22.82	31 590	40 010	44 220	56 860	367 570	465 590	514 590	661 620
5	16.25	27 610	34 970	38 650	49 690	259 120	328 220	362 780	466 420
	19.50	32 290	40 890	45 200	58 110	311 540	394 600	436 150	560 760
	25.60	40 540	51 360	56 760	72 980	414 690	525 270	580 570	746 440
$5\frac{1}{2}$	21.90	39 860	50 490	55 810	71 750	344 780	436 720	482 690	620 600
	24.70	44 320	56 140	62 050	79 780	391 280	495 630	547 800	704 310

* Based on the shear strength equal to 57.7% of minimum yield strength.

† Torsional data based on 20% uniform wear on outside diameter and tensile data based on 20% uniform wear on outside diameter.

TABLE 2.3(b) Used drillpipe collapse and internal pressure data API Premium Class. (Courtesy of API[4])

1	2	3	4	5	6	7	8	9	10
Size OD (in)	Nominal weight thds and couplings (lb/ft)	Collapse pressure based on minimum values (psi)*				Internal pressure at minimum yield strength (psi)*			
		E	X95	G105	S135	E	X95	G105	S135
$2\frac{3}{8}$	4.85	8550	10 150	10 900	12 920	9600	12 160	13 440	17 280
	6.65	13 380	16 950	18 730	24 080	14 150	17 920	19 810	25 470
$2\frac{7}{8}$	6.85	7670	9000	9620	11 210	9060	11 470	12 680	16 300
	10.40	14 220	18 020	19 910	25 600	15 110	19 140	21 150	27 200
$3\frac{1}{2}$	9.50	7100	8270	8800	10 120	8710	11 030	12 190	15 680
	13.30	12 020	15 220	16 820	21 630	12 620	15 980	17 660	22 710
	15.50	14 470	18 330	20 260	26 050	15 390	19 500	21 550	27 710
4	11.85	5730	6490	6820	7470	7860	9960	11 000	14 150
	14.00	9040	10 780	11 610	13 870	9900	12 540	13 860	17 820
	15.70	10 910	13 820	15 180	18 630	11 400	14 440	15 960	20 520
$4\frac{1}{2}$	13.75	4710	5170	5340	5910	7230	9150	10 120	12 010
	16.60	7550	8850	9460	10 990	8990	11 380	12 580	16 180
	20.00	10 980	13 900	15 340	18 840	11 470	14 520	16 050	20 640
	22.82	12 660	16 040	17 720	22 790	13 330	16 890	18 670	24 000
5	16.25	4510	4920	5060	5670	7100	9000	9950	12 790
	19.50	7070	8230	8760	10 050	8690	11 000	12 160	15 640
	25.60	11 460	14 510	16 040	20 540	12 000	15 200	16 800	21 600
$5\frac{1}{2}$	19.20	3760	4140	4340	4720	6630	8400	9290	11 940
	21.90	5760	6530	6860	7520	7880	9980	11 030	14 180
	24.70	7670	9000	9620	11 200	9050	11 470	12 680	16 300
$6\frac{5}{8}$	25.20	2930	3250	3350	3430	5980	7570	8370	10 760

* Data are based on minimum wall of 80% nominal wall. Collapse pressures are based on uniform OD wear. Internal pressures are based on uniform wear and nominal OD.

Approximate weight of tool joint and drillpipe assembly

The drillpipe weight given in Tables 2.2–2.5 is described as nominal weight and is used mainly for the purpose of identification when ordering. The exact weight of drillpipe is difficult to determine and, instead, an approximate weight is usually calculated.

The calculation of the approximate weight of a drillpipe and tool joint assembly involves the determination of adjusted weights of drillpipe and tool joint (reference should be made to Figure 2.6 and Tables 2.2–2.5).

(a) Approximate adjusted weight of drillpipe

$$\text{plain end weight} + \frac{\text{upset weight}}{29.4}$$

(b) Approximate adjusted weight of tool joint

$$= 0.222 \times L(D^2 - d^2) + 0.167(D^3 - D_{TE}^3) - 0.501d^2(D - D_{TE})$$

where: L = combined length of pin and box (in); D = outside diameter of pin (in); d = inside diameter of pin (in); D_{TE} = diameter of box at elevator upset (in).

(c) Approximate weight of tool joint and drillpipe assembly

$$= \frac{\text{approx. adjusted wt drillpipe} \times 29.4\text{ ft} + \text{approx. wt tool joint}}{29.4 + \text{tool joint adjusted length}}$$

where

$$\text{tool joint adjusted length} = \frac{L + 2.253(D - D_{TE})}{12}\text{ ft}$$

Example 2.1

Calculate the approximate weight of tool joint and drillpipe assembly for 5 in OD, 19.5 lbm/ft Grade E drillpipe having a $6\frac{3}{8}$ in OD/$3\frac{1}{2}$ in ID, NC50 tool joint. Assume the pipe to be internally–externally upset (IEU) and weight increase due to upsetting to be 8.6 lb.

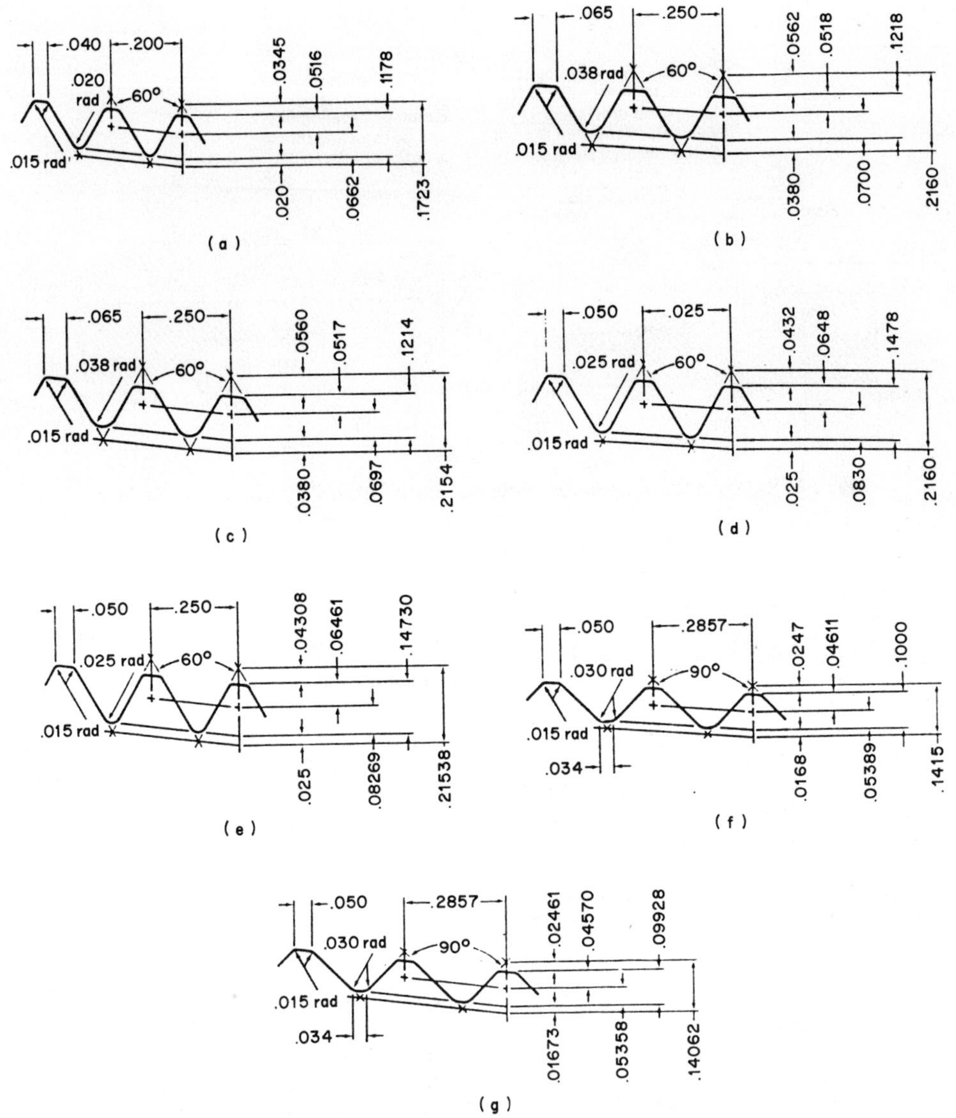

Fig. 2.7. Oilfield thread forms[1]: (a) V-0.040: 3 in taper per foot (TPF) on diameter; 5 threads per inch (TPI). (b) V-0.038R: 2 in TPF; on diameter 4 TPI. (c) V-0.038R: 3 in TPF on diameter; 4 TPI. (d) V-0.050: 2 in TPF on diameter; 4 TPI. (e) V-0.050: 3 in TPF on diameter; 4 TPI. (f) H-90: 2 in TPF on diameter; $3\frac{1}{2}$. TPI. (g) H-90: 3 in TPF on diameter: $3\frac{1}{2}$ TPI. (Courtesy of Drilco)

Solution

Referring to Table 2.6, NC50, $6\frac{3}{8}$ in OD/$3\frac{1}{2}$ in ID tool joint for a 19.5 lbm/ft nominal weight drillpipe is available in Grade X95 only (see columns 4, 5 and 6 of Table 2.6). Thus, $L = 17$ in, $D_{\text{TE}} = 5\frac{1}{8}$ in, $D = 6\frac{3}{8}$ in, $d = 3\frac{1}{2}$ in.

Approximate adjusted weight of tool joint

$$= 0.222 \times 17((6\tfrac{3}{8})^2 - (3\tfrac{1}{2})^2)$$

$$+ 0.167((6\tfrac{3}{8})^3 - (5\tfrac{1}{8})^3)$$

$$- 0.501 \times (3\tfrac{1}{2})^2(6\tfrac{3}{8} - 5\tfrac{1}{8})$$

$$= 107.15 + 20.79 - 7.67$$

$$= 120.27 \text{ lb}$$

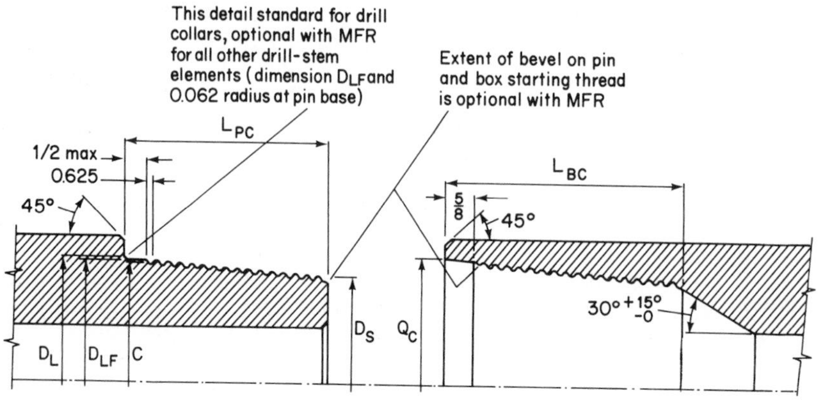

Fig. 2.8. Rotary shouldered connections[3] (see Table 2.7). (Courtesy of API)

TABLE 2.4(a) Used drillpipe torsional and tensile data API Class 2. (Courtesy of API[4])

1	2	3	4	5	6	7	8	9	10
Size OD (in)	*New wt. nom. wt./ thds and couplings (lb/ft)*	*Torsional yield strength based on eccentric wear (ft-lb)*†*				*Tensile data based on uniform wear load at minimum yield strength (lb)†*			
		E	*X95*	*G105*	*S135*	*E*	*X95*	*G105*	*S135*
$2\frac{3}{8}$	4.85	3150	3990	4410	5670	76 880	97 380	107 640	138 380
	6.65	4150	5260	5810	7470	107 620	136 330	150 680	193 730
$2\frac{7}{8}$	6.85	5340	6770	7480	9620	106 950	135 470	149 730	192 510
	10.40	7680	9720	10 750	13 820	166 500	210 900	233 100	299 700
$3\frac{1}{2}$	9.50	9350	11 840	13 090	16 830	153 000	193 800	214 200	275 400
	13.30	12 310	15 590	17 230	22 160	212 250	268 850	297 150	382 050
	15.50	14 010	17 750	19 620	25 220	250 500	317 300	350 700	450 900
4	11.85	12 860	16 290	18 000	23 140	182 020	230 560	254 840	327 640
	14.00	15 410	19 520	21 580	27 740	224 180	283 960	313 850	403 520
	15.70	17 110	21 670	23 950	30 790	253 880	321 580	355 430	456 980
$4\frac{1}{2}$	13.75	17 090	21 650	23 930	30 760	213 220	270 080	298 510	383 800
	16.60	20 370	25 800	28 520	36 660	260 100	329 460	364 140	468 180
	20.00	24 460	30 980	34 240	44 030	322 950	409 070	452 130	581 310
	22.82	26 590	34 400	38 020	48 880	367 570	465 590	514 590	661 620
5	16.25	23 110	29 280	32 360	41 600	259 120	328 220	362 780	466 420
	19.50	27 210	34 460	38 090	48 970	311 540	394 600	436 150	560 760
	25.60	34 650	43 900	48 520	62 380	414 690	525 270	580 570	746 440
$5\frac{1}{2}$	21.90	33 480	42 410	46 870	60 260	344 780	436 720	482 690	620 600
	24.70	37 410	47 380	52 370	67 330	391 280	495 630	547 800	704 310

* Based on the shear strength equal to 57.7% of minimum yield strength.

† Torsional data based on 35% eccentric wear on outside diameter and tensile data based on 20% uniform wear on outside diameter.

TABLE 2.4(b) Used drillpipe collapse and internal pressure data API Class 2. (Courtesy of API[4])

1	2	3	4	5	6	7	8	9	10
Size OD (in)	*Nominal weight thds and couplings (lb/ft)*	*Collapse pressure based on minimum values (psi)**				*Internal pressure at minimum yield strength (psi)**			
		E	*X95*	*G105*	*S135*	*E*	*X95*	*G105*	*S135*
$2\frac{3}{8}$	4.85	6020	6870	7240	8030	7800	9880	10 920	14 040
	6.65	11 480	14 540	16 080	20 630	11 490	14 560	16 090	20 690
$2\frac{7}{8}$	6.85	5270	5900	6150	6610	7360	9320	10 300	13 250
	10.40	12 250	15 520	17 160	22 060	12 280	15 550	17 190	22 100
$3\frac{1}{2}$	9.50	4790	5270	5450	6010	7080	8960	9910	12 740
	13.30	10 250	12 420	13 450	16 310	10 250	12 990	14 350	18 450
	15.50	12 480	15 810	17 480	22 470	12 510	15 840	17 510	22 510
4	11.85	3620	4020	4210	4550	6390	8090	8940	11 500
	14.00	6440	7410	7850	8840	8040	10 190	11 260	14 480
	15.70	8560	10 150	10 910	12 930	9260	11 730	12 970	16 670
$4\frac{1}{2}$	13.75	2960	3290	3400	3480	5870	7440	8220	10 570
	16.60	5170	5770	6010	6490	7300	9250	10 220	13 140
	20.00	8660	10 280	11 050	13 120	9320	11 800	13 040	16 770
	22.82	10 830	13 720	14 950	18 320	10 830	13 720	15 170	19 500
5	16.25	2850	3150	3240	3300	5770	7310	8080	10 390
	19.50	4760	5230	5410	5970	7060	8940	9880	12 710
	25.60	9420	11 270	12 160	14 590	9750	12 350	13 650	17 550
$5\frac{1}{2}$	19.20	2440	2610	2650	2650	5390	6830	7540	9700
	21.90	3640	4040	4230	4580	6400	8110	8960	11 520
	24.70	5260	5890	6140	6610	7360	9320	10 300	13 250
$6\frac{5}{8}$	25.20	1870	1900	1900	1900	...	...	...	...

* Data are based on minimum wall of 65% nominal wall. Collapse pressures are based on uniform OD wear. Internal pressures are based on uniform wear and nominal OD.

Approximate adjusted weight of drillpipe

$$= \text{plain-end weight} + \frac{\text{upset weight}}{29.4}$$

$$= \frac{\pi}{4}(5^2 - 4.276^2) \times \frac{1}{144} \times 489.5 + \frac{8.6}{29.4}$$

$$= 17.93 + 0.293 = 18.22 \text{ lbm/ft}$$

Adjusted length of tool joint

$$= \frac{L + 2.253(D - D_{TE})}{12}$$

$$= \frac{17 + 2.253(6\tfrac{3}{8}\text{ in} - 5\tfrac{1}{8}\text{ in})}{12}$$

$$= 1.651 \text{ ft}$$

Hence, approximate weight of tool joint and drillpipe assembly

$$= \frac{18.22 \times 29.4 + 120.26}{1.651 + 29.4}$$

$$= 21.12 \text{ lbm/ft}$$

(*Note:* Figure 2.9 shows the various types of upsets and Table 2.8 gives the calculated nominal weight for various types of upsets and sizes of drillpipe. Column 6, Table 2.8 gives the weight of upset.)

DRILL COLLARS

Drill collars are used to provide weight on bit and to keep drillpipe in tension. Drillpipe has a comparatively low stiffness, which makes it susceptible to buckling when under compression. Repeated buckling will eventually lead to drillpipe failure (see Chapter 8, page 167.)

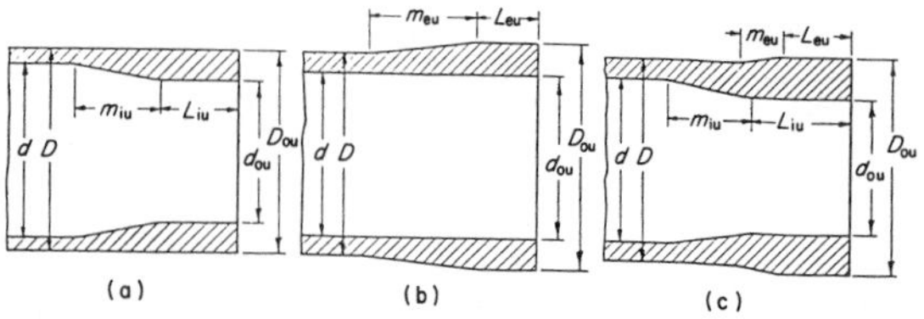

Fig. 2.9. Upset drillpipe for weld-on tool joints[5]: (a) internal upset; (b) external upset; (c) internal–external upset (see Table 2.8). (Courtesy of API)

Since elastic members can only buckle in compression, fatigue failure of drillpipe can be eliminated by keeping it in constant tension. In practice, 85% of the buoyant weight of drill collars (giving a safety factor of 1.15) is used as weight-on-bit, to ensure that the drillpipe is kept in tension. This practice also places the neutral point (point of zero tension and zero compression) in the drill collars.

Table 2.9 lists the sizes and weight per foot of available drill collars. From this table, the reader can observe that the inside diameter of drill collars is quite small, e.g. for a 5 in OD drill collar, the ID varies from $1\frac{1}{2}$ to $2\frac{1}{2}$ in. The reduced diameter of a drill collar results in a high pressure loss during the circulation of drilling mud and, for this reason, drilling engineers often select the largest available bore to limit pressure losses. Pressure losses will be discussed in Chapters 5 and 7.

Drill collars are normally manufactured in an average length of 31 ft. Owing to the large thickness of pipe body, drill collars are not provided with tool joints and, instead, pins and boxes are cut from the pipe body. The drill collar connection has a tapered, threaded jack screw[1] that forces the shoulders of individual drill collar joints together to form a metal seal (as shown in Figure 2.10) at the shoulders.

The drill collar threads are normally strengthened by cold-working the thread roots. Cold-working involves the prestressing of the thread root with a roller driven by a hydraulic ram. This results in a

TABLE 2.5(a) Used drillpipe torsional and tensile data API Class 3. (Courtesy of API[4])

1	2	3	4	5	6	7	8	9	10
Size OD (in)	New wt. nom. wt./ thds and couplings lb/ft	Torsional yield strength based on eccentric wear (ft-lb)*†				Tensile data based on uniform wear load at minimum yield strength (lb)†			
		E	X95	G105	S135	E	X95	G105	S135
$2\frac{3}{8}$	4.85	2690	3400	3760	4830	59 140	74 910	82 800	106 460
	6.65	3540	4480	4960	6370	82 050	103 930	114 870	147 690
$2\frac{7}{8}$	6.85	4550	5770	6380	8200	82 580	104 600	115 610	148 640
	10.40	6550	8290	9170	11 780	126 600	160 360	177 240	227 880
$3\frac{1}{2}$	9.50	7970	10 090	11 150	14 340	118 050	149 530	165 270	212 490
	13.30	10 490	13 290	14 690	18 890	162 220	205 480	227 120	292 000
	15.50	11 950	15 140	16 730	21 510	190 500	241 300	266 700	342 900
4	11.85	10 950	13 880	15 340	19 720	140 630	178 130	196 880	253 130
	14.00	13 140	16 640	18 390	23 650	172 580	218 600	241 600	310 640
	15.70	14 580	18 470	20 420	26 250	195 000	247 000	273 000	351 000
$4\frac{1}{2}$	13.75	14 560	18 440	20 380	26 210	164 330	208 150	230 060	295 790
	16.60	17 360	21 990	24 300	31 240	200 180	253 560	280 240	360 320
	20.00	20 850	26 410	29 190	37 530	247 720	313 780	346 820	445 900
	22.82	23 160	29 330	32 420	41 680	280 720	355 580	393 000	505 290
5	16.25	19 690	24 940	27 570	35 440	200 180	253 560	280 250	360 320
	19.50	23 180	29 370	32 460	41 730	240 300	304 380	336 420	432 540
	25.60	29 550	37 420	41 360	53 180	317 550	402 230	444 570	571 590
$5\frac{1}{2}$	21.90	28 530	36 130	39 940	51 350	266 480	337 540	373 070	479 660
	24.70	31 880	40 380	44 630	57 380	301 420	381 800	422 000	542 560

* The torsional yield strength is based on a shear strength of 57.7% of the minimum yield strength (following the maximum shear strain energy theory of yielding).

† Torsional data based on 45% eccentric wear on outside diameter. Tensile data based on $37\frac{1}{2}$% uniform wear on outside diameter.

TABLE 2.5(b) Used drillpipe collapse and pressure data API Class 3. (Courtesy of API[4])

1	2	3	4	5	6	7	8	9	10
Size OD (in)	*Nominal weight thds and couplings (lb/ft)*	*Collapse pressure based on minimum values (psi)**				*Internal pressure at minimum yield strength (psi)**			
		E	*X95*	*G105*	*S135*	*E*	*X95*	*G105*	*S135*
$2\frac{3}{8}$	4.85	4260	4590	4810	5350	6600	8360	9240	11 880
	6.65	10 030	12 050	13 040	15 760	9730	12 320	13 620	17 510
$2\frac{7}{8}$	6.85	3600	4010	4190	4530	6230	7890	8720	11 210
	10.40	10 800	13 680	14 880	18 230	10 390	13 160	14 540	18 700
$3\frac{1}{2}$	9.50	3230	3650	3790	4000	5990	7580	8380	10 780
	13.30	8040	9480	10 160	11 930	8670	10 990	12 140	15 610
	15.50	11 010	13 950	15 410	18 960	10 580	13 410	14 820	19 050
4	11.85	2570	2790	2840	2850	5400	6840	7560	9720
	14.00	4630	5070	5230	5810	6810	8620	9530	12 250
	15.70	6490	7480	7920	8940	7840	9930	10 970	14 110
$4\frac{1}{2}$	13.75	2090	2170	2170	2170	4970	6290	6960	8940
	16.60	3520	3930	4110	4420	6180	7830	8650	11 120
	20.00	6580	7590	8040	9100	7880	9990	11 040	14 190
	22.82	9140	10 900	11 750	14 040	9170	11 610	12 830	16 500
5	16.25	1990	2050	2050	2050	4880	6190	6840	8790
	19.50	3210	3630	3770	3960	5970	7570	8360	10 750
	25.60	7250	8460	9020	10 410	8250	10 450	11 550	14 850
$5\frac{1}{2}$	19.20	1640	1640	1640	1640	5100	6460	7140	9180
	21.90	2580	2810	2860	2870	5420	6860	7580	9750
	24.70	3600	4000	4190	4520	6230	7890	8720	11 210
$6\frac{5}{8}$	25.20	1170	1170	1170	1170	4110	5210	5750	7400

* Data are based on minimum wall of 55% nominal wall. Collapse pressures are based on uniform OD wear. Internal pressures are based on uniform wear and nominal OD.

thread surface with a greater resistance to cyclic stress reversals.

Drill collar features

Drill collars possess the following features[6].

Fishing necks

Large-sized drill collars with OD in excess of 8 in are manufactured with reduced diameters near the pin and box ends (Figure 2.11a). This feature is introduced to allow drill collars to be fitted with smaller connections which can be properly torqued with the available rig tongs and line pulls.

As outlined by Wilson[6], the name ‘fishing neck’ comes from the original use, which was to receive an overshot and grapple in case of joint failure in hole which demanded a fishing job.

Stepped-bore drill collars

In small and medium-sized drill collars where large bores are required, to reduce pressure losses, the pin strength can be increased by stepping its bore as shown in Figure 2.11(b). For example, in a $6\frac{1}{4}$ in OD drill collar with a $2\frac{13}{16}$ in bore, the pin bore is reduced to $2\frac{1}{4}$ in, to increase its strength.

Slip and elevator recesses

Slip and elevator recesses are introduced to allow drill collars to be handled like drillpipe by simply changing the drillpipe elevators and slips. Figure 2.11(c) shows a drill collar with slip and elevator recesses.

Spiral grooving

Differential sticking is more prevalent with drill collars than with drillpipe. A drill collar with spiral

TABLE 2.6 Tool joint dimensions for Grade E, X, G and S drillpipe (all dimensions in inches). (Courtesy of API[3])

1	2	3	4	5	6	7	8	9	10	11	12	13	14
Tool joint designation[1]	Drill pipe			Tool joint									Torsional ratio, pin to drill pipe
	Size and style	Nom. wt.[2] (lb/ft)	Grade	Outside dia. of pin and box, ±1/32	Inside dia. of pin[3], +1/64 −1/32	Bevel dia. of pin and box shoulder, ±1/64	Total length tool joint pin, $+\frac{1}{4}$ $-\frac{3}{8}$	Pin tong space, $\pm\frac{1}{4}$	Box tong space, $\pm\frac{1}{4}$	Combined length of pin and box, $\pm\frac{1}{2}$	Dia. of pin at elevator upset, max.	Dia. of box at elevator upset, max.	
				D	d	D_F	L_F	L_{PE}	L_B	L	D_{PE}	D_{TE}	
NC26($2\frac{3}{8}$IF)	$2\frac{3}{8}$ EU	6.65	E75	$3\frac{3}{8}$*	$1\frac{3}{4}$*	$3\frac{17}{64}$	10	7	8	15	$2\frac{9}{16}$	$2\frac{9}{16}$	1.10
			X95	$3\frac{3}{8}$*	$1\frac{3}{4}$*	$3\frac{17}{64}$	10	7	8	15	$2\frac{9}{16}$	$2\frac{9}{16}$	0.87
			G105	$3\frac{3}{8}$*	$1\frac{3}{4}$*	$3\frac{17}{64}$	10	7	8	15	$2\frac{9}{16}$	$2\frac{9}{16}$	0.79
NC31($2\frac{7}{8}$IF)	$2\frac{7}{8}$ EU	10.40	E75	$4\frac{1}{8}$*	$2\frac{1}{8}$*	$3\frac{61}{64}$	$10\frac{1}{2}$	7	9	16	$3\frac{3}{16}$	$3\frac{3}{16}$	1.03
			X95	$4\frac{1}{8}$*	2	$3\frac{61}{64}$	$10\frac{1}{2}$	7	9	16	$3\frac{3}{16}$	$3\frac{3}{16}$	0.90
			G105	$4\frac{1}{8}$*	2	$3\frac{61}{64}$	$10\frac{1}{2}$	7	9	16	$3\frac{3}{16}$	$3\frac{3}{16}$	0.82
			S135	$4\frac{3}{8}$	$1\frac{5}{8}$	$3\frac{61}{64}$	$10\frac{1}{2}$	7	9	16	$3\frac{3}{16}$	$3\frac{3}{16}$	0.82
NC38[4]	$3\frac{1}{2}$ EU	9.50	E75	$4\frac{3}{4}$*	3	$4\frac{37}{64}$	$11\frac{1}{2}$[4]	8	$10\frac{1}{2}$	$18\frac{1}{2}$	$3\frac{7}{8}$	$3\frac{7}{8}$	0.91
NC38($3\frac{1}{2}$IF)	$3\frac{1}{2}$ EU	13.30	E75	$4\frac{3}{4}$*	$2\frac{11}{16}$*	$4\frac{37}{64}$	12	8	$10\frac{1}{2}$	$18\frac{1}{2}$	$3\frac{7}{8}$	$3\frac{7}{8}$	0.98
			X95	5	$2\frac{9}{16}$	$4\frac{37}{64}$	12	8	$10\frac{1}{2}$	$18\frac{1}{2}$	$3\frac{7}{8}$	$3\frac{7}{8}$	0.87
			G105	5	$2\frac{7}{16}$	$4\frac{37}{64}$	12	8	$10\frac{1}{2}$	$18\frac{1}{2}$	$3\frac{7}{8}$	$3\frac{7}{8}$	0.86
			S135	5	$2\frac{1}{8}$	$4\frac{37}{64}$	12	8	$10\frac{1}{2}$	$18\frac{1}{2}$	$3\frac{7}{8}$	$3\frac{7}{8}$	0.80
		15.50	E75	5	$2\frac{9}{16}$	$4\frac{37}{64}$	12	8	$10\frac{1}{2}$	$18\frac{1}{2}$	$3\frac{7}{8}$	$3\frac{7}{8}$	0.97
			X95	5	$2\frac{7}{16}$	$4\frac{37}{64}$	12	8	$10\frac{1}{2}$	$18\frac{1}{2}$	$3\frac{7}{8}$	$3\frac{7}{8}$	0.83
			G105	5	$2\frac{1}{8}$	$4\frac{37}{64}$	12	8	$10\frac{1}{2}$	$18\frac{1}{2}$	$3\frac{7}{8}$	$3\frac{7}{8}$	0.90
NC40(4FH)	$3\frac{1}{2}$ EU	15.50	S135	$5\frac{1}{2}$	$2\frac{1}{4}$	$5\frac{1}{64}$	$11\frac{1}{2}$	7	10	17	$3\frac{7}{8}$	$3\frac{7}{8}$	0.87
	4 IU	14.00	E75	$5\frac{1}{4}$*	$2\frac{13}{16}$*	$5\frac{1}{64}$	$11\frac{1}{2}$	7	10	17	$4\frac{3}{16}$	$4\frac{3}{16}$	1.01
			X95	$5\frac{1}{4}$*	$2\frac{11}{16}$	$5\frac{1}{64}$	$11\frac{1}{2}$	7	10	17	$4\frac{3}{16}$	$4\frac{3}{16}$	0.86
			G105	$5\frac{1}{2}$	$2\frac{7}{16}$	$5\frac{1}{64}$	$11\frac{1}{2}$	7	10	17	$4\frac{3}{16}$	$4\frac{3}{16}$	0.93
			S135	$5\frac{1}{2}$	2	$5\frac{1}{64}$	$11\frac{1}{2}$	7	10	17	$4\frac{3}{16}$	$4\frac{3}{16}$	0.87
NC46(41F)	4 EU	14.00	E75	6*	$3\frac{1}{4}$*	$5\frac{23}{32}$	$11\frac{1}{2}$	7	10	17	$4\frac{1}{2}$	$4\frac{1}{2}$	1.43
			X95	6*	$3\frac{1}{4}$*	$5\frac{23}{32}$	$11\frac{1}{2}$	7	10	17	$4\frac{1}{2}$	$4\frac{1}{2}$	1.13
			G105	6*	$3\frac{1}{4}$*	$5\frac{23}{32}$	$11\frac{1}{2}$	7	10	17	$4\frac{1}{2}$	$4\frac{1}{2}$	1.02
			S135	6*	3	$5\frac{23}{32}$	$11\frac{1}{2}$	7	10	17	$4\frac{1}{2}$	$4\frac{1}{2}$	0.94
	$4\frac{1}{2}$ IU	13.75	E75	6	$3\frac{3}{8}$	$5\frac{23}{32}$	$11\frac{1}{2}$	7	10	17	$4\frac{11}{16}$	$4\frac{11}{16}$	1.20
	$4\frac{1}{2}$ IEU	16.60	E75	$6\frac{1}{4}$*	$3\frac{1}{4}$*	$5\frac{23}{32}$	$11\frac{1}{2}$	7	10	17	$4\frac{11}{16}$	$4\frac{11}{16}$	1.09
			X95	$6\frac{1}{4}$*	3	$5\frac{23}{32}$	$11\frac{1}{2}$	7	10	17	$4\frac{11}{16}$	$4\frac{11}{16}$	1.01
			G105	$6\frac{1}{4}$*	3	$5\frac{23}{32}$	$11\frac{1}{2}$	7	10	17	$4\frac{11}{16}$	$4\frac{11}{16}$	0.91
			S135	$6\frac{1}{4}$*	$2\frac{3}{4}$	$5\frac{23}{32}$	$11\frac{1}{2}$	7	10	17	$4\frac{11}{16}$	$4\frac{11}{16}$	0.81
	$4\frac{1}{2}$ IEU	20.00	E75	$6\frac{1}{4}$*	3	$5\frac{23}{32}$	$11\frac{1}{2}$	7	10	17	$4\frac{11}{16}$	$4\frac{11}{16}$	1.07
			X95	$6\frac{1}{4}$*	$2\frac{3}{4}$	$5\frac{23}{32}$	$11\frac{1}{2}$	7	10	17	$4\frac{11}{16}$	$4\frac{11}{16}$	0.96
			G105	$6\frac{1}{4}$*	$2\frac{1}{2}$	$5\frac{23}{32}$	$11\frac{1}{2}$	7	10	17	$4\frac{11}{16}$	$4\frac{11}{16}$	0.96
			S135	$6\frac{1}{4}$*	$2\frac{1}{4}$	$5\frac{23}{32}$	$11\frac{1}{2}$	7	10	17	$4\frac{11}{16}$	$4\frac{11}{16}$	0.81

$4\frac{1}{2}$ FH**	$4\frac{1}{2}$ IEU	16.60	E75	6*	3*	$5\frac{23}{32}$	11	7	10	17	$4\frac{11}{16}$	$4\frac{11}{16}$	1.12
			X95	6*	3*	$5\frac{23}{32}$	11	7	10	17	$4\frac{11}{16}$	$4\frac{11}{16}$	0.89
			G105	6*	3*	$5\frac{23}{32}$	11	7	10	17	$4\frac{11}{16}$	$4\frac{11}{16}$	0.81
			S135	$6\frac{1}{4}$	$2\frac{1}{2}$	$5\frac{23}{32}$	11	7	10	17	$4\frac{11}{16}$	$4\frac{11}{16}$	0.81
	$4\frac{1}{2}$ IEU	20.00	E75	6*	3*	$5\frac{23}{32}$	11	7	10	17	$4\frac{11}{16}$	$4\frac{11}{16}$	0.95
			X95	6*	$2\frac{1}{2}$	$5\frac{23}{32}$	11	7	10	17	$4\frac{11}{16}$	$4\frac{11}{16}$	0.95
			G105	6*	$2\frac{1}{2}$	$5\frac{23}{32}$	11	7	10	17	$4\frac{11}{16}$	$4\frac{11}{16}$	0.86
NC50($4\frac{1}{2}$IF)	$4\frac{1}{2}$ EU	13.75	E75	$6\frac{1}{4}$	$3\frac{7}{8}$	$5\frac{59}{64}$	$11\frac{1}{2}$	7	10	17	5	5	1.32
	$4\frac{1}{2}$ EU	16.60	E75	$6\frac{3}{8}$*	$3\frac{3}{4}$*	$5\frac{59}{64}$	$11\frac{1}{2}$	7	10	17	5	5	1.23
			X95	$6\frac{3}{8}$*	$3\frac{3}{4}$*	$5\frac{59}{64}$	$11\frac{1}{2}$	7	10	17	5	5	0.97
			G105	$6\frac{3}{8}$*	$3\frac{3}{4}$*	$5\frac{59}{64}$	$11\frac{1}{2}$	7	10	17	5	5	0.88
			S135	$6\frac{3}{8}$*	$3\frac{1}{2}$	$5\frac{59}{64}$	$11\frac{1}{2}$	7	10	17	5	5	0.81
	$4\frac{1}{2}$ EU	20.00	E75	$6\frac{3}{8}$*	$3\frac{5}{8}$*	$5\frac{59}{64}$	$11\frac{1}{2}$	7	10	17	5	5	1.02
			X95	$6\frac{3}{8}$*	$3\frac{1}{2}$	$5\frac{59}{64}$	$11\frac{1}{2}$	7	10	17	5	5	0.96
			G105	$6\frac{3}{8}$*	$3\frac{1}{2}$	$5\frac{59}{64}$	$11\frac{1}{2}$	7	10	17	5	5	0.86
			S135	$6\frac{5}{8}$	3	$5\frac{59}{64}$	$11\frac{1}{2}$	7	10	17	5	5	0.87
	5 IEU	19.50	E75	$6\frac{3}{8}$*	$3\frac{3}{4}$*	$5\frac{59}{64}$	$11\frac{1}{2}$	7	10	17	$5\frac{1}{8}$	$5\frac{1}{8}$	0.92
			X95	$6\frac{3}{8}$*	$3\frac{1}{2}$	$5\frac{59}{64}$	$11\frac{1}{2}$	7	10	17	$5\frac{1}{8}$	$5\frac{1}{8}$	0.86
			G105	$6\frac{1}{2}$	$3\frac{1}{4}$	$5\frac{59}{64}$	$11\frac{1}{2}$	7	10	17	$5\frac{1}{8}$	$5\frac{1}{8}$	0.89
			S135	$6\frac{5}{8}$	$2\frac{3}{4}$	$5\frac{59}{64}$	$11\frac{1}{2}$	7	10	17	$5\frac{1}{8}$	$5\frac{1}{8}$	0.86
	5 IEU	25.60	E75	$6\frac{3}{8}$*	$3\frac{1}{2}$	$5\frac{59}{64}$	$11\frac{1}{2}$	7	10	17	$5\frac{1}{8}$	$5\frac{1}{8}$	0.86
			X95	$6\frac{1}{2}$	3	$5\frac{59}{64}$	$11\frac{1}{2}$	7	10	17	$5\frac{1}{8}$	$5\frac{1}{8}$	0.86
			G105	$6\frac{5}{8}$	$2\frac{3}{4}$	$5\frac{59}{64}$	$11\frac{1}{2}$	7	10	17	$5\frac{1}{8}$	$5\frac{1}{8}$	0.87
$5\frac{1}{2}$ FH	5 IEU	19.50	E76	7*	$3\frac{3}{4}$	$6\frac{23}{32}$	13	8	10	18	$5\frac{1}{8}$	$5\frac{1}{8}$	1.53
			X95	7*	$3\frac{3}{4}$	$6\frac{23}{32}$	13	8	10	18	$5\frac{1}{8}$	$5\frac{1}{8}$	1.21
			G105	7*	$3\frac{3}{4}$	$6\frac{23}{32}$	13	8	10	18	$5\frac{1}{8}$	$5\frac{1}{8}$	1.09
			S135	$7\frac{1}{4}$	$3\frac{1}{2}$	$6\frac{23}{32}$	13	8	10	18	$5\frac{1}{8}$	$5\frac{1}{8}$	0.98
	5 IEU	25.60	E75	7*	$3\frac{1}{2}$	$6\frac{23}{32}$	13	8	10	18	$5\frac{1}{8}$	$5\frac{1}{8}$	1.21
			X95	7*	$3\frac{1}{2}$	$6\frac{23}{32}$	13	8	10	18	$5\frac{1}{8}$	$5\frac{1}{8}$	0.95
			G105	$7\frac{1}{4}$	$3\frac{1}{2}$	$6\frac{23}{32}$	13	8	10	18	$5\frac{1}{8}$	$5\frac{1}{8}$	0.99
			S135	$7\frac{1}{4}$	$3\frac{1}{4}$	$6\frac{23}{32}$	13	8	10	18	$5\frac{1}{8}$	$5\frac{1}{8}$	0.83
	$5\frac{1}{2}$ IEU	21.90	E75	7*	4	$6\frac{23}{32}$	13	8	10	18	$5\frac{11}{16}$	$5\frac{11}{16}$	1.11
			X95	7*	$3\frac{3}{4}$	$6\frac{23}{32}$	13	8	10	18	$5\frac{11}{16}$	$5\frac{11}{16}$	0.98
			G105	$7\frac{1}{4}$	$3\frac{1}{2}$	$6\frac{23}{32}$	13	8	10	18	$5\frac{11}{16}$	$5\frac{11}{16}$	1.02
			S135	$7\frac{1}{2}$	3	$7\frac{3}{32}$	13	8	10	18	$5\frac{11}{16}$	$5\frac{11}{16}$	0.96
	$5\frac{1}{2}$ IEU	24.70	E75	7*	4*	$6\frac{23}{32}$	13	8	10	18	$5\frac{11}{16}$	$5\frac{11}{16}$	0.99
			X95	$7\frac{1}{4}$	$3\frac{1}{2}$	$6\frac{23}{32}$	13	8	10	18	$5\frac{11}{16}$	$5\frac{11}{16}$	1.01
			G105	$7\frac{1}{4}$	$3\frac{1}{2}$	$6\frac{23}{32}$	13	8	10	18	$5\frac{11}{16}$	$5\frac{11}{16}$	0.92
			S135	$7\frac{1}{2}$	3	$7\frac{3}{32}$	13	8	10	18	$5\frac{11}{16}$	$5\frac{11}{16}$	0.86

* Denotes standard OD or standard ID.

** Obsolescent connection.

[1]The tool joint designation (Col. 1) indicates the size and style of the applicable connection. See special note.

[2]Nominal weights, threads and couplings, (Col. 3) are shown for the purpose of identification in ordering.

[3]The inside diameter (Col. 6) does not apply to box members, which are optional with the manufacturer.

[4]Length of pin thread reduced to $3\frac{1}{2}$ in ($\frac{1}{2}$ in short) to accommodate 3 in ID.

Special Note: IF denotes internal flush design in which the pin bore is approximately equal to the internal diameter of the non-upset section of drillpipe. This type is used on externally upset drillpipe. FH denotes full hole design in which the pin bore is equal to the internal diameter of the internal upset section of the drillpipe.

TABLE 2.7 Product dimensions of rotary shouldered connections (see Figure 2.8). Italicised connections are tentative. All dimensions in inches. (Courtesy of API[3])

1	*2*	*3*	*4*	*5*	*6*	*7*	*8*	*9*	*10*	*11*
Conn. number or size*	*Thread form*	*Threads per inch*	*Taper (in/ft on dia.)*	*Pitch dia. of thread at gauge point* C	*Large dia. of pin* D_L	*Dia. of flat on pin*[1], $\pm\frac{1}{64}$ D_{LP}	*Small dia. of pin* D_S	*Length of pin,* $+0$ $-\frac{1}{8}$ L_{PC}	*Depth*[2] *of box,* $+\frac{3}{8}$ -0 L_{BC}	*Box counter-bore,* $+\frac{1}{32}-\frac{1}{64}$ Q_c
				Number (NC) style						
NC10	V-0.055	6	$1\frac{1}{2}$	1.063	1.190	...	1.002	$1\frac{1}{2}$	$2\frac{1}{8}$	1.204†
NC12	V-0.055	6	$1\frac{1}{2}$	1.265	1.392	...	1.173	$1\frac{3}{4}$	$2\frac{3}{8}$	1.406†
NC13	V-0.055	6	$1\frac{1}{2}$	1.391	1.518	...	1.299	$1\frac{3}{4}$	$2\frac{3}{8}$	1.532†
NC16	V-0.055	6	$1\frac{1}{2}$	1.609	1.736	...	1.517	$1\frac{3}{4}$	$2\frac{3}{8}$	1.751†
NC23	V-0.038R	4	2	2.355 00	2.563	2.437	2.063	3	$3\frac{5}{8}$	$2\frac{5}{8}$
NC26**	V-0.038R	4	2	2.668 00	2.876	2.750	2.376	3	$3\frac{5}{8}$	$2\frac{13}{16}$
NC31**	V-0.038R	4	2	3.183 00	3.391	3.266	2.808	$3\frac{1}{2}$	$4\frac{1}{8}$	$3\frac{29}{64}$
NC35	V-0.038R	4	2	3.531 00	3.739	3.625	3.114	$3\frac{3}{4}$	$4\frac{3}{8}$	$3\frac{15}{16}$
NC38**	V-0.038R	4	2	3.808 00	4.016	3.891	3.349	4	$4\frac{5}{8}$	$4\frac{5}{64}$
NC40**	V-0.038R	4	2	4.072 00	4.280	4.156	3.530	$4\frac{1}{2}$	$5\frac{1}{8}$	$4\frac{11}{32}$
NC44	V-0.038R	4	2	4.417 00	4.625	4.499	3.875	$4\frac{1}{2}$	$5\frac{1}{8}$	$4\frac{11}{16}$
NC46**	V-0.038R	4	2	4.626 00	4.834	4.709	4.084	$4\frac{1}{2}$	$5\frac{1}{8}$	$4\frac{23}{32}$
NC50**	V-0.038R	4	2	5.041 70	5.250	5.135	4.500	$4\frac{1}{2}$	$5\frac{1}{8}$	$5\frac{5}{16}$
NC56	V-0.038R	4	3	5.616 00	5.876	5.703	4.626	5	$5\frac{5}{8}$	$5\frac{15}{16}$
NC61	V-0.038R	4	3	6.178 00	6.438	6.266	5.063	$5\frac{1}{2}$	$6\frac{1}{8}$	$6\frac{1}{2}$
NC70	V-0.038R	4	3	7.053 00	7.313	7.141	5.813	6	$6\frac{5}{8}$	$7\frac{3}{8}$
NC77	*V-0.038R*	*4*	*3*	*7.741 00*	*8.000*	*7.828*	*6.376*	$6\frac{1}{2}$	$7\frac{1}{8}$	$8\frac{1}{16}$
				Regular (Reg.) style						
$2\frac{3}{8}$REG	V-0.040	5	3	2.365 37	2.625	2.515	1.875	3	$3\frac{5}{8}$	$2\frac{11}{16}$
$2\frac{7}{8}$REG	V-0.040	5	3	2.740 37	3.000	2.890	2.125	$3\frac{1}{2}$	$4\frac{1}{8}$	$3\frac{1}{16}$
$3\frac{1}{2}$REG	V-0.040	5	3	3.239 87	3.500	3.390	2.562	$3\frac{3}{4}$	$4\frac{3}{8}$	$3\frac{9}{16}$
$4\frac{1}{2}$REG	V-0.040	5	3	4.364 87	4.625	4.515	3.562	$4\frac{1}{4}$	$4\frac{7}{8}$	$4\frac{11}{16}$
$5\frac{1}{2}$REG	V-0.050	4	3	5.234 02	5.520	5.410	4.333	$4\frac{3}{4}$	$5\frac{3}{8}$	$5\frac{31}{64}$
$6\frac{5}{8}$REG	V-0.050	4	2	5.757 80	5.992	5.882	5.159	5	$5\frac{5}{8}$	$6\frac{1}{16}$
$7\frac{5}{8}$REG	V-0.050	4	3	6.714 53	7.000	6.890	5.688	$5\frac{1}{4}$	$5\frac{7}{8}$	$7\frac{3}{32}$
$8\frac{5}{8}$REG	V-0.050	4	3	7.666 58	7.952	7.840	6.608	$5\frac{3}{8}$	6	$8\frac{3}{64}$
				Full-hole (FH) style						
$5\frac{1}{2}$FH	V-0.050	4	2	5.591 00	5.825	...	4.992	5	$5\frac{5}{8}$	$5\frac{5}{8}$

*The number of the connection in the number style is the pitch diameter of the pin thread at the gauge point, rounded to units and tenths of inches.

**Connections in the number (NC) style are interchangeable with connections having the same pitch diameter in the FH and IF styles.

†Box counterbore (Q_c) tolerance is ± 0.005 in on Connections NC10, NC12, NC13 and NC16.

[1] Dimension D_{LF} and the 0.062 in radius at the pin base are standard for drill collars and optional with the manufacturer for other drill stem elements.

[2] The length of perfect threads in the box shall not be less than the maximum pin length (L_{PC}) plus $\frac{1}{8}$ in.

grooving has a much reduced contact area; this greatly reduces the magnitude of the differential sticking force, as outlined in Chapter 12. Spiral grooving only reduces the weight of the drill collar joint by 4% (Figure 2.11d).

Square drill collars

Squared-section drill collars are used for special drilling purposes such as reducing deviation in crooked hole formation and to maintain the existing hole direction in directional drilling.

TABLE 2.8 Upset drillpipe for weld-on tool joints. (Courtesy of API[5])

1	2	3	4	5	6	7	8	9	10	11	12		13
Pipe size: outside dia. (in)	Nominal wt.[1] (lb/ft)	Wall thickness (in)	Inside diameter (in)	Calculated weight		Upset dimensions (in)[5]							
				Plain end (lb/ft)	Upset[4] (lb)	Outside diameter[2], $+\frac{1}{8}$, $-\frac{1}{32}$	Inside diameter at end of pipe[3], $\pm\frac{1}{16}$	Length of internal upset, $+1\frac{1}{2}$ $-\frac{1}{2}$	Length of internal taper, min.	Length of external upset min.	Length of external taper		Length end of pipe to taper fadeout, max.
											min.	max.	
D		t	d	W_{PE}	e_w	D_{ou}	d_{ou}	L_{iu}	M_{iu}	L_{eu}	M_{eu}		$L_{eu}+M_{eu}$
						Internal upset drillpipe							
$2\frac{7}{8}$	10.40	0.362	2.151	9.72	3.20	2.875	$1\frac{5}{16}$	$1\frac{3}{4}$	$1\frac{1}{2}$	...	...	...	...
$3\frac{1}{2}$	9.50	0.254	2.992	8.81	4.40	3.500	$2\frac{1}{4}$	$1\frac{3}{4}$	...	...	...	...	...
$3\frac{1}{2}$	13.30	0.368	2.764	12.31	4.40	3.500	$1\frac{15}{16}$	$1\frac{3}{4}$	$1\frac{1}{2}$	...	...	...	...
$3\frac{1}{2}$	15.50	0.449	2.602	14.63	3.40	3.500	$1\frac{15}{16}$	$1\frac{3}{4}$	$1\frac{1}{2}$	...	...	...	...
4*	11.85	0.262	3.476	10.46	4.20	4.000	$2\frac{15}{16}$	$1\frac{3}{4}$	...	...	...	...	...
4	14.00	0.330	3.340	12.93	4.60	4.000	$2\frac{3}{4}$	$1\frac{3}{4}$	2	...	...	...	...
$4\frac{1}{2}$*	13.75	0.271	3.958	12.24	5.20	4.500	$3\frac{3}{8}$	$1\frac{3}{4}$	...	...	...	...	...
5*	16.25	0.296	4.408	14.87	6.60	5.000	$3\frac{3}{4}$	$1\frac{3}{4}$	...	...	...	...	...
						External upset drillpipe							
$2\frac{3}{8}$	6.65	0.280	1.815	6.26	1.80	2.656	1.815	...	...	$1\frac{1}{2}$	$1\frac{1}{2}$	...	4
$2\frac{7}{8}$	10.40	0.362	2.151	9.72	2.40	3.219	2.151	...	...	$1\frac{1}{2}$	$1\frac{1}{2}$	...	4
$3\frac{1}{2}$	9.50	0.254	2.992	8.81	2.60	3.824	2.992	...	...	$1\frac{1}{2}$	$1\frac{1}{2}$	...	4
$3\frac{1}{2}$	13.30	0.368	2.764	12.31	4.00	3.824	2.602	$2\frac{1}{4}$	2	$1\frac{1}{2}$	$1\frac{1}{2}$	...	4
$3\frac{1}{2}$	15.50	0.449	2.602	14.63	2.80	3.824	2.602	...	...	$1\frac{1}{2}$	$1\frac{1}{2}$	...	4
4*	11.85	0.262	3.476	10.46	5.00	4.500	3.476	...	...	$1\frac{1}{2}$	$1\frac{1}{2}$	...	4
4	14.00	0.330	3.340	12.93	5.00	4.500	3.340	...	...	$1\frac{1}{2}$	$1\frac{1}{2}$	...	4
$4\frac{1}{2}$*	13.75	0.271	3.958	12.24	5.60	5.000	3.958	...	...	$1\frac{1}{2}$	$1\frac{1}{2}$	...	4
$4\frac{1}{2}$	16.60	0.337	3.826	14.98	5.60	5.000	3.826	...	...	$1\frac{1}{2}$	$1\frac{1}{2}$	...	4
$4\frac{1}{2}$	20.00	0.430	3.640	18.69	5.60	5.000	3.640	...	...	$1\frac{1}{2}$	$1\frac{1}{2}$	...	4
						Internal–external upset drillpipe							
$4\frac{1}{2}$	16.60	0.337	3.826	14.98	8.10	4.656	$3\frac{5}{32}$	$2\frac{1}{2}$	...	$1\frac{1}{2}$	1	$1\frac{1}{2}$	...
$4\frac{1}{2}$	20.00	0.430	3.640	18.69	8.60	4.781	3	$2\frac{1}{4}$	2	$1\frac{1}{2}$	1	$1\frac{1}{2}$	...
5	19.50	0.362	4.276	17.93	8.60	5.188	$3\frac{11}{16}$	$2\frac{1}{4}$	2	$1\frac{1}{2}$	1	$1\frac{1}{2}$	...
5	25.60	0.500	4.000	24.03	7.80	5.188	$3\frac{7}{16}$	$2\frac{1}{4}$	2	$1\frac{1}{2}$	1	$1\frac{1}{2}$	...
$5\frac{1}{2}$	21.90	0.361	4.778	19.81	10.60	5.563	4	$2\frac{1}{4}$	2	$1\frac{1}{2}$	1	$1\frac{1}{2}$	...
$5\frac{1}{2}$	24.70	0.415	4.670	22.54	9.00	5.563	4	$2\frac{1}{4}$	2	$1\frac{1}{2}$	1	$1\frac{1}{2}$	...

[1] Nominal weights (Col. 2), are shown for the purpose of identification in ordering.
[2] The ends of internal upset drillpipe shall not be smaller in outside diameter than the values shown in Col. 7, including the minus tolerance. They may be furnished with slight external upset, within the tolerance specified.
[3] Maximum taper on inside diameter of internal upset and internal–external upset is $\frac{1}{4}$ in/ft on diameter.
[4] Weight gain or loss due to end finishing.
[5] The specified upset dimensions do not necessarily agree with the bore and OD dimensions of finished weld-on assemblies. Upset dimensions were chosen to accommodate the various bores of tool joints and to maintain a satisfactory cross-section in the weld zone after final machining of the assembly.
* These sizes and weights are tentative.

TABLE 2.9 Drill collar weight (steel: lb/ft). (Courtesy of API[3])

1	*2*	*3*	*4*	*5*	*6*	*7*	*8*	*9*	*10*	*11*	*12*	*13*	*14*
Drill collar OD (in)	*Drill collar ID (in)*												
	1	$1\frac{1}{4}$	$1\frac{1}{2}$	$1\frac{3}{4}$	2	$2\frac{1}{4}$	$2\frac{1}{2}$	$2\frac{13}{16}$	3	$3\frac{1}{4}$	$3\frac{1}{2}$	$3\frac{3}{4}$	4
$2\frac{7}{8}$	19	18	16										
3	21	20	18										
$3\frac{1}{8}$	22	22	20										
$3\frac{1}{4}$	26	24	22										
$3\frac{1}{2}$	30	29	27										
$3\frac{3}{4}$	35	33	32										
4	40	39	37	35	32	29							
$4\frac{1}{8}$	43	41	39	37	35	32							
$4\frac{1}{4}$	46	44	42	40	38	35							
$4\frac{1}{2}$	51	50	48	46	43	41							
$4\frac{3}{4}$			54	52	50	47	44						
5			61	59	56	53	50						
$5\frac{1}{4}$			68	65	63	60	57						
$5\frac{1}{2}$			75	73	70	67	64	60					
$5\frac{3}{4}$			82	80	78	75	72	67	64	60			
6			90	88	85	83	79	75	72	68			
$6\frac{1}{4}$			98	96	94	91	88	83	80	76	72		
$6\frac{1}{2}$			107	105	102	99	96	91	89	85	80		
$6\frac{3}{4}$			116	114	111	108	105	100	98	93	89		
7			125	123	120	117	114	110	107	103	98	93	84
$7\frac{1}{4}$			134	132	130	127	124	119	116	112	108	103	93
$7\frac{1}{2}$			144	142	139	137	133	129	126	122	117	113	102
$7\frac{3}{4}$			154	152	150	147	144	139	136	132	128	123	112
8			165	163	160	157	154	150	147	143	138	133	122
$8\frac{1}{4}$			176	174	171	168	165	160	158	154	149	144	133
$8\frac{1}{2}$			187	185	182	179	176	172	169	165	160	155	150
9			210	208	206	203	200	195	192	188	184	179	174
$9\frac{1}{2}$			234	232	230	227	224	220	216	212	209	206	198
$9\frac{3}{4}$			248	245	243	240	237	232	229	225	221	216	211
10			261	259	257	254	251	246	243	239	235	230	225
11			317	315	313	310	307	302	299	295	291	286	281
12			379	377	374	371	368	364	361	357	352	347	342

K-Monel drill collars

K-Monel drill collars are manufactured from non-magnetic steel alloys, and are used to shield directional survey instruments from the magnetic effects of normal steel drill collars.

DRILL STRING ACCESSORIES

Heavy-wall drillpipe (HWDP)

Field experience[7] has shown that fatigue failure of drillpipe can be greatly reduced when the section modulus ratio at the interface of two different-sized drill stem members (i.e. at drillpipe/drill collar interface) is limited to 5.5 or below. The section modulus is defined as the ratio of the moment of inertia (I) to the external radius of the pipe.

In a drill string consisting of drill collars and standard drillpipe the ratio of section modulus of drill collar to section modulus of drillpipe at the interface is much larger than 5.5. Thus, to limit the ratio to 5.5 or less, a heavy-walled drillpipe section is inserted between the drill collars and drillpipe. The HWDP has the same OD as standard drillpipe but a much reduced ID, and has an extra-long tool joint. The use of HWDP between standard drillpipe and drill collars

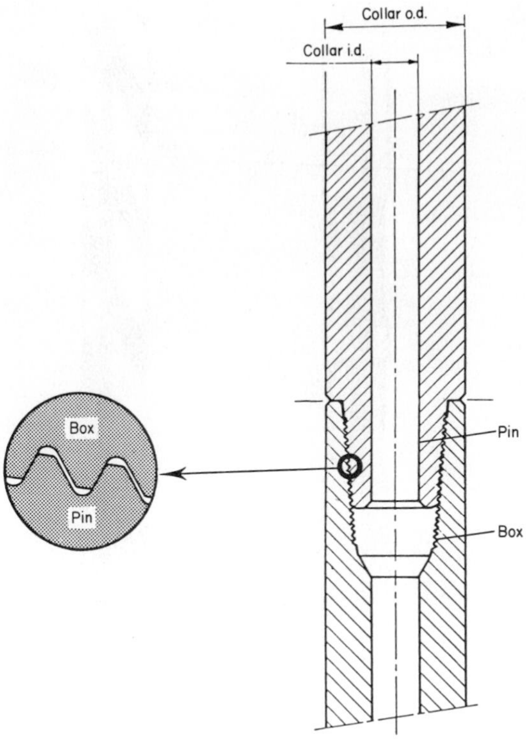

Fig. 2.10. Drill collar connection.

results in a much reduced section modulus at the interfaces.

HWDP is manufactured in four sizes, from $3\frac{1}{2}$ in OD to 5 in OD, and in lengths of 30.5 ft (9.3 m)[1]. HWDP can be distinguished by an integral centre wear pad which acts as a stabiliser, thereby increasing the overall stiffness of the drill string. Figure 2.12 gives a schematic view of an HWDP.

The HWDP is also used to ensure that the drillpipe is kept in constant tension (to avoid buckling) by having the neutral point of zero tension and zero compression in the length of HWDP. The HWDP can be safely run in compression, owing to its high stiffness compared with that of drillpipe.

In directional wells HWDP is also used to provide weight-on-bit in addition to the weight supplied by drill collars. The number of joints usually run varies from 30 to 70.

Drilling jar

A drilling jar may be defined[8] as a mandrel which slides within a sleeve, as shown in Figure 2.13. The free end of the mandrel is shaped in the form of a hammer, to provide a striking action against the face of the anvil. Jarring action is required during operations to free stuck pipe when an upward or downward pull is required to free the pipe.

Within the drill string, the mandrel end of the jar is connected to one end of the string, while the sleeve is connected to the other end[8]. Referring to Figure 2.13, upward movement of the mandrel causes the hammer to strike (or jar) against the anvil, producing an upward force on the drill string portion below the jar. A downward force may be obtained by reversing the position of the jar within the drill string. Jars are recommended to run in tension such that they are placed above the neutral point. Jars run in compression may trip while drilling if too much weight is accidentally applied on the bit[8].

In general, there are three types of jar: mechanical; hydraulic; and mixed, or hydromechanical.

Stabilisers

Stabilisers are tools placed above the bit and along the bottom hole assembly in order to control hole deviation, minimise dog-leg severity and prevent differential sticking. They achieve these functions by centralising and providing extra stiffness to the bottom hole assembly. Stabilisation also allows the bit to rotate perpendicular to the hole bottom and consequently improves its performance.

There are two basic types of stabiliser.

Rotating stabilisers

Rotating stabilisers include the integral blade stabiliser (Figure 2.14a), the sleeve stabiliser (Figures 2.14b and c) and the welded blade stabiliser (Figure 2.14d).

Integral blade stabilisers are machined from a solid piece of high-strength steel alloy. The blade faces are dressed with sintered tungsten carbide inserts. Integral blade stabilisers are available with either straight or spiral blades (Figure 2.14a).

A sleeve stabiliser comprises a body and a replaceable sleeve. Two different designs of sleeve stabilisers are available. In the first design the sleeve is screwed onto a one-piece body and made up with rotary tongs (Figure 2.14b). In the second design the sleeve is shrunk-fit to the main body (Figure 2.14c).

The welded blade stabiliser is particularly useful in soft formations where balling up of the mud and cuttings on the drill collar string may be a problem. The blades are welded to the body of the stabiliser and hard-faced with tungsten carbide to resist chipping and lengthen blade life (Figure 2.14d). The blades have

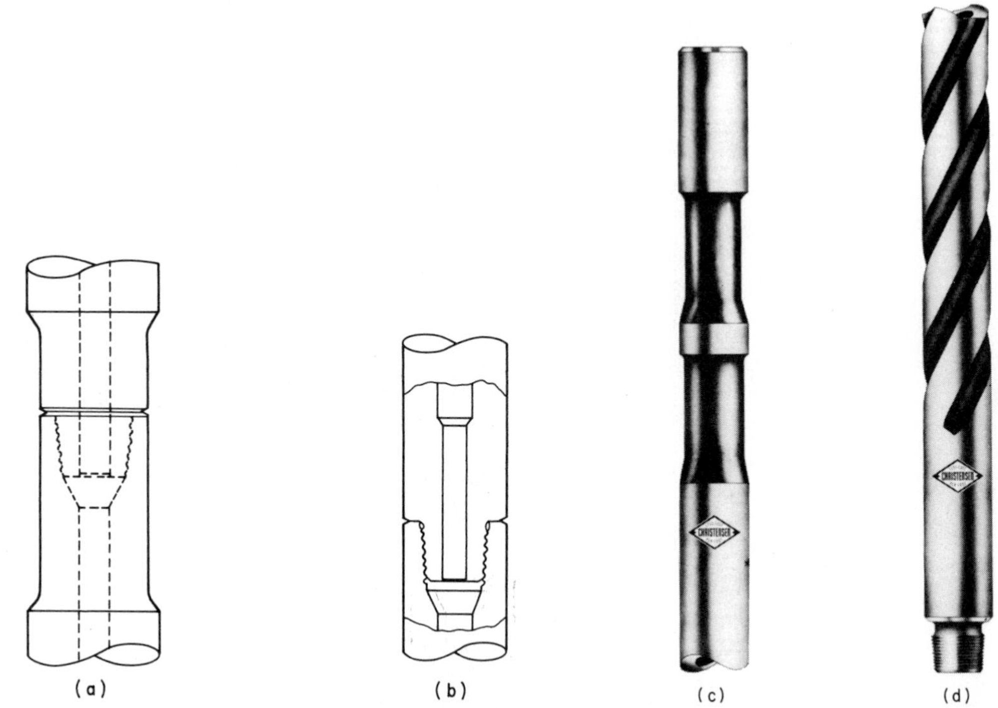

Fig. 2.11. Special features available on drill collars: (a) fishing neck; (b) stepped bore; (c) slip and elevator recesses; (d) spiral grooving. (Courtesy of Christensen Drilling Products)

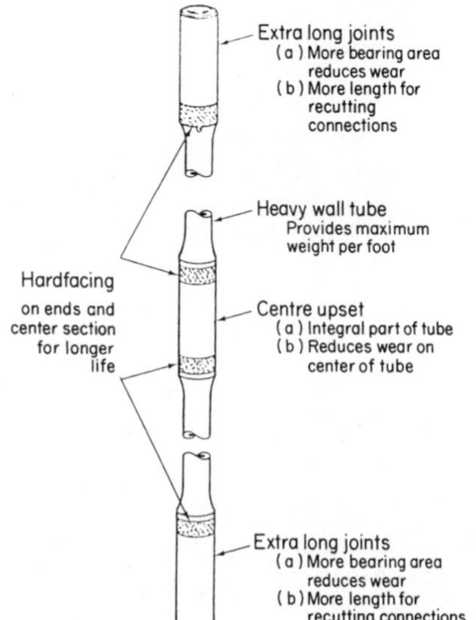

Fig. 2.12. Features of heavy walled drillpipe[1]. (Courtesy of Drilco)

bevelled edges to allow the stabiliser easy movement through casing and key seats.

Non-rotating stabilisers

A non-rotating stabiliser comprises a rubber sleeve and a mandrel (Figure 2.14e). The sleeve is designed to remain stationary while the mandrel and the drill string are rotating. This type is used to prevent reaming of the hole wall during drilling operations and to protect the drill collar from wall contact wear.

Reamers

Rotary reamers are provided with tungsten carbide rollers set vertically in the body of the reamer. Reamers can have three or six reaming cutters, as shown in Figure 2.15. Reamers are usually run immediately behind the bit to provide a gauge hole when drilling hard and abrasive rocks.

Shock subs

A shock sub, or vibration dampener, is a tool placed between the bit and drill collars. It is used to absorb

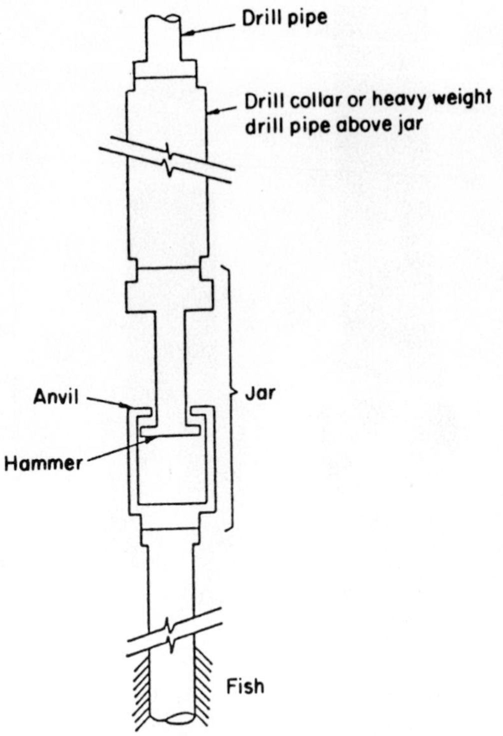

Fig. 2.13. Put simply, a drilling jar consists of a mandrel that slides within a sleeve. Upward motion of the mandrel causes its hammer to jar against the anvil, producing an upward force on the drill string below. Downward mandrel motion produces the opposite effect. (Courtesy of *World Oil*[7])

vibrations and impact loads caused by the movement of the drill string. Shock subs employ steel or rubber springs designed to absorb torsional and vertical vibration. Vibration occurs as the drill bit bounces off the formations during drilling, or when the drill string is rotating at its critical speed(s). Downhole vibration can result in drill collar failure, reduced bit life and limitation on the amount of weight or rotary speed that can be used[1]. A shock sub is used to eliminate these vibrations and increase drill string life. A shock sub is normally placed directly above the bit, to act as a large shock absorber. However, certain stabilisation requirements dictate the placement of a shock sub at 30 or 60 ft above the bit (see Chapter 8).

DRILL STRING DESIGN

Drill string design involves the determination of the lengths, weights and grades of drillpipe which can be used during drilling, coring or any other operation. Drill string design depends on several factors[3], including: hole depth and hole size; mud weight; desired safety factor in tension and/or margin of overpull; length and weight of drill collars; and desired drillpipe sizes and inspection class.

The following design criteria will be used to select a suitable drill string: (a) tension; (b) collapse; (c) shock loading; (d) torsion.

Tension

Prior to deriving any equation, it should be observed that only submerged weights are considered, since all immersed bodies suffer from lifting or buoyancy forces. Buoyancy force reduces the total weight of the body and its magnitude is dependent on fluid density. Buoyancy effects will also be discussed in Chapter 10.

Referring to Figure 2.16, the total weight, P, carried by the top joint of drillpipe at JJ is given by

$$P = (\text{weight of drillpipe in mud}) + (\text{weight of drill collars in mud})$$

(*Note:* Weight of drill bit and other accessories is normally included within the weight of the drill collars.)

$$P = [(L_{dp} \times W_{dp} + L_{dc} \times W_{dc})] \times \text{BF} \qquad (2.1)$$

where L_{dp} = length of drillpipe; W_{dp} = weight of drillpipe per unit length; L_{dc} = length of drill collars; W_{dc} = weight of drill collars per unit length; and BF = buoyancy factor. The buoyancy factor is given by

$$\text{BF} = \left(1 - \frac{\rho_m}{\rho_s}\right) = \left(1 - \frac{\gamma_m}{\gamma_s}\right)$$

where γ_m = specific gravity of mud; ρ_m = density of mud; γ_s = specific gravity of steel = 7.85; and ρ_s = density of steel = 489.5 lb/ft^3 (7850 kg/m^3). (*Note:* The BF relationship will be derived in Chapter 10.)

As can be seen from Tables 2.2–2.5, drillpipe strength is expressed in terms of yield strength. This is defined as the load at which deformation occurs. Under all conditions of loading, steel elongates initially linearly in relation to the applied load until the elastic limit is reached. Up to this limit, removal of applied load results in the steel pipe recovering its original dimensions. Loading a steel pipe beyond the elastic limit induces deformation which cannot be recovered, even after the load is removed. This deformation is described as yield and results in a reduction in pipe strength.

In practice, the condition of pipe and its service life are taken into account when a drill string is designed.

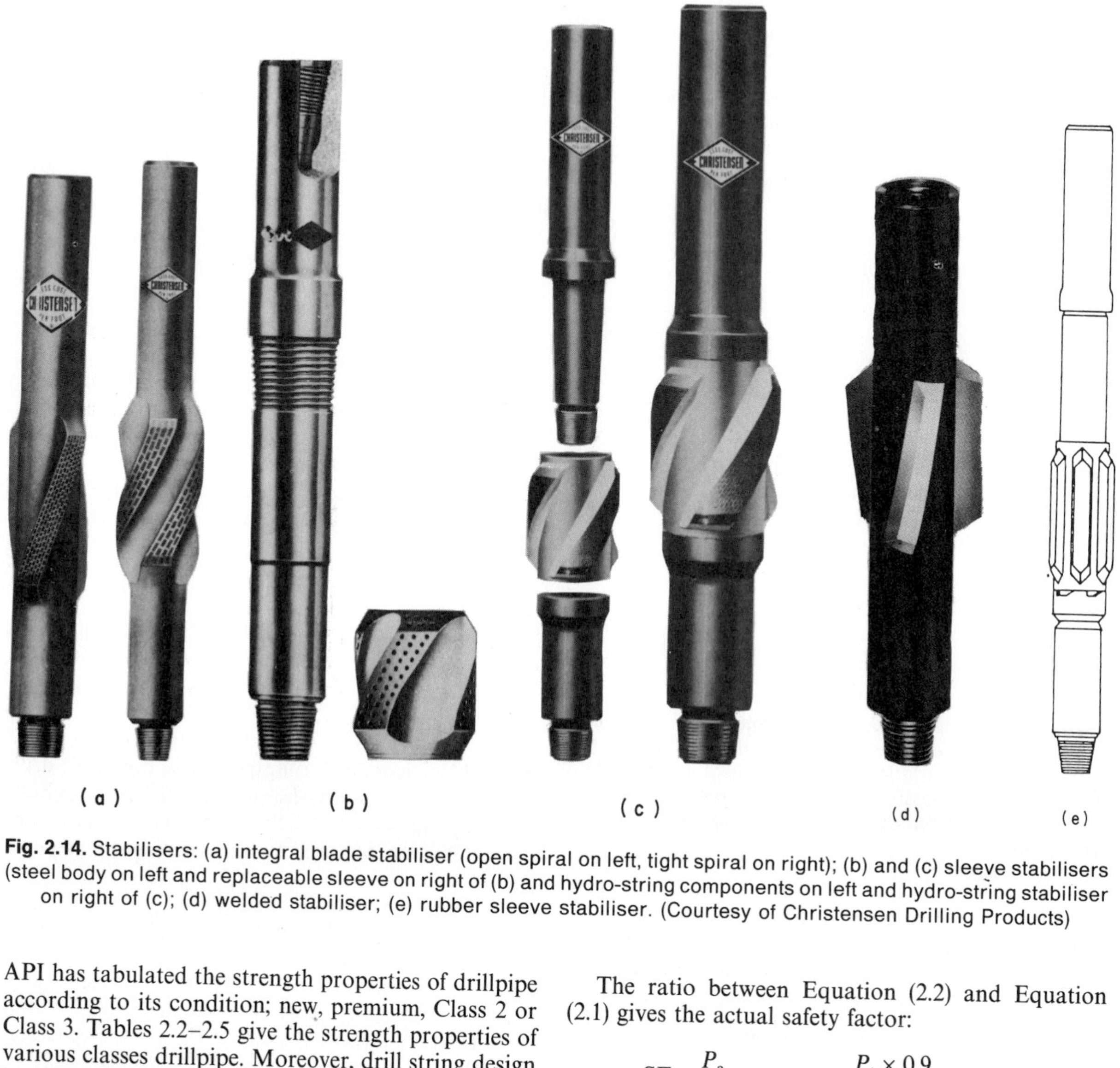

Fig. 2.14. Stabilisers: (a) integral blade stabiliser (open spiral on left, tight spiral on right); (b) and (c) sleeve stabilisers (steel body on left and replaceable sleeve on right of (b) and hydro-string components on left and hydro-string stabiliser on right of (c); (d) welded stabiliser; (e) rubber sleeve stabiliser. (Courtesy of Christensen Drilling Products)

API has tabulated the strength properties of drillpipe according to its condition; new, premium, Class 2 or Class 3. Tables 2.2–2.5 give the strength properties of various classes drillpipe. Moreover, drill string design is never based on the tabulated yield strength value but, instead, on 90% of the yield strength, to provide an added safety in the resulting design. Thus, maximum tensile design load, P_a, = theoretical yield strength (taken from API Tables 2.2–2.5) × 90% or

$$P_a = P_t \times 0.9 \tag{2.2}$$

where P_t is drillpipe yield strength.

The difference between Equations (2.2) and (2.1) gives the margin of overpull, MOP:

$$\text{MOP} = P_a - P \tag{2.3}$$

The design values of MOP normally range from 50 000–100 000 lb.

The ratio between Equation (2.2) and Equation (2.1) gives the actual safety factor:

$$\text{SF} = \frac{P_a}{P} = \frac{P_t \times 0.9}{(L_{dp} \times W_{dp} + L_{dc} \times W_{dc})\,\text{BF}} \tag{2.4}$$

The choice of an appropriate value of safety factor is dependent on overall drilling conditions, including hole drag and the likelihood of pipe becoming stuck. Dynamic loading, which arises from arresting the drillpipe by slips, must also be considered.

Simplifying Equation (2.4) further gives:

$$L_{dp} = \frac{P_t \times 0.9}{\text{SF} \times W_{dp} \times \text{BF}} - \frac{W_{dc}}{W_{dp}} L_{dc} \tag{2.5}$$

Equation (2.5) can also be expressed in terms of MOP instead of the SF term by combining Equation (2.1) with Equation (2.3), to obtain

Fig. 2.15. Roller reamers. (a) Three-point reamer. (b) Six-point reamer. (c) Types of cutter: bottom, soft formation; middle, hard formation; top, hard-abrasive formation.

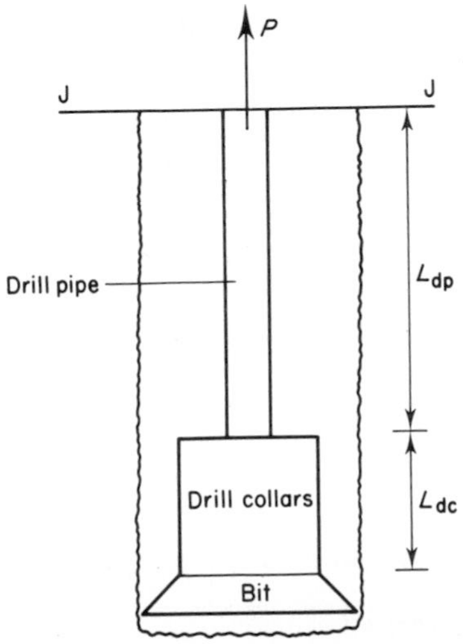

Fig. 2.16. Drillpipe design.

$$L_{dp} = \frac{P_t \times 0.9 - \text{MOP}}{W_{dp} \times \text{BF}} - \frac{W_{dc}}{W_{dp}} \times L_{dc} \qquad (2.6)$$

The term L_{dp} is sometimes expressed as L_{max}, to refer to the maximum length of a given grade of drillpipe which can be selected for a given loading situation.

Equations (2.5) and (2.6) can also be used to design a tapered string, consisting of different grades and sizes of drillpipe. In this case the lightest available grade is first considered and the maximum useable length selected as a bottom section. Successive heavy grades are then considered in turn to determine the usable length from each grade along the hole depth. Example 2.6 illustrates the design of a tapered string.

Collapse

Collapse pressure may be defined as the external pressure required to cause yielding of drillpipe or casing.

In normal drilling operations the mud columns inside and outside the drillpipe are both equal in height and are of the same density. This results in zero differential pressure across the pipe body and, in turn, zero collapse pressure on the drillpipe. In some cases, as in drill stem testing (DST), the drillpipe is run partially full, to reduce the hydrostatic pressure exerted against the formation. This is done to encourage formation fluids to flow into the well bore, which is the object of the test. Once the well flows, the collapsing effects are small, as the drillpipe is now full of fluids.

Thus, maximum differential pressure, Δp, across the drillpipe exists prior to the opening of the DST tool, and can be calculated as follows:

$$\Delta p = \frac{L\rho_1}{144} - \frac{(L - Y) \times \rho_2}{144} \qquad (2.7)$$

where Y = depth to fluid inside drillpipe (see Figure 2.17); L = total depth of well (ft); ρ_1 = fluid density outside the drillpipe (pcf); ρ_2 = fluid density inside the drillpipe (pcf).

When densities are expressed in ppg, Equation (2.7) becomes

$$\Delta p = \frac{L\rho_1}{19.251} - \frac{(L - Y)\rho_2}{19.251} \qquad (2.8)$$

Other variations of Equation (2.7) include the following:

(a) Drillpipe is completely empty, $Y = 0$, $\rho_2 = 0$:

$$\Delta p = \frac{L\rho_1}{144} \qquad (2.9)$$

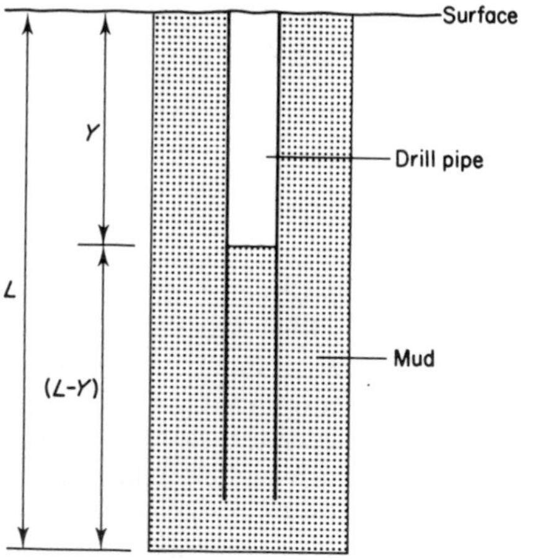

Fig. 2.17. Collapse considerations in drill string design.

(b) Fluid density inside drillpipe is the same as that outside drillpipe, i.e. $\rho_1 = \rho_2 = \rho$, and Equation (2.7) becomes

$$\Delta p = \frac{Y\rho}{144} \tag{2.10}$$

where ρ = density of mud (pcf).

Once the collapsing pressure, Δp, is calculated, it can then be compared with the theoretical collapse resistance of the pipe as determined by the manufacturer (see Tables 2.2–2.5). A safety factor in collapse can be determined as follows:

$$\text{SF} = \frac{\text{collapse resistance}}{\text{collapse pressure } (\Delta p)} \tag{2.11}$$

An SF of $1\frac{1}{8}$ is normally used.

It should be observed that the collapse resistance reported in API or manufacturers' tables is that value under zero load. Normally, drillpipe is under tension resulting from its own weight and the weight of drill collars. The combined loading of tension and collapse is described as biaxial loading. During biaxial loading, the drillpipe stretches and its collapse resistance decreases. The corrected collapse resistance of drillpipe can be determined as follows:

(1) Determine tensile stress at joint under consideration by dividing the tensile load by the pipe cross-sectional area.

(2) Determine the ratio between tensile stress and the average yield strength.

(3) Using Figure 2.18, determine the percentage reduction in collapse resistance corresponding to the ratio calculated in Step 2.

Shock loading

Shock loading arises whenever slips are set on moving drillpipe and can contribute to parting of pipe in marginal designs.

It can be shown that the additional tensile force, F_s,

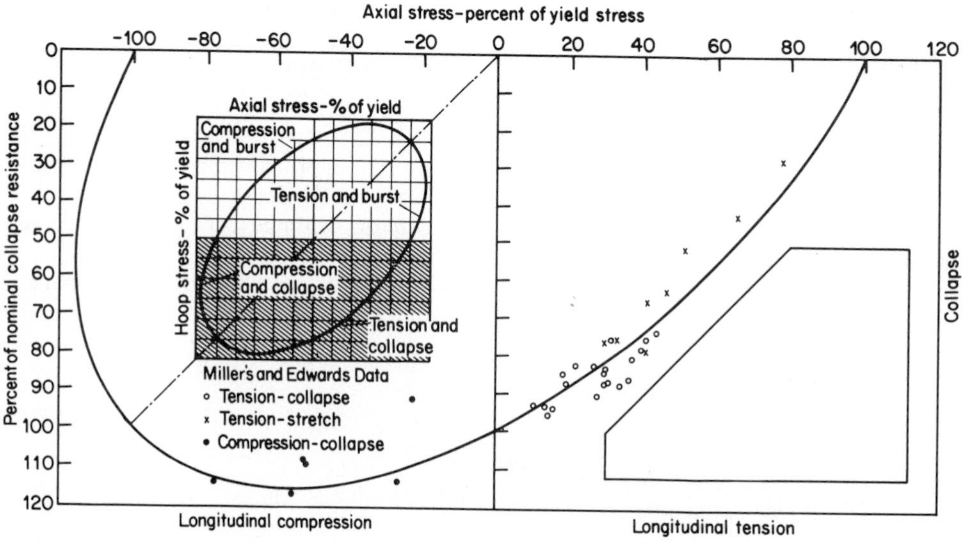

Fig. 2.18. Ellipse of biaxial yield stress or maximum shear–strain energy diagram[4]. (Courtesy of API)

generated by shock loading is calculated from

$$F_s = 3200 W_{dp} \quad (2.12)$$

where W_{dp} = weight of drill pipe per unit length. The derivation of Equation (2.12) is given in Chapter 10.

Torsion

It can be shown that the drill pipe torsional yield strength when subjected to pure torsion is given by

$$Q = \frac{0.096\,167 J Y_m}{D} \quad (2.13)$$

where Q = minimum torsional yield strength (lb-ft); Y_m = minimum unit yield strength (psi); J = polar moment of inertia = $\Pi/32(D^4 - d^4)$ for tubes = $0.098\,175(D^4 - d^4)$; D = outside diameter (in); d = inside diameter (in).

When drillpipe is subjected to both torsion and tension, as is the case during drilling operations, Equation (2.13) becomes

$$Q_\tau = \frac{0.096\,167 J}{D}\sqrt{Y_m^2 - \frac{P^2}{A^2}} \quad (2.14)$$

where Q_τ = minimum torsional yield strength under tension (lb-ft); J = polar moment of inertia = $\Pi/32(D^4 - d^4)$ for tubes = $0.098\,175(D^4 - d^4)$; D = outside diameter (in); d = inside diameter (in); Y_m = minimum unit yield strength (psi); P = total load in tension (lb); A = cross-sectional area (in^2).

It should be noted that Equations (2.13) and (2.14) can be used to calculate the maximum allowable make-up torque before the minimum torsional yield strength of the drill pipe body is exceeded.

Examples

In the following examples the nominal weight of drillpipe is used in order to reduce the length of calculations. For actual designs, the engineer is advised to use the method presented in Example 2.1 for calculating the approximate weight of the drillpipe and joint assembly, and use this figure in the calculations.

Example 2.2

A drill string consists of 600 ft of $8\frac{1}{4}$ in × $2\frac{13}{16}$ in drill collars and the rest is a 5 in, 19.5 lbm/ft Grade X95 drillpipe. If the required MOP is 100 000 lb and mud weight is 75 pcf (10 ppg), calculate the maximum depth of hole that can be drilled when (a) using new drillpipe and (b) using Class 2 drillpipe having a yield strength (P_t) of 394 600 lb.

Solution

(a) Weight of drill collar per foot is

$$A \times 1\text{ ft} \times \rho_s = \frac{\pi}{4}((8\tfrac{1}{4})^2 - (2\tfrac{13}{16})^2) \times 1\text{ ft} \times 489.5 \times \frac{1}{144}$$

$$= 160.6 \text{ lbm/ft}$$

where ρ_s = density of steel = 489.5 lbm/ft; A = cross-sectional area (in). (*Note:* From Table 2.9 weight of drill collar = 161 lbm/ft.) From Equation (2.6),

$$L_{dp} = \frac{P_t \times 0.9 - \text{MOP}}{W_{dp} \times \text{BF}} - \frac{W_{dc}}{W_{dp}} L_{dc}$$

From Table 2.2,

$$P_t = 501\,090 \text{ lb (for Grade X95 new pipe)}$$

$$\text{BF} = \left(1 - \frac{\rho_m}{\rho_s}\right) = \left(1 - \frac{75}{489.5}\right) = 0.847$$

and

$$\text{MOP} = 100\,000 \text{ lb}$$

Therefore,

$$L_{dp} = \frac{501\,090 \times 0.9 - 100\,000}{19.5 \times 0.847} - \left(\frac{160.6}{19.5}\right) \times 600$$

$$= 16\,309 \text{ ft}$$

Therefore, maximum hole depth that can be drilled with a new drillpipe of Grade X95 under the given loading condition is 16 309 + 600 = 16 909 ft.

(b) Now P_t = 394 600 lb:

$$L_{dp} = \frac{394\,600 \times 0.9 - 100\,000}{19.5 \times 0.847} - \left(\frac{160.6}{19.5} \times 600\right)$$

$$= 10\,506 \text{ ft}$$

Max. hole depth = 10 506 + 600 = 11 106 ft.

Example 2.3

If 10 000 ft of the drillpipe in Example 2.1 is used, determine the maximum collapse pressure that can be encountered and the resulting safety factor. The mud density is 75 pcf (10 ppg). If the fluid level inside the drillpipe drops to 6000 ft below the rotary table, determine the new safety factor in collapse.

Solution

(a) Maximum collapse pressure, Δp, occurs when the drillpipe is 100% empty. Thus,

$$\Delta p = \frac{L \times \rho_m}{144} = \frac{10\,000 \times 75}{144} = 5208 \text{ psi}$$

From Table 2.2, collapse resistance of new pipe of Grade X95 is 12 010 psi. Hence,

$$SF = \frac{12\,010}{5208} = 2.3$$

(b) When the mud level drops to 6000 ft below the surface, Equation (2.7) or (2.10) can be used. Thus,

$$\Delta p = \frac{6000 \times 75}{144}$$

$$= 3125 \text{ psi}$$

$$SF = \frac{12\,010}{3125} = 3.8$$

(*Note:* In this example the safety factor in collapse is calculated for the bottom joint only, which suffers no biaxial loading, since it carries no load (see Example 2.4).

Example 2.4

A drill string consists of 10 000 ft of drillpipe and a length of drill collars weighing 80 000 lb. The drillpipe is 5 in OD, 19.5 lb, Grade S135, premium class.

(a) Determine the actual collapse resistance of the bottom joint of drillpipe.

(b) Determine the safety factor in collapse.

(Assume collapse resistance = 10 050 psi and mud weight = 75 pcf.)

Solution

For a 5 in OD new drillpipe, the nominal ID is 4.276 in (i.e. thickness = 0.376 in). For a premium drillpipe, only 80% of the pipe thickness remains. Thus, reduced wall thickness for premium pipe = 0.8 × 0.362 = 0.2896 in and reduced OD for premium pipe

$$= \text{nominal ID} + 2 \times (\text{premium thickness})$$

$$= 4.276 + (0.2896) \times 2$$

$$= 4.8552 \text{ in}$$

Cross-sectional area (CSA) of premium pipe

$$= \frac{\pi}{4}(\text{OD}^2 - \text{ID}^2)$$

$$= \frac{\pi}{4}((4.855\,2)^2 - (4.276)^2)$$

$$= 4.153\,8 \text{ in}^2$$

Tensile stress at bottom joint of drillpipe

$$= \frac{\text{tensile load } (= \text{weight of drill collars})}{CSA}$$

$$= \frac{80\,000 \text{ lb}}{4.1538 \text{ in}^2} = 19\,259 \text{ psi}$$

The average yield strength of Grade S135 is 145 000 psi.

$$\text{Tensile ratio} = \frac{\text{tensile stress}}{\text{average yield strength}} = \frac{19\,260}{145\,000}$$

$$= 0.1328 = 13.3\%$$

From Figure 2.18 of biaxial loading, a tensile ratio of 13.3% reduces the nominal collapse resistance to 53%.

(a) Thus, collapse resistance of bottom joint of drillpipe

$$= 0.93 \times \text{collapse resistance under zero load}$$

$$= 0.93 \times 10\,050$$

$$= 9347 \text{ psi}$$

(b)

$$SF = \frac{\text{collapse resistance}}{\text{collapse pressure}}$$

For worst conditions, assume drillpipe to be 100% empty. Hence,

$$\text{collapse pressure} = \frac{75 \times 10\,000}{144}$$

$$= 5208 \text{ psi}$$

$$SF = \frac{9347}{5208} = 1.8$$

Example 2.5

Assuming that 10 000 ft of Grade X95 has been selected from Example 2.2, determine: (a) the safety factor during drilling; (b) the magnitude of shock loading; and (c) the safety factor when shock loading is included.

Solution

(a) Total weight carried by drillpipe

$$= (L_{dp} \times W_{dp} + L_{dc} W_{dc}) \times BF$$

$$= (10\,000 \times 19.5 + 600 \times 160.4) \times 0.847$$

$$= 246\,680 \text{ lb}$$

$$SF = \frac{\text{yield strength} \times 0.9}{\text{load carried by top joint}}$$

$$= \frac{501\,090 \times 0.9}{246\,680} = 1.83$$

(b) Force due to shock loading

$$= 3200 \times W_{dp}$$

$$= 3200 \times 19.5 = 62\,400 \text{ lb}$$

(c) Total load at top joint

$$= 246\,680 + 62\,400$$

$$= 309\,080 \text{ lb}$$

$$SF = \frac{501\,050 \times 0.9}{309\,080} = 1.46$$

Example 2.6

An exploration rig has the following grades of drillpipe to be run in a 15 000 ft deep well:

Grade E: 5/4.276 in; 19.5 lbm/ft; yield strength = 395 600 lb (1760 kN)

Grade G: 5/4.276 in; 19.5 lbm/ft; yield strength = 553 830 lb (2464 kN)

If the total length and weight of drill collars plus heavy-wall drillpipe is 984 ft (300 m) and 157 374 lb (700 kN), respectively, calculate: (a) the maximum length that can be used from each grade of drillpipe if an MOP of 50 000 lb is to be maintained for the lower grade; and (b) the MOP of the heavier grade. The maximum expected mud weight at 15 000 ft is 100 pcf (13.4 ppg, or 1.6 kg/l).

Solution

(a)

$$BF = \left(1 - \frac{\rho_m}{\rho_s}\right) = \left(1 - \frac{100}{489.5}\right) = 0.796$$

and

$$L_{dp} = \frac{P_t \times 0.9 - \text{MOP}}{W_{dp} \times BF} - \left(\frac{W_{dc}}{W_{dp}}\right) \times L_{dc}$$

The lightest grade (Grade E) should be used for the bottom part of the hole, while the highest grade should be used at the topmost section. Thus, Grade E will carry the weight of drill collars and heavy-wall drillpipe and the term W_{dc} should include the combined weight of these two items. Hence,

$$W_{dc}L_{dc} = \text{weight of drill collars} + \text{weight of heavy-wall drillpipe}$$

$$= 157\,374 \text{ lb}$$

$$L_{dp} = \frac{395\,600 \times 0.9 - 50\,000}{19.5 \times 0.796} - \frac{157\,374}{19.5}$$

$$= 11\,646 \text{ ft}$$

Accumulated length of drill string will consist of:

drill collar and heavy-wall drillpipe =	984 ft
Grade E drillpipe	= 11 646 ft
	Total = 12 630 ft

The top part of the well will consist of Grade G of length 15 000 − 12 630 = 2370 ft. Check Grade G for suitability:

This grade will carry a combined weight of Grade E (11 646 ft) + drill collars and heavy-wall drillpipe

$$= 11\,646 \times 19.5 + 157\,374$$

$$= 384\,471 \text{ lb}$$

and

$$L_{dp} = \frac{553\,830 \times 0.9 - 50\,000}{19.5 \times 0.796} - \frac{384\,471}{19.5}$$

$$= 9175 \text{ ft}$$

Hence, under the existing loading conditions, 9175 ft of Grade G can be used as a top section. In our example only 2370 ft are required.

(b) $\text{MOP} = P_t \times 0.9 - P$

P = buoyant weight carried by top joint
= weight of drillpipe of Grades E and G and weight of drill collars and heavy-wall drillpipe

$$= (2370 \times 19.5 + 11\,646 \times 19.5 + 157\,374) \times BF$$

$$= 430\,686 \times 0.796$$

$$= 342\,826 \text{ lb}$$

$$\text{MOP} = 553\,830 \times 0.9 - 342\,826$$

$$= 155\,621 \text{ lb}$$

Example 2.7

The following data refer to a drill string stuck at the drill collars:

drillpipe: 10 000 ft, 5/4.276 in, Grade E 19.5 lbm/ft, Class 2
drill collars: 600 ft, total weight 80 000 lb
make-up torque for drillpipe tool joints = 20 000 lb-ft
100% free point = 9900 ft

Determine the maximum torque that can be applied at the surface without exceeding the minimum torsional yield strength of drillpipe.

Solution

Using Equation (2.14),

$$Q_T = \frac{0.096\,167J}{D}\sqrt{Y_m^2 - \frac{P^2}{A^2}}$$

Since drillpipe is 100% free at 9900 ft, the total tensile load $(P) = 9900 \times 19.5 = 193\,050$ lb. (*Note:* Nominal weight of drillpipe is used.)

$$J = \frac{\pi}{32}(D^4 - d^4) = \frac{\pi}{32}(5^4 - (4.276)^4)$$

$$= 28.5383 \text{ in}^4$$

$$A = \frac{\pi}{4}(D^2 - d^2) = \frac{\pi}{4}(5^2 - (4.276)^2) = 5.27 \text{ in}^2$$

Tensile strength = 311 540 lb (from Table 2.4).
Minimum unit yield strength (Y_m)

$$= \frac{311\,540}{A} = \frac{311\,540}{5.27}$$

$$= 59\,116 \text{ psi}$$

$$Q_T = \frac{0.096\,167 \times 28.5383}{5}\sqrt{(59\,116)^2 - \frac{(193\,050)^2}{(5.27)^2}}$$

$$= 25\,468 \text{ lb-ft}$$

(maximum allowable torque for pipe body).

Since the make-up torque for the drillpipe tool joint is 20 000 ft-lb, the maximum allowable torque should be based on tool joint torque and not pipe body torque. Note that this information will be required in back-off operations, as will be discussed in Chapter 12.

STRETCH OF DRILLPIPE

Drillpipe stretches under the action of the weight of drill collars and its own weight.

Stretch due to weight carried

Assume that changes in drillpipe diameter are very small, such that drillpipe extension is only significant in the axial direction. If the weight of drill collars is P, (Figure 2.19) then, using Hookes' Law, we obtain

$$E = \frac{P/A}{e_1/L} \tag{2.15}$$

where E = Young's modulus of steel = 30×10^6 psi (210×10^9 N/m^2); e_1 = extension of drillpipe; and A = plain-end area of drillpipe. Rearranging Equation (2.15) gives

$$e_1 = \frac{PL}{AE} = \frac{PL}{\frac{\pi}{4}(\text{OD}^2 - \text{ID}^2) \times E} \tag{2.16}$$

Equation (2.16) can be simplified by replacing E by its numerical value and plain-end area by weight per unit length, W_{dp}:

$$W_{dp} = (\text{plain-end area}) \times 1 \text{ ft} \times \text{density of steel}$$

$$= \frac{\pi}{4}\frac{(\text{OD}^2 - \text{ID}^2)}{144} \times 1 \text{ ft} \times 489.5$$

where OD = outside diameter of drillpipe (in); ID = inside diameter of drillpipe (in). Therefore,

$$W_{dp} = 3.3993 \times \frac{\pi}{4}(\text{OD}^2 - \text{ID}^2) \tag{2.17}$$

Substituting the values of E and W_{dp} in Equation (2.16) gives:

$$e_1 = \frac{P \times L}{\frac{W_{dp}}{3.3993} \times 30 \times 10^6} \tag{2.18}$$

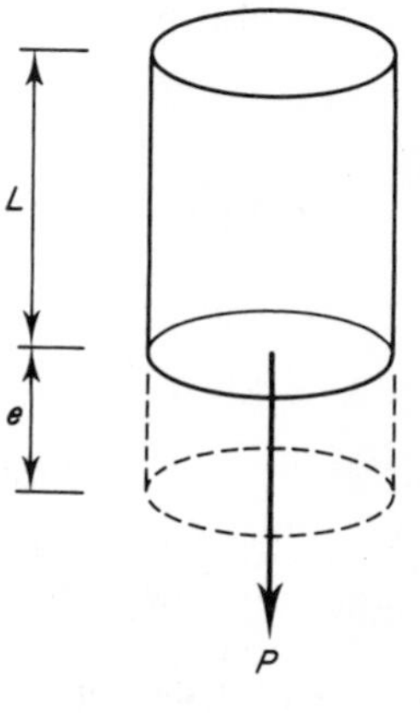

Fig. 2.19. Drillpipe stretch due to weight carried.

Normally, drillpipe extension is expressed in inches and Equation (2.18) simplifies to

$$e_1 = \frac{12 \times P \times L}{\frac{W_{dp}}{3.3993} \times 30 \times 10^6} = \frac{P \times L}{735\,444 W_{dp}} \text{ in} \quad (2.19)$$

where L is in ft, P is in lb and W_{dp} is in lbm/ft.

In metric units Equation (2.16) becomes:

$$e_1 = 373{\cdot}8 \times 10^{-10}\left(\frac{PL}{W_{dp}}\right) \quad (2.20)$$

where P = weight in N; L = drillpipe length in metres; and W_{dp} = weight per metre in kg/m.

Stretch due to own weight

Drillpipe also stretches under its own weight when fully suspended in mud. API[3] gives the following formula for calculating drillpipe extension in inches, due to its own weight:

$$e_2 = \frac{L^2}{72 \times 10^7}(489.5 - 1.44\rho_{m_1})$$

where ρ_{m_1} is in lbm/ft^3, or

$$e_2 = \frac{L^2}{9.625 \times 10^7}(65.44 - 1.44\rho_{m_2}) \quad (2.21)$$

where ρ_{m_2} is in lb/gal, L = drillpipe length (ft).

In metric units, the equivalent of Equation (2.20) is

$$e_2 = 2.346 \times 10^{-8} L^2(7.85 - 1.44\, P_m) \quad (2.20a)$$

where L is in m and Pm is in kg/l.

Example 2.8

A $3\frac{1}{2}$ in drillpipe, 13.3 lbm/ft, Grade S135 premium class, is used to run a $4\frac{1}{2}$ in OD liner to 21 000 ft. If the length of drillpipe is 17 500 ft, the mud weight is 120 pcf and the total weight of the liner is 50 000 lb, calculate the total stretch in the drillpipe.

Solution

e_1 = stretch due to weight carried

$$= \frac{P \times L}{735\,444 W_{dp}} = \frac{50\,000 \times 17\,500}{735\,444 \times 13.3} = 89.5 \text{ in}$$

e_2 = stretch due to suspended weight of drillpipe

$$= \frac{L^2}{72 \times 10^7}(489.5 - 1.44\rho_m)$$

$$= \frac{(17\,500)^2}{72 \times 10^7}(489.5 - 1.44 \times 120)$$

$$= 134.7 \text{ in}$$

$$\text{Total stretch} = e_1 + e_2$$

$$= 89.5 + 134.7 = 224.2 \text{ in}$$

$$= 18.7 \text{ ft}$$

CRITICAL ROTARY SPEEDS

The component of the drill string can vibrate in three different modes: (1) axial or longitudinal; (2) torsional; and (3) transverse or lateral.

According to Dareing[9], axial vibrations can be recognized at the surface by the kelly bounce and whipping of the drilling line. Torsional vibrations cannot be seen, since the rotary table tends to control the angular motion at the surface. Transverse motion is only possible with the drillpipe, as drill collars are only free to move axially and torsionally.

Vibration of the drill string occurs when the frequency of the applied force equals the natural free vibration frequency of the drill stem. This condition is normally described as 'resonance'. Hence, the drill bit must be rotated at a speed (or rpm) which is different from the natural frequency of the drill string.

Rotation of the drill string at its natural resonant frequency results in excessive wear, fatigue failure and rapid deterioration of drillpipe.

Two different methods for the calculation of critical rotary speeds exist in the literature. The first relies on total drill string length and drillpipe dimensions for calculating critical rotary speeds. The second uses the length of drill collars only for estimating critical rotary speed.

Calculation of critical rotary speeds based on total drill string length and drillpipe dimensions

API[3] recommends that critical longitudinal speeds be calculated using Equation (2.22). Thus,

$$\text{rpm} = \frac{258\,000}{L} \text{(longitudinal vibration)} \quad (2.22)$$

where L = total length of drill string (ft).

Secondary and higher harmonic vibrations will occur at 4, 9, 16, 25, 36, etc., times the rpm of Equation 2.22.

The critical transverse speed is calculated using the following equation:

$$\text{rpm} = \frac{4\,760\,000}{l^2}(D^2 + d^2)^{1/2} \text{ (transverse vibration)} \quad (2.23)$$

where l = length of one drillpipe joint (in); D = OD of drillpipe (in); and d = ID of drillpipe (in).

Example 2.9

Determine the critical rotary speeds for a 10 000 ft drill string consisting of 5 in OD/4.276 in ID drillpipe and 800 ft 8 in OD/3 in ID drill collars.

Solution

$$\text{Longitudinal rpm} = \frac{258\,000}{10\,000} = 25.8$$

$$\text{Transverse rpm} = \frac{4\,760\,000}{(30 \times 12)^2}(5^2 + 4.276^2)^{1/2} = 242$$

where l (from Equation 2.23) = 30 × 12 in.

Drill collar length

According to Dareing[9], critical rotary speeds should be based on the length of drill collars only. The drill collars are assumed to be fixed at the drill bit and free at the drill collar/drillpipe interface. Data collected over 25 years showed[9] that drill bit displacement frequencies are consistently three cycles for every bit revolution for three-cone bits. Thus, frequency, f, of vibration at the bit is given by

$$f = 3N \times \left(\frac{1}{60}\right) \text{cycles/s} \quad (2.24)$$

where N = rotary speed of drill bit.

Equation (2.24) is applicable to three-cone bits only. Simplifying Equation (2.24) gives the critical rotary speed, N:

$$N = 20 \times f \text{ rpm} \quad (2.25)$$

Drill collars are assumed to vibrate in two modes, longitudinal and torsional. The natural frequency of longitudinal vibrations of drill collars is given by

$$f_1 = \frac{4212}{L} \text{ c/s} \quad (2.26)$$

where L = length of drill collars. The natural torsional frequency of drill collars is given by

$$f_2 = \frac{2662}{L} \text{ c/s} \quad (2.27)$$

Dareing's method requires that the bit be rotated at a speed (N) less or greater than the natural frequencies (f_1 and f_2) of the drill collars.

Example 2.10

Determine the critical rotary speed of the drill string described in Example 2.8.

Solution

From Equation (2.26) natural longitudinal frequency, f_1, for 800 ft of drill collars is:

$$f_1 = \frac{4212}{800} = 5.265 \text{ c/s}$$

Also, natural torsional frequency, f_2, is given by Equation (2.27):

$$f_2 = \frac{2662}{800} = 3.3275 \text{ c/s}$$

Thus, the drill bit critical rotary speeds, N, are:

$$N_1(\text{longitudinal}) = 20 \times f_1 = 20 \times 5.265 = 105 \text{ rpm}$$

and

$$N_2(\text{transverse}) = 20 \times f_2 = 20 \times 3.327 = 67 \text{ rpm}$$

Hence, the drill bit must be rotated below 67 rpm, between 67 and 105 rpm or above 105 rpm, to avoid vibration.

Comparison of the results of Examples 2.9 and 2.10 shows that the two methods produce widely different values. The longitudinal critical frequency using the API[3] method is 25.8 rpm, compared with 105 rpm as calculated by the drill collar method[9]. Similar variations exist in the calculation of transverse frequencies. More work is required to compare the two cited methods. The author recommends the use of the drill collar method for its simplicity and sound theoretical background for determining the critical rotary speeds of tricone bits. For other bit types, the API method should be followed.

Drill string vibration can be reduced by the following methods:

(1) Changing the natural frequency: (a) by using a shock absorber; (b) by increasing the drill collar length; or (c) by using HWDP, or by increasing the length of the existing section of HWDP.
(2) Increasing or applying mechanical damping.

References

1. Drilco (1977). *Drilco Drilling Assembly Handbook.* Drilco (Smith International Division, Inc.) Publications.
2. Varco Oil Tools (1981). *Rotary Equipment Care and Maintenance Handbook.* Varco Publications.
3. API Spec. 7 (1984). *Specification for Rotary Drilling Equipment.* API Production Department.
4. API RP7G (1981). *API Recommended Practice for Drillstem Design and Operating Limits.* API Publications Department.
5. API Spec. 5A (1982). *Specification for Casing, Tubing and Drillpipe.* API Publications Department.
6. Wilson, G. (1976). How to drill a usable hole. *World Oil.*
7. Cheng, J. T. (1982). How to make heavy wall drillpipe last longer. *World Oil*, Oct.
8. Schmid, J. T., Jr. (1982). Designing BHAs for better drilling jar performance. *World Oil*, Oct.
9. Dareing, D. W. (1984). Drillcollar length is a major factor in vibration control. *JPT*, April, 639–644.

Problems

Problem 1

Determine the approximate weight of drillpipe, including tool joint, having 5 in OD, 19.5 lbm/ft nominal weight, Grade G, with a $6\frac{5}{8}$ in OD/$2\frac{3}{4}$ in ID, NC50 tool joint.

Answer: 22.55 lbm/ft

Problem 2

Design a tapered string to run a $4\frac{1}{2}$ in OD liner to 19 000 ft using the following data:

length of liner = 6000 ft
weight of liner = 90 600 lb
mud weight = 97 pcf
available drillpipe:
9000 ft, $3\frac{1}{2}$ in drillpipe, 13.3 lb/ft, Grade E (new)
5000 ft, $3\frac{1}{2}$ in drillpipe, 15.5 lb/ft, Grade E (new)
10 000 ft, $3\frac{1}{2}$ in drillpipe, 15.5 lb/ft, Grade G (new)

An MOP of 80 000 lb is to be maintained for the lower grade.

Answer: 6727 ft of Grade E, 13.3 lbm/ft as bottom section; 3273 ft of Grade E, 15.5 lbm/ft as middle section; Grade G 15.5 lbm/ft as top section.

Chapter 3

Hoisting

The main function of the hoisting equipment is to get the drill string and other necessary equipment in and out of the hole safely and efficiently. Equipment normally run in hole includes drill string assembly, casing, logging tools, deviation tools, and testing and completion equipment.

The main components of a typical hoisting system include:

Draw-works
Hoisting tackle, including crown and travelling blocks, hooks and elevators
Deadline anchors
Drilling line
Derrick

Figure 3.1 presents a schematic drawing of the main hoisting equipment.

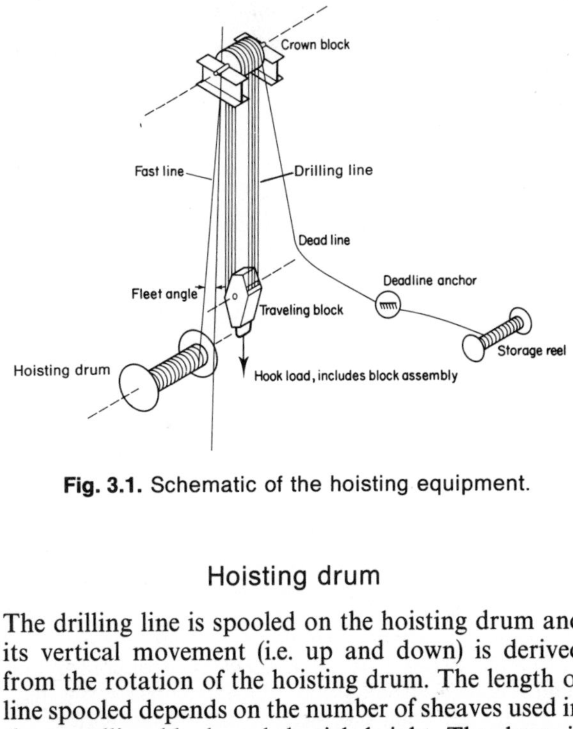

Fig. 3.1. Schematic of the hoisting equipment.

DRAW-WORKS

The draw-works represent the heart of the rig, enabling equipment to be run in and out of the hole and also providing power for making or breaking pipe joints. The principal components of a draw-work comprise drumshaft group; catshaft and coring reel group; main drive shaft and jackshaft group; rotary countershaft group; and controls (refer to Figure 3.2).

Drumshaft group

As can be seen from Figure 3.2, the main components of the drumshaft group are as follows.

Hoisting drum

The drilling line is spooled on the hoisting drum and its vertical movement (i.e. up and down) is derived from the rotation of the hoisting drum. The length of line spooled depends on the number of sheaves used in the travelling block and derrick height. The drum is powered from the main drive shaft through a jackshaft assembly using roller chains and sprockets.

The input power is used to rotate the drum during hoisting by engagement of high or low air clutches, causing the drilling line to spool onto the drum. For lowering, the drum is allowed to freewheel and the weight of the travelling block and other components will cause the drilling line to unreel.

A special design feature of the hoisting drum is a

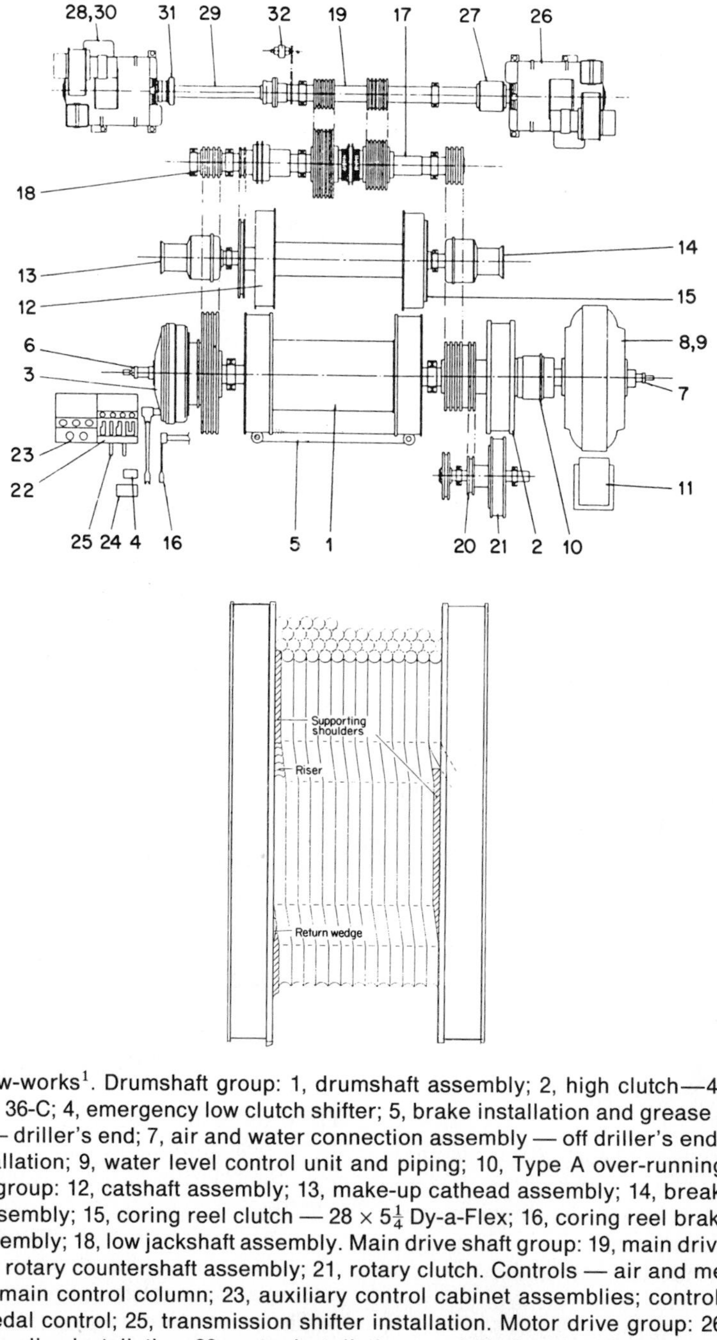

Fig. 3.2. (Top) The draw-works[1]. Drumshaft group: 1, drumshaft assembly; 2, high clutch—40 × 10 Dy-a-Flex; 3, low clutch — National Type 36-C; 4, emergency low clutch shifter; 5, brake installation and grease system; 6, air and water connection assembly — driller's end; 7, air and water connection assembly — off driller's end; 8, brake installation; 9, hydromatic brake installation; 9, water level control unit and piping; 10, Type A over-running clutch; 11, water level control units. Catshaft group: 12, catshaft assembly; 13, make-up cathead assembly; 14, break-out cathead assembly; 12, coring reel shaft assembly; 15, coring reel clutch — 28 × $5\frac{1}{4}$ Dy-a-Flex; 16, coring reel brake installation. Jackshaft group: 17, jackshaft assembly; 18, low jackshaft assembly. Main drive shaft group: 19, main driveshaft assembly. Rotary countershaft group: 20, rotary countershaft assembly; 21, rotary clutch. Controls — air and mechanical: air and water piping installation; 22, main control column; 23, auxiliary control cabinet assemblies; control for independent rotary drive brake; 24, foot pedal control; 25, transmission shifter installation. Motor drive group: 26, motor installations — rotary end; 27, motor coupling installation; 28, motor installations — driller's side; 29, floating drive shaft assembly; 30, motor installation with inertia brake — driller's side; 31, inertia brake installation. Miscellaneous: 32, oil lubrication installation; rope rollers and brackets; jackshaft shim installation. (Bottom): Hoisting drum with two-step spiral grooving.

two-step Spirallel* grooving, in which the drum grooves are parallel to the drum end flanges and kick over half a pitch in 45° every 180°. This design means that the wraps of drilling line on the drum are mostly parallel and allows the line to spool correctly at high speed. Grooves also provide a much better support for the first layer of the drum than is possible with a smooth drum.

As a rule of thumb, the draw-works should have 1 h.p. for each 10 ft to be drilled[2].

Brakes

The draw-works main brakes are used to stop the movement of the drill string prior to the arresting of the drill string in the rotary table by means of slips and for allowing drilling line to unreel while drilling ahead to maintain weight on bit. Most draw-works use friction-type mechanical brakes, which provide a braking effect by means of physical contact between the rims of the hoisting drum flanges and a brake band. All hoisting drums are provided with two brake flanges, one on each side of the hoisting drum.

Each brake band consists of a flexible steel band with a number of friction-material brake blocks bolted to it. The brake bands are wrapped around the brake rims of the hoisting drum. One end is anchored to each end of an equaliser beam, which pivots at its centre and connects both bands to ensure that braking is shared equally by each band. The other end of each band is movable and is connected through a magnification linkage to the driller's brake lever (see Figure 3.3).

*Trademark of National Supply Company.

Braking is achieved by lowering the brake lever, which, in turn, pushes the brake blocks against the rims of the rotating drum. As the brake blocks drag along the rim, the applied tension is magnified towards the fixed end. Hence, the fixed end of the brake band has more tension than the movable end. A driller of average build can apply a large braking force by merely putting pressure on the brake lever, owing to the magnification effect in the brake linkage mechanism.

Cooling system

The braking action caused by the brake blocks contacting the rims of the drum flanges generates a great deal of heat. This heat must be dissipated quickly, to avoid damage to the rims and brake blocks.

All draw-works manufacturers employ a water-cooling system for removing the heat generated during braking. Figure 3.4 gives a schematic diagram of a typical water cooling system. Important features of Figure 3.4 include a motor pump which circulates water through a piping system to the water-jacketed rims and to the auxiliary brake. Finally, the water is returned to the suction tank.

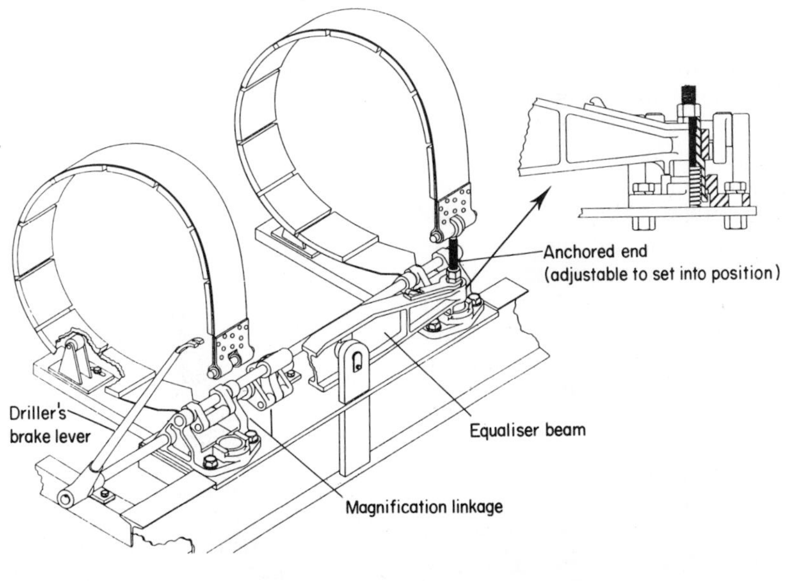

Fig. 3.3. Schematic of the brake system.

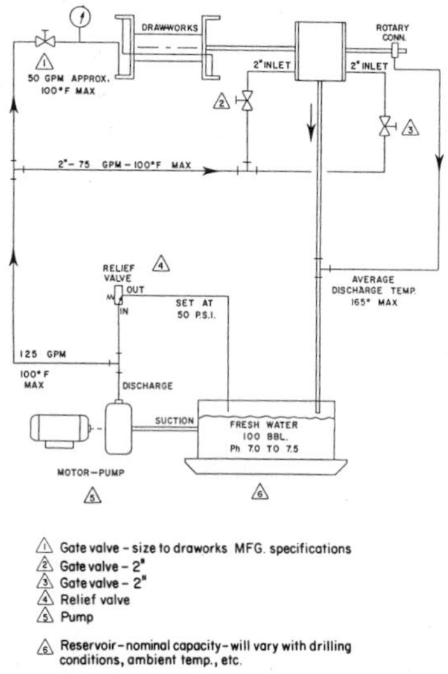

Fig. 3.4. Typical water cooling system for draw-works and eddy current brake[1]: 1, gate valve — size to draw-works manufacturing specifications; 2, 3, gate valves — 2 in; 4, relief valve; 5, pump; 6, reservoir — nominal capacity will vary with drilling conditions, ambient temperature, etc. 40–50 psi recommended by draw-works manufacturer. (Courtesy of National Supply Company)

Auxiliary brakes

During tripping-out operations, the drillpipe is removed in stands (≈93 ft, depending on size of derrick etc.) in such a way that the pipe is decelerated during the last 5–10 ft before being arrested in the rotary table with the slips. Thus, during tripping out operations, the main brakes are not required to absorb excessive energy.

During tripping-in (or RIH) operations without the aid of an auxiliary brake, the main brake must be applied continuously in order to lower the load at a controlled speed. As each stand of pipe is added, the load becomes greater. To use the band brake for this almost continuous operation would quickly cause fading and overheating. In practice, the main brakes are assisted by an auxiliary brake which is only engaged during tripping-in operations. The auxiliary brake provides a continuous retarding torque to control speed of lowering. The main brake is used for final stopping of each stand before being held in rotary table by the slips.

The auxiliary brake is cradle-mounted on the draw-works frame and is connected to the hoisting drum shaft by means of a sliding coupling. Two types of auxiliary brakes are in use: hydrodynamic and eddy current.

Hydrodynamic (or 'Hydromatic') auxiliary brake The hydrodynamic brake is built somewhat like a centrifugal pump using water to provide a cushioning effect, rather than pumping and producing pressure. The principal components of the hydrodynamic brake comprise a rotor assembly connected to the brake shaft and a stator assembly (Figure 3.5a). The brake shaft connects with the draw-works drum shaft through a sliding coupling. A braking effect is produced by the resistance to agitation of water circulated between the veined pockets of the rotor and stator elements[4]. The mechanical energy of the rotor is transferred to water in the form of heat energy. The magnitude of mechanical energy absorbed and, in turn, braking effect produced is proportional to the velocity of water in the brake chamber. Also, the braking effect is dependent on the water level inside the brake chamber, with the greatest braking effect being possible when the chamber is completely full of water[2]. Braking effect can be varied by changing the water level inside the chamber.

The hoisting drum friction brake cooling water is discharged through a passageway in the auxiliary brake shaft. A separate supply branch provides continuous flow of cool water to the auxiliary brake thereby preventing water in the chamber from overheating.

Eddy current auxiliary brake[4] The eddy current brake uses magnetic forces to slow the rotation of the hoisting drum. It consists of an iron rotor connected to the brake shaft, the rotor being surrounded by a stationary member stator which provides a controlled variable magnetic field (Figure 3.5b). The magnetic field of the stator is provided by coils excited by an outside source of d.c. electricity. The stator induces a magnetic force in the rotor which opposes the rotor

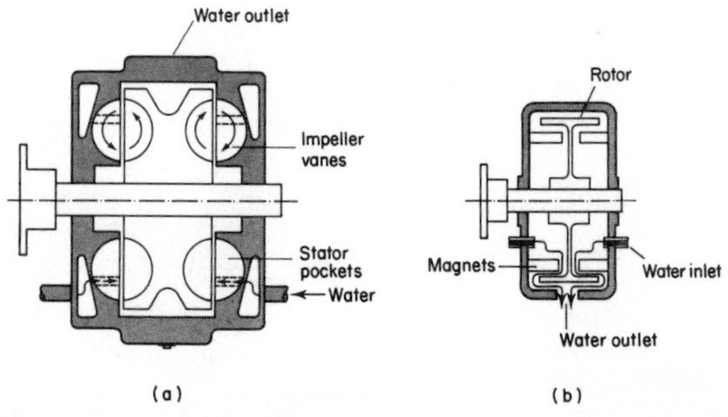

Fig. 3.5. Schematic of auxiliary brakes: (a) hydromatic brake; (b) eddy current brake[5]. (Courtesy of *Oil and Gas Journal*)

motion and, in turn, provides a braking effect. The magnitude of the braking effect can be varied by varying the intensity of the electromagnetic field in the stator.

The induced magnetic field also produces, within the rotor, eddy currents which, in turn, generate heat. This heat is removed by a continuous flow of cool water provided by the brake cooling system (shown in Figure 3.4).

Comparison of hydromatic and eddy current brake With the hydromatic brake the braking effect increases exponentially with speed of rotation. Thus, in lowering stands of pipe, the braking effect increases as the pipe tries to drop faster. As the band brake is applied to slow the stand near the rotary table, the hydromatic brake has less effect, until all braking is being done by the band brake. Hydromatic brakes are generally used on rigs where electrical power supply is limited. Where electrical power is readily available, as on a diesel–electric rig, it is normal to use an eddy current brake, which has the advantage that the braking effect is dependent upon the intensity of the electromagnetic field and is controlled by the driller through a small operating lever. Thus, the driller has more precise control of the braking effect.

Catshaft and coring reel group

The catshaft and coring reel group comprises the catheads, the catshaft assembly and the coring reel drum (Figure 3.2).

The catheads are spool-shaped, rotating drums powered by the jackshaft assembly. They consist of friction and mechanical rotating heads.

The friction catheads are used to transport heavy objects around the rig floor by means of a manila rope.

The mechanical catheads comprise the make-up cathead on the driller's side and the break-out cathead on the opposite side. The mechanical catheads are spooled with a suitable length of wire line which is connected to tongs. The line coming out of the mechanical catheads on the driller's side is termed the make-up line and the corresponding tong is described as the make-up tong. The make-up tong is used to tighten the drillpipe joints when the drill string is being run into the hole.

The mechanical cathead on the opposite side of the driller is described as the break-out cathead; the line and tong are also described as break-out line and break-out tong, respectively. The break-out tong is used to break the drillpipe joints during the removal of pipe from the hole.

Coring reel drum

The coring reel drum contains sufficient small-diameter (usually $\frac{9}{16}$ in) wire line to reach the bottom of the hole. It is used for lowering and retrieving any device to the hole bottom. Deviation recorders are one example. At any time, a deviation recorder can be lowered down the centre of the drill string to come to rest sitting above the drilling bit. After a preset time delay, a clockwork mechanism is activated to punch a pinhole in a paper disc, and then the deviation recorder is lifted out on the coring reel. The position of the pinhole shows the deviation of the hole from the vertical.

Main drive shaft and jackshaft group

On many modern rigs the prime movers are of the diesel–electric type, in which diesel engines are used to generate electricity. Electric cables are used to deliver power to motors attached to the main drive shaft (Figure 3.2), rotary table and mud pumps. The main drive shaft is equipped with two sprockets connected by roller chains to high- and low-drive sprockets on the jackshaft. The jackshaft is connected to the cat-shaft and drumshaft through roller chains and sprockets, as depicted in Figure 3.2.

Engagement of the high- or low-drive sprockets, catshaft or hoisting drumshaft is achieved by sliding-gear clutches.

With the combination of high- and low-drive sprockets on the jackshaft plus high- and low-air clutches on the drumshaft, the hoisting drum is, in effect, driven from a four-speed gearbox. Reverse is obtained by reversing the rotation of the d.c. electric motors. Speed control is by a foot or hand throttle at the driller's position.

Rotary countershaft group

When the rotary table is powered directly from the draw-works, a rotary countershaft is required. The rotary countershaft group comprises all the components required to transfer rotary motion to the rotary table. The equipment includes the rotary countershaft drive-chain and sprockets, air clutch, inertia brake and controls. In most modern rigs, however, the rotary table is independently powered by a separate d.c. motor and drive shaft assembly.

HOISTING TACKLE

The great weight of the drill string can easily be handled by employing a block-and-tackle system in which a continuous line is wound around a number of fixed and travelling pulleys. The line segments between sets of pulleys act to multiply the single pull exerted by the hoisting drum, thereby allowing many thousands of pounds of drill string weight or casing to be lowered into or removed from the hole. Ignoring friction at the pulleys, the pull required at the hoisting drum is approximately equal to the total weight handled, divided by the number of line segments between the sets of pulleys.

The main components of the travelling assembly include the following.

Crown block

The crown block (shown in Figure 3.6) provides a means of taking wire line from the hoisting drum to the travelling block. The crown block is stationary and is firmly fastened to the top of the derrick. It has a number of sheaves mounted on a stationary shaft, with each sheave acting as an individual pulley. The drilling line is reeved successively round the crown block and travelling block sheaves, with the end coming from the crown block sheave going to an anchoring clamp called a 'dead line anchor'.

The line section connecting the hoisting drum to the crown block is described as the 'fast line'. Thus, during hoisting operations the drum spools much more line than the distance travelled by the travelling block. The speed of the fast line is equal to the number of lines strung times the speed of the travelling block.

The crown block must be positioned such that the fast-line sheave is close to the centre line of the hoisting drum. The angle formed by the fast line and the vertical is described as the fleet angle (Figure 3.1). Fleet angle should be less than 1.5°, to minimise poor drum spooling and line wear[3].

The crown block is normally a steel framework with the sheaves mounted parallel on a shaft. The sheaves are individually mounted on double-row tapered roller bearings, to minimise friction. A sheave for the line from the coring reel drum is also on the crown block. Small sheaves for the manilla rope from the friction catheads may also be on the crown block.

Travelling block and drilling hook

The travelling block is somewhat similar to the crown block in that it has several independently mounted sheaves or pulleys. The sheaves are manufactured of high-quality steel and each is mounted on large-diameter, anti-friction bearings. The bearings are lubricated with grease through the stationary shaft which supports the sheaves. In general, the crown and travelling block sheave diameter should be 30–35 times the wire-line diameter, to prevent excessive wear and to increase the fatigue life of the drilling line. The sheave grooves[3] have tapering faces such that the sheave supports 150° of rope circumference. The travelling block is manufactured[3] to be: (a) short and slim, so that less head room is used at the top of the hoisting travel (also, it must have a low centre of gravity to resist tipping moments caused by efficiency differences in the line reeving); (b) heavy, to overcome the drilling line friction quickly, especially when falling empty; and (c) free of protrusions and sharp edges for the safety of workers.

The travelling block is often combined with the hook into one unit, known as a 'Hook-Block' as shown in Figure 3.6. The hook is used to connect the travelling block to the swivel and the rest of the drill string.

DEADLINE ANCHOR AND WEIGHT INDICATOR

The deadline anchor (Fig. 3.7) consists of a base and a slightly rotatable drum attached to the rig floor or rig substructure. A deadline anchor provides a means of securing the deadline and of measuring the hook load.

Measurement of hook load is achieved by installing a sensitive load cell or a pressure transformer between the lever arm of the drum and the stationary extension of the base. The load being run in or pulled out of the hole creates tension in the deadline, described as deadline load. This load causes torque on the deadline drum which is resisted by the load cell. The load is converted to hydraulic pressure within the load cell or the pressure transformer. A pressure signal representing the deadline load is sent to the rig floor through a fluid-filled hose connected to a weight indicator gauge at the driller's control position.

When drilling ahead, some of the total hook load is applied to the drilling bit, to give the optimum weight on the bit for the size and type of bit being used and the formation being drilled. The weight indicator has two pointers: one shows total hook load and the other shows weight on bit.

A pad-type load cell, mounted under a mast leg, may also be used to give total hook load. Once again, the load exerted on the load cell or pressure transformer is transmitted hydraulically to the driller's weight indicator via a high-pressure hose.

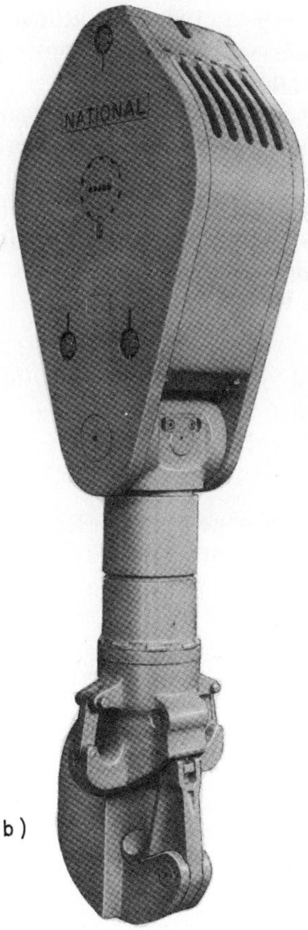

Fig. 3.6. Crown block and travelling block[1]: (a) crown block; (b) travelling block and hook. (Courtesy of National Supply Company)

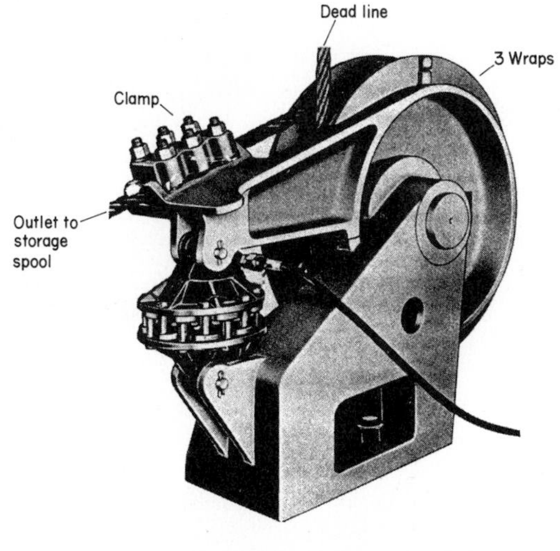

Fig. 3.7. Deadline anchor[1]. (Courtesy of National Supply Company)

DRILLING LINE (BLOCK LINE)

A drilling line is basically a wire rope made up of a number of strands wound around a steel core. Each strand contains a number of small wires wound around a central core. There are several types of wire rope: (a) round-strand (drilling line is a typical example); (b) flattened-strand; (c) locked-coil; (d) half-locked; and (e) multi-strand. The principal differences between them are their internal construction, their weight per unit length, the breaking strength, the number of wires in each strand, and the number of strands and type of core. In oil well drilling only round-strand wires are used, and, hence, discussion will be limited to this type.

Round-strand ropes

Round-strand ropes are the most widely used in most hoisting operations, including oil well and mining applications. They are more versatile and economical than the other types, but require more maintenance.

A round-strand rope consists of six strands wound over a fibre core or, sometimes, over a small wire rope. A round-strand rope can be simple, with one central wire, or compound, with several central wires.

A wire rope is described by the number of strands, the wire ropes making up the strand and the type of central core. For example, the ropes illustrated in Figure 3.8 can be described as:

Type A: either 6 × 9/9/1, meaning 6 strands each consisting of 9 outer wires, 9 inner wires and one central core, or 6 × 19, meaning 6 strands each containing 19 filler wires.
Type C: either 6 × 10/5/5/1 or 6 × 21.

A round-strand rope is also described by the type of lay of the main strands and filler wires. In general, there are two types of lay.

Lang's lay

In the Lang's lay type, wires in the strands and the strands in the rope are twisted in the same direction or same hand. Hence, a Lang's lay can be right lay or left lay, depending on whether the strands are twisted towards the left- or right-hand direction as one looks away from one end of the line. This type of twist increases the wire rope's resistance to wear although it tends to unwind if the ends are released unless they are pre-formed to their final shape before being twisted into the strands.

Ordinary (or regular) lay

In the ordinary lay the wires in the strands and the strands in the rope are twisted in opposite directions. This type has a small wearing surface, with the consequence of a shorter working life. The main advantage of this type is that it is easier to install and handle than the Lang's lay type.

The drilling line: design considerations

A drilling line is of the round-strand type with a Lang's lay and typically has a 6 × 19 construction, with independent wire rope core (IWRC). Size of drilling line varies from $\frac{1}{2}$ in to 2 in (51 mm). Figure 3.8

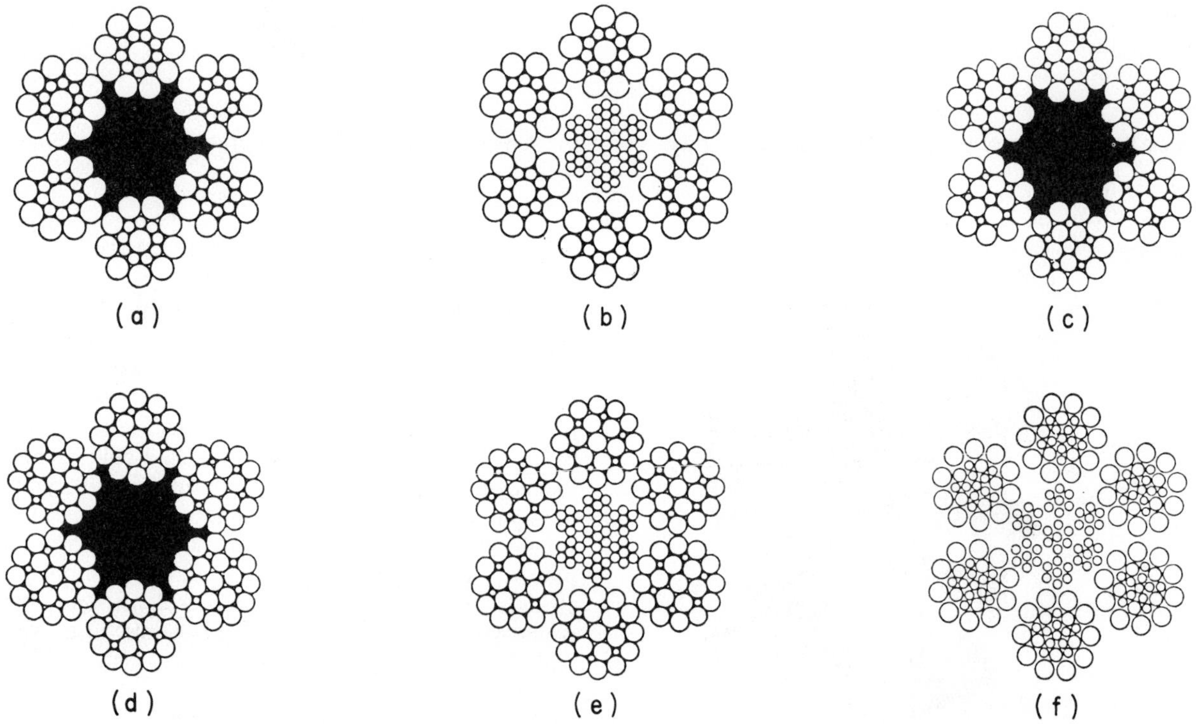

Fig. 3.8. Typical wire rope constructions, 6 × 19 classification[6]: (a) 6 × 19 Seale with fibre core; (b) 6 × 19 Seale with independent wire rope core; (c) 6 × 21 filler wire with fibre core; (d) 6 × 25 filler wire with fibre core; (e) 6 × 25 filler wire with independent wire rope core; (f) 6 × 25 Warrington Seale with independent wire rope core. (Courtesy of API)

TABLE 3.1 6 × 19 classification wirerope, bright (uncoated) or drawn-galvanized wire, independent wirerope core. (Courtesy of API[7])

1	2	3	4	5	6	7	8	9	10
Nominal diameter		*Approximate mass*		*Nominal strength*					
				Improved plough steel			*Extra improved plough steel*		
in	*mm*	*lb/ft*	*kg/m*	*lb*	*kN*		*lb*	*kN*	
$\frac{1}{2}$	13	0.46	0.68	23 000	102		26 600	118	
$\frac{9}{16}$	14.5	0.59	0.88	29 000	129		33 600	149	
$\frac{5}{8}$	16	0.72	1.07	35 800	159		41 200	183	
$\frac{3}{4}$	19	1.04	1.55	51 200	228		58 800	262	
$\frac{7}{8}$	22	1.42	2.11	69 200	308		79 600	354	
1	26	1.85	2.75	89 800	399		103 400	460	
$1\frac{1}{8}$	29	2.34	3.48	113 000	503		130 000	578	
$1\frac{1}{4}$	32	2.89	4.30	138 800	617		159 800	711	
$1\frac{3}{8}$	35	3.50	5.21	167 000	743		192 000	854	
$1\frac{1}{2}$	38	4.16	6.19	197 800	880		228 000	1010	
$1\frac{5}{8}$	42	4.88	7.26	230 000	1020		264 000	1170	
$1\frac{3}{4}$	45	5.67	8.44	266 000	1180		306 000	1360	
$1\frac{7}{8}$	48	6.50	9.67	304 000	1350		348 000	1550	
2	51	7.39	11.0	344 000	1530		396 000	1760	

gives the various types of wire rope of API 6 × 19 classification.

Besides construction details, a wire rope is also described by its nominal diameter, its mass per unit length and its nominal strength. Table 3.1 gives the design data for a 6 × 19 construction wire rope with an independent core. This type is the most widely used in the oil industry. For other types the reader is advised to consult API Specification 9A[7].

Static and dynamic crown (or derrick) loading

Analysis of static and dynamic crown loading is greatly simplified if the sheaves of both the crown block and the travelling block are considered as single pulleys.

Consider the hoisting system shown in Figure 3.9, in which the crown block has two sheaves and the travelling block has one. (*Note:* The travelling block always has one sheave less than the crown block.)

The fast-line load is the load carried by the hoisting drum line when in motion. The deadline load is the pull in the deadline anchor. The hook load is the total load carried by the travelling block including the weight of the block.

Under static conditions the hook load, W, will be supported by the two lines and each will carry $W/2$. Also, the fast-line load and the dead-line loads will be $W/2$.

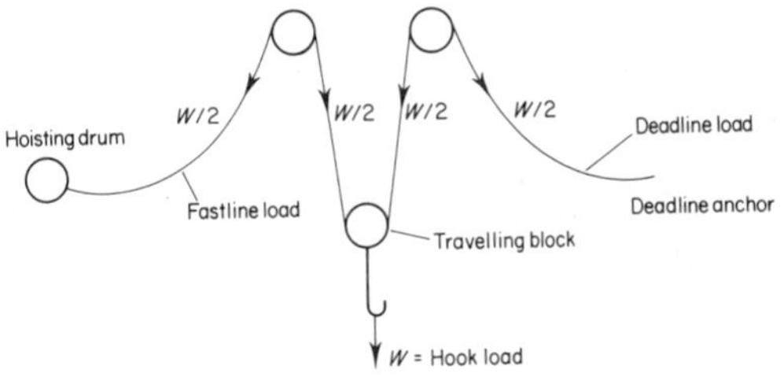

Fig. 3.9. Static and dynamic crown loading — hoisting system where crown block has two sheaves.

Static crown load (SCL)

$$= \text{fast-line load} + \text{hook load} + \text{deadline load}$$

$$= \frac{W}{2} + W + \frac{W}{2} = 2W$$

Similarly, if the crown block has three sheaves, as shown in Figure 3.10, then the static crown loading is given by

$$\text{SCL} = \frac{W}{4} + W + \frac{W}{4} = \frac{3}{2}W$$

In general, under static conditions,

$$\text{fast-line load} = \frac{W}{N}$$

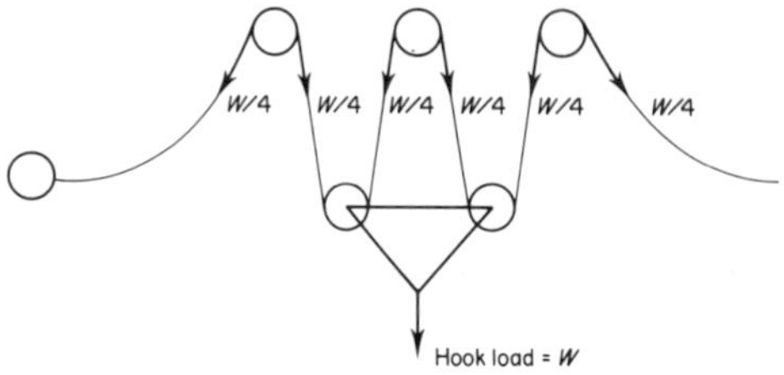

Fig. 3.10. Static and dynamic crown loading — hoisting system where crown block has three sheaves.

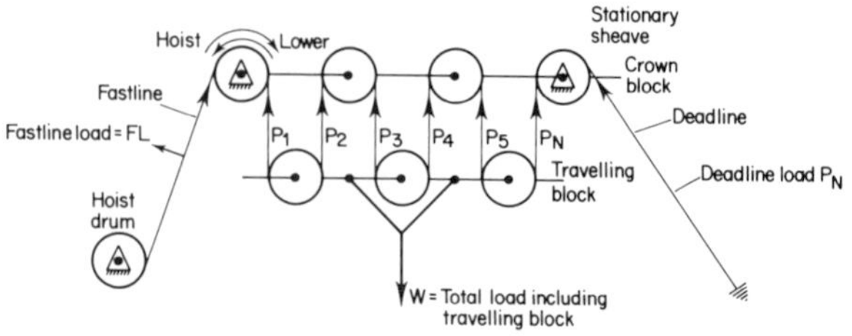

Fig. 3.11. Effects of friction on the efficiency of the hoisting system.

and

$$\text{deadline load} = \frac{W}{N}$$

where N is the number of lines strung between the travelling and crown blocks.

Thus, for N lines, the static crown load is given by

$$\text{SCL} = \frac{W}{N} + W + \frac{W}{N} = W\left(1 + \frac{2}{N}\right)$$

$$= W\left(\frac{N+2}{N}\right) \tag{3.1}$$

Under dynamic conditions, i.e. when the line is moving, the dynamic crown loading will be given by

dynamic crown load = fast-line load + hook load + deadline load

where the fast-line load is now magnified, owing to the effects of sheave efficiency due to movement of lines.

Efficiency of the hoisting system

In a block-and-tackle system where the wirerope is reeved over a number of sheaves, the line pull exerted by the hoisting drum is gradually reduced towards the deadline due to the losses caused by friction in the sheaves and in the bending of the rope around the sheaves[4]. The efficiency of the hoisting system is further reduced by internal friction in wirerope and hole friction.

For hoisting operations, an expression for the efficiency of the block-and-tackle system can be developed with reference to Figure 3.11.

Let

EF = Block-and-tackle efficiency factor

K = Sheave and line efficiency per sheave

N = Number of lines strung to travelling block

Or

N = Number of working sheaves

FL = Fast-line tension

DL = Deadline tension

Starting with a hoisting fast-line pull of FL, the friction from the first block sheave reduces the line pull in the first travelling line from FL to P_1, where P_1 is given by:

$$P_1 = FL \times K$$

Similarly, the line pull in the second travelling line will be reduced to P_2 where P_2 is given by:

$$P_2 = P_1 \times K$$

Or

$$P_2 = FL \times K^2$$

Similarly

$$P_N = FL \times K^N \tag{3.2}$$

If N is the number of lines supporting the hook load W, then:

$$W = P_1 + P_2 + P_3 + \cdots + P_N$$
$$= FL \times K + FL \times K^2 + FL \times K^3 + \cdots + FL \times K^N$$
$$= FL(K + K^2 + K^3 + \cdots + K^N)$$

The terms in brackets[7] form a geometric series, the sum of which is given by:

$$\frac{K(1-K^N)}{(1-K)}$$

Hence

$$W = \frac{FL \times K(1-K^N)}{(1-K)}$$

Or

$$FL = \frac{W(1-K)}{K(1-K^N)} \tag{3.3}$$

In the absence of friction,

$$FL = P_1 = P_2 = \cdots = P_N \qquad (3.3)$$

and hook load W is given by

$$W = P_{AV} \times N$$

Or

$$P_{AV} = \frac{W}{N} \qquad (3.4)$$

P_{AV} = average line pull on block-and-tackle system

Hence the efficiency factor (EF) of the hoisting system is the ratio of equation (3.4) to equation (3.3), i.e.:

$$EF = \frac{P_{AV}}{FL}$$

$$EF = \frac{K(1 - K^N)}{N(1 - K)} \qquad (3.5)$$

Example 3.1

Calculate the efficiency factor for a hoisting system employing 8 strung lines. Assume the value of K to be 0.9615.

Solution

$$EF = \frac{K(1 - K^N)}{N(1 - K)}$$

$$= \frac{0.9615(1 - 0.9615^8)}{8(1 - 0.9615)}$$

$$= 0.842$$

Using equation (3.5), Table 3.2 can be constructed for different numbers of lines strung between the crown and travelling blocks.

During lowering of the hook load it can be shown that the efficiency factor and fast-line load are given by[5A]:

$$(EF)_{\text{Lowering}} = \frac{NK^N(1 - K)}{(1 - K^N)} \qquad (3.6)$$

TABLE 3.2 Block and tackle efficiency factors for $K = 0.9615$.

Number of lines strung	*Efficiency factor*
6	0.874
8	0.842
10	0.811
12	0.782

4 0,907

$$(FL)_{\text{Lowering}} = \frac{WK^{-N}(1 - K)}{(1 - K^N)} \qquad (3.7)$$

Hook, fast-line and deadline loads

Referring to Figure 3.11, the hook load, HL, is given by

$$\text{HL} = W = \text{weight of drill string (or casing) in mud} + \text{weight of travelling block, hook, etc.} \qquad (3.8)$$

The hook load is supported by N lines, and, in the absence of friction, the fast-line load, FL, is given by

$$\text{FL} = \frac{\text{hook load}}{\text{number of lines supporting the hook load}}$$

$$= \frac{\text{HL}}{N}$$

Owing to friction, the fast-line load required to hoist the hook load is increased by a factor equal to the efficiency factor. Thus,

$$\text{FL} = \frac{\text{HL}}{N \times \text{EF}} \qquad (3.9)$$

Under static conditions the deadline load is given by HL/N. During motion the effects of sheaves friction must be considered and the deadline load is given by

$$\text{DL} = \frac{\text{HL} \times K^N}{N \times \text{EF}} \qquad (3.10)$$

If the breaking strength of the drilling line is known, then a design or design factor, DF, may be calculated, as follows:

$$\text{DF} = \frac{\text{nominal strength of wire rope (lb)}}{\text{maximum load carried by a single line (lb)}}$$

or

$$\text{DF} = \frac{\text{nominal strength of wire rope (lb)}}{\text{fast-line load (lb)}}$$

The design factor for a wire rope is applied to compensate for wear and shock loading, and its magnitude is dependent on the kind of work performed. API[7] recommends a minimum DF of 3 for hoisting operations and of 2 for setting casing or when pulling on stuck pipe.

Power requirements

The power requirements of the draw-works can be determined by consideration of the fast-line load and fast-line velocity.

For a given distance travelled by the hook load (say

1 unit), the fast-line has to move N times to provide the travelling block with the appropriate rope length, where N is the number of lines strung. Thus,

$$V_f = N \times V_L \tag{3.11}$$

where V_f = velocity of fast-line load; and V_L = velocity of travelling block or any of its lines. Therefore,

$$\text{power at drum} = \text{FL} \times V_f$$

From Equations (3.9) and (3.11) we obtain

$$\text{power} = \left(\frac{\text{HL}}{N \times \text{EF}}\right) \times N \times V_L$$

or

$$P = \frac{(\text{HL}) \times V_L}{\text{EF}} \tag{3.12}$$

In the Imperial system, power is quoted in horsepower, and Equation (3.12) becomes:

$$\text{Drumpower output} = \frac{\text{HL} \times V_L}{\text{EF} \times 33\,000}\ \text{horsepower} \tag{3.13}$$

where HL is in pounds force and V_L is in ft/min.

The input power to the draw-works is calculated by taking into account the efficiency of the chain drives and shafts inside the draw-works.

Example 3.2

The following data refer to a $1\frac{1}{2}$ in block line with 10 lines of extra improved plough steel wire rope strung to the travelling block:

hole depth	= 10 000 ft (3048 m)
drillpipe	= 5 in OD/4.276 in ID, 19.5 lb/ft (29.02 kg/m)
drill collars	= 500 ft, 8 in/$2\frac{13}{16}$ in, 150 lb/ft (223 kg/m)
mud weight	= 75 pcf
Line and sheave efficiency coefficient	= 0.961 5

Calculate:

(1) weight of drill string in air and in mud;
(2) hook load, assuming weight of travelling block and hook to be 23 500 lb
(3) deadline and fast-line loads, assuming an efficiency factor of 0.81;
(4) dynamic crown load;
(5) wireline design during drilling if breaking strength of wire is 228 000 lb (1010 kN) (see Table 3.1);
(6) design factor when running 7 in casing of 29 lb/ft (43.2 kg/m).

Solution

(1) Weight of string in air

= weight of drillpipe + weight of drill collars

= (10 000 − 500) × 19.5 + 500 × 150

= 260 250 lb

(*Note:* Weight of string in air is also described as pipe setback load.)

Weight of string in mud

= buoyancy factor × weight in air

$$= \left(1 - \frac{75}{489.5}\right) \times 260\,250$$

= 220 432 lb

$$\left(\textit{Note: } \text{BF} = 1 - \frac{\rho_{\text{mud}}}{\rho_{\text{steel}}} = 1 - \frac{75}{489.5} = 0.847\right)$$

(2) Hook load

= weight of string in mud

+ weight of travelling block, etc.

= 220 432 + 23 500

= 243 932 lb

(3) Deadline load

$$= \frac{\text{HL}}{N} \times \frac{K^{10}}{\text{EF}} = \frac{243\,932}{10} \times \frac{(0.9615)^{10}}{0.81}$$

= 20 336 lb

$$\text{Fast-line load} = \frac{\text{HL}}{N \times \text{EF}} = \frac{243\,932}{10 \times 0.81}$$

= 30 115 lb

(4) Dynamic crown load

= DL + FL + HL

= 20 336 + 30 115 + 243 932

= 294 383 lb

$$\text{(5) Design factor} = \frac{\text{breaking strength}}{\text{fast-line load}}$$

$$= \frac{228\,000}{30\,115} = 7.6$$

(6) Weight of casing in mud

= 10 000 × 29 × BF

= 245 630 lb

$$\text{HL} = \text{weight of casing in mud} + \text{weight of travelling block, etc.}$$
$$= 245\,630 + 23\,500$$
$$= 269\,130 \text{ lb}$$

$$\text{FL} = \frac{\text{HL}}{N \times \text{EF}} = \frac{269\,130}{10 \times 0.81} = 33\,226 \text{ lb}$$

$$\text{DF} = \frac{228\,000}{33\,226} = 6.9$$

Ton-miles (megajoules) of a drilling line

The total service life of a drilling line may be evaluated by taking into account the work done by the line during drilling, fishing, coring, running casing, etc., and by evaluating the stresses imposed by acceleration and deceleration loadings, vibration stresses and bending stresses when the line is in contact with the drum and sheave surfaces (API RP 9B: Section 4)[8]. Owing to the large number of variables involved, API recommends that the service life of a drilling line be computed by evaluating the work done in raising and lowering loads, in making round trips and in the operation of drilling, coring and setting casing. Derivation of most equations required to determine the service life of a drilling line is based on API RP 9B standards[8].

Round trip operations Typical round trip operations include running in and pulling drillpipe during drilling (see Chapter 1).

(a) *Work done by travelling assembly* In running the drill string, the travelling assembly, including the travelling block, hook, links and elevator of weight M, moves a distance approximately equal to twice the length of the stand, $2L_s$, for each stand. Hence, work done, WD, by travelling assembly

$$= \text{force} \times \text{distance}$$
$$= M \times 2L_s$$

and for N stands

$$\text{WD} = N \times M \times 2L_s$$

In pulling the string out of hole a similar amount of work is done. Hence, work done by the travelling block assembly during one complete round trip is equal to $2(NM \times 2L_s) = 4ML_sN$.

For a hole depth of D, assume the drillpipe to extend to the bottom of the hole, such that $L_sN = D$, where D is hole depth. Therefore,

$$\text{WD by travelling assembly} = 4MD \qquad (3.14)$$

(b) *Work done by drill string* WD = average weight lowered × depth

Average weight

$$= \frac{\text{initial load} + \text{final load}}{2}$$

$$= \frac{\text{weight of 1 stand} + \text{weight of } N \text{ stands}}{2}$$

The average weight must be adjusted for the effects of buoyancy, and this results in a buoyant average weight of

$$\frac{W_eL_s + W_eL_sN}{2}$$

where W_e is the effective weight per foot (or metre) of drillpipe in mud. Therefore,

$$\text{WD by drill string} = \tfrac{1}{2}(W_eL_s + W_eL_sN) \times D$$
$$= \tfrac{1}{2}(W_eL_s + W_eD) \times D$$

Assuming the frictional effects to be the same going into the hole as coming out, the WD in pulling the drillpipe is the same as in lowering it. Hence, for a round trip,

$$\text{WD by drilling line} = 2 \times \tfrac{1}{2}(W_eL_s + W_eD)D$$

$$\text{WD} = W_e(L_s + D)D \qquad (3.15)$$

(c) *Work done by drill collars* A correction for the weight of drill collars should be made for the above equations, since it was assumed that drillpipe extends to the bottom of hole.

$$\text{WD in raising drill collars and subs} = 2CD \qquad (3.16)$$

where C = effective weight of drill collar assembly in mud minus effective weight of the same length of drillpipe in mud, as shown in Figure 3.12

$$C = (LW_{dc} - LW_{dp}) \times \text{BF}$$

where W_{dc} = weight of drill collar in air; W_{dp} = weight of drillpipe in air, and L = length of drill collars.

(d) *Work done in round trip operations*
Total WD in making a round trip = Equations (3.14) + (3.15) + (3.16)

$$T_r = 4MD + W_e(L_s + D)D + 2CD \qquad (3.17)$$

In Imperial units, Equation (3.17) becomes

$$T_r = \frac{D(L_s + D)W_e}{10\,560\,000} + \frac{D(M + C/2)}{2\,640\,000} \text{ ton-miles}$$

where M = mass of travelling assembly (lb); L_s = length of each stand (ft); and D = hole depth (ft). (3.17a)

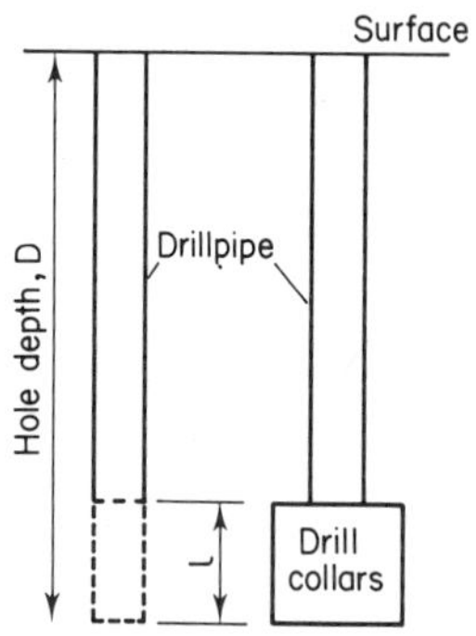

Fig. 3.12. Correction for weight of drill collars.

In metric units M is in kN; D and L_s are in metres; and W_e and C are in kg/m. Hence, in field units, Equation (3.17) becomes

$$T_r = 10^{-6}\left[D(L_s + D) \times 9.81 W_e + 4D\left(M + \frac{9.81C}{2}\right)\right] \text{MJ} \qquad (3.17b)$$

Drilling operations The ton-mile (or megajoule: MJ) service performed by a drilling line during drilling operations is expressed in terms of the work performed in making round trips, since there is a direct relationship, as shown in the following cycle of drilling operations[8] (see also Chapter 1, page 9):

(1) drill ahead a length of kelly;
(2) pull up length of kelly;
(3) ream ahead a length of kelly;
(4) pull up a length of kelly;
(5) put kelly in rathole;
(6) pick up a single (or double);
(7) lower drill string in hole; and
(8) pick up kelly and drill ahead.

Operations (1) and (2) give 1 round trip of WD; operations (3) and (4) give 1 round trip of WD; operation (7) gives $\frac{1}{2}$ round trip of WD; and operations (5), (6) and (8) give approximately $\frac{1}{2}$ round trip of WD. Therefore,

$$\text{total WD} = 3 \text{ round trips} = 3T_r$$

However, in drilling a length of section from depth d_1 to depth d_2, the work done, T_d, is given by

$$T_d = 3T_r = 3(T_r \text{ at } d_2 - T_r \text{ at } d_1)$$
$$= 3(T_2 - T_1) \qquad (3.18)$$

Coring operations The following cycle is normally adopted:

(1) core ahead a length of core barrel;
(2) pull up length of kelly;
(3) put kelly in rathole;
(4) pick up a single joint of drillpipe;
(5) lower drill string in hole; and
(6) pick up kelly.

Operations (1) and (2) give 1 round trip of WD; operation (5) gives $\frac{1}{2}$ round trip of WD; and operations (3), (4) and (6) give $\frac{1}{2}$ round trip of WD. Therefore,

total WD in coring = 2 round trips to bottom

or

$$T_c = 2(T_2 - T_1) \qquad (3.19)$$

where T_2 = WD for 1 round trip at depth d_2, where coring stopped before coming out of the hole; and T_1 = WD for 1 round trip at depth d_1, where coring started.

Setting casing operations The total WD in this case is the same as that for tripping of drillpipe but the effective weight, W_e, is that of the casing. Also, running casing is a one-way operation and the WD for tripping will have to be multiplied by $\frac{1}{2}$. The extra weight for drill collars, C, is not required in this case. Therefore,

$$\text{WD in setting casing} = T_s = \tfrac{1}{2}(4MD + W_{cs}(L_s + D)D) \qquad (3.20)$$

where W_{cs} = effective weight per unit length of casing in mud; and L_s = length of a casing joint.

In Imperial units, Equation (3.20) becomes

$$T_s = \frac{1}{2}\left[\frac{D(L_s + D) \times W_{cs}\ \text{ft(ft)}\ \frac{\text{lbm}}{\text{ft}}}{\frac{5280\ \text{ft}}{\text{mile}}\ \frac{2000\ \text{lbm}}{\text{ton}}} + \frac{4MD\ \text{lbm} \times \text{ft}}{\frac{2000\ \text{lbm}}{\text{ton}}\ \frac{5280\ \text{ft}}{\text{mile}}}\right]$$

$$= \frac{1}{2}\left(\frac{D(L_s + D) \times W_{cs}}{10\,560\,000} + \frac{MD}{2\,640\,000}\right) \text{ton-miles}$$

In metric units, Equation (3.20) becomes

$$T_s = \tfrac{1}{2} \times 10^{-6}(D(L_s + D) \times 9.81 \times W_{cs} + 4MD) \text{ MJ}$$

Evaluation of total service and cut-off practice

Grand total of work performed by drilling line = sum of Equations (3.17) + (3.18) + (3.19) + (3.20).

The service life of a drilling line can be increased[8] by frequently cutting off a suitable length in order to avoid highly stressed portions being constantly left in the same positions. Highly stressed points of the drilling line are normally found on the top of the crown block sheaves, on the bottom of the travelling block sheaves and at cross-over points on the hoisting drum.

The frequency at which the wire rope is cut off is determined by the total work done during drilling, tripping, etc. The frequency of cut-off is dependent on type and diameter of wire and on the local conditions. For the Gulf Coast in the USA, API recommends the use of Table 3.3 for calculating the cut-off frequencies in terms of ton-miles for the given wire rope sizes.

The cut-off length is related to the hoisting drum diameter and the derrick height. API[8] recommends Table 3.4 to be used for the calculation of cut-off length of line in terms of drum laps. Drum laps can be converted to lengths in feet by use of the following relationship:

$$\text{length} = \text{number of laps} \times \text{drum circumference}$$
$$= \text{number of laps} \times \Pi \times D$$

where D = drum diameter. For drums with two-step 'spirallel' grooving design, a quarter-lap should be added to the value obtained from Table 3.4, while for all other types of grooving a half-lap should be added to the cut-off length.

It should be observed that the frequency between cut-off in terms of ton-miles, as presented in Table 3.3, is only valid for DF = 5. For other safety factors the ton-mile frequency should be modified as shown in the following example.

TABLE 3.3 Frequency of cut-off for DF = 5, extracted from API[8].

Wire rope diameter (inches)	*Ton-miles between cut-off*
1	600
$1\frac{1}{8}$	800
$1\frac{1}{4}$	1100
$1\frac{3}{8}$	1900
$1\frac{1}{2}$	2600

Example 3.3

Given the following:

drum diameter = 30 in two-step spirallel grooving
derrick height = 136 ft
wire rope size = $1\frac{1}{4}$ in
design factor = 3.5

determine: (a) ton-miles between cuts; (b) cut-off length. (Assume that the data in Table 3.3 are applicable.)

Solution

(a) From Table 3.3, for a $1\frac{1}{4}$ in wire rope, the ton-miles between cuts are 1100 for DF = 5. For DF = 3.5,

$$\text{ton-miles} = \left(\frac{3.5}{5}\right) \times 1100$$
$$= 770$$

(b) From Table 3.4, cut-off length in terms of drum laps for a derrick height of 136 ft and drum diameter of 30 in is 10.5. Since the drum has a 'spirallel'

TABLE 3.4 Recommended cut-off lengths in terms of drum laps. (Courtesy of API[8])

1	*2*	*3*	*4*	*5*	*6*	*7*	*8*	*9*	*10*	*11*	*12*	*13*	*14*	*15*
Derrick or mast height (ft)	*Drum diameter (in)*													
	11	*13*	*14*	*16*	*18*	*20*	*22*	*24*	*26*	*28*	*30*	*32*	*34*	*36*
	Number of drum laps per cut-off													
187										$15\frac{1}{2}$	$14\frac{1}{2}$	$13\frac{1}{2}$	$12\frac{1}{2}$	$11\frac{1}{2}$
142, 143, 147								$13\frac{1}{2}$	$12\frac{1}{2}$	$11\frac{1}{2}$	$11\frac{1}{2}$	$10\frac{1}{2}$		
133, 136, 138						$15\frac{1}{2}$	$14\frac{1}{2}$	$12\frac{1}{2}$	$11\frac{1}{2}$	$11\frac{1}{2}$	$10\frac{1}{2}$	$9\frac{1}{2}$		
126, 129, 131				$17\frac{1}{2}$	$15\frac{1}{2}$	$14\frac{1}{2}$	$12\frac{1}{2}$	$12\frac{1}{2}$	$11\frac{1}{2}$	$10\frac{1}{2}$	$9\frac{1}{2}$	$9\frac{1}{2}$		
94, 96, 100		$19\frac{1}{2}$	$17\frac{1}{2}$	$14\frac{1}{2}$	$12\frac{1}{2}$	$11\frac{1}{2}$	$10\frac{1}{2}$	$9\frac{1}{2}$	$9\frac{1}{2}$	$8\frac{1}{2}$				
87		$17\frac{1}{2}$	$14\frac{1}{2}$	$12\frac{1}{2}$	$11\frac{1}{2}$									
66	$12\frac{1}{2}$	$11\frac{1}{2}$												

grooving design, a $\frac{1}{4}$ lap should be added. Hence, cut-off length in terms of drum laps

$$= 10.5 + 0.25 = 10.75$$

Length to be cut for every 770 ton-mile service

$$= 10.75 \times \pi \frac{30 \text{ in}}{12 \text{ in/ft}}$$

$$= 84.4 \text{ ft}$$

Example 3.4

Using the data given in Example 3.2, determine: (a) round trip ton-miles at 10 000 ft; (b) casing ton-miles if one joint of casing = 40 ft (12.2 m); (c) design factor of the drilling line when the $9\frac{5}{8}$ ft casing is run to 10 000 ft; (d) the ton-miles when coring from 10 000 ft to 10 180 ft; and (e) the ton-miles when drilling from 10 000 ft to 10 180 ft.

Solution

(a) From Equation (3.17a):

$$T_r = \frac{D(L_s + D)W_e}{10\,560\,000} + \frac{D(M + C/2)}{2\,640\,000}$$

$$M = 23\,500 \text{ lb}$$

$$C = (LW_{dc} - LW_{dp}) \times \text{BF}$$

$$= (500 \times 150 - 500 \times 19.5) \times 0.847$$

$$= 55\,267 \text{ lb}$$

$$D = 10\,000 \text{ ft}$$

$$L_s = 93 \text{ ft}$$

$$W_e = 19.5 \times \text{BF} = 16.52 \text{ lb/ft}$$

Therefore,

$$T_r = \frac{10\,000(93 + 10\,000) \times 16.52}{10\,560\,000} + \frac{10\,000(23\,500 + 55\,267/2)}{2\,640\,000}$$

$$= 157.9 + 193.7$$

$$= 351.6 \text{ ton-miles}$$

(b) $$T_s = \frac{1}{2}\left(\frac{D(L_s + D)W_{cs}}{10\,560\,000} + \frac{DM}{2\,640\,000}\right)$$

$$W_{cs} = \text{weight of casing in air} \times \text{BF}$$

$$= 29 \times 0.847 = 24.56 \text{ lb/ft}$$

and

$$L_s = 40 \text{ ft}$$

$$T_s = \frac{1}{2}\left(\frac{10\,000(40 + 10\,000) \times 24.56}{10\,560\,000} + \frac{10\,000 \times 23\,500}{2\,640\,000}\right)$$

$$= \frac{1}{2}(233.5 + 89.0)$$

$$= 161.3 \text{ ton-miles}$$

(c) DF = 5.6 (see Example 3.2).

(d) $$T_c = 2(T_2 - T_1)$$

where T_2 = round trip time at 10 180 ft, where coring stopped; and T_1 = round trip time at 10 000 ft, where coring started. Therefore,

$$T_2 = \frac{10\,180(93 + 10\,180) \times 16.52}{10\,560\,000} + \frac{10\,180(23\,500 + 55\,267/2)}{2\,640\,000}$$

$$= 163.6 + 197.2$$

$$= 360.8 \text{ ton-miles}$$

$$T_1 = 351.6 \text{ (from Part a)}$$

Therefore,

$$T_c = 2(360.8 - 351.6)$$

$$= 18.4 \text{ ton-miles}$$

(e) $$T_d = 3(T_2 - T_1)$$

$$= 3(360.8 - 351.6)$$

$$= 27{\cdot}6 \text{ ton-miles}$$

DERRICKS

A derrick is a structure of square cross-section constructed of special structural steel, having a yield strength of greater than 33 000 psi. It consists of four legs, connected by horizontal structural members described as girts. The derrick is further strengthened by bracing members connecting the girts.

A derrick is also equipped with a substructure (derrick floor) on which the drilling equipment is mounted. The derrick substructure is composed of derrick supports and rotary supports. The derrick supports consist of four posts and exterior bracing between the supports. The rotary supports consist of beams, posts and braces to support the rotary table and pipe setback load (weight of string standing in the derrick). The substructure height above the ground level varies according to the size of the substructure base. For a base size of 24–26 ft the height of the substructure is 7.25 ft. For a base size of 30 ft the height of the substructure can be 7.25, 10 or 14 ft (API[9]). The rating of derricks is based on pipe setback load and wind velocity as detailed in API standard 4A.[9].

The height of the substructure is often dependent on the size and pressure rating of the wellhead equipment and blow-out prevention equipment which has to be installed at the top of the hole, below the rotary table. Thus, a deep high-pressure well may require a high substructure to give room for the wellhead and BOPs.

Referring to Figure 3.13, a standard derrick can be designated by the following[9]:

The *height* is the vertical distance along the neutral axis of the derrick leg from the top of the derrick floor joists to the bottom of the water table beams.

The *base square* is the distance from heel to heel of adjacent legs at the top of the base plate.

The *window opening* distance is measured parallel to the centre line of the derrick from the top of the base plate. Figure 3.14 illustrates the different types of derrick window.

The *water-table opening* is an opening in the top of the derrick in which the crown block fits.

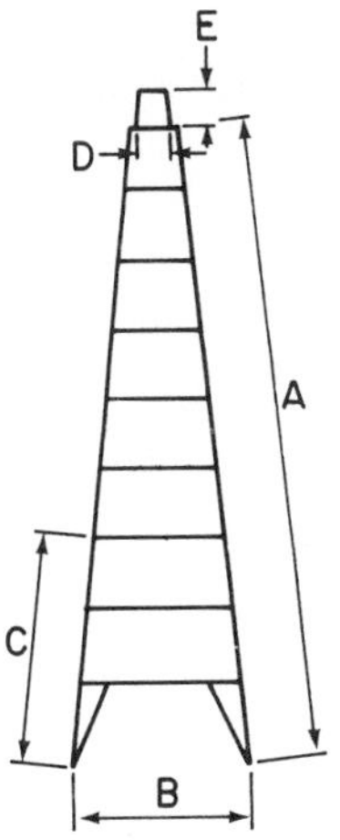

Fig. 3.13. Standard derrick dimensions[8]: A = height — top of floor joists to bottom of water-table beams; B = base square; C = window openings; D = water-table opening; E = gin-pole clearance. (Courtesy of API)

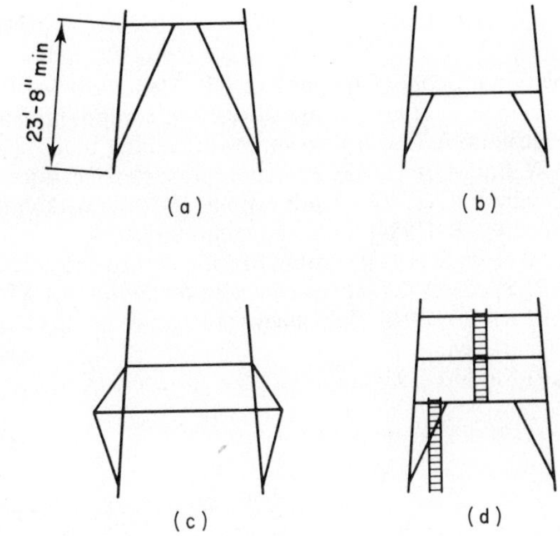

Fig. 3.14. Derrick windows[8]: (a) V-window; (b) and (c) draw-works windows; (d) ladder window. (Courtesy of API)

The *gin pole* is used to hoist the crown block to its place at the water-table opening. In Figure 3.13 the gin pole clearance, *E*, is measured as the distance between the header of the gin pole and the top of the crown support beam.

Types of steel derrick

Steel derricks are of two types.

Standard derrick

The standard type is a bolted structure that must be assembled part by part. It is currently used on offshore platforms, where the derrick is kept static during the drilling of the specified number of wells.

Portable derrick (or mast)

The portable type is pivoted at its base and is lowered to the horizontal by use of the draw-works after completion of the hole. It dismantles into a number of pin-jointed sections, each of which is one truck load. This type is normally used in land drilling where the complete rig must be moved to a different location after completion of each well. In the Middle East, for example, such locations are normally between 5 and 20 miles apart. At the new location the sections are quickly pinned together and the mast raised to the vertical with wire line from the draw-works.

References

1. National Supply Company (1983). *Manufacturing Catalogue.* Cheadle Heath, Stockport, England.
2. McGhee, E. (1958). Draw-works and compound transmission. *Oil and Gas Journal.*
3. Fredhold, A. (1955). Crown and travelling block. *Petroleum Engineer*, Dec.
4. API Bull. D10 (1982). *Procedure for Selecting Rotary Drilling Equipment.* API Production Department.
5. Cordrey, R. (1980). Depth ratings of draw-works mean more than horsepower. *Oil and Gas Journal*, June 9.

5a. Cordrey, R. (1984). Personal communication.

6. Crake, W. S. (1973). Fitting drilling rigs to their job — whether rig is new or old. SPE Reprint Series.
7. API Spec. 9A (1976). *Specification for Wirerope.* API Production Department.
8. API RP 9B (1980). *Recommended Practice on Application, Care and Use of Wirerope for Oilfield Service.* API Production Department.
9. API Std. 4A (1967). *Specification for Steel Derricks.* API Production Department.

Problems

Problem 1

Given the following data for a deep well:

hole depth	= 12 000 ft (3658 m)
drilling line	= $1\frac{1}{4}$ in (35 mm), 8 lines with efficiency factor of 0.841, and K = 0.9615
mud weight	= 15.9 ppg (1.9 kg/l)
drum diameter	= 30 in (761 mm), 2 spiralled grooving
derrick height	= 142 ft (43 m)
derrick capacity	= 1 000 000 lb (4448 kN)
breaking strength of line	= 65 tons (59 000 kg)
drillpipe	= 19.5 lbm/ft (29.02 kg/m)
drill collars	= 656 ft (200 m) long, 302 lb/ft (449 kg/m)
travelling block, hook, etc.	= 23 000 lb (102.3 kN)
casing dimensions	= $13\frac{3}{8}$ in, 54.5 lbm/ft (339.7 mm, 81.2 kg/m)
length of stand	= 93 ft (28 m)

calculate: (a) design factor when drilling at 12 000 ft; (b) dynamic crown load; (c) round-trip ton-miles (MJ) at 12 000 ft; (d) pipe setback load (i.e. weight of drill string in air); (e) casing ton-miles (MJ) if the length of one joint of casing = 40 ft (12.2 m); (f) design factor of the drilling line when the given casing is run to 12 000 ft and the dynamic crown loading at this depth; (g) if the design factor in Part (f) is less than 2, what can be done to increase it? (h) ton-miles between cuts; and (i) cut-off length. (Assume that the data of Table 3.3 are applicable.)

Answer: 2.6; 427 983 lb; 626 ton-miles; 385 288 lb; 334.5 ton-miles; 1.7 and 647 081 lb; increase number of lines; 616 ton-miles and 92.3 ft.

Problem 2

A block line for a new desert rig is to be ordered for an expected operation of 4000 MJ. The engineer decided on a $1\frac{1}{4}$ in (31.75 mm) line and a total of 8 lines to be wound round the crown block. The derrick height is 38 m above the rig floor and the drum diameter is 660 mm. From previous experience, it is known that the length of cut for $1\frac{1}{4}$ in line is 16 ton-miles/ft (229 MJ/m).

Calculate the length of line that should be ordered and the weight of this line. Assume the sheave diameter = 30 × block line diameter and a minimum of 10 laps of wire is to be left on the hoisting drum. (A $1\frac{1}{4}$ in line weighs 4.30 kg/m.)

Answer: 433 m; 18.27 kN.

Chapter 4

Rotary Drilling Bits

The drill bit represents the heart of the drill string, and therefore its proper selection and use cannot be overemphasised. The drill bit crushes the rock under the combined action of weight on bit and rotary speed. The resulting chippings are flushed away by the circulating fluid, to allow the bit to attack a new surface of rock. The process of cutting the rock and flushing the drill cuttings results in a drill hole.

This chapter will be confined to the discussion of rotary drilling bits of the roller cone type. Diamond and polycrystalline diamond compact bits will be briefly reviewed.

ROLLER CONE BITS (ROCK BITS)

As the name implies, a roller cone bit employs cones which rotate about their own axis or about the bit axis. Roller cone bits are the most widely used in oil-well drilling. They are also used in mining and civil engineering drilling. These bits were first used in 1920, and have gradually developed such that 95% of today's oil-well drilling employs roller cone bits[1].

As will be discussed later, the bit cones can have milled teeth cut from the body of the cone or tungsten carbide buttons inserted into the cones. Milled teeth bits are suitable for soft formations, while insert-type bits can drill medium to hard formations.

There are three types of roller bit:

(1) The two-cone bit is currently manufactured as a milled tooth bit only, which restricts its use for soft formations.
(2) The three-cone bit is the most widely used, employing either milled teeth or tungsten carbide inserts. The remaining discussion will be devoted to three-cone bits. Figures 4.1 and 4.2 show a milled-tooth bit and a tungsten carbide insert bit, respectively.

Fig. 4.1. Milled tooth bit[1]. (Courtesy of Smith Tool Company)

Fig. 4.2. Tungsten carbide insert bit[1]. (Courtesy of Smith Tool Company)

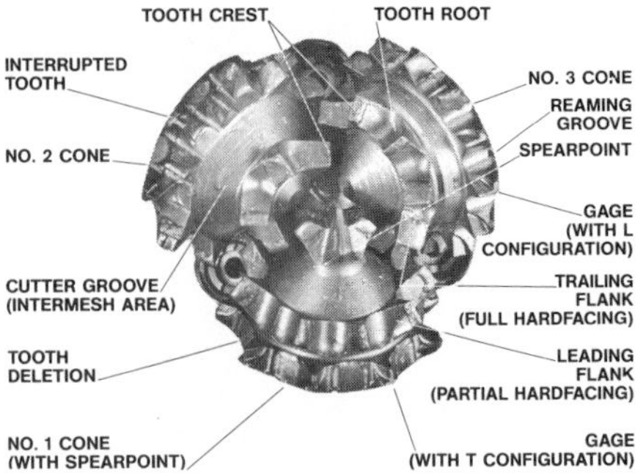

Fig. 4.3. Milled tooth bit nomenclature[1]. (Courtesy of Smith Tool Company)

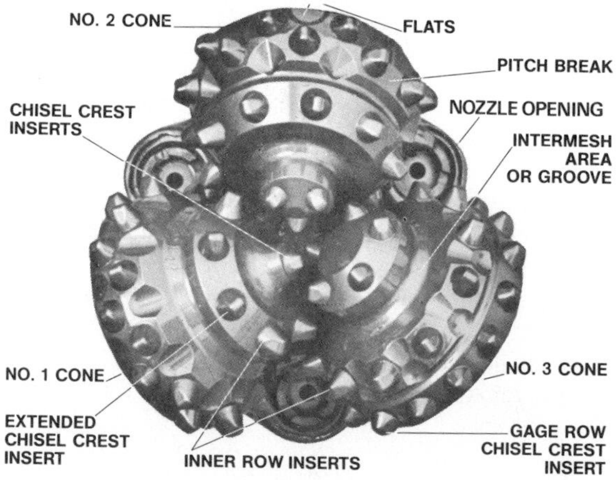

Fig. 4.4. Tungsten carbide insert bit nomenclature[1]. (Courtesy of Smith Tool Company)

(3) The four-cone bit is currently only manufactured as a milled tooth bit and is used for drilling large hole sizes, e.g. 26 in (660.4 mm) and larger.

THE THREE-CONE BIT

Principal features

As the name implies, a three-cone bit employs three cutting cones, each fitted on a leg, with a suitable bearing arrangement. Figures 4.3 and 4.4 give the nomenclature for milled tooth and insert bits, respectively.

Briefly, a three-cone bit consists of three equal-sized cones and three identical 'legs'. Each cone is mounted on bearings which run on a pin that forms an integral part of the bit leg (see Figure 4.5). The three legs are welded together and form the cylindrical section, which is threaded to make a pin connection to provide a means of attachment to the drill string (Figures 4.1 and 4.2). Each leg is provided with an opening (for fluid circulation), the size of which can be reduced by fitting nozzles of different sizes (Figure 4.4). Nozzles are used to provide constriction in order to obtain high jetting velocity for efficient hole cleaning. Mud pumped through the drill string passes through the pin bore and through the three nozzles, with each nozzle accommodating one-third of the flow, if all the nozzles are of the same size.

The factors influencing the design of a roller bit include the type and hardness of formation and the size of hole to be drilled. Formation hardness dictates the type and property of the material used for the manufacture of the cutting elements. The steel selected has a high content of nickel and is further strengthened by the addition of molybdenum[2].

Design factors

The design of the various bit parts is largely dictated by the formation properties and size of hole. The three legs and journals are identical, but the shape and the distribution of cutters on the three cones differ[2]. The design should also ensure that the three legs are equally loaded, to avoid the excessive loading of one leg only. The following factors are normally considered when designing and manufacturing of soft and hard three-cone bits: (a) journal angle; (b) amount of offset; (c) teeth; (d) bearings; and (e) interrelationship between (c) and (d).

Important bit nomenclature is presented in Figures 4.3 and 4.4.

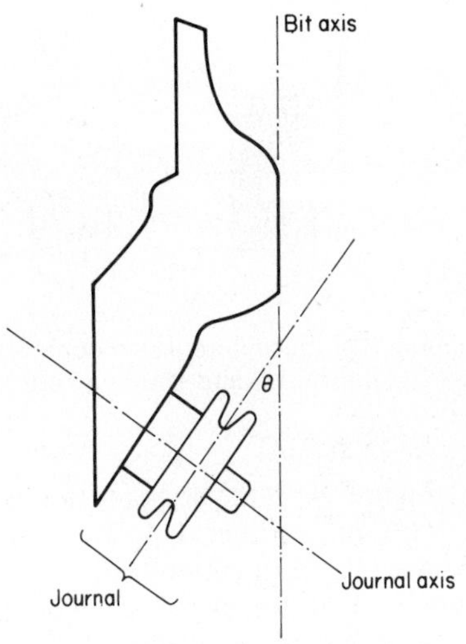

Fig. 4.6. Definition of journal angle.

Journal angle

The bit journal is the bearing load-carrying surface, as shown in Figures 4.5 and 4.6. The journal angle is defined as the angle formed by a line perpendicular to the axis of the journal and the axis of the bit. Figure 4.6 is a section through one leg of a three-cone bit. Angle θ in Figure 4.6 is the journal angle.

The magnitude of the journal angle directly affects the size of the cone. An increase in journal angle will result in a decrease in the basic angle of the cone and, in turn, cone size. Figure 4.7 shows how the cone size decreases as the journal angle increases from 0° to 45°. At a journal angle of 45° the cutters can theoretically become truly rolling.

The smaller the journal angle the greater the gouging and scraping action by the three cones[1]. As the journal angle increases from zero, the cutters (or cones) must be shaped (by removal of excess steel, as

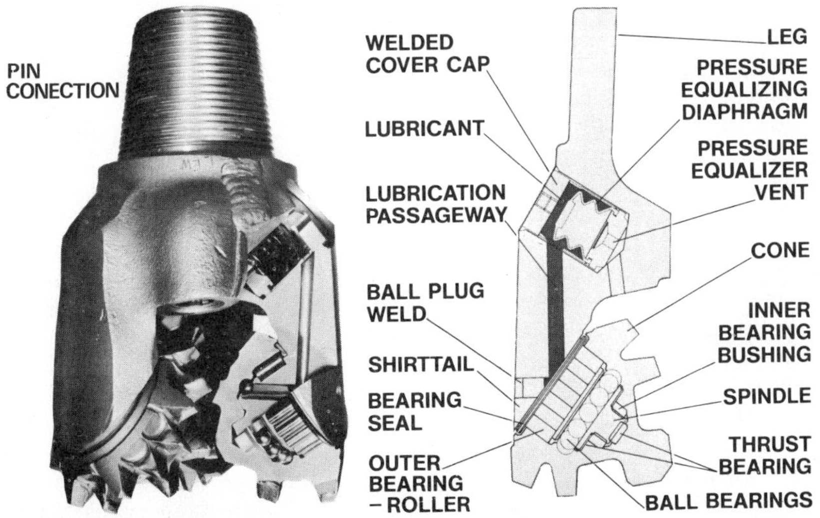

Fig. 4.5. Sealed bearing milled tooth bit nomenclature[1]. (Courtesy of Smith Tool Company)

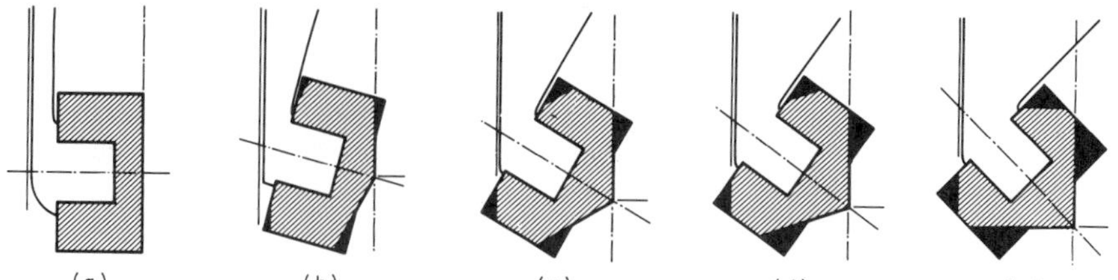

Fig. 4.7. Influence of journal angle on cone size[2]: (a) 0° journal; (b) 15° journal; (c) 30° journal; (d) 36° journal; (e) 45° journal. Solid shading represents sections removed. (Courtesy of Reed Rock Bits)

shown in Figure 4.7) to prevent the three cones interfering with one another. It follows that the journal angle influences both cutter size and cutter shape.

Optimum journal angles for soft and hard roller bits are 33° and 36°, respectively.

Cone offset

The degree of offset is defined as the horizontal distance between the axis of the bit and a vertical plane through the axis of the journal[3], as shown in Figure 4.8. To understand the effect of cone offset and cone design on rock cutting, consider the sketches shown in Figure 4.8. The single cone shown in Figure 4.8a. has its apex at the center of bit rotation and will move in a circle centered at the apex. This movement produces a true rolling action. In Figure 4.8b., the cone is modified so that it incorporates two basic cone angles, neither of which has its apex at the center of bit rotation. In this case, the conical heel surface tends to rotate about its theoretical apex, and the inner cone surface about its own apex. Since the cone is forced to rotate about the centerline of the bit, it slips as it rotates producing a tearing, gouging action. Practical experience has shown that soft rocks can be drilled efficiently by a scraping, gouging action. This action can be further increased by offsetting the centerlines of the cones from the center of bit rotation as shown in Figure 4.8c. The amount of offset is directly related to the strength of rock being drilled. For soft formations, a three-cone bit is made with a large offset so that the cones slip while rolling on the bottom of the hole.

Hard rocks are characterised by brittleness and high strength, and can be efficiently drilled by use of crushing and chipping actions. The bit applies sufficient force to overcome the compressive strength of the rock immediately beneath the tooth, creating fines and chips. Scraping and twisting actions, therefore, are not required for hard rock bits and the amount of offset is zero.

For rocks of medium hardness the skew angle can be up to 2°.

Teeth

The length and geometry of the teeth are directly related to the strength of the rock to be drilled. Tooth depth is also limited by the size of the cone and bearing structure.

The criteria employed in tooth design include:

(a) *spacing and interfitting of teeth*, which are governed by tooth strength, depth and included-angle requirements[3] (Interfitting of cone teeth (Figure 4.9) permits meshing of teeth for cleaning and, in turn, efficient drilling action.); and (b) *shape and length of teeth*, which are often dictated by formation characteristics.

Long, slender and widely spaced teeth are used for drilling soft formation rocks. Soft formations, being relatively yielding, allow longer teeth to be used, which results in the breaking of a greater volume of rock at any one contact between bit and rock. The wide spacing allows the easy removal of the resulting drill cuttings, and self-cleaning of the bit. The included angle for a soft bit tooth ranges from 39° to 42°.

For hard formations the teeth are manufactured shorter, heavier and more closely spaced, to withstand the high compressive loads required to break the rock. The teeth are not intended to penetrate the rock, but simply to fracture it by application of high compressive loads.

A medium-hard formation would have a moderate number of teeth, having 43–45° included angles[2]. The included angle for the hard bit teeth is 45°–50°.

(c) *Types of teeth.* The three-cone bit teeth can be milled type or insert type. Milled teeth are cut from the cone body (Figures 4.1 and 4.3), one side of the tooth being hard-faced with a resistant material such as tungsten carbide to provide a self-sharpening effect. As the unfaced side of the tooth wears away, it leaves a sharp edge suitable for efficient drilling. Additional tooth life can be obtained[2] by facing one side completely with tungsten carbide while partially facing the opposite side. This type of design reduces the initial tooth wear.

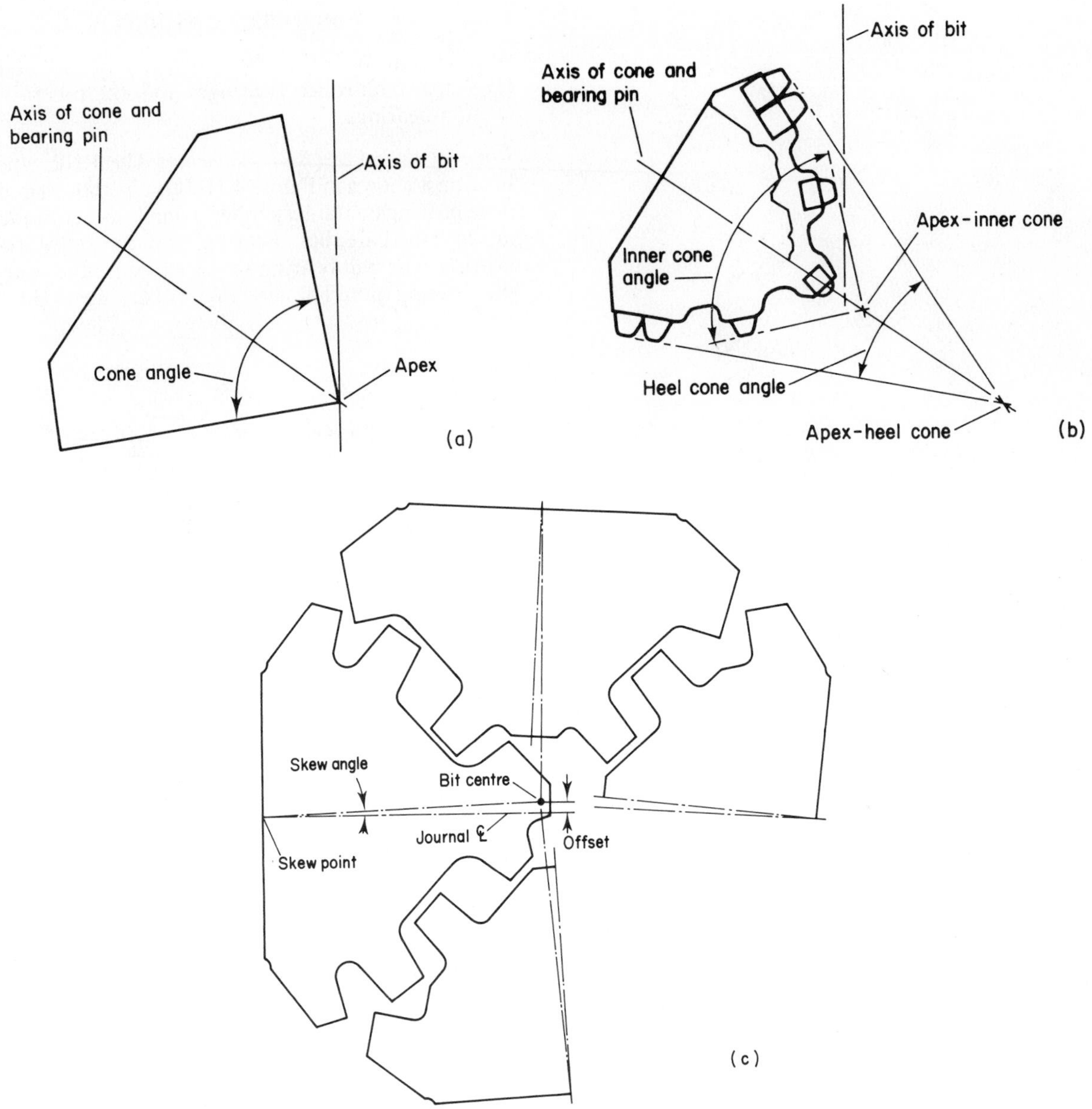

Fig. 4.8. Cone design (a) rolling cone design; (b) soft formation design; (c) cutter offset—skew angle[2]. (Courtesy of Reed Rock Bits)

Milled-tooth bits are only suitable for drilling very soft to soft formations in which moderate weights are required to fracture the rock.

For hard rocks the cutting elements of the cone are of the insert type. Inserts (or buttons) are made of tungsten carbide and are cold-pressed into holes already drilled in the cone shell.

There are several shapes of inserts, each being designed to suit the hardness of rock being drilled, as shown in Figure 4.10. Chisel-shaped inserts are used to drill soft rocks, while round or hemispherical inserts are used to drill medium to hard rocks. Figure 4.2 shows an insert-type bit with chisel-shaped buttons.

Bearings

Bit bearings are used to perform the following functions: (a) support radial loads; (b) support thrust or

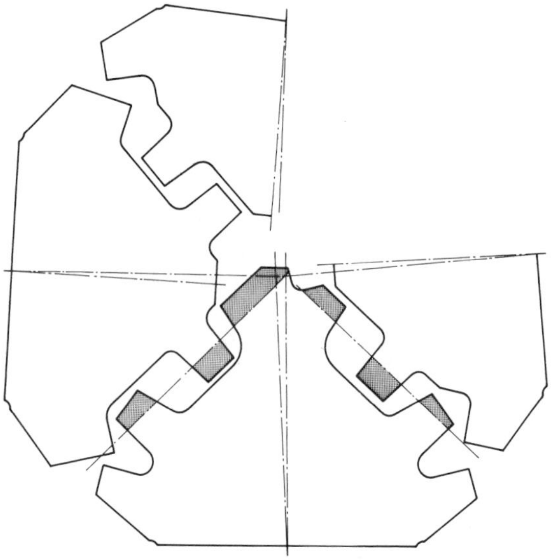

Fig. 4.9. Interfitting teeth[2]. (Courtesy of Reed Rock Bits)

axial loads; and (c) secure the cones on the legs. Function (a) is performed by the outer and nose bearing; functions (b) and (c) are performed by ball bearings and friction thrust faces.

Currently, two different bearing arrangements are in use: (1) anti-friction bearings; and (2) friction bearings.

Anti-friction bearings

Anti-friction or roller bearings are of two types: (1) roller–ball–roller bearings; and (2) roller–ball–friction bearings.

Roller–ball–roller (RBR) bearings The RBR type of bearing is shown in Figure 4.11. It is characterised by a nose bearing containing rollers (small, solid cylinders), an intermediate ball bearing and an outer roller bearing. The ball bearing serves to secure the cone on the bearing pin. The size of the bearing is directly

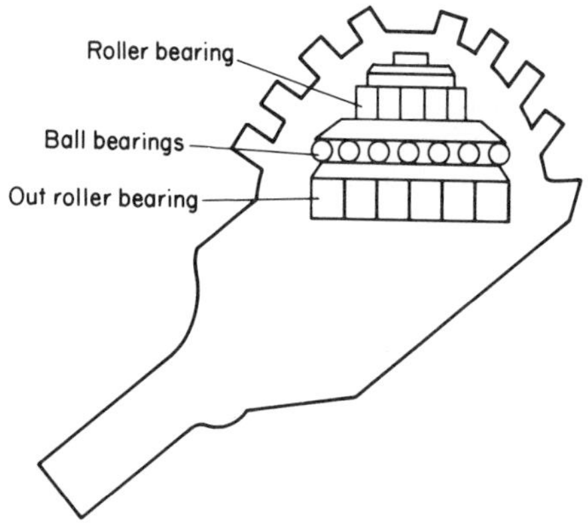

Fig. 4.11. Schematic of roller–ball–roller bearing.

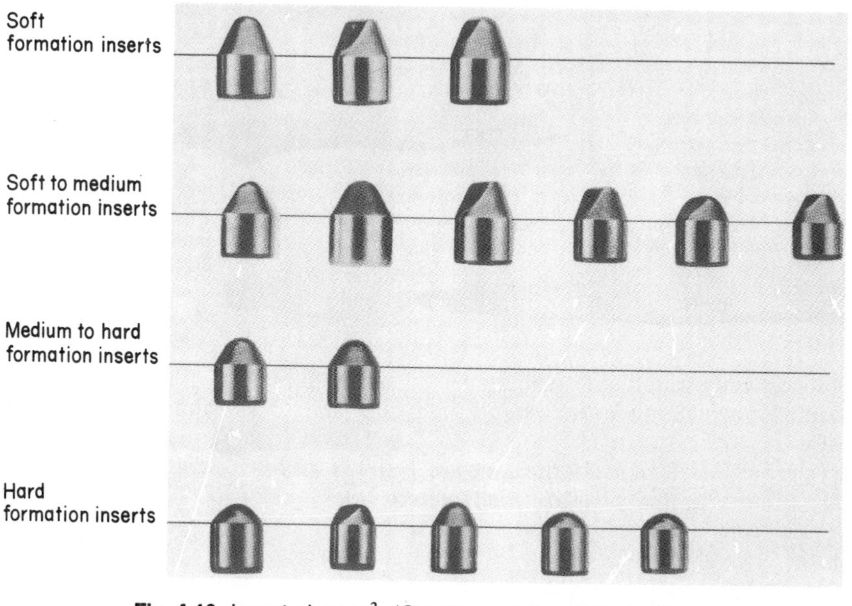

Fig. 4.10. Insert shapes[2]. (Courtesy of Reed Rock Bits)

influenced by the journal angle and by the cone size. In general, a compromise is achieved between the size of the bearings, cone shell thickness, and roller and ball sizes, for optimum strength of each component.

RBR bearings suffer from spalling of the races on the load side, which results from the high unit stress imposed by the rollers. Drill bits with RBR bearings have a shorter life compared to those with the more rugged design of friction-type bearings.

The roller–ball–roller assembly is normally used in bits larger than $12\frac{1}{4}$ in where adequate space is available and in situations where high rotary speeds are required.

Roller–ball–friction (RBF) bearings The RBF bearing is characterised by a friction-type bearing at the nose section, as shown in Figure 4.5. The inner ball bearing and outer roller bearing are the same as in RBR bearings. The friction bearing consists of a special case-hardened bushing pressed into the nose of the cone. The surface of the pin is also hard-faced with a special alloy ('Stellite') so that rotation of the bushing on the pin displays a low coefficient of friction and resists seizure and wear[4].

Friction-type bearings were introduced to the drilling industry to overcome the shortcomings of RBR bearings of spalling in the races of the pin. Other advantages of replacing rollers with friction-type bearings include: (a) maximum cone strength is obtained through thicker cone sections; and (b) maximum pin strength is obtained through larger pin diameter[4].

The roller–ball–friction design is common in bit sizes up to $12\frac{1}{4}$ in.

Friction bearings

Another variation in bearing design is to replace the rollers from both the nose and outer bearing with friction-type bearings. This, again, allows the journal diameter and length to be increased, which results in a more rugged design of bearing. A friction–ball–friction or solid journal bearing is shown in Figure 4.12. There is another variation (Hughes) wherein the ball bearing is replaced by a steel ring.

Bearing lubrication

Bearings are classified as non-sealed or sealed. Non-sealed bearings are lubricated by the mud system, while sealed bearings are lubricated by a custom-made system built within the leg body. Lubrication by the mud is not generally recommended, as mud contains abrasive solids (sand, barite, etc.) which reduce the working life of the bit. Non-sealed bearings are lubricated by mud entering through the face where the cone meets the journal.

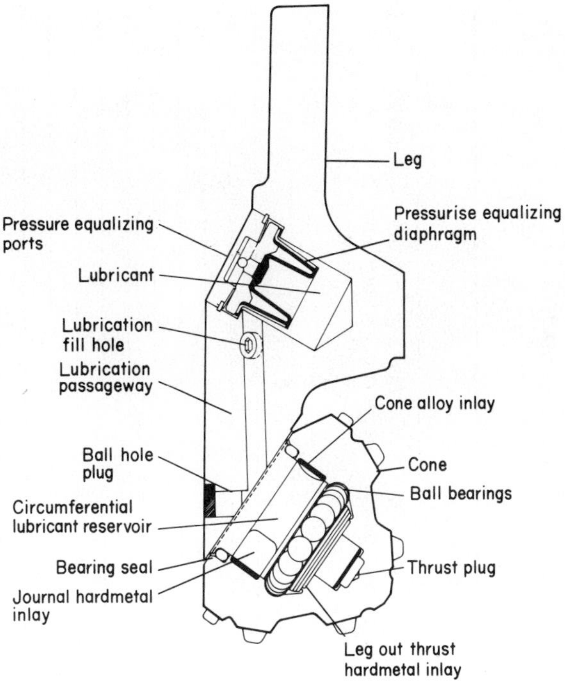

Fig. 4.12. Solid journal (or friction–ball–friction) bearing[1]. (Courtesy of Smith Tool Company)

A sealed bearing consists of bearing, seal, reservoir and pressure compensator (Figure 4.12). The seal is an O-ring type placed at the contact between the cone and the bearing lowermost point. The O-ring provides a positive sealing, preventing mud from entering the bearing or grease from leaving the bearing past the seal face. The reservoir provides a lubricant (a special grease) to the bearing through a connecting passageway, as shown in Figure 4.12. The movement of the lubricant is controlled by a pressure compensating system.

The pressure compensator has a flexible diaphragm which operates within a metal protector and is retained by a vented steel cap[4]. The pressure compensator maintains equal pressure inside and outside the bearing. The pressure compensating mechanism is also equipped with a pressure relief valve. The latter protects the bearing seal and compensator from damage when heat causes breakdown of grease into gaseous components, causing a build-up of interior pressures[4].

ROCK BIT CLASSIFICATION

From the preceding section, it is apparent that various designs of milled-tooth or insert bits can be manufactured by using different combinations of size, shape

TABLE 4.1 Form for classification of rock (roller) bits. (Courtesy of API[9])

MANUFACTURER

	Series	Formations	Types	*Features:* *Standard* *1*	*'T' gauge* *2*	*Gauge insert* *3*	*Rllr. seal bearing* *4*	*Seal brg. and gauge* *5*	*Friction seal brg.* *6*	*Friction brg. and gauge* *7*	*Other* *8*	*Other* *9*
Milled tooth bits	1	Soft formations having low compressive strength and high drillability	1									
			2									
			3									
			4									
	2	Medium to medium hard formations with high compressive strength	1									
			2									
			3									
			4									
	3	Hard semi-abrasive or abrasive formations	1									
			2									
			3									
			4									
	4	For future use	1									
			2									
			3									
			4									
Insert bits	5	Soft to medium formations with low compressive strength	1									
			2									
			3									
			4									
	6	Medium hard formations of high compressive strength	1									
			2									
			3									
			4									
	7	Hard semi-abrasive and abrasive formations	1									
			2									
			3									
			4									
	8	Extremely hard and abrasive formations	1									
			2									
			3									
			4									

NOTE: Bit classifications are general and are to be used only as simple guides. All bit types will drill effectively in formations other than those specified. This chart shows the relationship between the specific bit types.

and type of tooth, amount of offset, bearing type and lubrication mechanism. Further, there are several bit manufacturers, each adding their own modifications to the basic bit design. Thus, for the same rock type, one is faced with many different designs from several bit manufacturers.

The International Association of Drilling Contractors (IADC) recognised this difficulty and, in 1972, they prepared a comparison chart for easy reference. According to the IADC chart (Table 4.1), each bit is distinguished by a three-code system.

The first code or digit defines the series classification relating to the cutting structure. For milled-tooth bits, the first code carries the numbers 1 to 3, which describe soft, medium and hard rocks, respectively. Soft rocks (number 1) require long, slim and widely spaced teeth for efficient drilling. Medium formations (number 2) require short and less widely spaced teeth in order to withstand the high compressive loads. Hard formations (number 3) require very short and closely spaced teeth for maximum bit life and efficient drilling. Insert bits carry as their first code the numbers 5 to 8; again, these numbers relate to increasing rock hardness, as depicted in Table 4.1.

The second code relates to formation hardness subdivision within each group and carries the numbers 1 to 4. These numbers signify formation hardness, from softest to hardest within each series.[2]

The third code defines the mechanical features[2] of the bit, such as non-sealed or sealed bearings.

Fig. 4.13. Drag bits.

Tables 4.2 and 4.3 give comparison charts for milled-tooth and insert bits, respectively, using data from four different manufacturers.

As an example of the use of Tables 4.2 and 4.3, consider a bit having the code 134. From Table 4.2 it can be seen that a bit code of 134 indicates a milled-tooth bit suitable for a soft formation (Class 3). This bit type is also characterised by a sealed bearing. It can be ordered from the four cited manufacturers as follows:

Hughes	:	XIG
Security	:	S44
Reed	:	S13
Smith	:	SDG

An insert bit having the code 627 can be ordered from Hughes as J55, from Reed as FP62 and from Smith as F5.

POLYCRYSTALLINE DIAMOND COMPACT (PDC) BITS

A polycrystalline diamond compact bit is a new generation of the old drag or fishtail bit (Figure 4.13) and employs no moving parts (e.g. there are no bearings). The new bit is designed to break the rock in shear and not in compression, as is the case with roller cone bits, or by a ploughing/grinding action, as is the case with diamond bits. Breakage of rock in shear requires significantly less energy than in compression. Thus, less weight on bit can be used, resulting in less wear and tear on the rig and drill string[6]. Figure 4.14 shows two different types of PDC bit. PDC bits are also known as 'stratapax' bits.

The mechanics of rock breakage by a diamond compact (or stratapax) bit, a three-cone bit and a conventional diamond bit is depicted in Figure 4.15. The fact that a PDC bit fails the rock in shear limits its application to the drilling of rocks of soft and medium hardness. Shear failure also requires that the bit be self-sharpening for efficient rock cutting.

A PDC bit employs a large number of cutting elements, each called a drill blank. The drill blank is made by bonding a layer of polycrystalline man-made diamond to a cemented tungsten carbide substrate in a high-pressure, high-temperature process to produce an integral blank. This process produces a blank (or compact) having the hardness and wear resistance of diamond complemented by the impact resistance of the cemented tungsten carbide layer[5]. The diamond layer is composed of many tiny diamonds which are grown together at random orientation[6] for maximum

TABLE 4.2 Milled tooth bit comparison chart

*IADC Code**		*Smith*					*Hughes*				
		Standard Roller Bearings 1	*Roller Bearing Gage* 3	*Sealed Roller Bearings* 4	*Sealed Roller Brg. Gage* 5	*Sealed Journal Bearing* 6	*Standard Roller Bearings* 1	*Roller Bearing Gage* 3	*Sealed Roller Bearings* 4	*Sealed Roller Brg. Gage* 5	*Sealed Journal Bearing* 6
1	1	DS		SDS		FDS	R1		X3A		J1
	2	DT	DTT	SDT		FDT	R2				J2
	3	DG	DGT	SDG	SDGH	FDG	R3		XIG		J3
2	1	V2	V2H	SV	SVH	FV	R4				J4
3	1	L4	L4H	SL4	SL4H		R7				J7
	2						R8				
		Reed					*Security*				
1	1	Y11		S11			S3S		S33S	S33SG	S33SF
	2	Y12	Y12T	S12		FP12	S3	S3T	S33		S33F
	3	Y13	Y13T-Y13G	S13	S13G	FP13	S4	S4T	S44	S44G	
2	1	Y21	Y21G	S21	S21G	FP21	M4N		M44N	M44NG	M44NF
	2	Y22					M4				
	3				S23G		M4L		M44L		M44LF
3	1	Y31	Y31G		S31G		H7	H7T	H77		H77F
	3							H7SG		H77SG	
	4								H77C		H77CF

strength and wear resistance. The blanks are bonded to specially shaped tungsten carbide studs and are then attached to the bit body using a low-temperature brazing method or by interference fitting. Cutting efficiency is maximised by precise positioning and angling of the cutters[5]. Figure 4.16 shows different types of cutter.

During drilling the compact provides a continuous sharp cutting edge, owing to continuous microchipping of the diamond surface resulting from wear. This feature is necessary for efficient rock cutting.

PDC bit design is influenced by nine variables[6]: (1) bit body material; (2) bit profile; (3) gauge protection; (4) cutter shape; (5) number or concentration of cutters; (6) locations of cutters; (7) cutter exposure; (8) cutter orientation; and (9) hydraulics (see Chapter 7).

Bit body material. Two types of body materials are in use: (1) heat-treated alloy steel, as used in roller cone bits; and (2) a tungsten carbide matrix, as used in natural diamond bits.

Steel body bits are less durable and less resistant to erosion by the drilling fluid than are the matrix body bits. Steel body bits use stud-type cutters (Figure 4.16) which are attached to the body by interference or shrink fitting. The steel body is also provided with three or more nozzles for fluid passage.

Tungsten carbide matrix body bits are manufactured (or cast) in a mould similar to the process of manufacturing diamond bits. This allows more complex profiles to be obtained. Owing to the high temperature required when casting a matrix body, it is not possible to insert compact blanks until after the furnacing of the body, as the compact diamond will be destroyed. The bonding between the small diamond crystals is destroyed at about 750°C.

Bit profile affects cleaning and stability of the hole and gauge protection. Two bit profiles are in common use; double-cone and shallow-cone (see Figure 4.17). The double-cone profile allows more cutters to be placed near the gauge and will control hole deviation. The shallow-cone profile affords less area for cleaning but the bit drills faster than the bit of the double-cone profile owing to the more direct loading of the cutters on the bit face by the weight applied by the drill collars.

Gauge protection in steel body bits is provided by

TABLE 4.3 Insert bit comparison chart

*IADC Code**		*Smith*			*Hughes*			*Reed*			*Security*		
		Roller Bearing Air 2	*Sealed Roller Bearing 5*	*Sealed Journal Bearing 7*	*Roller Bearing Air 2*	*Sealed Roller Bearing 5*	*Sealed Journal Bearing 7*	*Roller Bearing Air 2*	*Sealed Roller Bearing 5*	*Sealed Journal Bearing 7*	*Roller Bearing Air 2*	*Sealed Roller Bearing 5*	*Sealed Journal Bearing 7*
4	3			F1			J11						
	1		2JS	F2			J22			FP51		S84	S84F
	2			F27					S52	HS51			
5										FP52			
	3		3JS	F3	HH33	X33	J33		S53	FP53		S86	S86F
	4			F37						HPSM	S8JA	S88	S88F
	1	4JA	4JS	F4						FP62			
				F45	HH44	X44	J44			FP62X			M84F
	2	5JA	5JS	F47			J44C			HPM			M84CF
6				F5			J55R	Y62BJA	S62B	FP62B			M89TF
	3												M89F
				F57	HH55		J55	J63JA	S63	HPMH			M90F
	4								S64	FP64			
	2			F6				Y72JA	S72	FP72			
	3	7JA		F7	HH77		J77	Y73JA	S73	HPH	H8JA	H88	H84F
3													H88F
	4								S74	FP74			
8	1				HH88						H9JA	H99	H99F
	3	9JA		F9	HH99		J99	Y83JA	S83	FP83	H10JA	H100	H100F

(a)

(b)

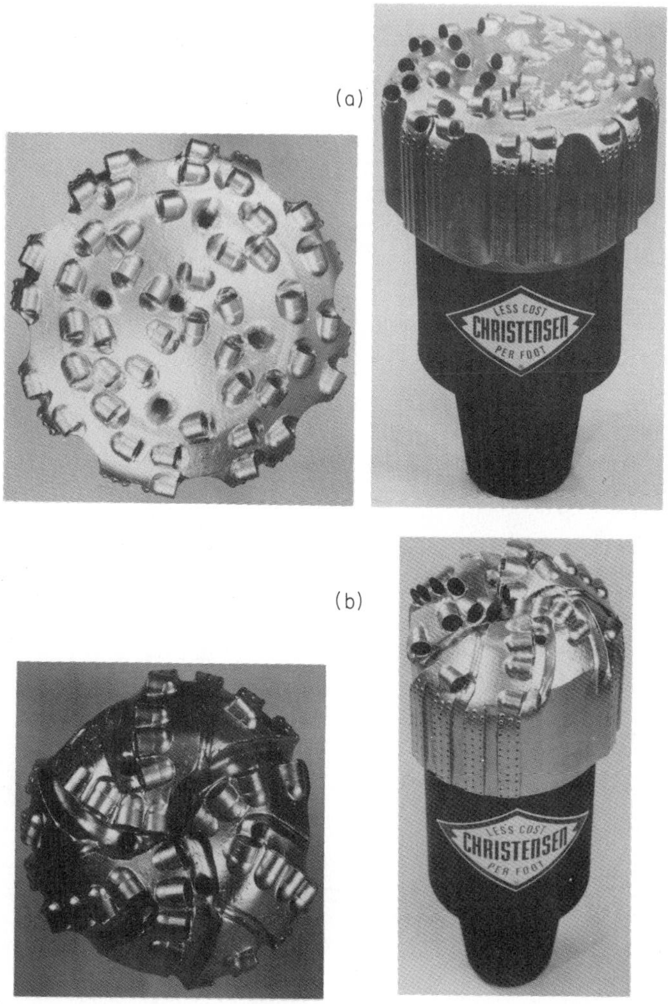

Fig. 4.14. Polycrystalline diamond compact bits[6]: (a) shallow cone; (b) double cone. (Courtesy of Christensen Drilling Products)

tungsten carbide inserts placed near the edges, while the matrix body bit utilises natural diamonds for gauge protection.

Cutter shape. Polycrystalline diamond blanks are produced in three basic shapes: (1) the standard cylindrical shape; (2) the chisel (or parabolic) shape; and (3) the convex shape.

Concentration of cutters. Longer bit life is generally obtained with greater concentration of cutters. However, the penetration rate decreases with increasing concentration, owing to the difficulty of cleaning the areas between the cutters.

Location of cutters. Field experience and fracture mechanics models are used to locate cutters for maximum cutting and minimum wear and torque.

Cutter exposure. Penetration rate increases with increased cutter exposure; however, greater exposure makes the cutter more vulnerable to breakage (Figure 4.18).

Cutter orientation is described by back and side rake angles, as shown in Figure 4.19. The back rake angle varies between 0° and 25°, and its magnitudes directly affects the rate of penetration. As the rake angle increases, the penetration rate decreases, but the resistance to cutting edge damage increases, as the load is now spread over a larger area (Figures 4.15 and 4.19). The side rake angle assists hole cleaning by mechanically directing cuttings towards the annulus.

Hydraulics (see Chapter 7). PDC bits require optimum hydraulics for efficient hole cleaning, and, in

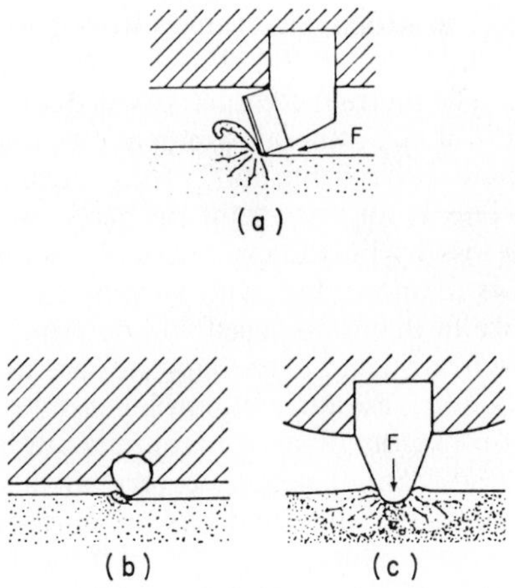

Fig. 4.15. Rock cutting mechanics[6]: (a) diamond compact bit (shearing); (b) diamond bit (ploughing/grinding); (c) roller cone bit (crushing). (Courtesy of Christensen Drilling Products)

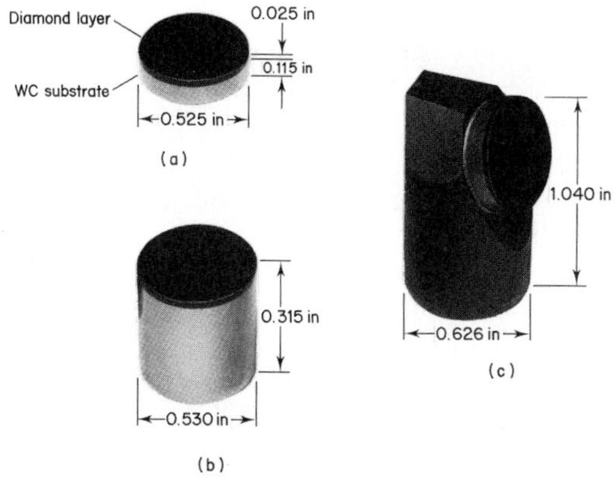

Fig. 4.16. Polycrystalline diamond compacts[6]: (a) diamond compact; (b) long cylinder cutter; (c) stud cutter. (Courtesy of Christensen Drilling Products)

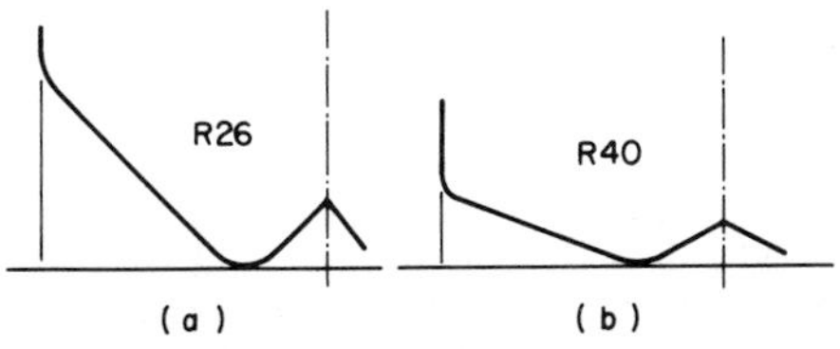

Fig. 4.17. Bit profiles[6]: (a) double cone; (b) shallow cone. (Courtesy of Christensen Drilling Products)

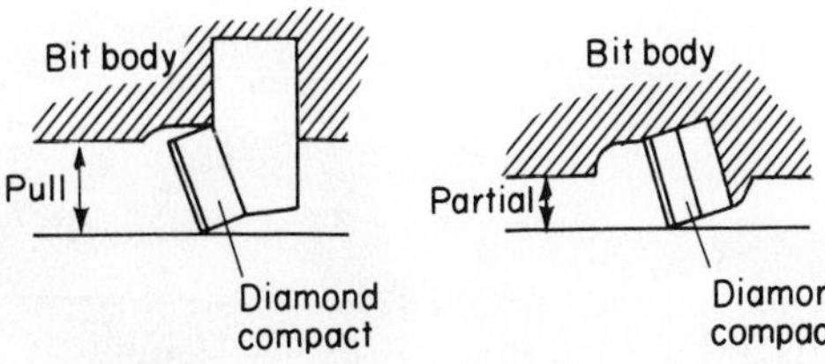

Fig. 4.18. Cutter exposure[6]. (Courtesy of Christensen Drilling Products)

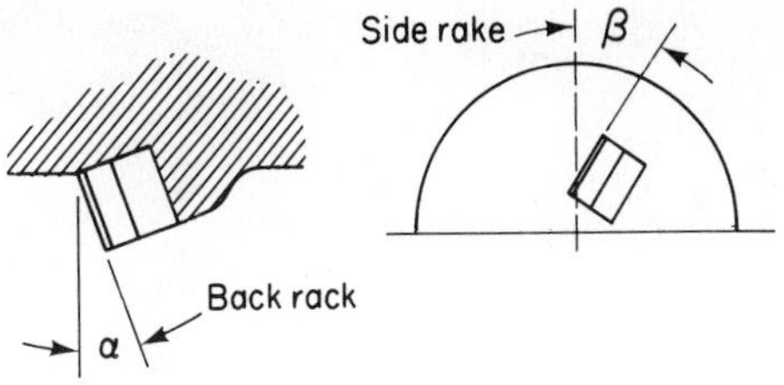

Fig. 4.19. Cutter orientation[6]. (Courtesy of Christensen Drilling Products)

turn, efficient hole making. Also, since the bit nozzles are close to hole bottom, maximum jetting speed will result in improved cleaning and high penetration rates.

Selection of nozzles and calculation of jetting speed are presented in detail in Chapter 7. It should be noted that PDC bits may have more than three nozzles and, in addition, the nozzles may not be round, as with the roller cone bit. Hydraulic calculations are, therefore, based on total flow area (TFA), and manufacturers' charts should be used to determine the size of nozzles corresponding to the calculated TFA.

PDC bits are also used in coring applications.

DIAMOND BITS

The cutting elements of a diamond bit consist of a large number of small-sized diamonds geometrically distributed across a tungsten carbide body. The bit does not employ moving parts and is normally used for hard and abrasive rock drilling and when longer bit runs are required in order to reduce trip time, e.g. in deep wells, and in offshore drilling, where rig costs are very high. Diamond bits are manufactured as either drilling or coring bits. A coring bit is used in conjunction with a core barrel to obtain samples of the formation in the form of cores.

Diamond is the hardest mineral known to man and has a value of 10 on the Mohs scale of mineral

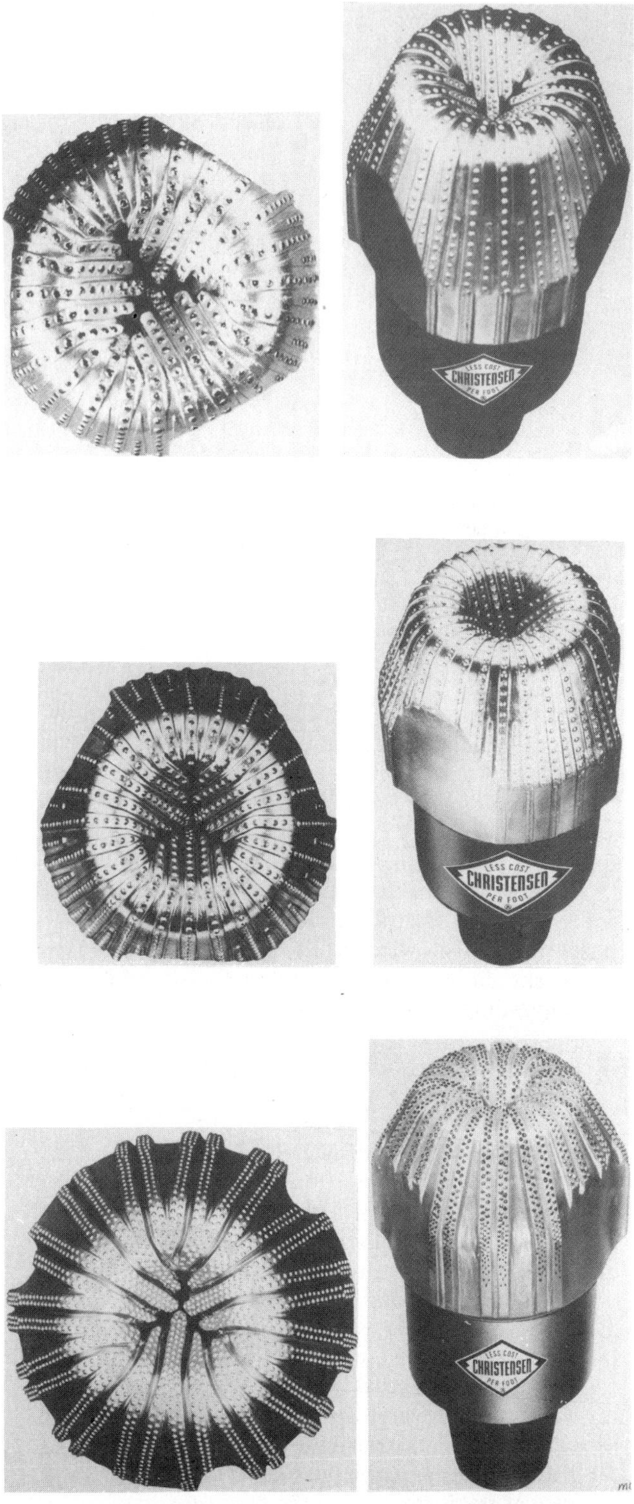

Fig. 4.20. Diamond bits with different cone profiles[8]. (Courtesy of Christensen Drilling Products)

hardness. The Mohs scale ranges from 1 for very soft rock, i.e. talc, to 10 for the hardest mineral, i.e. diamond. The thermal conductivity of diamond is also the highest of any mineral, which allows the diamond to dissipate heat from its cutting structures very quickly. This is an important property in protecting diamond loss by burning or thermal fracture. Figure 4.20 shows diamond bits with various cone profiles.

The size of diamonds used will determine the type of rock to be drilled. Large diamonds are used to drill soft rocks, since these rocks can be easily penetrated. For hard rocks, small-sized diamonds are used, since the diamonds cannot penetrate very far in such rock.

The majority of diamond bits are currently manufactured as coring bits, as the compact bits have taken over a share of the market previously occupied by diamond bits. A diamond coring bit contains a central hole corresponding to the size of core required. A typical coring assembly consists of a diamond coring bit, a core barrel, drill collars and drill pipe to surface.

A full discussion of diamond bits is beyond the scope of this book and the interested reader is advised to consult References 6 and 7 for further details.

BIT SELECTION

The following methods can be used to select the most appropriate bit type for a particular formation: (a) cost per foot; (b) specific energy; (c) bit dullness; and (d) offset well bit records and geological information. The information presented in this section is an abbreviated version of a detailed discussion presented in an SPE paper[10].

Cost per foot

The criterion for bit selection is normally based on cost per foot, C, which is determined by use of the following equation:

$$C = \frac{B + (T + t)R}{F} \text{ \$/ft} \tag{4.1}$$

where B = bit cost (\$); T = trip time (h); t = rotating time (h); R = rig cost per hour (\$/h); and F = length of section drilled (ft).

Equation (4.1) shows that cost/ft is controlled by five variables, and for a given bit cost, B, and hole section, F, cost/ft will be highly sensitive to changes in rig cost per hour, R, trip time, T, and rotating time, t. The trip time may not always be easy to determine unless a straight running in (RIH) and pulling out (POH) of hole is made. In this case, T is the sum of RIH and POH times. If the bit is pulled out for some reason, say to casing shoe for a wiper trip, such duration, if added, will influence the total trip time and, in turn, cost/ft. Bit performance, therefore, can be changed by some arbitrary factor. Rotation time is straightforward and is directly proportional to cost/ft, assuming that other variables remain constant.

The rig cost, R, will greatly influence the value of cost/ft. For a given hole section in a field that is drilled by different rigs, having different values of R, the same bit will produce different values of cost/ft, assuming that the same rotating hours are used in all rigs. It should be pointed out that if the value of R is taken as arbitrary (say 900 \$/h), then Equation (4.1) will yield equivalent values of cost/ft for all rigs. The value of cost/ft in this case is not a real value and does not relate to actual or planned expenditure.

The criterion for selection of bits on the basis of cost/ft is to choose the bit that consistently produces the lowest value of C in a given formation or hole section (see example 4.1).

Performance of a bit in the different parts of a given hole section can be determined, while the bit is drilling, by a cumulative cost per foot (CCF) method. In this method Equation (4.1) is used to determine cost/ft for, say, every 10 ft by assuming a reasonable figure for round trip time, T. When the value of CCF starts to increase, Reference 11 suggests that it is time to pull the current bit out of hole. In other words, the CCF is being used as a criterion for determining the depth at which the current bit becomes uneconomical. The drawbacks with the use of the CCF method are: (a) accurate measurement and prediction of F, t and T are necessary; and (b) the CCF may have suddenly increased, owing to the drilling of a hard streak of formation and may decrease once the bit has passed this streak (see example 4.1).

Because of these uncertainties, pulling out of bit on the evidence of one value of CCF may prove to be premature. Vargo[12] suggests the determination of several increasing values of CCF and also the corresponding incremental cost per foot (ICF) for each, say, 10 ft interval. A probabilistic test for trend is then performed on the difference between ICF and CCF to confirm that there is an upward trend for CCF.

Specific energy

The specific energy method provides a simple and practical method for the selection of drill bits. Specific energy, SE, may be defined as the energy required to remove a unit volume of rock and may have any set of consistent units. The specific energy equation can be derived by considering the mechanical energy, E, expended at the bit in one minute. Thus,

$$E = W \times 2\pi R \times N \text{ in-lb} \tag{4.2}$$

where W = weight on bit (lb); N = rotary speed (rpm); and R = radius of bit (in).

The volume of rock removed in 1 min is:

$$V = (\pi R^2) \times \text{PR in}^3 \tag{4.3}$$

where PR = penetration rate in ft/h.

Dividing Equation (4.2) by Equation (4.3) gives specific energy in terms of volume as

$$\text{SE} = \frac{E}{V}$$

$$= \frac{W \times 2\pi R \times N}{\pi R^2 \times \text{PR}} \quad \frac{\text{lb} \times \text{in} \times \dfrac{1}{\text{min}}}{\left[\left(\text{in}^2 \times \dfrac{\text{ft}}{\text{h}} \times \dfrac{\text{h}}{60\text{ min}} \times \dfrac{12\text{ in}}{\text{ft}}\right)\right]}$$

$$\text{SE} = 10 \frac{WN}{R \times \text{PR}} \quad \frac{\text{lb} \times \text{in}}{\text{in}^3} \tag{4.4}$$

Replacing R by $D/2$, where D is the hole diameter, Equation (4.4) becomes

$$\text{SE} = 20 \frac{WN}{D \times \text{PR}} \text{ in-lb/in}^3 \tag{4.5}$$

In metric units, Equation (4.5) becomes

$$\text{SE} = 2.35 \frac{WN}{D\text{PR}} \text{ MJ/m}^3 \tag{4.5a}$$

where W is in kg; D is in mm; and PR is in m/h.

Since penetration rate, PR, is equal to footage, F, divided by rotating time, t, Equation (4.5) becomes

$$\text{SE} = \frac{(20WN)}{(DF)} t \tag{4.6}$$

In Reference 13 it was concluded that SE is not a fundamental intrinsic property of rock. It is highly dependent on type and design of bit. This means that, for a formation of a given strength, a soft formation bit will produce an entirely different value of SE from that produced by a hard formation bit. This property of SE, therefore, affords accurate means for selection of appropriate bit type. The bit that gives the lowest value of SE in a given section is the most economical bit.

Equation (4.5) also shows that, for a given bit type in a formation of constant strength, SE can be considered constant under any combination of WN values. This is because changes in WN usually lead to increased values of PR (under optimum hydraulics) and this maintains the balance of Equation (4.5). The penetration rate is, however, highly influenced by changes in *WN*, and for a particular bit type an infinite number of PR values exist for all possible combinations of *WN* values. It follows that specific energy is a direct measure of bit performance in a particular formation and provides an indication of the interaction between bit and rock. The fact that SE, when compared with the penetration rate, is less sensitive to changes in *WN* makes it a practical tool for bit selection.

Example 4.1

The data given in Table 4.4 represent a complete run using an insert-type bit. The run has been divided into 10 sections. The cumulative cost per foot and specific energy at each section is calculated using Equations (4.1) and (4.5), respectively, and is given in Table 4.4. Figure 4.21 shows the CCF and SE in graphical form.

An interesting feature of Figure 4.21 shows that the CCF continues to decrease up to the depth at which the bit was pulled out of hole. Specific energy does not, however, exhibit such a fixed trend, which indicates that the bit performance in each interval is controlled by rock hardness of the interval in question. Also, the CCF of each point along the trend is influenced by the bit performance of the previous points. Hence, if the CCF assumes a sudden minimum or maximum value in a given interval, the actual bit performance is not evaluated, but an average value of the CCF between this interval and previous intervals. By contrast, the SE graph gives a 'spot' value of bit performance in each interval. No cumulative figures enter into the calculations of SE apart from the use of cumulative depth for plotting SE, as shown in Figure 4.21.

From the above discussion, it is seen that the performances of different bit types can be compared on the basis of their specific energies, irrespective of rig cost and trip time. This independence of SE from rig cost and trip time presents the drilling engineer with a convenient method of bit selection.

Example 4.2

Figure 4.22 summarises the drilling data from 43 wells drilled to an average depth of 8700 ft (2650 m). The specific energies of four drill bits used in this field were determined and plotted against depth, as shown in Figure 4.22. The bit types given in Figure 4.22 are the most frequently used and are produced by different manufacturers. Bit types F2 and J22 are of the insert type, while types J3 and S21 are milled-tooth bits. It is required to find the best bit types for various sections of the hole for drilling future holes.

Solution

With the minimum value of SE being used as a criterion for bit selection, Figure 4.22 indicates that bit J22 should be used for the top section (2500–5750 ft), bit F2 for the middle section (5750–7800 ft) and bit J3 for the last section (7800–8600 ft).

TABLE 4.4 Breakdown of a complete run using an insert bit type

Daily footage (ft)	*Cumulative footage (ft)*	*Depth out (ft)*	*WN (× 1000)*	*Trip time (h)*	*Rotating time (h)*		*Cumulative cost/foot (\$/ft)*	*Specific energy (in lb/in³) (× 1000)*
					Daily	*Cumul.*		
325	325	5793	45 × 60	4.35	12	12	76.9	163
63	388	5856	45 × 60	4.40	4	16	73.8	280
374	762	6230	60 × 60	4.70	23.5	39.5	65.7	369
249	1011	6479	60 × 60	4.90	14.5	54	62.6	342
171	1182	6650	60 × 60	4.99	7.5	61.5	59.3	258
331	1513	6981	40 × 60	5.20	16.5	78	56.3	212
159	1672	7140	40 × 65	5.40	7	85	54.8	187
369	2041	7509	45 × 60	5.60	22	107	54.7	263
378	2419	7887	45 × 60	5.90	23	130	54.8	268
251 (bit pulled out of hole)	2670	8138	45 × 80	6.10	14	144	54.4	328

Note: Trip time is a calculated value using a 6 h round trip time per 8000 ft as a basis. Cumulative cost per foot is calculated using $R = 900$ \$/h, $B = \$10\,260$ and the formula

$$CCF = \frac{B + (T + t)}{F}$$

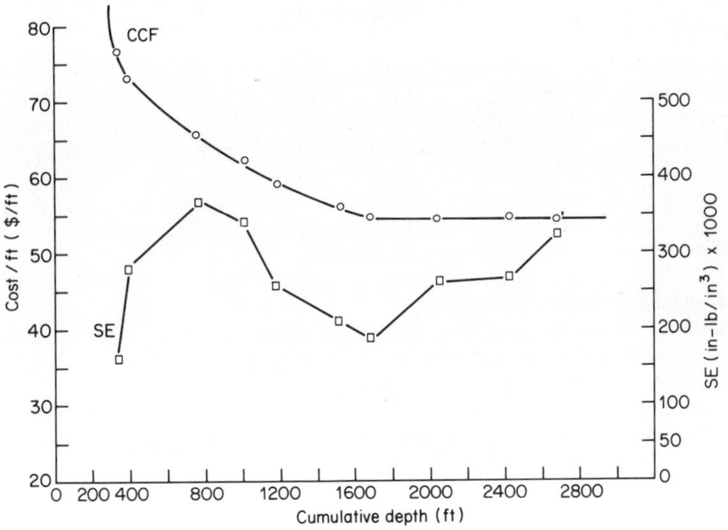

Fig. 4.21. CCF and SE plotted against cumulative depth.

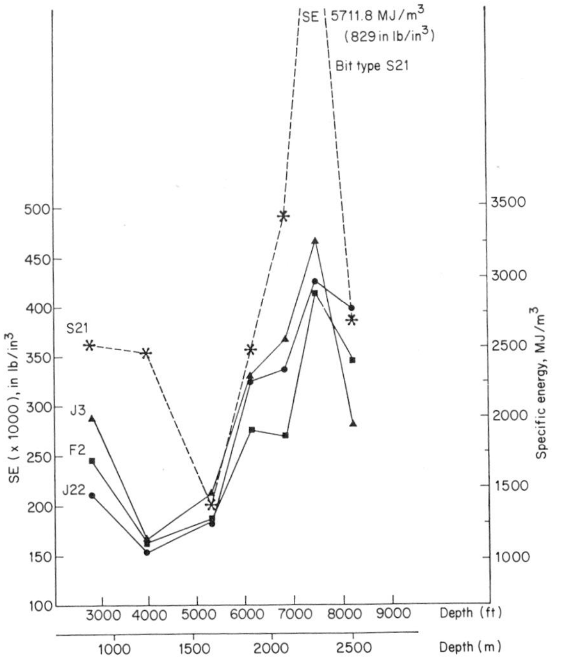

Fig. 4.22. Comparison of the performance of four bit types in terms of specific energy: ●, bit type J22; ■, bit type F2; ▲, bit type J3; *, bit type S21.

Bit dullness

The degree of dullness can be used as a guide for selecting a particular bit. Bits that wear too quickly are obviously less efficient and have to be pulled out of the hole more frequently, which increases total drilling cost. Bit dullness is described by tooth wear and bearing condition.

Tooth wear is reported as the total height remaining and is given a code from T1 to T8. T1 indicates that $\frac{1}{8}$ of the tooth height has gone; T4 indicates that $\frac{1}{2}$ of the tooth height has gone. Similarly, bearing life is described by eight codes, from B1 to B8. The number B8 indicates that the bearing life has gone or the bearing is locked. A bit that shows high tooth wear and low bearing life is, therefore, not suitable for the particular formation selected. If such a bit were a 1-1-1 type, then the use of a bit with a higher numerical code could reduce wear and bearing deterioration. A bit type 1-2-4 may be chosen; the code 2 is selected for the high rock strength, reducing tooth wear, and the code 4 is for a sealed bearing. Code 1 indicates that the bit is a milled-tooth type.

Well bit records and geological information

Drilling data from offset wells and geological information can provide useful guides for the selection of drill bits. Sonic logs from such wells can also be used to provide an estimate of rock strength, which, in turn, provides a guide for selecting the proper bit type.

References

1. Smith Tool (1979). *Drilling and Bit Technology Seminar.*
2. Reed Rock Bits Manual. *Rock Bit Design.*
3. Bentson, H. G. (1981). *Rock-Bit Design, Selection and Evaluation.* Smith Tool Publication.
4. Cook, J. H. and McElya, F. H. (1973). Development and application of journal bearing bits. *IADC Rotary Drilling Conference*, Houston, Texas, Feb. 28–March 2.
5. Madigan, J. and Caldwell, R. H. (1980). Applications for Stratapax blank bits from analysis of carbide insert and steel tooth bit performance. *55th Annual Fall Technical Conference*, Sept. 21–24.
6. Christensen (1982). *Christensen Diamond Compact Bit Manual.*
7. Christensen (1982). *Christensen Field Handbook.*
8. Christensen (1982). *Diamond Drill Bit Technology.*
9. API (1984). *Recommended Practice for Drill Stem Design and Operating Limits.* API Production Department.
10. Rabia, H. (1983). Specific energy as a criterion for bit selection. SPE No. 12355.
11. Moore, P. L. (1974). *Drilling Practices Manual.* Penn Well Books, Tulsa.
12. Vargo, L. (1982). On the optimal time to pull a bit under conditions of uncertainty. *JPT.* Dec., 2903–2904.
13. Rabia, H. (1982). Specific energy as a criterion for drill performance prediction. *Int. J. Rock Mech. Min. Sci.*, 19, No. 1, Feb.

Chapter 5

Fundamentals of Fluid Flow

This chapter aims to give a brief introduction to the fundamentals of fluid flow in pipes and annuli. The discussion will be limited to Bingham plastic and power-law models, these models being the most widely used in the drilling industry. Full details of the derivation of the various equations are given to help the reader appreciate the limitations of these equations.

This chapter will cover the following topics:

Fluid flow
Viscosity
Types of flow
Criteria for type of flow
Types of fluid
Viscometers
Derivation of laminar flow equations
  Bingham plastic model
  Power-law model
Turbulent flow
Flow through nozzles

FLUID FLOW

A fluid flowing along a conduit of any cross-section has a stationary layer adjacent to the conduit wall. The velocity of this layer is zero and the velocities of adjacent layers increase progressively until a maximum velocity is attained at the centre of the conduit, as shown in Figure 5.1.

This progression from zero velocity at the pipe wall to maximum velocity at the centre of the conduit results in the sliding of layers past one another; a high-velocity layer slides past its adjacent low-velocity layer, and so on. Fluid flow is, therefore, a result of this sliding action and, in order to maintain this flow, a continuous supply of energy is required. For example, a pump is required to lift water from a deep well to a surface tank.

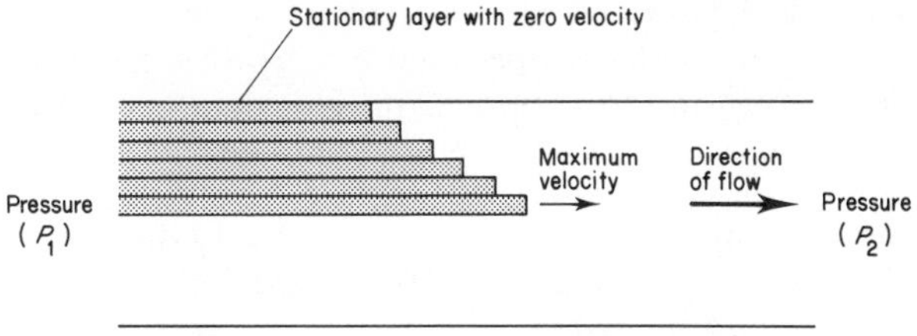

Fig. 5.1. Fluid flow through a pipe.

The sliding action of fluid layers is accompanied by shear stress (or frictional drag) which is highly dependent on the velocity and viscosity of the fluid.

VISCOSITY

Viscosity is a property which controls the magnitude of the shear stress that develops when one layer of fluid slides over another. Viscosity is, therefore, a measure of the strength of the internal resistance offered by the cohesive forces between the fluid molecules when motion is induced. Viscosity is also dependent on the type and the temperature of fluid. Temperature largely affects the intermolecular distances. For liquids the distance between the molecules is increased with increasing temperature, which reduces the magnitude of the cohesive forces and, in turn, the fluid viscosity. For gases the increased temperature causes the vibrational forces of the molecules to increase and the cohesive forces to decrease. In practice, the vibrational forces of gas exceed the cohesive forces, which results in increased viscosity with increasing temperature.

For a drilling mud composed of water and solids the viscosity is controlled by quantity, size and shape of solids.

The viscosity of fluids may be related to measurable parameters by considering the deformation of an elemental cube as shown in Figure 5.2. The cube is subjected to a force, F, applied parallel to the surface labelled 1, having a cross-sectional area A. The resulting shear stress, τ, is given by

$$\tau = \frac{F}{A} \tag{5.1}$$

This shear stress results in deformation of the fluid layer from a cubic shape (Figure 5.2a) to a rhombic shape, as shown in Figure 5.2(b). This deformation is analogous to the elongation or strain in elastic solids. Fluid deformation is referred to as shear strain and is described by the ratio between the velocity difference between the top and bottom of the deformed cube and the height of the cube.

Hence,

$$\text{shear strain} \qquad \gamma = \frac{(V + \mathrm{d}V) - V}{\mathrm{d}r}$$

$$= \frac{\mathrm{d}V}{\mathrm{d}r}\,\mathrm{s}^{-1} \tag{5.2}$$

Experimental evidence indicates that τ is related to γ either linearly or non-linearly. Fluids exhibiting a linear relationship between τ and γ are referred to as Newtonian fluids and, for this type, viscosity, μ, is expressed as

$$\tau = \mu\gamma \tag{5.3}$$

Substitution of Equations (5.1) and (5.2) in Equation (5.3) gives

$$\frac{F}{A} = \mu\left(\frac{-\mathrm{d}V}{\mathrm{d}r}\right) \tag{5.4}$$

The negative sign is included to indicate that velocity decreases away from the centre as the distance, dr, increases, to allow for the fact that a stationary layer exists at the pipe wall. The reader should note that the term 'viscosity' in Equations (5.3) and (5.4) and the rest of the book refers to dynamic viscosity.

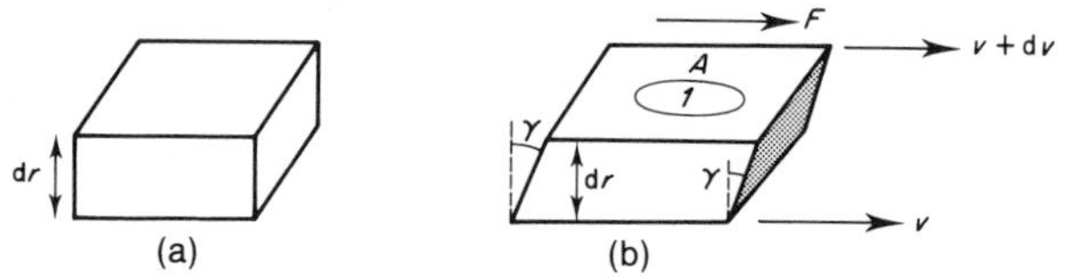

Fig. 5.2. Elemental cube of fluid: (a) before shear; (b) after shear.

Units of viscosity

In both Imperial and metric units viscosity is normally expressed in units of poise (P) or centipoise (cP).

The consistent units of viscosity in the Imperial system are

$$\frac{\text{lb s}}{\text{ft}^2} \text{ or } \frac{\text{lbm}}{\text{ft s}}$$

In metric units, from Equation (5.4),

$$\mu = \frac{F/A}{\mathrm{d}v/\mathrm{d}r} = \frac{\mathrm{N/m^2}}{\left(\dfrac{\mathrm{m/s}}{\mathrm{m}}\right)}$$

$$= \frac{\mathrm{N\,s}}{\mathrm{m}^2}$$

$$= \frac{(\mathrm{kg\,m/s^2})\,\mathrm{s}}{\mathrm{m}^2} = \frac{\mathrm{kg}}{\mathrm{m\,s}}$$

$$= \frac{\mathrm{g} \times 10^3}{\mathrm{cm} \times 10^2\,\mathrm{s}} = \frac{10\,\mathrm{g}}{\mathrm{cm\,s}}$$

By definition,

$$1\ \mathrm{P} = 1\,\frac{\mathrm{g}}{\mathrm{cm\,s}}$$

and 1 P = 100 cP. Hence,

$$1\ \mathrm{cP} = 10^{-3}\,\frac{\mathrm{kg}}{\mathrm{m\,s}}$$

Also,

$$1\ \mathrm{cP} = 2.0886 \times 10^{-5}\,\frac{\mathrm{lb\,s}}{\mathrm{ft}^2}$$

Or

$$1\ \mathrm{cP} = 6.719 \times 10^{-4}\,\frac{\mathrm{lbm}}{\mathrm{ft\,s}}$$

Hence

$$1\ \mathrm{P} = 2.089 \times 10^{-3}\,\frac{\mathrm{lb\,s}}{\mathrm{ft}^2}$$

and

$$\frac{\text{lb s}}{\text{ft}^2} = 47\,886 \text{ cP}$$

TYPES OF FLOW

Generally speaking, two types of flow can be recognised: laminar flow and turbulent flow.

Laminar flow

In laminar flow the flow pattern is smooth, with fluid layers travelling in straight lines parallel to the conduit axis. The velocity of each layer increases towards the middle of the stream until some maximum velocity is reached.

In laminar flow shear resistance is caused by the sliding action only and is independent of the roughness of the pipe. Laminar flow develops at low velocities and there is only one component of fluid velocity: a longitudinal component.

A special type of laminar flow with a flat centre portion is called a plug flow (Figure 5.3). In the flat portion there is no shear of fluid layers, and this is the reason that they are moving at the same velocity. It should be observed that plug flow occurs only with yield stress materials.

In oil-well drilling plug flow occurs at low velocities and when the mud thickness (viscosity) is large.

Turbulent flow

In turbulent flow the flow pattern is random in both time and space. The chaotic and disordered motion of fluid particles in turbulent flow results in two components of velocity: a longitudinal and a transverse

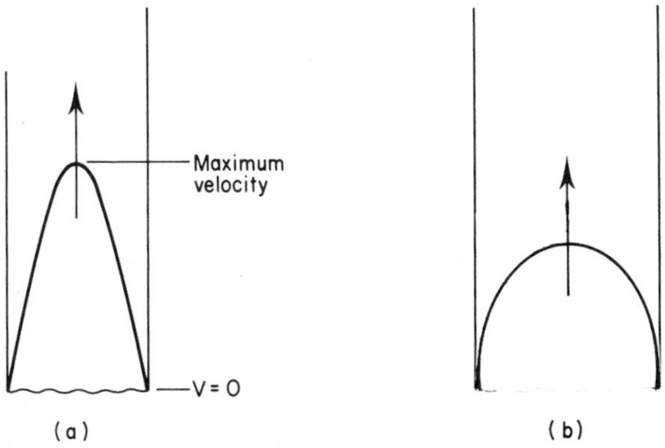

Fig. 5.3. (a) Laminar flow; (b) plug flow.

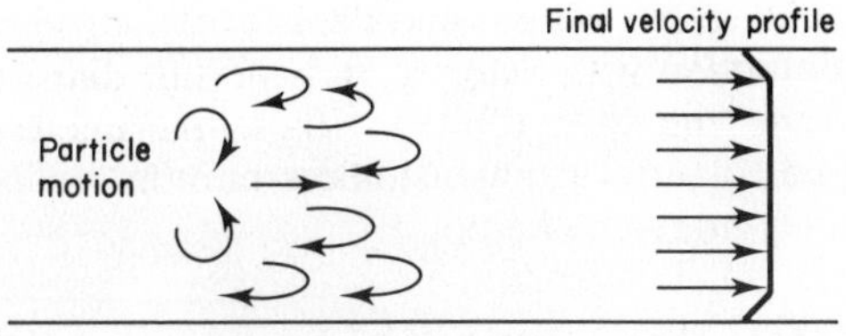

Fig. 5.4. Turbulent flow.

component. The longitudinal velocity attempts to make the fluid flow parallel to the conduit axis, while the transverse component attempts to move the fluid in a direction normal to the pipe axis.

The motion of particles in a direction normal to the longitudinal direction generates another shear resistance in addition to the laminar shear resistance. As previously discussed, the laminar shear resistance develops as a result of the sliding of one layer over another. In a fully developed turbulent flow the shear resistance due to turbulence can be many times the laminar shear resistance.

Despite turbulence, the final velocity profile tends to be a uniform one (Figure 5.4); this is largely attributed to the mixing of fluid particles which leads to interchange of momentum between high-velocity and low-velocity particles, which gives rise to a fairly flat profile.

Even in turbulent flow, particle fluctuation near the conduit wall dies out and the flow pattern in this region is essentially laminar. This region is normally called the laminar sublayer, and its thickness depends on the degree of turbulence. The relationship between the thickness of this laminar sublayer and the degree of turbulence is an inverse one.

In oil-well drilling turbulent flow is to be avoided as far as possible, since turbulence can cause severe hole erosion. Pressure losses also increase with degree of turbulence. However, in cementing turbulence is deliberately initiated to help to displace the mud cake from the walls of the hole, thus allowing the cement to contact the fresh surfaces of the formation. This will then result in a better cement job, as will be discussed in Chapter 11.

CRITERIA FOR TYPE OF FLOW

From the previous discussion it is apparent that in laminar flow shear resistance is dependent solely on the sliding action of layers. In turbulent flow the additional turbulent shear resistance is dependent on the magnitude of the transverse velocity. Thus, it is convenient to use fluid velocity as a criterion for determining the type of flow. Two other fluid properties, namely viscosity and density, can also be used in conjunction with velocity and conduit diameter to determine the type of flow. These parameters are grouped to form a dimensionless number called the Reynolds number, (Re):

$$(Re) = \frac{DV\rho}{\mu} \tag{5.5}$$

where D = conduit diameter; V = fluid velocity; ρ = fluid density; and μ = fluid viscosity.

It has been experimentally established that at a certain critical value of (Re) the flow pattern changes from laminar to turbulent. The magnitude of this critical value depends on many factors, including pipe wall roughness, viscosity of fluid and proximity of vibration. In most applications, however, fully turbulent flow develops at (Re) values of greater than 3000. For (Re) values of less than 2000, the flow is always laminar. In the transitional flow, where Re is between 2000 and 3000, the flow is often described as 'plug flow'. In plug flow a central portion exists where shear resistance is zero (as shown in Figure 5.3).

Reynolds number using field units

Parameter	*Metric unit*	*Imperial unit*
Diameter	mm	in
Velocity	m/s	ft/min
Density	kg/l	lbm/gal
Viscosity	cP	cP

Metric units:

$$(Re) = \frac{DV\rho}{\mu}$$

$$\text{Constant} = \frac{\left(\text{mm} \times \dfrac{1\ \text{m}}{10^3\ \text{mm}}\right)\left(\dfrac{\text{m}}{\text{s}}\right)\left(\dfrac{\text{kg}}{\text{l}} \times \dfrac{1000\ \text{l}}{\text{m}^3}\right)}{\left(\text{cP} \times \dfrac{10^{-3}\ \text{Kg/m s}}{\text{cP}}\right)}$$

$$(Re) = \frac{1000\,DV\rho}{\mu} \tag{5.5a}$$

Imperial units:

$$(Re) = \frac{DV\rho}{\mu}$$

Constant

$$= \frac{\left(\text{in} \times \dfrac{1\ \text{ft}}{12\ \text{in}}\right)\left(\dfrac{\text{ft}}{\text{min}} \times \dfrac{\text{min}}{60\ \text{s}}\right)\left(\dfrac{\text{lbm}}{\text{gal}} \times \dfrac{7.48\ \text{gal}}{1\ \text{ft}^3}\right)}{\text{cP} \times \dfrac{\left(2.0886 \times 10^{-5}\dfrac{\text{lb s}}{\text{ft}^2}\right)}{\text{cP}}}$$

$$= 497.4092\frac{\text{ft.lbm}}{\text{s}^2\text{.lb}}$$

$$(Re) = 15.46\frac{DV\rho}{\mu} \tag{5.5b}$$

TYPES OF FLUID

Newtonian fluid

A Newtonian fluid is defined by a straight-line relationship between τ and γ with a slope equal to the dynamic viscosity of the fluid, i.e. $\tau = \mu\gamma$. In this type of fluid, viscosity is constant and is only influenced by changes in temperature and pressure (as shown in Figure 5.5). Examples include oil and water.

Non-Newtonian fluid

A non-Newtonian fluid is a fluid that does not show a linear relationship between τ and γ, i.e. μ is not a constant. The viscosity of this type of fluid is proportional to the magnitude of shear stress or the duration of shear. Examples include drilling mud and cement slurries.

In general three major types of non-Newtonian fluid can be recognised.

Bingham plastic fluid (time-independent)

In a Bingham plastic fluid, deformation takes place after a minimum value of shear stress is exceeded. This minimum value is referred to as the yield stress or

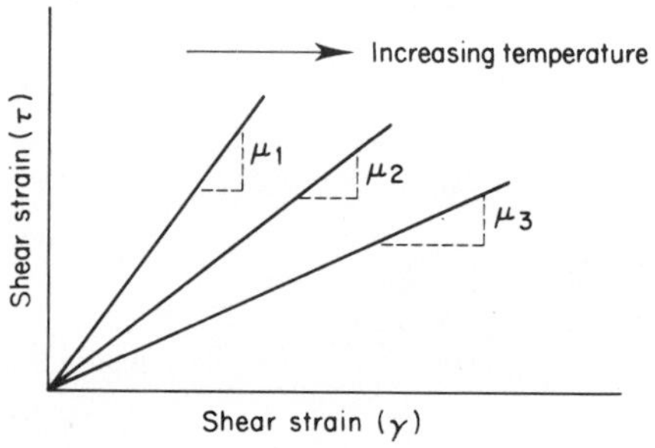

Fig. 5.5. Newtonian fluid.

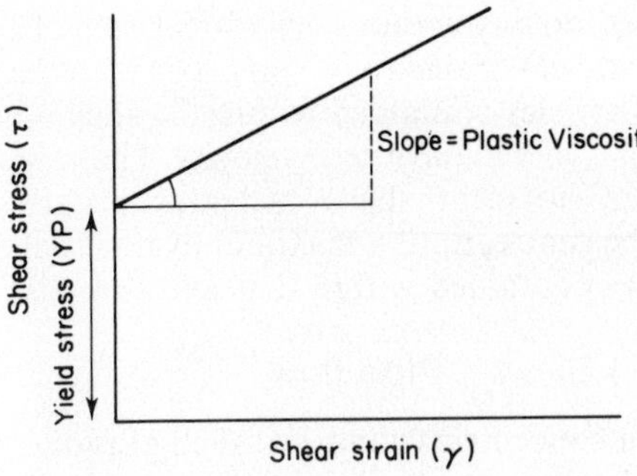

Fig. 5.6. Bingham plastic flow.

'yield point' (YP) (Figure 5.6). Beyond YP the relationship between τ and γ is linear, with a constant value of viscosity known as plastic viscosity (PV). Plastic viscosity is, again, dependent on temperature and pressure.

Hence, for a Bingham plastic fluid,

$$\tau = \mathrm{YP} + (\mathrm{PV})\,\gamma$$

or

$$\tau = \mathrm{YP} + (\mathrm{PV})\left(\frac{-\mathrm{d}v}{\mathrm{d}r}\right) \tag{5.6}$$

The yield point or yield stress is normally measured in lb/100 ft^2 using a viscometer, as detailed in Chapter 6. In metric units the yield point is expressed in $\mathrm{N/m}^2$.

Power-law fluid (time-independent)

In a power-law fluid τ and γ are related by the following expression:

$$\tau = K(\gamma)^n$$

or

$$\tau = K\left(\frac{-\mathrm{d}v}{\mathrm{d}r}\right)^n \tag{5.7}$$

where n = flow behaviour index, which varies between 0 and 1; and k = consistency index. A power-law fluid is shown graphically in Figure 5.7.

It should be noted that when $n = 1$, Equation (5.7) reduces to

$$\tau = K\gamma$$

where $K = \mu$, and the relationship reduces to that of a Newtonian fluid. The value of n gives an indication of the degree of non-Newtonian behaviour, and k refers to the consistency of the fluid. Large values of k mean that the fluid is very thick.

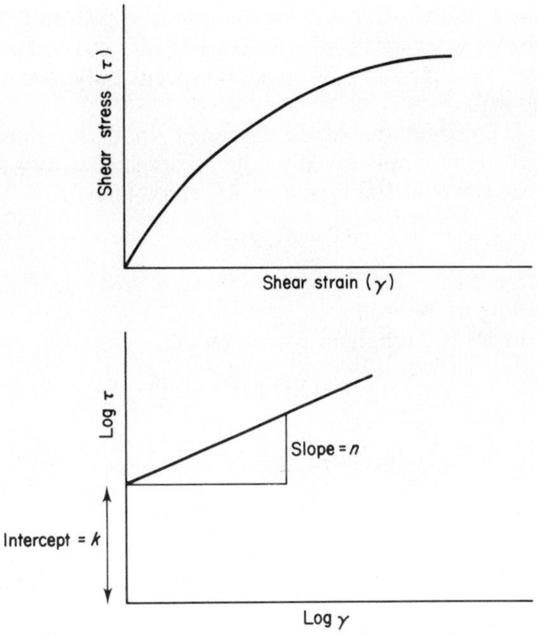

Fig. 5.7. Power-law fluid: (a) power-law relationship on a linear scale; (b) power-law relationship on a log–log scale.

Time-dependent fluid

Bingham plastic and power-law fluids are referred to as time-independent fluids, in which the magnitude of viscosity is not affected by the duration of shear. A time-dependent fluid is one whose apparent viscosity at a fixed value of shear rate and temperature changes with the duration of shear. Two types of time-dependent fluids can be recognised:

(1) A *thixotropic fluid* exhibits a decrease in shear stress with duration of shear at constant shearing rate. A thixotropic fluid gels when it is static but returns to the liquid state upon agitation. Examples of thixotropic fluids include paint, grease and solutions of polymers.

(2) A *rheopectic fluid* exhibits an increase in shear stress with duration of shear at a given shearing rate and constant temperature. True rheopectic fluids are rare. Gypsum and bentonite suspensions are examples.

VISCOMETERS

The important rheological properties of fluids are normally measured with a viscometer (or a rheometer). The rotational type is typically designed to rotate at two different speeds, namely 600 and 300 rpm, or at six speeds of 3, 6, 100, 200, 300 and 600 rpm. Most field instruments have two speeds only: 300 and 600 rpm.

The apparatus is so designed that the plastic viscosity is simply given by the difference between the shear stress at 600 rpm and 300 rpm. Thus,

$$\text{PV} = \theta_{600} - \theta_{300}$$

where θ_{600} = dial reading at 600 rpm; and θ_{300} = dial reading at 300 rpm.

From the Bingham plastic model,

$$\tau = \text{YP} + (\text{PV})\gamma \qquad (5.6)$$

At 300 rpm, $\tau = \theta_{300}$ and $\gamma = \gamma_{300}$. Therefore, Equation (5.6) becomes

$$\theta_{300} = \text{YP} + (\text{PV})\gamma_{300}$$

or

$$(\text{PV})\gamma_{300} = \theta_{300} - \text{YP} \qquad (5.6a)$$

Similarly, at 600 rpm, Equation (5.6) becomes

$$(\text{PV})\gamma_{600} = \theta_{600} - \text{YP} \qquad (5.6b)$$

and

$$(\text{PV})2\gamma_{300} = \theta_{600} - \text{YP} \qquad (5.6c)$$

where $2\gamma_{300} = \gamma_{600}$. Dividing Equation (5.6c) by Equation (5.6a) yields

$$2 = \frac{\theta_{600} - \text{YP}}{\theta_{300} - \text{YP}}$$

or

$$\text{YP} = 2\theta_{300} - \theta_{600}$$

$$\text{YP} = \theta_{300} - \text{PV} \qquad (5.8)$$

(*Note:* $\text{PV} = \theta_{600} - \theta_{300}$.)

DERIVATION OF LAMINAR FLOW EQUATIONS

Bingham plastic model

Fluid flow equations for pipes are normally derived using the following assumptions:

(1) Fluid velocity at the pipe wall is zero. This assumption effectively means that there is no slippage at the pipe wall.

(2) The magnitude of viscosity is independent of time or duration of shear. In other words, the fluid considered is a time-independent one and shear stress is a function of shear strain only.

(3) Particles moving within a cylindrical shell of a finite thickness travel parallel to the pipe axis with the same velocity. Particles contained within the shell adjacent to the pipe wall will have zero velocity. The velocity of particles in adjacent shells increases progressively towards the centre, until a maximum value is attained by particles contained within the central shell.

Pipe flow

Consider a concentric cylindrical shell of radius r and length L, as shown in Figure 5.8. During steady state flow of fluid, i.e. when the fluid is not accelerating, the following forces act on the shell: (a) differential pressure, $(P_1 - P_2)$, which causes the fluid to move with a steady speed of V; and (b) shear stress, resulting from the sliding action of particles within the shell over particles immediately outside this surface. The shear stress, τ, opposes the forward motion of the fluid particles within the shell, and for steady state flow an equilibrium between the forces exists such that

$$\text{forward force} = \text{opposing force}$$

$$(P_1 - P_2)\ \text{end area} = \tau(\text{surface area of shell})$$

But

$$\text{surface area of shell} = (2\pi r)L$$

Therefore,

$$(P_1 - P_2)\pi r^2 = 2\pi r L \times \tau$$

or

$$\tau = \frac{\Delta p r}{2L} \qquad (5.9)$$

where Δp is the frictional pressure loss $(P_1 - P_2)$ or the pressure drop.

Distribution of shear stress Equation (5.9) can be used to determine the shear stress distribution during fluid flow. At the pipe wall

$$r = D/2 \text{ and } \tau = \tau_w$$

where τ_w is the pipe wall shear stress. By use of these values, Equation (5.9) becomes

$$\tau_w = \frac{\Delta p D/2}{2L}$$

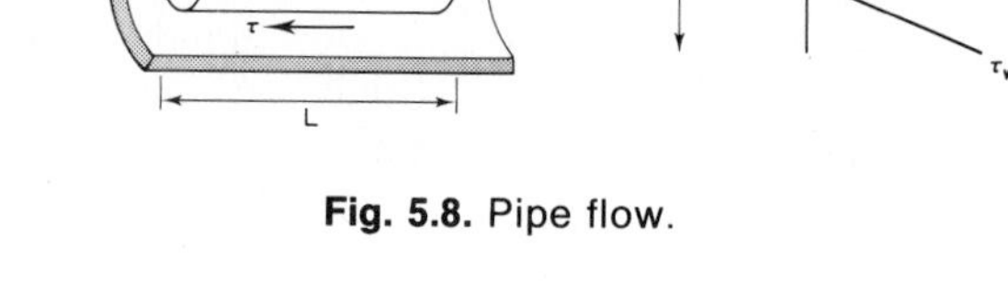

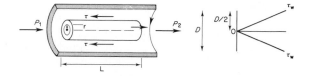

Fig. 5.8. Pipe flow.

or

$$\tau_w = \frac{\Delta p D}{4L} \tag{5.9a}$$

From Equations (5.9) and (5.9a) we obtain shear stress at any point in terms of τ_w and r as follows:

$$\tau = \frac{2r}{D}\tau_w \tag{5.9b}$$

where r is the radial distance from the centre of the pipe. Hence, from Equation (5.9b),

$$\text{at } r = 0,\ \tau = 0$$

The shear stress distribution is shown in Figure 5.8.

Laminar pipe flow equation The laminar fluid flow equation is developed by writing the Bingham plastic model, substituting Equation (5.9) into this model and integrating the resulting equation.

The Bingham plastic model states that

$$\tau = \text{YP} + (\text{PV})\left(-\frac{dv}{dr}\right) \tag{5.6}$$

Substitution of Equation (5.9) in Equation (5.6) gives

$$\frac{\Delta p r}{2L} = \text{YP} + (\text{PV})\left(-\frac{dv}{dr}\right)$$

Integrating with respect to r yields

$$\int \frac{\Delta p r}{2L}\,dr = \int \text{YP}\,dr - \int (\text{PV})\frac{dv}{dr}\,dr$$

$$\frac{\Delta p}{2L}\frac{r^2}{2} = \text{YP}r - \text{PV}\cdot V + C \tag{5.10}$$

At the pipe boundary

$$V = 0,\ r = R$$

where R is the radius of the pipe. Therefore,

$$C = \frac{\Delta p R^2}{4L} - \text{YP}\cdot R \tag{5.10a}$$

Substituting for C in Equation (5.10) yields

$$\frac{\Delta p}{4L}r^2 = \text{YP}\cdot r - (\text{PV})V + \left(\frac{\Delta p R^2}{4L} - \text{YP}R\right)$$

or

$$V = \frac{\Delta p}{4L(\text{PV})}(R^2 - r^2) + \frac{\text{YP}}{\text{PV}}(r - R) \tag{5.10b}$$

Since it is more convenient to express flow equations in terms of volume flow rate, Q, rather than velocity, Equation (5.10b) is further simplified.

Using

$$dQ = V\,dA = V\,2\pi\,rdr$$

and the value of V from Equation (5.10b), we obtain

$$Q = \int_0^R V\,2\pi\,rdr$$

$$= 2\pi \int_0^R \left\{\frac{\Delta p}{4L(\text{PV})}(R^2 - r^2) + \frac{\text{YP}}{\text{PV}}(r - R)\right\} rdr$$

$$= 2\pi\left[\frac{\Delta p}{4L(\text{PV})}\left(R^2\frac{r^2}{2} - \frac{r^4}{4}\right) + \frac{\text{YP}}{\text{PV}}\left(\frac{r^3}{3} - \frac{r^2 R}{2}\right)\right]_0^R$$

$$= 2\pi\left[\frac{\Delta p}{4L(\text{PV})}\left(\frac{R^4}{2} - \frac{R^4}{4}\right) + \frac{\text{YP}}{\text{PV}}\left(\frac{R^3}{3} - \frac{R^3}{2}\right)\right]$$

$$= \frac{\pi \Delta p}{8L(\text{PV})}R^4 - \frac{\pi(\text{YP})}{3(\text{PV})}R^3$$

(N.B. The reader should note that the above result is only an approximate solution. The correct result is obtained by splitting the integral into two regions: (1) $\tau < \text{YP}$, and (2) $\tau > \text{YP}$. The final result should include a $(\text{YP})^4$ term. In practice, the approximate solution was found to provide acceptable results.)

But average velocity

$$\bar{V} = \frac{Q}{A} = \frac{\dfrac{\pi \Delta p R^4}{8L(\text{PV})} - \dfrac{\pi(\text{YP})R^3}{3(\text{PV})}}{\pi R^2}$$

$$= \frac{\Delta p R^2}{8L(\text{PV})} - \frac{\text{YP}R}{3(\text{PV})}$$

Rearranging,

$$\Delta p = \frac{8L(\text{PV})\bar{V}}{R^2} + \frac{8L(\text{YP})}{3R} \tag{5.11}$$

Equation (5.11) is in consistent units and must be converted to oil field units. The conversion is carried out as follows.

Imperial units

First term

$$= \frac{8L(\text{PV})\bar{V}}{R^2}$$

$$= \frac{8(\text{ft})\left(\text{cP} \times \dfrac{2.0886 \times 10^{-5}\ \text{lb s}}{\text{cP}\quad \text{ft}^2}\right)\left(\dfrac{\text{ft}}{\text{min}} \times \dfrac{\text{min}}{60\ \text{s}}\right)}{\text{in}^2}$$

$$= \frac{16.7088 \times 10^{-5}}{60}\left(\frac{\text{lb}}{\text{in}^2}\right) = \frac{1}{359\,092}\ \text{psi}$$

$$\text{Second term} = \frac{8L(\text{YP})}{3R} = \frac{8(\text{ft})\left(\dfrac{\text{lb}}{100\ \text{ft}^2} \times \dfrac{\text{ft}^2}{144\ \text{in}^2}\right)}{3\ \text{in}\ \dfrac{\text{ft}}{12\ \text{in}}}$$

$$= \frac{1}{450}\ \text{psi}$$

Hence, Equation (5.11) becomes

$$\Delta p \simeq \frac{L(\mathrm{PV})\bar{V}}{90\,000D^2} + \frac{L(\mathrm{YP})}{225D} \text{ psi} \tag{5.12}$$

where $D = 2R$.

Metric units

$$\text{First term} = \frac{8L(\mathrm{PV})\bar{V}}{R^2} = \frac{8(\mathrm{m})(\mathrm{cP})\ \mathrm{m/s}}{\mathrm{mm}^2 \dfrac{1\ \mathrm{m}^2}{10^6\ \mathrm{mm}^2}}$$

But

$$1\ \mathrm{cP} = 10^{-3}\ \frac{\mathrm{kg}}{\mathrm{m\ s}}$$

Therefore,

$$\text{first term} = \frac{8(\mathrm{m})\left(\mathrm{cP} \times \dfrac{10^{-3}\ \mathrm{kg/m\ s}}{\mathrm{cP}}\right)\mathrm{m/s}}{\mathrm{mm}^2(1\ \mathrm{m}^2/10^6\ \mathrm{mm}^2)}$$

$$= 8000\ \mathrm{N/m^2}$$

$$\text{Second term} = \frac{8L(\mathrm{YP})}{3R} = \frac{8(\mathrm{m})(0.479\ \mathrm{N/m^2})}{3\ \mathrm{mm} \times \dfrac{1\ \mathrm{m}}{10^3\ \mathrm{mm}}}$$

It should be noted that the yield point of a drilling mud is measured in the field using a viscometer. The instrument gives the value of yield point directly in lb/100 ft^2 calculated as the difference between twice the reading at 300 rpm and the reading at 600 rpm (see Chapter 6). As there is no metric version of the viscometer to date, yield point is still measured in lb/100 ft^2 and later converted to N/m^2. The conversion factor from lb/100 ft^2 to N/m^2 is 0.479.

Therefore, the constant of the second term

$$= \frac{8 \times 0.479}{3 \times 10^3} = 1.277 \times 10^3\ \mathrm{N/m^2}$$

Hence, Equation (5.11) becomes:

$$\Delta p = \left(\frac{8000L(\mathrm{PV})\bar{V}}{R^2}\right) + \frac{1277L(\mathrm{YP})}{R}\ \mathrm{N/m^2}$$

Since diameter $D = 2R$, the above equation simplifies to

$$\Delta p = \frac{32\,000L(\mathrm{PV})\bar{V}}{D^2} + \frac{2554L(\mathrm{YP})}{D}\ \mathrm{N/m^2} \tag{5.13}$$

Pressure is normally expressed in bar (or kPa), and the above equation becomes

$$\Delta p = \frac{0.32L(\mathrm{PV})\bar{V}}{D^2} + \frac{0.025\,54L(\mathrm{YP})}{D}\ \mathrm{bar} \tag{5.13a}$$

or

$$\Delta p = \frac{32L(\mathrm{PV})\bar{V}}{D^2} + \frac{2.55L(\mathrm{YP})}{D}\ \mathrm{kPa} \tag{5.13b}$$

Critical velocity in pipe flow

Equations (5.12) and (5.13) are only applicable to laminar flow. In the previous sections it was indicated that laminar flow occurs at $(Re) < 2000$ and turbulent flow at $(Re) > 3000$. Hence, an expression for the critical velocity can be obtained by letting $(Re) = 3000$.

As discussed previously, a Newtonian fluid is one in which YP = 0. Thus, substituting YP = 0 in Equation (5.13) gives

$$\Delta p = \frac{32\,000\ L\mu_e\ \bar{V}}{D^2}\ \mathrm{N/m^2} \tag{5.14}$$

where μ_e is the effective viscosity. Equation (5.14) is also known as the Hagen–Poiseuille equation.

A value for effective viscosity (μ_e) may be used in Equation (5.14) so that the magnitude of Equation (5.14) is numerically equal to that of Equation (5.13). Thus,

$$\frac{32\,000L\mu_e V}{D^2} = \frac{3200L(\mathrm{PV})V}{D^2} + \frac{2554LYP}{D}$$

Therefore,

$$\mu_e = \mathrm{PV} + \frac{2554}{32\,000}\left(\frac{D}{V}\right)\mathrm{YP} \tag{5.15}$$

Also (for metric units),

$$(Re) = \frac{1000\bar{V}D\rho}{\mu_e} \tag{5.15a}$$

And for turbulent flow $Re = 3000$. Hence, Equation (5.15a) becomes

$$3000 = \frac{1000V_cD\rho}{\mu_e} \tag{5.16}$$

where V_c is the critical velocity at which turbulence takes place.

Substituting μ_e from Equation (5.15) in Equation (5.16), we obtain

$$3000 = \frac{\rho V_c D}{\mu_e} = \frac{\rho V_c D}{(\mathrm{PV}) + \dfrac{2554}{32\,000}\dfrac{(D\mathrm{YP})}{V_c}}$$

Simplifying,

$$\tfrac{1}{3}\rho D\ V_c^2 - (\mathrm{PV})V_c - 0.0798\ D\mathrm{YP} = 0$$

Since velocity cannot be negative, only the positive root of the above equation is useful. Therefore,

$$V_c = \frac{PV + \sqrt{PV^2 + 4\left(\frac{\rho D}{3}\right)(0.0798\, DYP)}}{\frac{2}{3}\rho D}$$

$$= 1.5\left(\frac{PV + \sqrt{PV^2 + 0.1064\rho D^2 YP}}{\rho D}\right) \text{m/s} \quad (5.17)$$

Hence, Equation (5.17) can be used as a criterion for determining whether the flow is laminar or turbulent by simply comparing velocity V_c with the average velocity of fluid as determined from $\bar{V} = Q/A$.

If $V_c < \bar{V}$, flow is turbulent, and if $V_c > \bar{V}$, flow is laminar and Equation (5.12) or Equation (5.13) is used to determine pressure drop.

In order to obtain V_c in Imperial units, the same procedure as outlined above is used to simplify Equation (5.12) to obtain

$$\mu_e = PV + 400\ YP\left(\frac{D}{V}\right) \quad (5.18)$$

and, for turbulent flow,

$$(Re) = 3000 = 15.46\frac{\rho V D}{\mu_e} \quad (5.19)$$

From Equations (5.18) and (5.19), we obtain

$$V_c = \frac{97\ PV + 97\sqrt{PV^2 + 8.2\ \rho D^2\ YP}}{\rho D} \text{ ft/min} \quad (5.20)$$

It should be observed that other forms of Equation (5.20) have appeared in the literature in which turbulence was assumed to start at Reynolds numbers of 2000 or 2500, viz.

$(Re) = 2000$

$$V_c = \frac{1.08\ PV + 1.08\sqrt{PV^2 + 12.3\ \rho D\ YP}}{\rho D} \text{ ft/s} \quad (5.20a)$$

$(Re) = 2500$

$$V_c = \frac{1.13\ PV + 1.13\sqrt{PV^2 + 8.8\ \rho D\ YP}}{\rho D} \text{ ft/s} \quad (5.20b)$$

Example 5.1

A drilling mud is pumped at the rate of 200 gal/min through a drillpipe of 4.5 in internal diameter and 400 ft in length. The fluid has a density of 9 lbm/gal, a plastic viscosity of 15 cP and a yield point of 10 lb/100 ft^2. Determine the type of flow and the magnitude of the pressure drop through the drillpipe.

Solution

Using Equation (5.20), determine the critical velocity of the fluid, V_c:

$$V_c = \frac{97 \times 15 + 97\sqrt{(15)^2 + 8.2 \times 9 \times (4.5)^2 \times 10}}{9 \times 4.5}$$

$$= 330.9 \text{ ft/min}$$

$$\text{Average velocity } \bar{V} = \frac{Q}{A}$$

or

$$\bar{V} = \frac{24.5Q}{D^2} \text{ ft/min}$$

$$\bar{V} = \frac{24.5 \times 200}{(4.5)^2} = 242 \text{ ft/min}$$

Since $V_c > \bar{V}$, flow is laminar, and Equation (5.12) is applicable:

$$\Delta p = \frac{L(PV) \times \bar{V}}{90\,000 D^2} + \frac{L\ YP}{225D}$$

$$= \frac{400 \times 15 \times 241.9}{90\,000 \times (4.5)^2} + \frac{400 \times 10}{225 \times (4.5)}$$

$$= 0.80 + 3.95$$

$$\Delta p = 4.8 \text{ psi}$$

Annular flow

Several versions of annular equations for laminar flow have appeared in the literature, each producing a different value of pressure drop for the same volume flow rate and fluid properties. In oil well drilling, laminar annular pressure losses are small, representing less than 10% of total pressure losses, and are normally incorporated within the turbulent losses[1]. In oil well drilling, annular flow is encountered when mud flows through the annulus between the drillpipe (or drill collars) and casing (or hole).

Without significant loss of accuracy, the annular geometry may be represented by a narrow slot between two infinite plates, as shown in Figure 5.9. Under steady state flows, the forces acting on the narrow slot

$$\Delta p\ W \times E = 2\tau_w(L \times E)$$

$$\tau_w = \frac{\Delta p\ W}{2L}$$

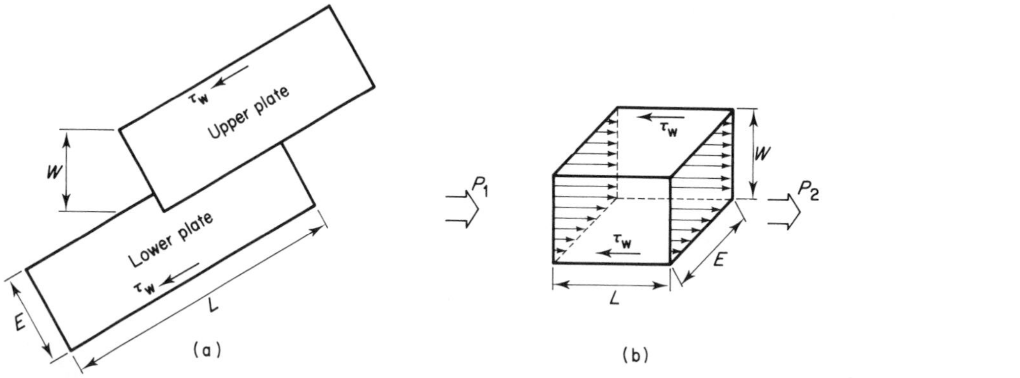

Fig. 5.9. Fluid in a narrow slot: (a) flow in a slot; (b) cubic element within the slot. W = vertical distance separating plates; L = length of plate (infinite); E = width of plate.

where $\Delta p = P_1 - P_2$; τ_w = shear stress at the pipe wall; E = width of plate; L = length of plate; and W = vertical distance between plates.

If the distance between the centre of this cube and the extreme boundary is taken as Y, then

$$W = 2Y$$

Hence,

$$\tau_w = \frac{\Delta p \, 2Y}{2L} = \frac{\Delta p \, Y}{L}$$

At any distance, y, from the centre, this equation may be modified to give shear stress, as follows:

$$\tau = \frac{\Delta p \cdot y}{L} \tag{5.21}$$

The Bingham plastic model states that

shear stress = yield stress + plastic viscosity × rate of shear

or

$$\tau = \text{YP} + \text{PV}\left(\frac{-\text{d}v}{\text{d}y}\right) \tag{5.6}$$

Substituting Equation (5.21) into Equation (5.6) yields

$$\frac{\Delta p y}{L} = \text{YP} + (\text{PV})\left(\frac{-\text{d}v}{\text{d}y}\right)$$

Integrating with respect to y,

$$\int \frac{\Delta p}{L} y \, \text{d}y = \int (\text{YP}) \, \text{d}y - \int (\text{PV}) \, \text{d}v$$

$$\left(\frac{\Delta p}{2L}\right) y^2 = \text{YP}\, y - (\text{PV})v + C \tag{5.22}$$

where C is a constant. At the boundary $y = Y$ and $v = 0$. Therefore,

$$C = \left(\frac{\Delta p}{2L}\right) Y^2 - \text{YP}\, Y \tag{5.22a}$$

Substituting Equation (5.22a) in Equation (5.22) and rearranging yields

$$v = \frac{\Delta p}{2L(\text{PV})}(Y^2 - y^2) + \frac{\text{YP}}{\text{PV}}(y - Y) \tag{5.22b}$$

But

$$\text{d}Q = \text{volume flow rate} = v \cdot \text{d}A$$

where A = slot area = WE = $E(2Y)$ (see Figure 5.9); $\text{d}A = E(2 \cdot \text{d}y)$. Substituting the value of v from Equation (5.22b) in the expression for $\text{d}Q$ we obtain

$$Q = \int_0^Y \left\{\frac{\Delta p}{2L(\text{PV})}(Y^2 - y^2) + \frac{\text{YP}}{(\text{PV})}(y - Y)\right\} E(2 \cdot \text{d}y)$$

$$= \left[\frac{\Delta p}{2L(\text{PV})}(Y^2 y - y^3/3) + \frac{\text{YP}}{\text{PV}}\left(\frac{y^2}{2} - yY\right)\right]_0^Y 2E$$

$$= \frac{\Delta p}{2L(\text{PV})}\left(Y^3 - \frac{Y^3}{3}\right) + \frac{\text{YP}}{\text{PV}}\left(\frac{Y^2}{2} - Y^2\right) 2E$$

$$= \left[\frac{\Delta p}{L(\text{PV})}\left(\frac{2}{3}Y^3\right) - \frac{\text{YP}}{\text{PV}} Y^2\right] E$$

and average velocity is given by

$$\bar{V} = \frac{Q}{A}$$

$$\bar{V} = \frac{\left(\dfrac{2\Delta p}{3L(\text{PV})} Y^3 - \dfrac{\text{YP}}{\text{PV}} Y^2\right) \cdot E}{2YE}$$

Therefore,

$$\bar{V} = \frac{\Delta p Y^2}{3L(\text{PV})} - \frac{\text{YP}Y}{2(\text{PV})}$$

Solving for Δp and using $Y = \frac{W}{2} = \frac{1}{2}\left(\frac{D_h - D_p}{2}\right)$

$$= \left(\frac{D_h - D_p}{4}\right)$$

where D_h = hole diameter; D_p = drillpipe (or collar) outside diameter yields

$$\Delta p = \frac{48L(\text{PV})\bar{V}}{(D_h - D_p)^2} + \frac{6L(\text{YP})}{(D_h - D_p)} \quad (5.23)$$

Imperial units Equation (5.23) can be converted to field units (in order to express Δp in units of psi) by the following method.

$$\text{First term} = \frac{48L(\text{PV})\bar{V}}{(D_h - D_p)^2}$$

$$= \frac{48\ (\text{ft})}{\text{in}^2}\left(\text{cP} \times \frac{2.0886 \times 10^{-5}\ \text{lb s}}{\text{cP ft}^2}\right) \times \left(\frac{\text{ft}}{\text{min}} \times \frac{\text{min}}{60\ \text{s}}\right)$$

$$\text{Constant of first term} = \frac{48 \times 2.0886}{60 \times 10^5} = \frac{1}{59\,848.7}\ \frac{\text{lb}}{\text{in}^2}$$

$$\simeq \frac{1}{60\,000}\ \text{psi}$$

$$\text{Second term} = \frac{6L(\text{YP})}{(D_h - D_p)} = \frac{6 \times \text{ft}}{\text{in}}\left(\frac{\text{lb}}{100\ \text{ft}^2}\right)$$

$$= \frac{6\left(\text{ft} \times \frac{12\ \text{in}}{\text{ft}}\right)}{100 \times \text{in}}(\text{lb})\left(\frac{1}{\text{ft}^2} \times \frac{\text{ft}^2}{144\ \text{in}^2}\right)$$

$$= \frac{6 \times 12}{14\,400}\ \frac{\text{lb}}{\text{in}^2} = \frac{1}{200}\ \text{psi}$$

Hence, Equation (5.23) becomes, in field units,

$$\Delta p = \frac{L(\text{PV})\bar{V}}{60\,000(D_h - D_p)^2} + \frac{L(\text{YP})}{200(D_h - D_p)}\ \text{psi} \quad (5.24)$$

Metric units

$$\text{First term} = \frac{48\ \text{m}\left(\text{cP} \times \frac{10^{-3}\ \frac{\text{kg}}{\text{ms}}}{\text{cP}}\right)(\text{m/s})}{\text{mm}^2 \times \frac{\text{m}^2}{10^6\ \text{mm}^2}}$$

$$= 48\,000\ \text{N/m}^2$$

$$\text{Second term} = \frac{6\ \text{m} \times (0.479\ \text{N/m}^2)}{\text{mm} \times \frac{\text{m}}{10^3\ \text{mm}}}$$

$$= 2874\ \text{N/m}^2$$

Hence, Equation (5.23) becomes

$$\Delta p = \frac{48000L(\text{PV})\bar{V}}{(D_h - D_p)^2} + \frac{2874L(\text{YP})}{(D_h - D_p)}\ \text{N/m}^2 \quad (5.25)$$

or

$$\Delta p = \frac{48L(\text{PV})\bar{V}}{(D_h - D_p)^2} + \frac{2.874L(\text{YP})}{(D_h - D_p)}\ \text{kPa} \quad (5.26)$$

or

$$\Delta p = \frac{0.48L(\text{PV})\bar{V}}{(D_h - D_p)^2} + \frac{0.0287L(\text{YP})}{(D_h - D_p)}\ \text{bar} \quad (5.27)$$

Example 5.2

Determine the annular pressure losses using the following data:

length of drill pipe	= 9000 ft
drill pipe	= OD/ID: 5 in/4.276 in
hole diameter	= 8.5 in
PV	= 20 cP
yield point	= 20 lb/100 ft²
flow rate	= 300 gpm
mud density	= 10 ppg

Solution

Equation (5.24)

$$\Delta p = \frac{L(\text{PV})\bar{V}}{60\,000(D_h - D_p)^2} + \frac{L(\text{YP})}{200(D_h - D_p)}$$

$$\bar{V} = \frac{24.5Q}{(D_h^2 - D_p^2)} = \frac{24.5 \times 300}{(8.5)^2 - (5)^2} = 155.6\ \text{ft/min}$$

(*Note:* The critical velocity is 402 ft/min, indicating that the flow is laminar.)

$$\Delta p = \frac{9000 \times 20 \times 155.6}{60\,000(8.5 - 5)^2} + \frac{9000 \times 20}{200(8.5 - 5)}$$

$$= 38.1 + 257.1 = 295.2\ \text{psi}$$

Critical velocity in annular flow

Using the same argument as given above (page 92), Equation (5.25) can be used to describe Newtonian fluid by letting YP = 0. Therefore,

$$\Delta p = \frac{48\,000\, L\mu_e v}{(D_h - D_p)^2} \tag{5.28}$$

where μ_e is the effective viscosity of the fluid.

A value of μ_e may be used in Equation (5.28) so that this equation is numerically equal to Equation (5.25). Hence,

$$\frac{48\,000\, L\mu_e \bar{v}}{(D_h - D_p)^2} = \frac{48\,000\, L\,(\text{PV})\,\bar{v}}{(D_h - D_p)^2} + \frac{2874\, L\,(\text{YP})}{(D_h - D_p)^2}$$

$$\mu_e = \text{PV} + \frac{2874}{48\,000}\frac{D_e\,\text{YP}}{\bar{v}} \tag{5.28b}$$

where $D_e = d_h - D_p$.

By use of $(Re) = \dfrac{1000\, vD_c}{\mu_e}$, $(Re) = 3000$ for turbulent flow and replacing $\bar{v}$ by V_c, Equation (5.28b) becomes

$$3000 = 1000 \frac{\rho V_c D_e}{\left[\text{PV} + \dfrac{2874}{48\,000}\dfrac{D_e(\text{YP})}{V_c}\right]}$$

Taking positive roots only, we obtain:

$$V_c = \frac{\text{PV} + \sqrt{\text{PV}^2 + 4\dfrac{\rho D}{3}(0.059) D_e(\text{YP})}}{2\rho D_e/3}$$

$$V_c = 1.5\left[\frac{\text{PV} + \sqrt{\text{PV}^2 + 0.079\rho D_e^2(\text{YP})}}{\rho D_e}\right] \text{m/s} \tag{5.29}$$

In Imperial units critical velocity may be determined by using the above procedure to arrive at

$$V_c = \frac{97\,\text{PV} + 97\sqrt{\text{PV}^2 + 6.2\,\rho D_e^2(\text{YP})}}{\rho D_e} \text{ft/min} \tag{5.30}$$

$$\left[\textit{Note: } \mu_e = \text{PV} + 300\,\text{YP}\frac{(D_h - D_p)}{v}\right]$$

Power-law model

The power-law model states that

$$\tau = K(\gamma)^n$$

or (5.31)

$$\log\tau = \log K + n \log \gamma$$

Hence, a graph of log τ against log γ is a straight line with a slope n and intercept of log K. Such a graph can be constructed by plotting the values of shear stress measured by a viscometer at 300 rpm and 600 rpm, as shown in Figure 5.10.

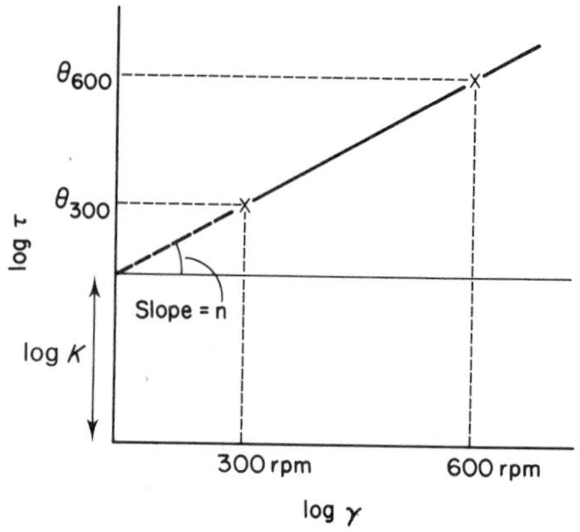

Fig. 5.10. Viscometer readings plotted on a log τ–log γ graph.

Hence,

$$\text{slope } (n) = \frac{\log\theta_{300} - \log K}{\log\gamma_{300}}$$

The viscometer is so designed that the shear strain at 300 rpm is equivalent to 511 s^{-1}. Therefore,

$$n = \frac{\log\theta_{300} - \log K}{\log 511}$$

$$K = \frac{\theta_{300}}{(511)^n} \tag{5.32}$$

Also,

$$n = \frac{\log\theta_{600} - \log\theta_{300}}{\log\gamma_{600} - \log\gamma_{300}}$$

where $\gamma_{600} = 1022$ s^{-1}

$$n = \frac{\log\left(\dfrac{\theta_{600}}{\theta_{300}}\right)}{\log 1022 - \log 511} = \frac{\log\left(\dfrac{\theta_{600}}{\theta_{300}}\right)}{\log\left(\dfrac{1022}{511}\right)}$$

or

$$n = 3.32 \log\left(\frac{\theta_{600}}{\theta_{300}}\right) \tag{5.33}$$

Pipe flow

The power-law model equation can be rewritten as

$$\tau = K\left(\frac{-\mathrm{d}v}{\mathrm{d}r}\right)^n \tag{5.34}$$

Shear stress, τ, in terms of Δp, r and L is also given by

$$\tau = \frac{\Delta p r}{2L} \tag{5.9}$$

Substituting Equation (5.9) in Equation (5.34) yields

$$\left(\frac{\Delta p r}{2L}\right) = K\left(\frac{-\mathrm{d}v}{\mathrm{d}r}\right)^n$$

$$\left(\frac{\Delta p}{2LK} r\right)^{1/n} = \left(\frac{-\mathrm{d}v}{\mathrm{d}r}\right)$$

$$\left(\frac{\Delta p}{2LK}\right)^{1/n} (r)^{1/n}\,\mathrm{d}r = -\mathrm{d}v$$

Integrating with respect to r gives

$$\left(\frac{\Delta p}{2LK}\right)^{1/n} \frac{r^{\left(\frac{n+1}{n}\right)}}{\left(\frac{n+1}{n}\right)} = -v + C \tag{5.34a}$$

where C is a constant.

At boundary $r = R$ and $v = 0$

$$C = \left(\frac{\Delta p}{2LK}\right)^{1/n} \times \frac{R^{[(n+1)/n]}}{\frac{n+1}{n}}$$

Hence, Equation (5.34a) becomes

$$\left(\frac{\Delta p}{2LK}\right)^{1/n} \frac{r^{[(n+1)/n]}}{\frac{n+1}{n}} = -v + \left(\frac{\Delta p}{2LK}\right)^{1/n} \frac{R^{[(n+1)/n]}}{\frac{n+1}{n}}$$

or

$$v = \left(\frac{\Delta p}{2LK}\right)^{1/n} \frac{n}{n+1} (R^{[(n+1)/n]} - r^{[(n+1)/n]})$$

Using

$$Q = v\mathrm{d}A = \int_0^R \left\{\left(\frac{\Delta p}{2LK}\right)^{1/n} \left(\frac{n}{n+1}\right)(R^{[(n+1)/n]} - r^{[(n+1)/n]})\right\} 2\pi r \mathrm{d}r$$

$$Q = \left(\frac{\Delta p}{2LK}\right)^{1/n} \left(\frac{n}{n+1}\right) \left[R^{[(n+1)/n]}\frac{r^2}{2} - \frac{r^{[(3n+1)/n]}}{\left(\frac{3n+1}{n}\right)}\right]_0^R 2\pi$$

$$\left[\textit{Note: } \int r^{[(n+1)/n]} r\,\mathrm{d}r = r^{[(2n+1)/n]} \cdot \mathrm{d}r = \frac{r^{[(2n+1)/n+1]}}{\left(\frac{2n+1}{n} + 1\right)} = \left(\frac{n}{3n+1}\right) r^{[(3n+1)/n]}\right]$$

$$Q = \left(\frac{\Delta p}{2LK}\right)^{1/n} \left(\frac{n}{n+1}\right) \left(\frac{R^{[(n+1)/n]} R^2}{2} - \frac{n}{3n+1} R^{[(3n+1)/n]}\right) 2\pi$$

Average velocity, $\bar{V}$, is given by

$$\bar{V} = \frac{Q}{A} = \frac{\left(\frac{\Delta p}{2LK}\right)^{1/n} \left(\frac{n}{n+1}\right) \times \left\{\frac{R^{[(n+1)/n]}}{2} R^2 - \frac{n}{3n+1} R^{[(3n+1)/n]}\right\} 2\pi}{\pi R^2}$$

Rearranging gives

$$\Delta p = \frac{2LK}{R}\left[\left(\frac{3n+1}{n}\right)\frac{\bar{V}}{R}\right]^n$$

But $R = D/2$, where D is the diameter of the pipe. Therefore,

$$\Delta p = \frac{4LK}{D}\left[2\left(\frac{3n+1}{n}\right)\frac{\bar{V}}{D}\right]^n \tag{5.35}$$

Field units of Equation (5.35)

Imperial units

$$K = \frac{\theta_{300}}{(511)^n} = \left(\frac{\text{lb}}{100\ \text{ft}^2} \times \frac{\text{ft}^2}{144\ \text{in}^2}\right)(\text{s}^{-1})^n$$

$$K = \frac{1}{14\,400}\ \text{psi}\ (\text{s}^{-1})^n$$

Therefore,

$$\Delta p = \text{LK}\frac{4\ \text{ft} \times \frac{1}{14\,400}\text{psi s}^n}{\text{in} \times \frac{\text{ft}}{12\ \text{in}}} \times \left[2\left(\frac{3n+1}{n}\right)\frac{(\text{ft/min})\frac{\text{min}}{60\ \text{s}}\bar{V}}{\left(\text{in} \times \frac{\text{ft}}{12\ \text{in}}\right)\bar{D}}\right]^n$$

$$\Delta p = \frac{1}{300D} KL\left[\frac{2}{5}\left(\frac{3n+1}{n}\right)\frac{\bar{v}}{D}\right]^n \qquad (5.36)$$

where $\bar{V}$ is in ft/min and Δp is in psi.

Metric units In SI units the various terms of Equation (5.35) have the following units: $K = \text{Pa}\cdot s^n$, $L = \text{m}$, $D = \text{m}$, $\bar{V} = \text{m/s}$. Hence, Equation (5.35) gives the pressure drop in Pa (N/m^2) directly.

In oil well drilling the units of length and diameter are given in metres and millimetres, respectively, while the consistency index K is determined from viscometer readings. The units of K as determined from this apparatus are in lb $\cdot$ s^n/100 ft^2 and must, therefore, be converted to Ns^n/m^2 or Pa $\cdot$ s^n. Hence, the power-law equations in this book will be modified to suit field requirements in order to produce the units of pressure in bars, N/m^2 or kPa.

Since

$$K = \frac{\theta_{300}}{(511)^n}\frac{\text{lb}}{100\ \text{ft}^2}\ \text{s}^n$$

where θ_{300} is measured in lb/100 ft^2; therefore, to convert the value of K to N/m$^2\cdot$ s^n multiply by a factor of 0.479.

Hence the metric field version of Equation (35) is:

$$\Delta p = \frac{4\ \text{m} \times 0.479 \dfrac{\text{N}}{\text{m}^2}\text{s}^n}{\text{mm} \times \dfrac{\text{m}}{10^3\ \text{mm}}}\frac{LK}{D} \times \left[2\left(\frac{3n+1}{n}\right)\text{mm} \times \frac{\text{m/s}}{10^3\ \text{mm}}\right]^n$$

$$\Delta p = \frac{1916KL}{D}\left[2000\left(\frac{3n+1}{n}\right)\frac{V}{D}\right]^n \text{N/m}^2 \qquad (5.37)$$

$$\Delta p = 19.16 \times 10^{-3}\frac{KL}{D}\left\{2000\left(\frac{3n+1}{n}\right)\frac{\bar{V}}{D}\right\}^n \text{bar} \qquad (5.37\text{a})$$

In Equations (5.37) and (5.37a) K is measured in lb s^n/100 ft^2.

Annular flow

Equation (21) states that

$$\tau = \frac{\Delta p y}{L}$$

Substituting Equation (5.21) in Equation (5.31), we obtain

$$\tau = \frac{\Delta p y}{L} = K(\gamma)^n = K\left(\frac{-\text{d}v}{\text{d}y}\right)$$

where $\gamma = (-\text{d}v)/\text{d}y$. Therefore,

$$\left(\frac{\Delta p}{LK}\right)^{1/n} y^{1/n} = \frac{-\text{d}v}{\text{d}y}$$

Integrating with respect to y gives

$$\left(\frac{\Delta p}{LK}\right)^{1/n}\frac{y^{[(n+1)/n]}}{\dfrac{n+1}{n}} = v + C$$

where C is a constant.

Boundary conditions give $v = 0$ at $y = Y$.

$$C = \left(\frac{\Delta p}{LK}\right)^{1/n}\left(\frac{n}{n+1}\right)Y^{[(n+1)/n]}$$

Therefore,

$$v = \left(\frac{\Delta p}{LK}\right)^{1/n}\left(\frac{n}{n+1}\right)\{Y^{[(n+1)/n]} - Y^{[(n+1)/n]}\}$$

Using $Q = v\text{d}A$ and the value of r from the above equation yields

$$Q = \int_0^Y \left\{\left(\frac{\Delta p}{LK}\right)^{1/n}\left(\frac{n}{n+1}\right)[Y^{[(n+1)/n]} - y^{[(n+1)/n]}]\right\} E(2\cdot \text{d}y)$$

where E = width of plate (see Figure 5.9).

$$Q = \left(\frac{\Delta p}{LK}\right)^{1/n}\left(\frac{n}{n+1}\right)\left[Y^{[(n+1)/n]}y - \left(\frac{n}{2n+1}\right)y^{[(2n+1)/n]}\right]_0^Y 2E$$

$$= \left(\frac{\Delta p}{LK}\right)^{1/n}\left(\frac{n}{n+1}\right)Y^{[(2n+1)/n]}\left[\frac{n+1}{(2n+1)}\right]2E$$

Average velocity, $\bar{V}$, is given by

$$\bar{V} = \frac{Q}{A} = \frac{\left(\dfrac{\Delta p}{LK}\right)^{1/n}\left(\dfrac{n}{n+1}\right)y^{[(2n+1)/n]}\left(\dfrac{n+1}{2n+1}\right)2E}{2YE}$$

or

$$\Delta p = \frac{KL}{Y}\left[\left(\frac{2n+1}{n}\right)\frac{\bar{V}}{Y}\right]^n$$

Referring to Figure 5.9,

$$Y = \frac{W}{2} = 1/2\left(\frac{D_h - D_p}{2}\right) = \frac{1}{4}(D_h - D_p)$$

$$\Delta p = \frac{4KL}{D_h - D_p}\left[4\frac{(2n+1)}{n}\frac{\bar{V}}{(D_h - D_p)}\right]^n \qquad (5.38)$$

Equation (5.38) is in consistent units and must be converted to field units.

Field units of Equation (5.38)

Imperial units Constant of Equation (5.38)

$$= \frac{4\left(\dfrac{1}{14\,400}\right)(\text{psi s}^n)(\text{ft})}{\text{in}\left(\dfrac{\text{ft}}{12\text{ in}}\right)}\left[4\frac{\text{ft}}{\text{min}}\times\frac{1\text{ min}}{60\text{ s}}\frac{1}{\text{in}\left(\dfrac{\text{ft}}{12\text{ in}}\right)}\right]^n$$

$$= \frac{1}{300}\text{ psi s}^n\left[\frac{48}{60}\frac{1}{\text{s}}\right]^n$$

Equation (5.38) becomes

$$\Delta p = \frac{KL}{300(D_h - D_p)}\left[\frac{4}{5}\left(\frac{2n+1}{n}\right)\frac{\bar{V}}{D_h - D_p}\right]^n \quad (5.39)$$

where Δp is in psi.

Metric units

$$\Delta p = \frac{4\times\text{m}\times 0.479\,(\text{N/m}^2)\,\text{s}^n}{\text{mm}\times\left(\dfrac{\text{m}}{1000\text{ mm}}\right)}$$
$$\times\left[4\left(\frac{2n+1}{n}\right)\frac{\text{m/s}}{\text{mm}\times\dfrac{\text{m}}{1000\text{ mm}}}\right]^n$$

$$\Delta p = 1916\times 10^{-3}\frac{KL}{(D_h - D_p)}$$
$$\times\left[\frac{4000}{D_h - D_p}\left(\frac{2n+1}{n}\right)\bar{V}\right]^n\ \text{N/m}^2$$

or

$$\Delta p = 19.16\times 10^{-3}\left[\frac{4000}{D_h - D_p}\left(\frac{2n+1}{n}\right)\bar{V}\right]^n\ \text{bar} \quad (5.40)$$

Critical velocities for power-law model

Pipe flow

Metric units In Equation (5.14) it was shown that for a Newtonian fluid (YP = 0), pressure loss, Δp, is given by

$$\Delta p = \frac{32\,000\,L\mu_e\bar{V}}{D^2}\ \text{N/m}^2$$

An effective viscosity, μ_e, may be substituted in Equation (5.14) such that Equation (5.14) is numerically equal to Equation (5.37), i.e.

$$\left(\frac{32\,000\,Lv}{D^2}\right)\mu_e = 1916\frac{KL}{D}\left[2000\left(\frac{3n+1}{n}\right)\frac{v}{D}\right]^n$$

$$\mu_e = \frac{1916}{32\,000}\frac{DK}{v}\left[2000\left(\frac{3n+1}{n}\right)\frac{v}{D}\right]^n \quad (5.41)$$

The equation for Reynolds Number in metric units is given by

$$(Re) = 1000\frac{DV\rho}{\mu_e} \quad (5.41\text{a})$$

and when $(Re) = 3000$ flow is turbulent and $v = V_c$ (critical velocity), hence, substituting Equation (5.41) in Equation (5.41a), we obtain

$$3000 = 1000\frac{DV_c\rho}{\left\{\dfrac{1916}{32\,000}\dfrac{DK}{V_c}\left[2000\left(\dfrac{3n+1}{n}\right)\dfrac{V_c}{D}\right]^n\right\}}$$

$$= \frac{DV_c\rho}{0.179\dfrac{DK}{V_c}\left[2000\left(\dfrac{3n+1}{n}\right)\dfrac{V_c}{D}\right]^n}$$

$$V_c = \left[\frac{0.179K}{\rho}\right]^{[1/(2-n)]}$$
$$\times\left[\frac{2000}{D}\left(\frac{3n+1}{n}\right)\right]^{[n/(2-n)]}\ \text{m/s} \quad (5.42)$$

Imperial units By the same procedures as for metric units, critical velocity in Imperial units may be shown to be

$$V_c = \left[\frac{5.82\times 10^4 K}{\rho}\right]^{[1/(2-n)]}$$
$$\times\left[\frac{1.6}{D}\left(\frac{3n+1}{4n}\right)\right]^{[n/(2-n)]}\ \text{ft/min} \quad (5.43)$$

Annular flow

From equations (39) and (28) and the Reynolds Number equation, the critical velocities for annular flow can be determined as detailed for pipe flow:

Metric system

$$V_c = \left[\frac{0.119K}{\rho}\right]^{[1/(2-n)]}$$
$$\times\left[\frac{4000}{D_h - D_p}\left(\frac{2n+1}{n}\right)\right]^{[n/(2-n)]}\ \text{m/s} \quad (5.44)$$

Imperial units

$$V_c = \left[\frac{3.878(10^4)K}{\rho}\right]^{[1/(2-n)]}$$
$$\times\left[\frac{2.4}{D_h - D_p}\left(\frac{2n+1}{3n}\right)\right]^{[n/(2-n)]}\ \text{ft/min} \quad (5.45)$$

The derivation of Equations (5.44) and (5.45) is left as an exercise for the reader.

TURBULENT FLOW

In the previous sections analytical techniques were used to describe laminar fluid flow and to derive expressions for pressure losses in pipes and annuli. Owing to the chaotic nature of fluid particle movement in turbulent flow, it is extremely difficult to arrive at an exact analytical method for determining pressure losses. A large number of workers in this field have arrived at the conclusion that pressure losses in turbulent flow are best determined from charts or by using the following equation:

$$\Delta p = \frac{2fL\rho \bar{V}^2}{D} \tag{5.46}$$

The friction factor, *f*

Solution of Equation (5.46) becomes straightforward once the value of the friction factor, f, is determined. In laminar flow the basic equation relating Δp to L, μ_e, $\bar{V}$ and D is given by Equation (5.11), when $PV = \mu_e$ and $YP = 0$, as

$$\Delta p = \frac{32\, L\mu_e V}{D^2}\ \text{N/m}^2 \tag{5.47}$$

where $D = 2R$. Equating Equations (5.46) and (5.47) yields

$$\frac{2fL\rho \bar{V}^2}{D} = \frac{32L\mu_e \bar{V}}{D^2}$$

$$f = \frac{16\mu_e}{\rho \bar{v} D}$$

$$= 16\frac{1}{\dfrac{\rho\bar{v}D}{\mu_e}}$$

or

$$f = \frac{16}{Re} \tag{5.48}$$

where $(Re) = \rho\bar{v}D/\mu_e$. In other words, for laminar flow the friction factor is inversely proportional to the Reynolds number only.

It was shown earlier that in turbulent flow the velocity profile is approximately uniform across the pipe cross-section and only layers adjacent to the pipe wall will feel the effects of wall roughness. If the pipe roughness is of small magnitude, such that the pipe protuberances penetrate a small part of the laminar sublayer, then the friction factor is independent of pipe roughness. As the length of the protuberances increases, the whole of the laminar sublayer and adjacent layers begin to feel the effects of pipe wall roughness. The pipe wall roughness, ε, is measured by the roughness height or protuberance length, and is expressed as a dimensionless quantity by dividing ε by the pipe diameter.

In drilling engineering it would be extremely difficult to routinely measure pipe wall roughness while the pipe is in service, owing to the hostile environment in which the equipment is used and to the practical difficulties in carrying out this measurement. Because of these difficulties and the need to arrive at simple and practical equations for determining turbulent pressure losses, the following relationships can be used to determine the friction factor, f.

Blasius equation Schlumberger (1984):

$$f = 0.057(Re)^{-0.2} \tag{5.49}$$

Prandtl equation:

$$1/\sqrt{f} = 4\log[(Re)\sqrt{f}] + 0.8 \tag{5.50}$$

Equation (5.49) is accurate up to $(Re) = 10^5$, which is more than sufficient for practical drilling engineering problems. Equation (5.50) is somewhat difficult to solve and normally only provides marginally accurate results for (Re) larger than 10^5.

For a fuller treatment of turbulent flow the reader is advised to consult the textbook by Skelland[1].

The turbulent flow equation will first be derived in consistent units and then converted to field units for the appropriate system.

Recall

$$(Re) = \frac{\rho VD}{\mu} \tag{5.5}$$

Combining Equations (5.5) and (5.49) yields

$$f = 0.057\left(\frac{\rho v D}{\mu}\right)^{-0.2} \tag{5.51}$$

Substituting Equation (5.51) in Equation (5.46) yields

$$\Delta p = 2(0.057)\left(\frac{\rho v D}{\mu}\right)^{-0.2}\frac{L\rho v^2}{D}$$

Simplifying further,

$$\Delta p = (2 \times 0.057)\frac{L\rho^{0.8} v^{1.8} \mu^{0.2}}{D^{1.2}} \tag{5.52}$$

The following substitution will be made in the above equation:

$$Q = vA,\ v = \frac{Q}{A} = \frac{Q}{\frac{\pi}{4}D^2}$$

or

$$v = \frac{4Q}{\pi D^2} \tag{5.53}$$

Also, turbulent viscosity, μ, is normally taken to be equal to PV/3.2. Making substitutions for v and μ in Equation (5.52) yields

$$p = (2 \times 0.057)\left(\frac{4}{\pi}\right)^{1.8}(3.2)^{-0.2} \times \frac{L\rho^{0.8}Q^{1.8}(\mathrm{PV})^{0.2}}{D^{4.8}} \tag{5.54}$$

Equation (5.54) is in consistent units and must be converted to field units.

Field units for Equation (5.54)

Metric units

$$p = (2 \times 0.057)\left(\frac{4}{\pi}\right)^{1.8}(3.2)^{-0.2}L\,(\mathrm{m}) \times \rho^{0.8}\left[\frac{\mathrm{kg}}{1} \times \left(\frac{1000\,1}{\mathrm{m}^3}\right)\right]^{0.8} \times Q^{1.8}\left(\frac{1}{\mathrm{min}} \times \frac{\mathrm{m}^3}{1000\,1} \times \frac{\mathrm{min}}{60\,\mathrm{s}}\right)^{1.8} \times pv^{0.2}\left(\mathrm{cP} \times \frac{10^{-3}\frac{\mathrm{kg}}{\mathrm{ms}}}{\mathrm{cP}}\right)^{0.2} \div D^{4.8}\left(\mathrm{mm} \times \frac{\mathrm{m}}{1000\,\mathrm{m}}\right)^{4.8}$$

Hence,

$$\text{constant} = 2 \times (0.057)\left(\frac{4}{\pi}\right)^{1.8}(3.2)^{-0.2} \times (1000)^{(0.8-1.8+4.8)} \times (60)^{-1.8} \times (10^{-3})^{0.2} = 5\,546\,778$$

$$\text{dimensions} = (\mathrm{m})^1(\mathrm{kg})^{0.8}(\mathrm{m})^{-2.4}(\mathrm{m})^{5.4}(\mathrm{s})^{-1.8}(\mathrm{kg})^{0.2} \times (\mathrm{m})^{-0.2}(\mathrm{s})^{-0.2}(\mathrm{m})^{-4.8} = \frac{\mathrm{kg}}{\mathrm{m\,s}} = \frac{1}{\mathrm{m}^2}\left(\frac{\mathrm{kg\,m}}{\mathrm{s}}\right) = \mathrm{N/m^2}\ (\text{or Pa})$$

Hence,

$$p = 5\,546\,778\frac{\rho^{0.8}Q^{1.8}(\mathrm{PV})^{0.2}L}{D^{4.8}}\ \mathrm{N/m^2}$$

$$\Delta p = 55.5\frac{\rho^{0.8}Q^{1.8}(pv)^{0.2}L}{D^{4.8}}\ \mathrm{bar} \tag{5.55}$$

Imperial units In Imperial units Equation (5.54) becomes

$$p = \frac{8.9 \times 10^{-5}\ \rho^{0.8}\ Q^{1.8}\ (\mathrm{PV})^{0.2}\ L}{D^{4.8}} \tag{5.56}$$

Units of Equation (5.56) are: ρ = ppg, Q = gpm, PV = cP, L = ft, D = in.

For annular flow, Equations (5.55) and (5.56) should be modified as follows.

From Equations (5.46) and (5.51),

$$\Delta p = 0.114\left(\frac{\rho v(D_h - D_p)}{\mu}\right)^{-0.2}\frac{L\rho V^2}{D_h - D_p} = 0.114\frac{\rho^{0.8}V^{1.8}(\mu)^{0.2}L}{D_h - D_p^{1.2}} \tag{5.57}$$

Substituting the following equations in Equation (5.57):

$$\mu = \frac{PV}{3.2}\ (\mu = \text{turbulent fluid viscosity})$$

and

$$V = \frac{Q}{A} = \frac{4}{\pi}\frac{Q}{(D_h^2 - D_p^2)} = \frac{4}{\pi}\frac{Q}{(D_h - D_p)(D_h + D_p)}$$

yields

$$\Delta p = 0.114\left(\frac{4}{\pi}\right)^{1.8}(3.2)^{-0.2}\rho^{0.8}(PV)^{0.2}Q^{1.8} \times \left[\frac{1}{(D_h - D_p)(D_h + D_p)}\right]^{1.8} \times \frac{1}{(D_h - D_p)^{1.2}} \times L$$

Therefore,

$$\Delta p = 0.114\left(\frac{4}{\pi}\right)^{1.8}(3.2)^{-0.2} \times \frac{\rho^{0.8}(PV)^{0.2}Q^{1.8}L}{(D_h - D_p)^3(D_h + D_p)^{1.8}} \tag{5.58}$$

Using metric field units and simplifying gives

$$\Delta p = 55.5\frac{\rho^{0.8}Q^{1.8}(pv)^{0.2}L}{(D_h - D_p)^3(D_h + D_p)^{1.8}} \tag{5.59}$$

(Δp is in bars.)

In Imperial field units

$$\Delta p = \frac{8.91 \times 10^{-5} \rho^{0.8} Q^{1.8} (pv)^{0.2}}{(D_h - D_p)^3 (D_h + D_p)^{1.8}} \text{ psi} \qquad (5.60)$$

FLOW THROUGH NOZZLES

Tri-cone bits are normally equipped with two or three nozzles to direct the drilling fluid at high velocity to the bottom of hole in order to lift the cuttings up the hole and also to clean the teeth of the bit. Owing to the small area of a nozzle, fluid velocity through the nozzle is normally high and the flow pattern is always turbulent.

Figure 5.11 gives a cross-section of a typical nozzle. Thus, a nozzle can be approximated as an orifice enabling Bernoulli's equation to be applied at points 1 and 2 of Figure 5.11:

$$\left(p_1 + \frac{1}{2}\rho v_1^2\right)_{\text{point 1}} = \left(p_2 + \frac{1}{2}\rho v_2^2\right)_{\text{point 2}}$$

$$(p_1 - p_2) = \frac{1}{2}\rho(v_2^2 - v_1^2)$$

When $A_2 < A_1$, $v_2^2 \gg v_1^2$ and the above equation simplifies to

$$\Delta p = \frac{1}{2}\rho v_2^2$$

where $\Delta p = p_1 - p_2$

$$v_2 = \sqrt{\frac{2\Delta p}{\rho}} \qquad (5.61)$$

Equation (5.61) does not consider frictional loss in the nozzle and a correction factor known as the coefficient of discharge, C_d, is normally introduced to take account of this frictional loss. Hence,

$$v_2 = C_d \sqrt{\frac{2\Delta p}{\rho}} \qquad (5.62)$$

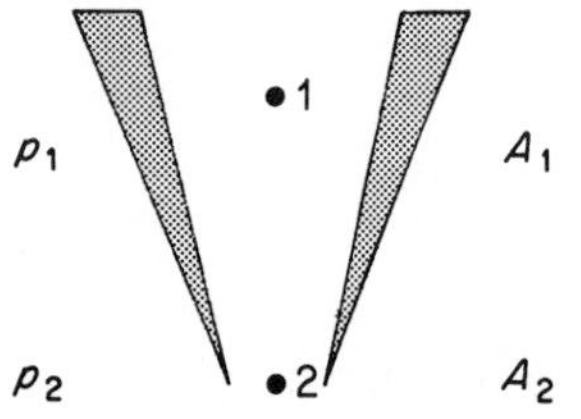

Fig. 5.11. Fluid flow through a nozzle.

The value of $C_d \simeq 0.95$, and Equation (5.62) becomes

$$V_n = 0.95\sqrt{\frac{2\Delta p}{\rho}} = 1.344\sqrt{\frac{\Delta p}{\rho}} \qquad (5.63)$$

where V_n = nozzle velocity or jet velocity.

Rearranging Equation (5.63) to express Δp in terms of V_n and ρ, Equation (5.64) is obtained:

$$\Delta p = \frac{V_n^2 \times \rho}{1.805} \qquad (5.64)$$

Equation (5.63) in Imperial units

$$V_n(\text{ft/s}) = 1.344 \sqrt{\frac{\Delta p\left(\dfrac{\text{lb}}{\text{in}^2}\dfrac{144\ \text{in}^2}{\text{ft}^2}\right)}{\rho\left(\dfrac{\text{lbm}}{\text{gal}} \times \dfrac{7.48\ \text{gal}}{\text{ft}^3}\right)}}$$

$$V_n = 5.897\sqrt{\text{ft}^3 \times \frac{1}{\text{ft}^3} \times 32\frac{\text{ft}^2}{\text{s}^2}\left(\frac{\Delta p}{\rho}\right)}$$

$$V_n = 33.3585\sqrt{\frac{\Delta p}{\rho}} \text{ ft/s} \qquad (5.65)$$

where Δp is in lbf/in^2 and ρ is in lbm/gal), or

$$\Delta p = \frac{\rho V_n^2}{1113} \text{ psi} \qquad (5.66)$$

Equation (5.63) in metric units

$$V_n = 1.344 \sqrt{\frac{\text{bar}\left(\dfrac{10^5\ \text{N/m}^2}{\text{bar}}\right)}{\dfrac{\text{kg}}{\text{l}}\left(\dfrac{1000\ \text{l}}{\text{m}^3}\right)}} \text{ m/s} \qquad (5.67)$$

$$V_n = 13.44\sqrt{\frac{\Delta p}{\rho}}$$

or

$$\Delta p = \frac{\rho V_n^2}{180.6} \text{ bar} \qquad (5.68)$$

Nozzle area

A tri-cone bit is normally equipped with three nozzles and the objective of the hydraulic programme is to determine the total area and size of nozzles. Using $Q = VA$,

$$A_T = \frac{Q}{V_n} \qquad (5.69)$$

where A_T = total area of nozzles.

Field units of Equation (5.69)

Imperial units

$$A_T(\text{in}^2) = \frac{Q\dfrac{\text{gal}}{\text{min}} \times \left(\dfrac{\text{ft}^3}{7.48\ \text{gal}}\right)\left(\dfrac{\text{min}}{60\ \text{s}}\right)}{V_n(\text{ft/s})}\left(\frac{144\ \text{in}^2}{\text{ft}^2}\right)$$

$$A_T = \frac{0.32Q}{V_n}\ \text{in}^2 \qquad (5.70)$$

where Q is in gal/min and V_n in ft/s. For three nozzles

$$A_T = 3A = 3\frac{\pi}{4}d^2$$

or

$$d = \sqrt{\frac{4A_T}{3\pi}} \qquad (5.71)$$

where d is size of nozzle in inches.

Normally, a nozzle size is expressed in multiples of $\frac{1}{32}$ of an inch; an open nozzle is given size 32 or 1 in (24.5 mm). Thus, Equation (5.71) can be modified to give nozzle sizes in multiples of $\frac{1}{32}$ as follows:

$$d_N = 32\sqrt{\frac{4A_T}{3\pi}} \qquad (5.72)$$

Metric units

$$A_T(\text{mm}^2) = \frac{\left(Q\dfrac{1}{\text{min}} \times \dfrac{\text{m}^3}{1000\ \text{l}} \times \dfrac{\text{min}}{60\ \text{s}}\right)}{V_n(\text{m/s})} \times \left(\frac{10^6\ \text{mm}^2}{\text{m}^2}\right)$$

$$A_T = \frac{1000Q}{60V_n}\ \text{mm}^2$$

where A_T is area of three nozzles. Also,

$$d = \sqrt{\frac{4A_T}{3\pi}}\ \text{mm} \qquad (5.73)$$

As manufacturers specify nozzle sizes in multiples of $\frac{1}{32}$ of an inch, Equation (5.73) should first be converted to inches and then multiplied by 32 to obtain multiples of 32. Thus,

$$d = \frac{1}{25.4}\sqrt{\frac{4A_T}{3\pi}}\ \text{in}$$

where 1 in = 25.4 mm, nozzle size, $d_N = d \times 32$ and

$$d_N = 1.2598\sqrt{\frac{4A_T}{3\pi}} \qquad (5.74)$$

Units of A in Equation (5.74) are still mm^2 but the size of nozzle is expressed in multiples of $\frac{1}{32}$ in.

Example 5.3

Determine the nozzle sizes in multiples of 32 to be used in a three-cone bit when 500 gpm of mud is circulated at a pressure drop of 1000 psi through the bit. The mud density is 10 ppg.

Solution

From Equation (5.66),

$$\Delta p = \frac{V_n^2}{1113}$$

or

$$V_n = 33.3585\sqrt{\frac{\Delta p}{\rho}}$$

$$= 33.3585\sqrt{\frac{1000}{10}}$$

$$V_n = 333.58\ \text{ft/s}\ (101.7\ \text{m/s})$$

From Equation (5.70),

$$A_T = \frac{0.32Q}{V_n} = \frac{0.32 \times 500}{333.58}$$

$$= 0.4796\ \text{in}^2\ (309.4\ \text{mm}^2)$$

From Equation (5.72),

$$d_N = 32\sqrt{\frac{4A_T}{3\pi}}$$

$$d_N = 14.44\ \text{in}\ (353.8\ \text{mm})$$

Nozzle sizes come only in integer values of $\frac{1}{32}$ in. The above result indicates that two nozzles of size 14 and one of size 15 should be used. This can be checked as follows:

$$A_T = 2 \times \text{area of}\ \frac{14}{32}\ \text{nozzle} + \text{area of}\ \frac{15}{32}\ \text{nozzle}$$

$$= 2\left[\frac{\pi}{4}\left(\frac{14}{32}\right)^2\right] + \frac{\pi}{4}\left(\frac{15}{32}\right)^2$$

$$A_T = 0.3007 + 0.1726$$

$$= 0.4733\ \text{in}^2\ (\text{approximately equal to } 0.4796)$$

If two 15/32 nozzles were used instead of one, then

the total area A_T becomes:

$$A_T = 2 \times \frac{\pi}{4}\left(\frac{15}{32}\right)^2 + \frac{\pi}{4}\left(\frac{14}{32}\right)^2$$

$$= 0.4954 \text{ in}^2$$

This is greater than 0.4796 in^2 calculated from Equation (5.72). Therefore, two size 14 and one size 15 should be used.

References

1. Skelland, A. H., 1967. *Non-Newtonian Flow and Heat Transfer*. John Wiley, New York.
2. Schlumberger, 1984. *Cementing Technology*. Dowell Schlumberger Publications

Problem

Prove that the effective viscosity in field units for an annular flow obeying a power-law model is given by

$$\mu_e = 200K(D_h - D_p)\left[\frac{0.8}{(D_h - D_p)}\left(\frac{2n+1}{n}\right)\right]^n V^{n-1} \text{ (Imperial units)}$$

and

$$\mu_e = 0.04K(D_h - D_p)\left[\frac{4000}{(D_h - D_p)}\left(\frac{2n+1}{n}\right)\right]^n V^{n-1} \text{ (metric units)}$$

Chapter 6

Mud Engineering

This chapter aims to provide a brief insight into the functions and composition of typical drilling muds. No attempt is made here to cover specific mud engineering topics in detail, as this would require a complete textbook. The interested reader is advised to consult the references given at the end of this chapter for full treatment of specific topics. The chapter, however, documents the latest ideas on the subject, as extracted from recently published literature.

The following topics will be discussed:

Functions of drilling mud
Drilling mud
Types of drilling mud
  water-base muds
  oil-base muds
  emulsion muds
Fundamental properties of mud
Mud calculations
Mud contaminants
Mud conditioning equipment

FUNCTIONS OF DRILLING MUD

A drilling mud is used to carry out the following functions: (a) cool the drill bit and lubricate its teeth; (b) lubricate and cool the drill string; (c) control formation pressure; (d) carry cuttings out of the hole; (e) stabilise the well bore to prevent it from caving in; and (f) help in the evaluation and interpretation of well logs. These functions will now be explained in detail.

Function (a)

One of the prime functions of mud is to cool the drill bit and lubricate its teeth. The drilling action requires a considerable amount of mechanical energy in the form of weight-on-bit, rotation and hydraulic energy. A large proportion of this energy is dissipated as heat, which must be removed to allow the drill bit to function properly. The mud also helps remove drill cuttings from the spaces between the teeth of the bit, thereby preventing bit balling.

Function (b)

A rotating drill string generates a considerable amount of heat which must be dissipated outside the hole. The mud helps to cool the drill string by absorbing this heat and releasing it, by convection and radiation, to the air surrounding the surface pit tanks. The mud also provides lubrication by reducing friction between drill string and borehole walls. Lubrication is normally achieved by the addition of bentonite, oil, graphite, etc.

Function (c)

For safe drilling, high-formation pressures must be contained within the hole to prevent damage to equipment and injury to drilling personnel. The drilling mud achieves this by providing a hydrostatic pressure just greater than the formation pressure. For effective drilling, the difference between the hydrostatic and formation pressures should be zero. In practice, an overbalance of 100–200 psi is normally used to provide an adequate safeguard against well-kick. The pressure overbalance is sometimes referred to as chip hold down pressure (CHDP), and its value directly influences penetration rate. In general, penetration rate decreases as the CHDP increases. When an abnormally pressured formation is encountered, the CHDP becomes negative and a sudden increase in penetration rate is observed. This is normally taken as an indication of a well kick (see Chapter 13).

Function (d)

For effective drilling, cuttings generated by the bit must be removed immediately. The drilling mud carries these cuttings up the hole and to the surface, to be separated from the mud. Hence, mud must also possess the necessary properties to allow cuttings to be separated at the surface and to be recirculated.

The carrying capacity of mud depends on several factors, including annular velocity, plastic viscosity and yield point of the mud and slip velocity of the generated cuttings.

In general, the resultant velocity (or lift velocity) of cuttings is the difference between annular velocity, V_r, and slip velocity, V_s. By use of the power-law model, the slip velocity of cuttings is given by

$$V_s = \frac{175 D_p(\rho_p - \rho_m)^{0.667}}{\rho_m^{0.333}\mu_e^{0.333}} \text{ ft/min} \tag{6.1}$$

where D_p = particle diameter (in); ρ_p = density of particle (lb/gal); ρ_m = density of mud (lbm/gal); μ_e = effective mud viscosity (cP) (see below, page 113).

In metric units the equivalent of Equation (6.1) is

$$V_s = \frac{15.23(\rho_s - \rho_m)^{0.667}}{\rho_m^{0.333}\mu_e^{0.333}} \text{ m/s} \tag{6.2}$$

where densities are measured in kg/l.

Annular velocity, V_r, is simply the volume flow rate divided by the annular area. Lift velocity, V, is given by

$$V = V_r - V_s \tag{6.3}$$

The mud must also be capable of keeping the cuttings in suspension when circulation is stopped, to prevent them from accumulating on the bottom of the hole and causing pipe sticking.

Function (e)

The formation of a good mud cake helps to stabilise the walls of the hole, somewhat similarly to the effect of adding a layer of plaster to interior house walls. The pressure differential between the hydrostatic pressure of mud and that of formation pressure is also instrumental in keeping the walls of the borehole stable.

Shale stability is largely dependent on the type of mud used. This subject will be discussed more fully in Chapter 12.

It should be noted that the best way to keep a hole stable is to reduce the time during which the hole is kept open.

Function (f)

Wire line logs are run in mud-filled holes in order to ascertain the existence and size of hydrocarbon zones. Open-hole logs are also run to determine porosity, boundaries between formations, locations of geo-pressured (or abnormally pressured) formations and the site for the next well[1]. Hence, the drilling mud must possess such properties that it will aid the production of good logs. A comprehensive treatment of well logging and the effects of mud on the generated logs can be found in Reference 2.

DRILLING MUD

Fresh water has a density of 62.3 lbm/ft^3 (1000 kg/m^3), which gives a pressure gradient of 0.433 psi/ft (0.0481 bar/m). Hence, for a 10 000 ft (3048 m) well, the bottom hole pressure due to a full column of water is 4330 psi (299 bar). At this depth normally pressured formations have a pressure gradient of 0.465 psi/ft (0.105 bar/m), giving a formation pressure at 10 000 ft of 4650 psi (321 bar). Hence, if water is used as the drilling fluid at this depth, formation fluids will invade the wall, causing a kick. In actual fact, the formation pressure gradient at 10 000 ft can often be greater than the 0.465 psi/ft given previously and, consequently, the pressure differential in the direction of the well will be even greater.

To provide more pressure, the mud density must be increased. This is achieved by the addition of high-gravity solids.

TYPES OF DRILLING MUD

Generally speaking, drilling mud may be defined as a suspension of solids in a liquid phase; the liquid can be water or oil. Three types of drilling mud are in common use: (1) water-base muds; (2) oil-base muds; and (3) emulsion muds.

Water-base muds

Water-base muds consist of four components: (1) liquid water (which is the continuous phase and is used to provide initial viscosity); (2) reactive fractions to provide further viscosity and yield point; (3) inert fractions to provide the required mud weight; and (4) chemical additives to control mud properties.

Components (2) and (3) represent the solid fraction of mud; the reactive fractions of mud are always low-gravity solids, while the inert fractions can be low- or high-gravity solids.

The reactive fraction of a water-base mud consists of clays such as bentonite and attapulgite. The inert solids include sand, barite, limestone, chert, etc.

Reactive fraction of mud solids

Clays (or low-gravity reactive solids) are added to water to provide the viscosity and yield point properties necessary to lift the drill cuttings or to keep them in suspension (see Equations (6.1) and (6.2)). The mechanism by which high viscosity and yield point are developed is very complex and is not yet completely understood. It is related to the internal structure of the clay particles and the electrostatic forces that act to hold them together when they are dispersed in water.

Full discussion of clays is beyond the scope of this book, and the interested reader is advised to consult references at the end of this chapter for a fuller treatment of the subject. References 3, 4 and 5 are particularly useful.

There are two types of clay currently in use for making water-base muds:

(a) *Bentonitic clay*: this is a member of the montmorillonite (smectite) group of clays, and can only be used with fresh water since high viscosity and yield point do not develop in saltwater.

(b) *Attapulgite* (or salt gel): this is a member of the palygorskite group of clays, and has the ability to develop the required high viscosity in both fresh and saltwater.

Nature of clays

API[6] defines clays as "natural, earthy, fine-grained materials that develop plasticity when wet". Clays form from the chemical weathering of igneous and metamorphic rocks. The major source of clays used commercially is volcanic ash, the glassy component of which weathers very readily, usually to bentonite. Ash layers occur interstratified with other sedimentary rocks, and are easily mined by opencast methods. The famous Wyoming bentonite is such a weathered ash layer.

The characteristic feature of clay minerals is that they possess an atomic structure in which the atoms form layers. There are three types of atomic layers that give clays their special properties:

(a) tetrahedral layers

These are made up of a flat honeycomb sheet of tetrahedra made up of a central silicon atom surrounded by four oxygens. The tetrahedra are linked to form a sheet by sharing three of their oxygen atoms with adjacent tetrahedra.

(b) octahedral layers

These are sheets composed of linked octahedra, each made up of an aluminium or magnesium atom surrounded by six oxygens. Again the links are made by shearing oxygen atoms between two or three neighbouring octahedra.

(c) exchangeable layers

These are layers of atoms or molecules bound loosely into the structure, which can be exchanged with other atoms or molecules, and are important in giving the clay its particular physical properties.

The nature of these layers and the way they are stacked together on top of one another define the type of clay mineral. A typical clay structure is shown in Figure 6.1. It is of the clay mineral pyrophyllite, which is related broadly to bentonite. In this structure, two tetrahedral layers are sandwiched together, joined by an octahedral layer that lies between them, which uses the unshared oxygen atoms of each tetrahedral layer to act as the oxygen atoms at the corner of the octahedra. The consequence of this sharing is that the sandwich of tetrahedral and octahedral layers is very tightly bonded together and forms a brittle thin sheet.

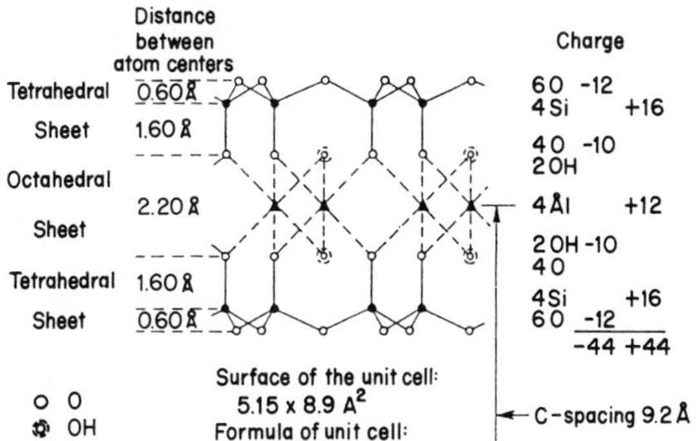

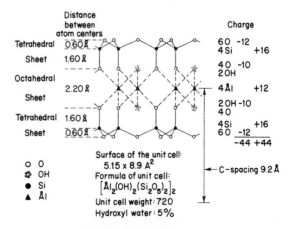

Fig. 6.1. Atom arrangement in the unit cell of a three-layer mineral (schematic)[1].

The sandwiches of tetrahedral and octahedral layers are joined in many clays by exchangeable layers. The bonding of these exchangeable layers is very weak – they may be composed of potassium, sodium, calcium or magnesium atoms or of water or organic molecules. The result is that clays split very easily along the exchangeable layers, which gives the characteristic platy shape to their individual particles.

The repeat distance between successive layers when they are stacked together varies from 9–15 Ångstrom units (0.9–1.5 nm, or about 4 one-hundred millionths of an inch).

In some clays, the exchangeable layer is relatively tightly bound in the structure, and requires a long period of weathering for chemical exchange to take place. In other clay families, such as those to which bentonite and attapulgite belong, the exchangeable layer can be removed or added to the structure simply by placing the clay in a solution of appropriate composition. A measure of how readily exchange can take place is the *cation exchange capacity* of the clay. A known weight of clay is dispersed in a solution of magnesium chloride, to replace as much as possible of the exchangeable layer with magnesium. It is then transferred to a solution of potassium or calcium chloride, and the amount of potassium or calcium absorbed by the clay is measured. This amount, expressed as milliequivalents per 100 grammes of dry clay is the cation exchange capacity (CEC). Typical values for bentonite are 70–130 and for attapulgite 5–99. Because the CEC is related to the ease with which water molecules enter the structure, CEC is used as a convenient guide to the quality of the clay.

Hydration of clays

Clays with a high CEC can exchange a large amount of water into the exchangeable layer, and also adsorb water onto the outer surfaces of individual plates of clay. It is this effect that gives the clays their desirable high viscosity and high yield point.

Adsorption of water into the exchangeable layer causes a very striking expansion of the clay structure, since the water molecules force the layers apart until there are four or more layers of water molecules in the space that previously only contained a single layer of sodium or calcium atoms. For sodium bentonite, the layer repeat distance (basal spacing) can increase from 9.8 Ångstroms before hydration to 40 Ångstroms or MURG after hydration (1 Ångstrom unit = 10^{-7} mm). For calcium bentonite the increase in basal spacing is from 12.1 Å to 17 Å.

The individual plates of clay can also adsorb water on their outer surfaces, where it is held by electrostatic attraction from the broken atomic bonds.

Overall hydration leads to the transformation of clay from a dry powder to a plastic slurry with an increase in volume of several hundred percent. The effectiveness of this process is measured by the clay yield, which is defined by API as the number of barrels of 15 CP mud which can be obtained from 1 ton (2000 lb) of dry clay.

In general the clay yield depends on:

(a) the purity of the clay:

(b) the nature of the atoms present in the exchangeable layer;

(c) the salinity of the water used.

Bentonite and attapulgite

Bentonite is a roll formation consisting primarily of montmorillonite. The name montmorillonite comes from the French town of Montmorillon, where the clay was first mined in 1874. The basic structure of montmorillonite is close to that of pyrophyllite shown in Figure 6.1, except that there are a small number of exchangeable ions, sodium, calcium and magnesium, as well as some water molecules in the exchangeable layer. The most common natural varieties of bentonite are those with sodium and calcium as exchangeable ions.

Figure 6.2 shows an electron micrograph of the minute platy crystals of which bentonite is made up. Each plate is about 1 μm in diameter and less than 0.1 μm thick, and is crinkled so that the aggregate somewhat resembles cornflakes.

Attapulgite belongs to a quite different family of clay minerals. In this family, the tetrahedra in the tetrahedral sheets of atoms do not all point the same way, but some tetrahedra in the sheet are inverted. Instead of crystallising as platy crystals, attapulgite forms needle-like crystals, shown in Figure 6.3. Attapulgite-based muds have excellent viscosity and yield strength, and retain these properties when mixed with salt water. However they have the disadvantage of suffering high water loss, giving poor sealing properties.

Dispersion, flocculation and deflocculation

When a suspension of clay and water is agitated, three modes of plate association occur: (1) edge to edge; (2) face to edge; and (3) face to face.

Depending on the type of plate association, four different states[1] can be distinguished: (1) dispersion; (2) aggregation; (3) flocculation; and (4) deflocculation.

When the individual plates are suspended in

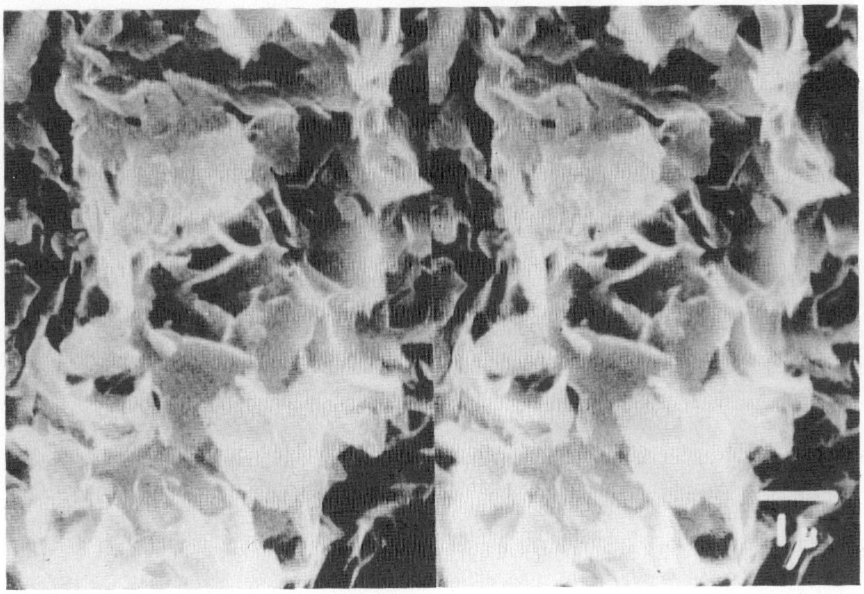

Fig. 6.2. Electron micrograph of Wyoming montmorrillonite (after Borst & Keller, 1969)[8].

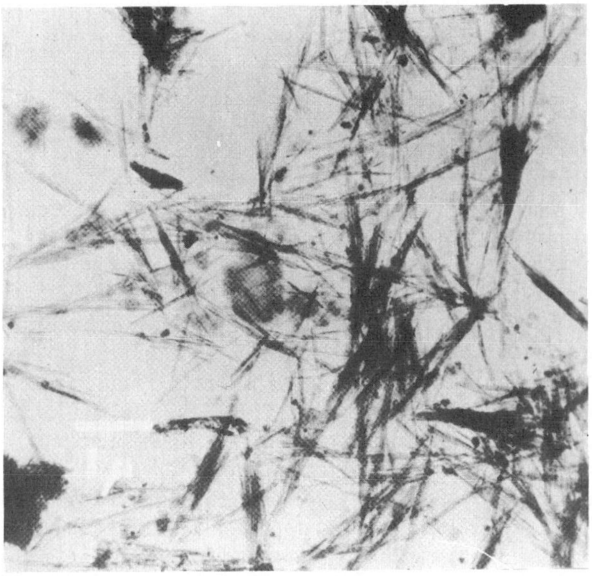

Fig. 6.3. Electron micrograph of Georgia attapulgite (after Borst & Keller, 1969)[8].

solution, with no face or edge association between individual plates, the suspension is said to be dispersed (Figure 6.4a). Dispersion results in an increase in viscosity and gel strength.

Face-to-face association of the individual plates results in an aggregated solution (Figure 6.4b). Aggregation results in a decrease in viscosity and gel strength.

When the clay plates join one another through the edges and faces, the resulting suspension is said to be flocculated (Figure 6.4c). Flocculation results in the formation of flocks or clusters of particles of clay. In drilling mud flocculation results in excessive gelation.

In a flocculated solution the edge-to-face association can be broken up by the use of a chemical thinner. The resulting suspension is described as deflocculated (Figure 6.4d).

Inert fractions of mud

The inert fractions of mud include low-gravity and high-gravity solids. Low-gravity solids include sand and chert.

High-gravity solids are added to increase the weight or density of mud. They are also referred to as weighting material, and mud containing such solids is termed 'weighted mud'. The following high-gravity solids are currently used.

Barite (or *barium sulphate*, $BaSO_4$) has a specific gravity of 4.2 and is used to prepare mud densities in excess of 10 ppg (1.19 kg/l). Barite is preferred to other weighting agents because of its low cost and high purity.

Lead sulphides such as galena are used as weighting

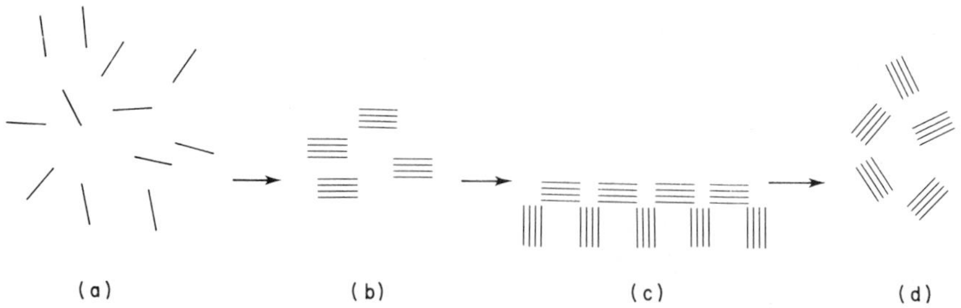

Fig. 6.4. Schematic presentation of the states of clay suspensions: (a) dispersed; (b) flocculated; (c) aggregated; (d) deflocculated (face-to-face association).

materials because of their high specific gravities (in the range 6.5–7) allowing mud weight of up to 35 ppg (4.16 kg/l) to be prepared.

Iron ores have a specific gravity of 5+ and are more erosive than other weighting materials. Iron ores also contain toxic materials.

Chemical additives

Chemical additives are used to control viscosity, yield point, gel and fluid loss properties of mud. Fluid loss additives will be discussed below (page 118) in conjunction with filtrate analysis.

Two types of chemical additives can be distinguished: thinners and thickeners.

Mud thinners

Mud thinners operate on the principle of reducing viscosity by breaking the attachment of the clay plates through the edges and faces. The thinner then attaches itself to the clay plates, preventing the maintenance of attractive forces between the sheets. The following is a list of the most widely used thinners.

Phosphates include sodium tetraphosphate and sodium acid pyrophosphate. They are suitable for any pH value but have a temperature limitation of 175°F (79°C).

Chrome lignosulphonate is the most commonly used thinner, but decomposes at 300°F (149°C). This chemical has the ability to deflocculate and disperse the clay particles, thereby reducing viscosity, yield point and water loss. The deflocculation is achieved by the attachment of the chrome lignosulphonates to the broken edges of clay plates. This reduces the attractive forces between clay sheets and results in a solution of reduced viscosity and gel strength.

Lignites decompose at 350°F (177°C). They can also be used as water-loss control agents.

Surfactants (surface tension-reducing agents) help thin mud and also reduce water loss. They can also be used as emulsifiers (see oil-based muds).

Mud thickeners

Mud thickeners include the following.

Lime or *cement*, which can be used to thicken mud or increase its viscosity. Viscosity increase is due primarily to flocculation of clay plates, resulting from the replacement of Na^+ cations by Ca^{++} cations.

Polymers

Polymers are chemicals consisting of large molecules made up of many repeated small units called monomers.

Polymers are used for filtration control, viscosity modification, flocculation and shale stabilisation. When added to mud, polymers cause little change in the solid content of the mud.

Polymer muds possess a high shear-thinning ability at high shear rates such as those encountered inside the drillpipe and across the bit. Shear thinning reduces the viscosity of mud and, in turn, reduces the frictional pressure losses across the bit and inside the pipes. Hence, a greater proportion of the available hydraulic horsepower can be expended at the bit, resulting in faster penetration rates and improved hydraulics. Certain polymer muds are also characterised by their insensitivity to contamination by salt, cement and gypsum.

In general, three types of polymers can be recognised:

(a) *Extenders*: include sodium polyacrylate (trade name BENEX) which increases viscosity by flocculating the bentonite;

(b) *Colloidal polymers*: include sodium carboxy methyl cellulose (CMC), Hydroxyethyl cellulose (HEC) and starch. (The term colloid comes from the Greek word 'kwλλd', which means glue[5].)

CMC is an anionic polymer produced by the treatment of cellulose with caustic soda and then monochloro acetate. Its molecular weight ranges between 50 000 and 400 000.

HEC is made by a similar process as that needed to manufacture CMC but with ethylene oxide after the caustic soda. Its main advantage is its ability to hydrate in all types of saltwaters.

Starch[1] is principally produced from corn or potatoes and is gelatinised by the action of heat and hydrochloric acid and is finally dried and ground. The molecular weight of starch can be up to 100 000. Starch is used to develop viscosity and to act as a filtration control agent. The latter property is attributed to the expansion of the starch particles in water and the formation of sponge-like bags[11] which wedge into the openings of the filter cake, resulting in a much reduced filtrate loss. The major disadvantage of starch is that it is amenable to bacterial attack at low pH values.

(c) *Long Chain polymers*: include the xanthan gum polymer. The xanthan gum polymer is a water soluble biopolymer produced by the action of bacteria on carbohydrates[9] and has a molecular weight of 5 000 000. The xanthan gum polymer is amenable to bacterial attack at temperatures above 300°F.

The main advantage of the xanthan gum polymer is that it can build viscosity in fresh, sea and salt water, without assistance from other additives. Additional increase in viscosity may also be obtained by cross-linking of the polymer molecules by the addition of chromic chloride at a concentration of 0.3 lb/lb of biopolymer[9]. The xanthan gum polymer also provides a high, but fragile, gel strength, thus allowing the mud to suspend the cuttings during non-circulation periods. A xanthan gum polymer mud can be formulated from any water type and is resistant to contamination by anhydrite, gypsum and salt.

Since polymers are specially prepared chemicals, they are inherently more expensive than bentonite and other viscosifying materials. In addition, polymers do not increase the solid content of the mud and, in turn, do not increase the density of the mud. In general, polymer muds have a weight limitation of 13 ppg (97 pcf, or 1.56 kg/l).

Types of water-base muds

The following is a summary of the most widely used types of water-base muds. For a full discussion of all types, the reader is advised to consult references (3, 4, 10 and 11).

Clear water

Fresh water and saturated brine can be used to drill hard (11), compacted and near-normally pressured formations.

Native mud

This mud is made by pumping water down the hole during drilling and letting it react with formations containing clays or shales. The water dissolves the clays and returns to the surface as mud. This mud is characterised by its high solids content and a high filter loss resulting in a thick filter cake.

Calcium muds

The swelling and hydration of clays and shales can be greatly reduced when calcium mud is used as the drilling fluid. Calcium muds are superior to freshwater muds when drilling massive sections of gypsum and anhydrite as they are not susceptible to calcium contamination.

When calcium is added to a suspension of water and bentonite, the calcium cations will replace the sodium cations on the clay platelets. Since the calcium cations will be strongly attached to the surfaces of the clay sheets, compared with the sodium cations, the platelets tend to be pulled closer together and the clay structure collapses forming aggregates. In general the hydration of clay is reduced with increasing concentration of calcium cations. The swelling of the clay is reduced by 50% when the calcium concentration in water is 150 ppm (parts per million).

Hence, when calcium mud comes into contact with shaly formations, the swelling of shale is greatly reduced due to the presence of the calcium cations. Also the disintegration of drill cuttings is greatly reduced in the presence of calcium cations.

The major advantage of calcium muds is their ability to tolerate a high concentration of drilled solids without these affecting the viscosity of mud.

Calcium muds are classified according to the percentage of soluble calcium in the mud.

(a) *Lime Mud*: this mud contains up to 120 ppm of soluble calcium. It is prepared by mixing bentonite, lime (calcium hydroxide $Ca(OH)_2$), thinner, caustic soda and an organic filtration control agent. The lime provides the inhibiting ion (Ca^{++}).

(b) *Gyp Mud*: this mud contains up to 1200 ppm of soluble calcium. It is similar to lime mud except that the lime is replaced by gypsum (calcium sulphate). Gyp muds also possess a greater temperature stability than lime muds.

Lignosulphonate mud

This mud type is considered to be suitable when:

(a) high mud densities are required (>14 lb/gal or 1.68 kg/l);

(b) working under moderately high temperatures of 250–300°F (121–149°C);
(c) high tolerance for contamination by drilled solids, common salt, anhydrite, gypsum and cement is required;
(d) low filter loss is required.

This type consists of freshwater or saltwater, bentonite, chrome or ferrochrome lignosulphonate, caustic soda, CMC or stabilised starch. Optional materials such as lignite, oil, lubricants, surfactants, may be used.

This mud type is not suitable for drilling shale sections due to adsorption of water from the mud by the clay surfaces with ultimate heaving of the shale section. The major disadvantage of this type of mud is the damage it causes to formation permeability. Clays of productive zones disperse in the presence of lignosulphonate causing the clay plates to move freely and ultimately block the pores of the production zone. Due to these disadvantages, the mud is now seldom used.

KCl/polymer muds

The basic components of potassium chloride (KCl)/polymer muds are:
(a) fresh water or sea water;
(b) KCl;
(c) inhibiting polymer;
(d) viscosity building polymer (such as the bacterially produced xanthane-type);
(e) stabilised starch or CMC;
(f) caustic soda or caustic potash; and
(g) lubricants, etc.

The KCl/polymer mud is suitable for drilling shale sections due to its superior sloughing-inhibiting properties resulting from the use of KCl and the inhibiting polymer. Inhibition by KCl is attributed to the replacement of the sodium ions in shales by potassium ions which are more tightly held. The inhibiting polymer is an anionic polymer which attaches itself to the positively-charged edges of the exposed shale sections and prevents the shale from coming in contact with the water.

This mud type is also suitable for drilling potentially productive sands (hydrocarbon zones) which are amenable to permeability damage by fresh water. Due to its low solids content, this mud is often described as non-dispersed mud.

Due to their low tolerance of solids, KCl/polymer muds require the use of efficient desanders and desilters to remove the very fine cuttings.

The advantages of this mud include:
(a) high shear thinning behaviour facilitating solids removal through the shale shaker, desanders and desilters;
(b) high true yield strength;
(c) improved borehole stability; and
(d) good bit hydraulics and reduced circulating pressure losses.

The disadvantage is their instability at temperatures above 250°F (121°C).

Salt-saturated muds

A salt-saturated mud is one in which the water phase is saturated with sodium chloride. Normally the concentration of salt is in excess of 315 000 ppm[11]. This mud is suitable for drilling salt domes and salt sections due to its ability to prevent hole enlargement through those sections which can be caused when using fresh water muds. When used with a polymer, salt-saturated mud can be used to inhibit the swelling of bentonitic shales. A salt-saturated mud consists of:
(a) fresh, brine or seawater;
(b) common salt (NaCl);
(c) encapsulating polymer;
(d) CMC or starch.

This type of mud is characterised by its low tolerance to high concentrations of low gravity solids (e.g. drilled solids) and its high requirement for filter-loss additives.

Oil-base muds

Oil-base muds are emulsions of water in oil in which crude or diesel oil is the continuous phase and water is the dispersed phase. Oil-base muds are sometimes described as 'invert emulsions', since water droplets are emulsified in a continuous phase of oil. True emulsions are those in which the continuous phase is water and oil is the dispersed phase.

Water is used mainly to give the emulsion the required properties of gel strength and barite suspension. Water and oil are emulsified by the use of suitable emulsifiers such as soaps and by agitation. Soaps made from monovalent ions such as sodium (Na^+) or divalent ions such as calcium (Ca^{2+}) are used as emulsifiers. A sodium soap molecule has two ends, a sodium end which is soluble in water, and an opposite end made up of a large organic group which is soluble in oil[11]. The soap molecules bridge together oil and water interfaces, which results in a stable emulsion[11]. Calcium soaps have two large organic groups attached to a central calcium ion (Ca^{2-}), the latter being soluble in water. These divalent soaps produce invert emulsions when mixed into oil and water.

Agitation is required to break the water into small droplets which can be easily dispersed within the oil phase. The oil/water ratio determines the final properties of the emulsion and its stability. In general, higher oil/water ratios give increased resistance to contamination and increased temperature stability[11].

Oil-base muds are used to drill holes with severe shale problems and to reduce torque and drag problems in deviated wells.

Oil-base muds tend to be more stable at high temperatures than are water-base muds.

From the completion standpoint, oil is an excellent drilling fluid, since it does not damage hydrocarbon-bearing zones, and thus preserves the natural permeability of the area around the well bore. Mud filtrate from water-base muds can penetrate the formation to a few feet and can drastically reduce the natural permeability of the formation.

The main disadvantages of oil-base muds are: (a) the environment is contaminated (especially in offshore operations); (b) flammability becomes a hazard; (c) drilled-solids removal from an oil-base mud is usually more difficult than from a water-base mud, owing to the high plastic viscosity of the emulsion[10]; and (d) electric logging is more difficult with oil-base muds[10].

Emulsion muds

Emulsion muds (or true emulsions) are those in which water is the continuous phase and oil is the dispersed phase. Oil is added to increase penetration rate, reduce filter loss, improve lubricity, reduce chances of lost circulation and reduce drag and torque in directional wells. An oil-emulsion mud normally contains 5–10% of oil by volume[11]. The emulsion can be formulated by the use of sodium soap emulsifiers as described above (page 112).

FUNDAMENTAL PROPERTIES OF MUD

The fundamental properties of mud include: (a) weight (or density); (b) rheological properties; (c) filtrate and filter cake; and (d) pH value.

Mud weight

Mud weight or, more precisely, mud density is defined as the mass of a given sample of mud divided by its volume. Mud weight is dependent largely upon the quantity of solids in the liquid phase, either in solution or suspended by the particles of the liquid phase.

In equation form, the density of mud, ρ_m, is given by

$$\rho_m = \frac{M_W + M_S}{V_W + V_S} \tag{6.4}$$

where M_W and M_S are the masses of water (or oil) and solids, respectively, and V_W and V_S are the volumes of water (or oil) and solids, respectively.

Equation (6.4) shows that mud density can be increased above the density of water by adding a material (normally solids) with a specific gravity greater than that of water. A mud density of less than that of water can be achieved by the addition of oil or by aerating the liquid phase. The use of Equation (6.4) in various forms is given in the section on 'Mud calculations'. Mud weight is measured in the field using a mud balance, as shown in Figure 6.5. A steel cup is filled with a freshly collected mud sample and then balanced on a knife-edge. The resulting reading is the mud weight (more accurately mud density) in pcf, ppg or Kg/m^3.

Rheological properties of mud

The most important rheological properties of mud are: (a) plastic viscosity; (b) yield point; and (c) gel strength.

Plastic viscosity

Viscosity is a property which controls the magnitude of shear stress which develops as one layer of fluid slides over another. It is a measure of the friction between the layers of the fluid and provides a scale for describing the thickness of a given fluid. Viscosity is largely dependent on temperature and, for liquids, decreases with increasing temperature; the reverse is true for gases. The concept of viscosity has been described in detail in Chapter 5.

Because viscosity is dependent on the velocity of the fluid and the pattern of flow, whether laminar or turbulent, absolute or effective viscosity is difficult to measure. In drilling engineering only changes in the annular viscosity are of concern to the drilling engineer, as such changes directly affect cuttings removal and annular pressure losses, which, in turn, affect the actual hydrostatic pressure of mud.

In practice, two models are used to describe the flow of fluids: the Bingham plastic and power-law models.

For the Bingham plastic model, it was shown in Chapter 5 that the annular effective viscosity may be determined from

$$\mu_e = PV + \frac{2874}{48\,000}(D_h - D_p)\frac{YP}{\bar{v}} \quad \text{metric units} \tag{6.5}$$

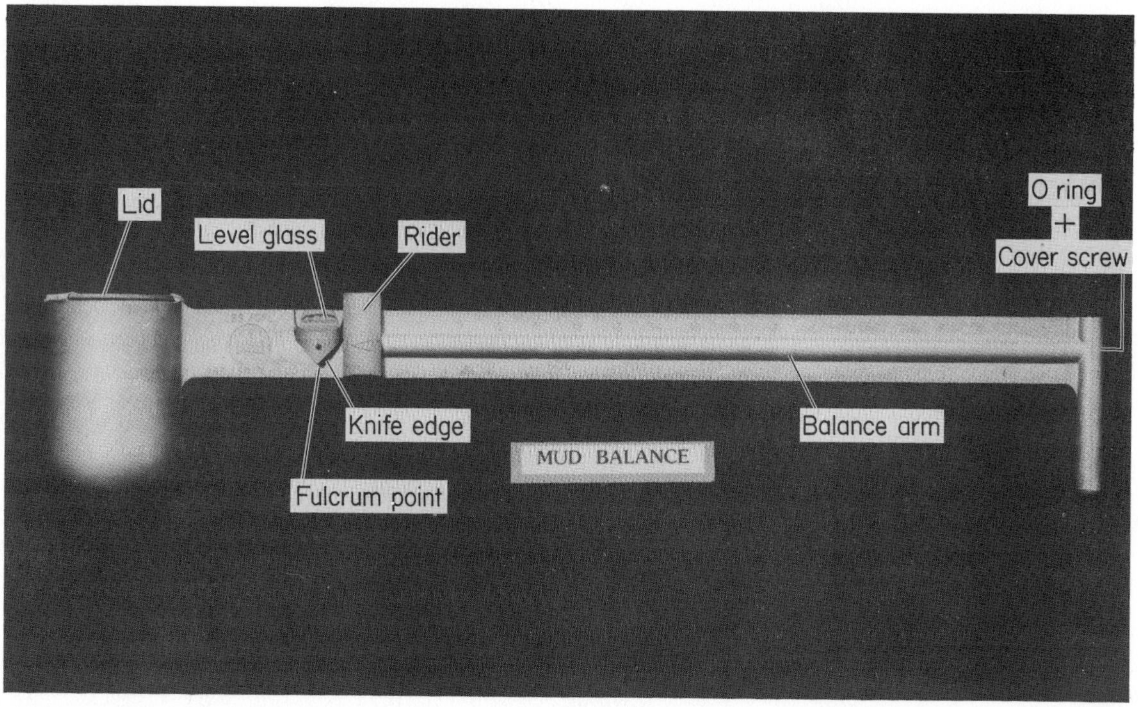

Fig. 6.5. Diagram of mud balance[1]. (Courtesy of Baroid/NL)

$$\mu_e = PV + \frac{300}{\bar{v}}(D_h - D_p)YP \text{ Imperial units} \quad (6.5a)$$

For the power-law model, the annular effective viscosity is given by

$$\mu_e = \frac{1916K(D_h - D_p)}{4800\bar{v}} \times \left[\frac{4000\bar{v}}{(D_h - D_p)}\frac{(2n+1)}{n}\right]^n \text{ metric units} \quad (6.6)$$

$$\mu_e = \left[\frac{2.4\bar{v}}{(D_h - D_p)}\frac{(2n+1)}{3n}\right]^n \times 200K\left(\frac{D_h - D_p}{\bar{v}}\right) \text{ Imperial units} \quad (6.6a)$$

In Equations (6.5) and (6.6) the significant terms are: μ_e = effective viscosity (theoretical); PV = plastic viscosity or Bingham viscosity as determined by a viscometer; YP = yield point; D_h = hole diameter; D_p = drillpipe outside diameter; $\bar{v}$ = average annular velocity; K and n are the indices of the power law model $\tau = K(\gamma)^n$; n = flow behaviour index = 3.32 log $(\theta_{600}/\theta_{300})$; K = consistency index = $\theta_{300}/(511)^n$; θ_{600} = viscometer reading at 600 rpm; and θ_{300} = viscometer reading at 300 rpm.

Measurements of plastic viscosity and yield point are made using a viscometer (see Figure 6.6). This instrument consists of two cylinders: the outer cylinder or rotor sleeve and an inner, stationary cylinder. A fresh sample of mud is placed in a cylindrical container in which the instrument head is immersed to a scribed line. Rotation of the rotor sleeve in the mud produces a torque on the inner cylinder. Movement of the inner cylinder is restricted by a high-precision torsion spring and the deflection of the latter is recorded on a dial. To determine PV and YP the rotor sleeve is made to run at 600 rpm and a stabilised dial reading, θ_{600}, is taken. Similarly, at 300 rpm, a stabilised reading, θ_{300}, of the torque is taken. The instrument constants are so arranged that plastic viscosity and yield point can be determined from the following equations:

$$PV = \theta_{600} - \theta_{300}$$

$$YP = \theta_{300} - PV$$

Another field measurement of viscosity is the Marsh funnel viscosity, which is determined using a Marsh Funnel (Figure 6.7). This measurement relies on the principle that thick or viscous mud flows more

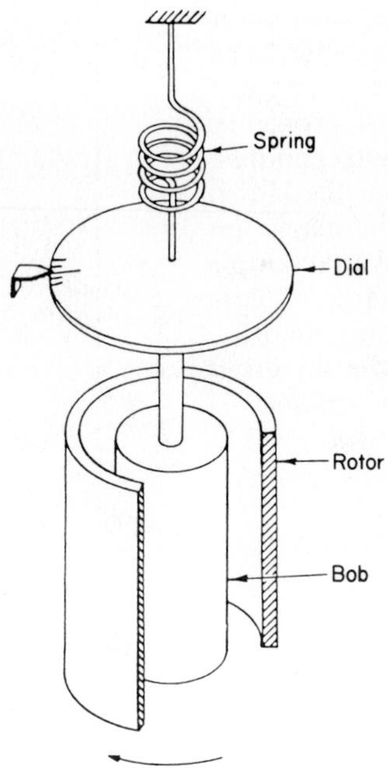

(a)

(b)

Fig. 6.6. (a) Diagram of concentric cylinder viscometer[7]; (b) apparatus for measuring viscosity, yield point and gel strength (rheometer). (Courtesy of API)[6] slowly than thin mud, and is normally used to monitor the changes in viscosity of the circulating mud. The Marsh funnel viscosity is easy to determine and can provide early warnings of potential fluid problems. However, it has no scientific foundation and cannot be used in quantitative analysis. To determine the Marsh Funnel viscosity, the funnel is filled to its maximum capacity of 1500 cc with a freshly collected mud sample and the time (in seconds) required to discharge 950 cc is noted. This time is taken as the Marsh funnel viscosity.

Usually the funnel is allowed to discharge into a container of approximately 950 cc in volume, as shown in Figure 6.7. The discharge time gives a comparative value of the static viscosity, and the value is usually higher than the plastic viscosity.

The marsh funnel viscosity is highly dependent on the rate of gelation and on the density which varies the hydrostatic head of the column of mud in the funnel (1). The measurement is also extremely sensitive to changes in the volume of discharge as shown in Fig. 6.8.

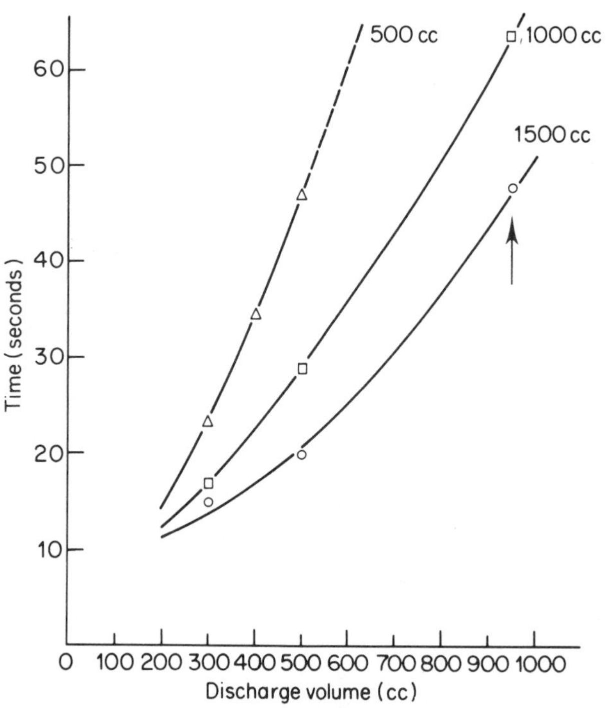

Fig. 6.8. Effects of varying discharge volume and funnel volume on discharge times: ○, volume of mud in funnel = 1500 cc; □, volume of mud in funnel = 1000 cc; △, volume of mud in funnel = 500 cc. Mud weight = 72 pcf.

Yield point

The yield point is a measure of the attractive forces between particles of mud resulting from the presence of positive and negative charges on the surfaces of these particles. The yield point is a measure of the forces that cause mud to gel once it is motionless, and it directly affects the carrying capacity of mud. The yielding of mud can be compared to the yielding of metals in that a minimum level of stress (the yield stress) must be provided before deformation of mud is achieved.

Fig. 6.7. Marsh Funnel Viscometer[1]. (Courtesy of Baroid/NL)

Example 6.1

Given the following well data:

$\theta_{600} = 36$
$\theta_{300} = 24$
hole size, $D_h = 12\frac{1}{4}$ in (311 mm)
drillpipe, $D_p = 5$ in (127 mm)
flow rate, $Q = 600$ gpm (2272 l/min)
hole depth = 7100 ft (2164 m)
drill collars = 8 in OD, 620 ft long
mud weight = 75 pcf

Determine: (a) thickness or effective viscosity, using the Bingham plastic and power-law models; and (b) the equivalent annular circulating density, using the Bingham plastic model only. (Assume an open hole from surface to 7100 ft, i.e. no surface casing.)

Solution

(a) *Effective viscosity*

Bingham plastic model

$$PV = \theta_{600} - \theta_{300} = 36 - 24 = 12 \text{ cP}$$

$$YP = \theta_{300} - PV = 24 - 12 = 12 \frac{\text{lb}}{100 \text{ ft}^2}$$

$$\bar{V} = \frac{24.5}{D_h^2 - D_p^2} = \frac{24.5 \times 600}{(12.25)^2 - 5^2} = 117.5 \text{ ft/min}$$

From Equation (6.5),

$$\mu_e = PV + \frac{300(D_h - D_p)YP}{\bar{v}}$$

$$= 12 + \frac{300(12.25 - 5) \times 12}{117.5} = 234 \text{ cP}$$

(In metric units, $\mu_e = 233$ cP.)

Power-law model

$$n = 3.32 \log \frac{\theta_{600}}{\theta_{300}}$$

$$= 3.32 \log \frac{36}{24} = 0.5846$$

$$K = \frac{\theta_{300}}{(511)^n} = \frac{24}{(511)^{0.5846}} = 0.626$$

$$D_h - D_p = 7.25 \text{ in}, \ \bar{V} = 117.5 \text{ ft/min}$$

$$\mu_e = \left[\frac{2.4}{(D_h - D_p)} \frac{(2n+1)}{3n}\right]^n \left[200K \frac{(D_h - D_p)}{\bar{v}}\right]$$

$$= \left[\frac{2.4 \times 117.5(2 \times 0.5846 + 1)}{7.25 \times 3 \times 0.5846}\right]$$

$$\times \left(\frac{200 \times 0.626 \times 7.25}{117.5}\right)$$

$$= 74 \text{ cP}$$

(In metric units, $\mu_e = 74$ cP.)

(b) *Equivalent annular circulating density* During circulation, the total pressure exerted on hole bottom is the sum of the hydrostatic pressure and annular pressure loss. This total pressure is also described as dynamic pressure, to distinguish it from the static bottom hole pressure exerted by a static column of mud. Thus,

$$\text{dynamic pressure} = \text{hydrostatic pressure}, H_s + \text{annular pressure loss}, \Delta P_a$$

$$\text{equivalent density} \times g \times \text{hole depth} = H_s + \Delta P_a \quad (6.7)$$

In Imperial units, Equation (6.7) becomes

$$\frac{\rho_e \times H}{144} = H_s + \Delta P_a$$

or

$$\rho_e = \frac{144}{H}(H_s + \Delta P_a) \quad (6.8)$$

where ρ_e = equivalent density (lb/ft^3); H = hole depth (ft); and H_s and ΔP_a are in psi.

In metric units, Equation (6.7) becomes

$$\rho_e = \frac{1}{0.0981H}(H_s + \Delta P_a) \quad (6.8a)$$

where ρ_e = equivalent density (kg/l); H = hole depth (m); and H_s and ΔP_a are in bars.

Pressure losses in the annulus are normally laminar in nature. Hence, annular losses around drillpipe,

$$\Delta P_{adp} = \frac{L \, PV \, \bar{V}}{60\,000(D_h - D_p)^2} + \frac{L \, YP}{200(D_h - D_p)}$$

where L = length of drillpipe = 7100 − 620 = 6480 ft; and $\bar{V}$ = 117.5 ft/min (from Part a).

$$\Delta P_{adp} = \frac{6480 \times 12 \times 117.5}{60\,000(12.25 - 5)^2} + \frac{6480 \times 12}{200(12.25 - 5)}$$

$$= 56.5 \text{ psi}$$

annular losses around drill collars,

$$\Delta P_{adc} = \frac{620 \times 12 \times 170.8}{60\,000(12.25 - 8)^2} + \frac{620 \times 12}{200(12.25 - 8)}$$

$$= 9.9 \text{ psi}$$

where L = 620 ft; and

$$\bar{v} = \frac{24.5 \times 600}{D_h^2 - OD_{dc}^2}$$

$$= \frac{24.5 \times 600}{12.25^2 - 8^2} = 170.8 \text{ ft/min}$$

where OD_{dc} = drill collar outside diameter. Total annular pressure losses,

$$\Delta P_a = \Delta P_{adp} + \Delta P_{adc} = 54.6 + 9.9$$

$$= 65 \text{ psi}$$

Therefore,

$$\text{dynamic pressure} = H_s + 65$$

$$= \frac{75 \times 7100}{144} + 65 = 3762.9 \text{ psi}$$

From Equation (6.8), equivalent circulating density

$$= \frac{144}{H}(H_s + \Delta P_a)$$

$$= \frac{144}{7100}(3762.9) = 76.3 \text{ pcf}$$

Gel strength

Gel strength is a measure of the ability of mud to develop and retain a gel structure. It is analogous to shear strength, and defines the ability of mud to hold solids in suspension. It also gives an indication of the thixotropic properties of mud and, consequently, the thickness of quiescent mud. Thixotropy refers to the ability of a suspension of fluid such as mud to develop a semi-solid structure when at rest, and to assume a liquid state when in motion.

The gel strength of mud can be determined using a viscometer, as shown in Figure 6.6. The sample of mud is stirred at high speed and then allowed to rest for either 10 s or 10 min. The torque reading at 3 rpm is normally taken as the value of the gel strength. Two readings, one after 10 s and another after 10 min, are taken to represent the initial gel strength and the 10 min gel strength, respectively. Gel strength is read in lb/100 ft^2 directly, and to convert to metric units, the value should be multiplied by a factor of 0.478.

Filtrate and filter cake

When a drilling mud comes into contact with porous rock, the rock acts as a screen allowing the fluid and small solids to pass through, retaining the larger solids. The fluid lost to the rock is described as 'filtrate'. The layer of solids deposited on the rock surface is described as 'filter cake'. It should be observed that filtration occurs only when there is a positive differential pressure in the direction of the rock.

The quality of mud is dependent on the volume of filtrate lost to the formation and the thickness, and strength of the filter cake. In general, the volume of filrate lost to the formation is dependent on the magnitude of differential pressure and mud solids characteristics of the filter cake. The initial volume of filtrate lost to the rock while the filter cake is forming is described as 'spurt loss'.

The volume of filtrate and thickness of filter cake of a mud sample can be determined using a filter press, Figure 6.9. The apparatus consists of a filter cell mounted in a frame, a pressure source and a graduated cylinder for receiving and measuring the volume of filtrate. The filter cell consists of a cylinder with an inside diameter of 3 in (76.2 mm) and a height of at least 2.5 in (64 mm)[12]. The cylinder is so fitted that a pressure medium can conveniently be admitted into and bled from the top. A fine screen is placed at the base of the cell and is then covered with a filter paper. The base of the cell is provided with a drain tube for filtrate discharge.

To perform a filtration test, a freshly collected mud sample is poured into the cylinder and is then subjected to a pressure of either 100 psi or 500 psi: and temperatures ranging from 200°F to 300°F.

The volume of filtrate is measured by placing a dry, graduated cylinder under the filter cell and the volume of filtrate collected is reported in cubic centimetres (or cc) per 30 minutes. The volume of filter cake deposited on the filter paper is then measured and is reported in 32-seconds of an inch. The cake is also noted as either soft, firm or tough[11].

A drilling mud with a high water loss is particularly damaging from the logging and production standpoints. A large invasion zone impairs the natural productivity of the formation and special type perforators or several acid washes may be required to remove this damage.

A number of additives may be used to reduce the volume of water loss. These include: bentonite, emulsified oil, dispersants, CMC and starch. When starch is used, the pH of mud must be greater than 11.5, otherwise the starch will be attacked by bacteria, reducing its ability to control water loss. Normally, a bactericide is used to prevent the fermentation of starch.

An ideal mud is one which has a small water loss and deposits a thin, tough filter cake on the surfaces of permeable formations.

The pH of mud

The acidity or alkalinity of any solution is normally described by the use of a pH value. The pH is defined as the negative logarithm of the hydrogen ion (H^+) content of the solution, i.e. $pH = -\log H^+$.

The measurement of pH is based on the principle that electrolytic solutions are capable of dissociating into hydrogen cations (H^+) and hydroxyl groups (OH^-). The hydrogen cations represent the acidic component of the solution, while the hydroxyl groups contribute to the alkalinity of the solution. It follows that the addition of materials that increase the concentration of H^+ results in a decrease in the pH values;

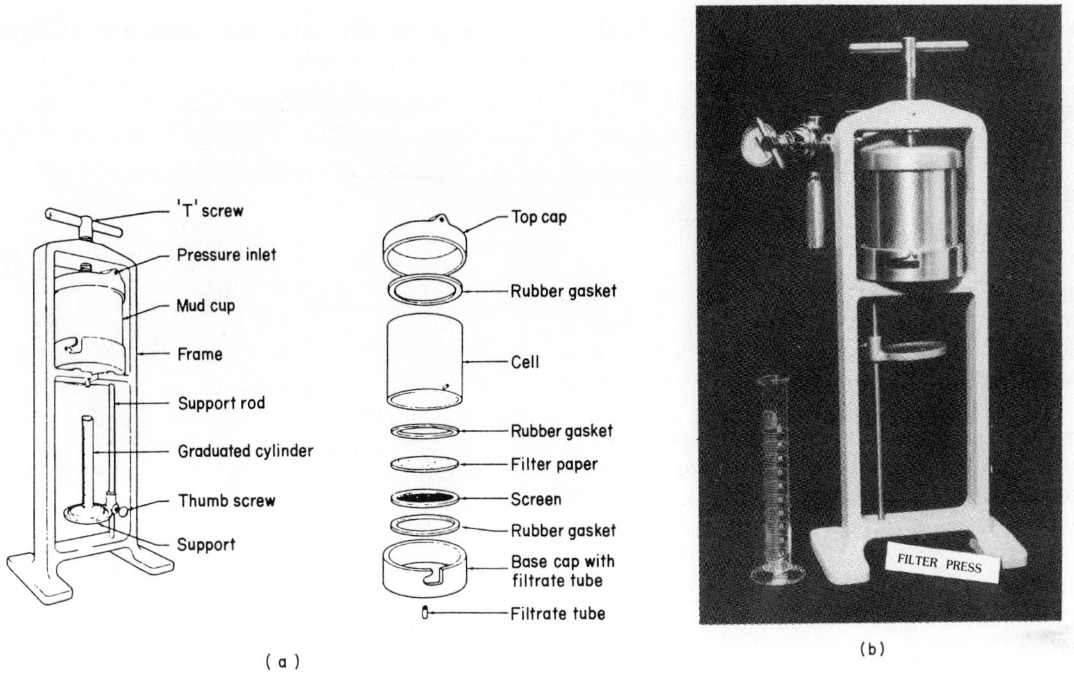

Fig. 6.9. Standard filter press[1]: (a) standard filter press and mud cell assembly; (b) Baroid standard filter press. (Courtesy of Baroid/NL)

conversely, the addition of materials that decrease the concentration of H^+ (i.e. increase OH^-) would result in an increase in the pH value of the solution.

The relationship between the concentrations of H^+ and of OH^- was found to be constant for most solutions and is given by

$$H^+ \times OH^- = 10^{-14}$$

For neutral solutions such as distilled water, the concentration of H^+ is equal to that of OH^-, giving $H^+ = OH^- = 10^{-7}$, and

$$pH = -\log H^+ = -\log 10^{-7} = +7$$

Hence, Table 6.1 can be constructed for various concentrations of H^+ and OH^-.

The pH of mud plays a major role in controlling the solubility of calcium. At high pH values calcium solubility is very limited, which makes high-pH muds suitable for drilling calcium-type formations which are susceptible to erosion by water.

Starch is used as a fluid loss additive. However, this additive is susceptible to bacterial attack at pH values below 11.5. Thus, close control of the pH of mud should be maintained when starch is used as an additive, unless a bactericide is used.

It has also been observed[13] that mud can be easily flocculated at pH values greater than 11 and less than 8. For pH values in the range 8–11 clay dispersion can be easily promoted.

The pH value is also an important indicator for controlling corrosion. A minimum value of 7 should

TABLE 6.1

H^+	OH^-	$pH = -\log H^+$	*Description*
10^0	10^{-14}	0	Acidic
10^{-7}	10^{-7}	7	Neutral
10^{-10}	10^{-4}	10	Basic
10^{-14}	10^0	14	Basic

TABLE 6.2

Solution	*pH*
Distilled water	7
Bentonite suspension	8
Caustic soda (NaOH) 10% concentration	13.3
Lignite (10%)	5
Sodium acid phosphate	3.9–4.2

always be maintained, to prevent corrosion to casing, drill pipe, etc.

The pH values of typical solutions are given in Table 6.2.

MUD CALCULATIONS

This section is intended to provide a basic knowledge of the way routine mud calculations are performed in the field. The equations used are widely accepted in Imperial units and certain modifications were made to adapt them to metric units.

In most mud calculations the objective is either to increase or to decrease mud weight. Density can be increased by the addition of solids and decreased by dilution with water, diesel oil, or by aerating the mud.

Mud weight increase calculations

Assume:

M_1 = mass of original mud

V_1 = volume of original mud

ρ_1 = density or weight of original mud

To increase mud density, solids should be added to the mud. The following particulars refer to solids:

M_2 = mass of added solids

V_2 = volume of added solids

ρ_2 = density of added solids

The addition of solids will produce a new mud having a mass M_3, a volume V_3 and a final density of ρ_3.

Hence, from the law of conservation of mass,

$$M_3 = M_1 + M_2 \tag{6.9}$$

and

$$V_3 = V_1 + V_2 \tag{6.10}$$

Equation (6.10) is only true for solids that do not dissolve in water. For soluble materials, Equation (6.10) does not apply until the liquid phase is saturated; thereafter, the addition of solids results in a proportional increase in volume.

From the basic definition of density and Equations (6.9) and (6.10), we obtain

$$\rho_3 = \frac{M_3}{V_3} = \frac{M_1 + M_2}{V_1 + V_2} \tag{6.11}$$

Keeping the term M_2 unchanged, and manipulating other terms, we arrive at

$$\rho_3 = \frac{V_1\rho_1 + M_2}{V_1 + \dfrac{M_2}{\rho_2}}$$

$$\rho_3 = \frac{\rho_1 + \dfrac{M_2}{V_1}}{1 + \left(\dfrac{M_2}{V_1}\right)\dfrac{1}{\rho_2}} \tag{6.11a}$$

The term M_2/V_1 refers to the mass of weighting material, normally barite, per volume of original mud. In practice, calculations are based on the mass of barite in lbm (or kg) per bbl (or m^3) of original mud.

Let

$$X = \frac{M_2}{V_1} \text{ lbm/bbl (kg/m}^3\text{)} \tag{6.12}$$

where X = mass of weighting material per unit volume. Substituting Equation (6.12) in Equation (6.11a) gives:

$$\rho_3 = \frac{\rho_1 + X}{1 + \dfrac{X}{\rho_2}}$$

Simplifying further,

$$X = \frac{\rho_2(\rho_3 - \rho_1)}{\rho_2 - \rho_3} = \frac{\rho_3 - \rho_1}{1 - \left(\dfrac{\rho_3}{\rho_2}\right)} \tag{6.13}$$

where ρ_2 is the density of added solids, e.g. barite or clay.

Equation (6.13) forms the basis for most mud calculations involving the determination of the mass of solids required to increase mud weight from ρ_1 to ρ_3. Since initial and final mud densities are known, the value of X can be easily determined. Equation (6.13) will be expressed in field units for the case of barite (density = 35.5 ppg or 4250 kg/m^3) only. For other types of solid Equation (6.13) can be easily modified by following the same procedure as outlined below.

Imperial units

From Equation (6.13),

$$X = \frac{\rho_3 - \rho_1}{1 - \left(\dfrac{\rho_3}{\rho_2}\right)}$$

Most calculations in Imperial units are made on the basis of the number of pounds of barite per barrel of initial mud.

$$X = \frac{M_2}{V_1} = \frac{M_2}{1\ \text{bbl}} = \frac{M_2}{42\ \text{gal}}$$

$$X = \frac{M_2}{42} = \frac{\rho_3 - \rho_1}{1 - \dfrac{\rho_3}{35.5}}$$

$$M_2 = \frac{42 \times 35.5(\rho_3 - \rho_1)}{(35.5 - \rho_3)} = 1491\left(\frac{\rho_3 - \rho_1}{35.5 - \rho_3}\right) \tag{6.14}$$

where ρ_3 and ρ_1 are in ppg and M_2 is in pounds.

Since each sack of barite contains 94 lbm, Equation (6.14) becomes

$$S = \frac{15.9(\rho_3 - \rho_1)}{(35.5 - \rho_3)} \tag{6.15}$$

where S is the number of sacks.

Metric version of Equation (6.13)

For an initial volume of 10 m^3, Equations (6.12) and (6.13) become

$$X = \frac{M_2}{V_1} = \frac{M_2}{10} \tag{6.12a}$$

Therefore,

$$\frac{M_2}{10} = \frac{\rho_3 - \rho_1}{1 - \dfrac{\rho_3}{4250}} \tag{6.13a}$$

or

$$M_2 = \frac{42\,500(\rho_3 - \rho_1)}{4250 - \rho_3} \tag{6.16}$$

where M_2 is the mass in kg of barite required to bring about a change in density from ρ_1 to ρ_3 (kg/m^3) and ρ_2 is density of barite = 4250 kg/m^3. Since each sack contains 42.64 kg,

$$\text{number of sacks} = \frac{M_2}{42.64}$$

$$= \frac{996.7(\rho_3 - \rho_1)}{4250 - \rho_3} \tag{6.17}$$

In Equation (6.17) ρ_1 and ρ_3 are in kg/m^3.

When densities are given in kg/l, Equation (6.16) becomes

$$M_2(\text{kg}) = 10\ \text{m}^3 \times \frac{1000\ \text{l}}{\text{m}^3}\ \frac{\rho_3 - \rho_1}{1 - \dfrac{\rho_3}{4.25}}$$

where ρ_3 and ρ_1 are in kg/l and 4.25 is the density of barite (kg/l).

$$M_2 = 10\,000 \times 4.25\left(\frac{\rho_3 - \rho_1}{4.25 - \rho_3}\right)$$

or

$$M_2 = 42\,500\frac{(\rho_3 - \rho_1)}{(4.25 - \rho_3)}\ \text{kg} \tag{6.18}$$

In sacks

$$S = 996.7\frac{(\rho_3 - \rho_1)}{(4.25 - \rho_3)} \tag{6.19}$$

Pit volume increase resulting from the addition of barite

Since

$$\text{volume} = \frac{\text{mass}}{\text{density}}$$

volume increase due to addition of barite = volume of added solids

$$= V_2 = \frac{M_2}{\rho_2} \tag{6.20}$$

Recall

$$X = \frac{M_2}{V_1} = \frac{\rho_3 - \rho_1}{1 - \left(\dfrac{\rho_3}{\rho_2}\right)} \tag{6.13}$$

or

$$M_2 = \frac{V_1\rho_2(\rho_3 - \rho_1)}{(\rho_2 - \rho_3)} \tag{6.21}$$

Combining Equations (6.20) and (6.21) yields

$$V_2 = \frac{V_1\rho_2(\rho_3 - \rho_1)}{(\rho_2 - \rho_1)} \div \rho_2$$

or

$$V_2 = \frac{V_1(\rho_3 - \rho_1)}{(\rho_2 - \rho_3)} \tag{6.22}$$

In metric units For 10 m^3 of initial volume Equation (6.22) becomes

$$V_2 = \frac{10(\rho_3 - \rho_1)}{(4250 - \rho_3)} \tag{6.23}$$

In Imperial units For an initial volume of 1 bbl, Equation (6.22) becomes

$$V_2 = \frac{42(\rho_3 - \rho_1)}{(35.5 - \rho_3)} \tag{6.24}$$

A summary of the above equations can be found in Tables 6.3 and 6.4.

Example 6.2

Determine the quantity of barite required to change the density of mud from 1.5 kg/l (12.53 ppg) to 2 kg/l (16.7 ppg). Calculate the increase in pit volume due to the addition of such a quantity of barite for an initial mud volume of 10 m^3 (63 bbl).

Solution

Metric units (based on an initial mud volume of 10 m^3)

$$M_2 = \frac{42\,500(\rho_3 - \rho_1)}{(4.250 - \rho_3)}$$

where ρ_3 and ρ_1 are in kg/l.

$$M_2 = \frac{42\,500(2 - 1.5)}{(4.25 - 2)} = 9444 \text{ kg}$$

Number of sacks

$$= \frac{9444}{42.64}$$

$$= 222$$

Pit volume increase

$$= \frac{9444 \text{ kg}}{4250 \text{ kg/m}^3}$$

$$= 2.222 \text{ m}^3 (2222 \text{ l})$$

Imperial units (based on an initial volume of 1 bbl, hence all results multiplied by 63)

$$M_2 = 1491\left(\frac{\rho_3 - \rho_1}{35.5 - \rho_3}\right) \times 63$$

$$= 1491\left(\frac{16.7 - 12.53}{35.5 - 16.7}\right) \times 63$$

$$M_2 = 20\,835 \text{ lb}$$

Number of sacks

$$= \frac{20\,835}{94}$$

$$= 222$$

Pit volume increase

$$= \frac{20\,835 \text{ lb}}{35.5 \text{ lb/gal}}$$

$$= 587 \text{ gal}$$

$$= 14 \text{ bbl}$$

Pit volume increase may also be obtained by use of Equation (6.23) or Equation (6.24).

Example 6.3

Determine the density of a water-base mud containing 5% bentonite by weight. The density of bentonite is 2500 kg/m^3 (20.8 ppg).

Solution

Recall Equation (6.11):

$$\rho_3 = \frac{M_1 + M_2}{V_1 + V_2}$$

In metric units:

$$M_1 = \text{mass of water} = 95\%, \text{ say } 95 \text{ kg}$$

$$M_2 = \text{mass of bentonite} = 5\% \text{ or } 5 \text{ kg}$$

$$V_1 = \text{volume of water} = \frac{M_1}{\rho_1} = \frac{95 \text{ kg}}{1000 \text{ kg/m}^3} \text{ m}^3$$

$$V_2 = \text{volume of bentonite} = \frac{M_2}{\rho_2} = \frac{5}{2500} \text{ m}^3$$

Therefore,

$$\rho_3(\text{kg/m}^3) = \frac{95 + 5}{\frac{95}{1000} + \frac{5}{2500}} = 1031 \text{ kg/m}^3$$

In Imperial units:

$$\rho_3 = \frac{95 \text{ lbm} + 5 \text{ lbm}}{\frac{95 \text{ lbm}}{8.33\left(\frac{\text{lbm}}{\text{gal}}\right)} + \frac{5 \text{ lbm}}{20.8\left(\frac{\text{lbm}}{\text{gal}}\right)}} = 8.59 \text{ lbm/gal}$$

where the density of water = 8.33 lbm/gal.

Mud weight decrease calculations

Mud weight may be decreased by the addition of water or diesel oil, or by aerating the mud. This section will be devoted to the calculation of the volume of liquid required to bring about a reduction in density from ρ_1 to ρ_3.

Let

$$X = \frac{\text{mass of water (or oil) required}}{\text{initial volume of mud}}$$

$$= \frac{M_2}{V_1} = \frac{\rho_2 V_2}{V_1}$$

Hence, Equation (6.13) becomes

$$X = \frac{\rho_2 V_2}{V_1} = \frac{\rho_3 - \rho_1}{1 - \left(\frac{\rho_3}{\rho_2}\right)}$$

or

$$V_2 = \frac{V_1 \rho_2}{\rho_2}\left(\frac{\rho_3 - \rho_1}{\rho_2 - \rho_3}\right)$$

$$V_2 = V_1 \frac{(\rho_3 - \rho_1)}{(\rho_2 - \rho_3)} \qquad (6.25)$$

Note that Equation (6.25) is similar to Equation (6.22) derived for pit volume increase calculation.

If water is added, $\rho_2 = 1000$ kg/m^3, then Equation (6.25) becomes, in metric units,

$$V_w = V_1 \frac{(\rho_3 - \rho_1)}{(1000 - \rho_3)}$$

where V_w is the volume of water required for density reduction. Since $\rho_1 > \rho_3$, this equation is conveniently written as

$$V_w = V_1 \frac{(\rho_1 - \rho_3)}{(\rho_3 - 1000)} \qquad (6.26)$$

where V_1 is the initial volume of mud.

Another version of Equation (6.26) which may be found in the literature is

$$\rho_3 = \frac{1 + \frac{1000 V_w}{V_1}}{1 + \frac{V_w}{V_1}} \qquad (6.27)$$

In Imperial units, Equation (6.25) becomes

$$V_w = V_1(\text{bbl}) \frac{\rho_3 - \rho_1}{\rho_2 - \rho_3} \qquad (6.28)$$

or

$$V_w = V_1 \left(\frac{\rho_3 - \rho_1}{8.33 - \rho_3}\right) \qquad (6.29)$$

where ρ_3 and ρ_1 are in ppg.

Variations of Equation (6.25)

From Equation (6.25), the following equations may be derived:

$$\frac{V_2}{V_1} = \frac{(\rho_1 - \rho_3)}{(\rho_3 - \rho_2)} \qquad (6.30)$$

Also, since $V_1 = V_3 - V_2$, Equation (6.30) becomes

$$\frac{V_2}{V_3 - V_2} = \frac{(\rho_1 - \rho_3)}{(\rho_3 - \rho_2)}$$

$$V_2(\rho_3 - \rho_2 + \rho_1 - \rho_3) = V_3(\rho_1 - \rho_3)$$

or

$$\frac{V_2}{V_3} = \frac{(\rho_1 - \rho_3)}{(\rho_1 - \rho_2)} \qquad (6.31)$$

Also, since $V_2 = V_3 - V_1$, Equation (6.30) becomes

$$\frac{V_1}{V_3} = \frac{(\rho_3 - \rho_2)}{(\rho_1 - \rho_2)} \qquad (6.32)$$

In practice, only Equation (6.25) should be used, and modifications may be made to the final results if required.

All equations required for mud weight decrease calculations are given in Tables 6.3 and 6.4.

Example 6.4

It is required to reduce mud weight from 25.1 ppg (3 kg/l) to 22.6 ppg (2.7 kg/l) in order to combat a lost circulation problem. Calculate the volumes of water and oil required to bring about this reduction. Also, if oil is used, what is the percentage of oil in mud if the initial volume of mud is 629 bbl (100 m^3). The density of oil is 6.87 ppg (823 kg/m^3).

Solution

In metric units:

From Equation (6.25),

$$V_2 = V_1 \frac{(\rho_3 - \rho_1)}{(\rho_2 - \rho_3)} = V_1 \frac{(\rho_1 - \rho_3)}{(\rho_3 - \rho_2)}$$

(a) Water:

$$\rho_2 = 1000 \text{ kg/m}^3,\ \rho_3 = 2.7 \text{ kg/l} = 2700 \text{ kg/m}^3$$

$$\rho_1 = 3000 \text{ kg/m}^3,\ V_1 = 100 \text{ m}^3$$

Therefore,

$$V_w = \frac{100(3000 - 2700)}{(2700 - 1000)} = 17.65 \text{ m}^3$$

TABLE 6.3 Equations for mud calculations: Imperial system

(1) To increase density from ρ_1 to ρ_3 per 1 bbl of initial mud volume.

(a) Mass of barite required is

$$M = 1491\frac{(\rho_3 - \rho_1)}{(35.5 - \rho_3)}$$

where mud densities are in ppg, where ppg = pcf/7.48

(b) Number of sacks of barite, S, is

$$S = \frac{15.9(\rho_3 - \rho_1)}{(35.5 - \rho_3)}$$

(c) Increase in pit volume in bbl is

$$V = \frac{42(\rho_3 - \rho_1)}{(35.5 - \rho_3)}$$

(2) To reduce mud weight from ρ_1 to ρ_3

(a) Volume of liquid (water or diesel oil), V_2, in bbl required is

$$V_2 = V_1\frac{(\rho_3 - \rho_1)}{(\rho_2 - \rho_3)}$$

where V_1 = initial volume of mud; ρ_1 = initial mud density; ρ_2 = density of added liquid; and ρ_3 = final required density. Other versions of the above equation are:

$$V_1 = V_3\left(\frac{\rho_3 - \rho_2}{\rho_1 - \rho_2}\right)$$

and

$$V_2 = V_3\left(\frac{\rho_1 - \rho_3}{\rho_1 - \rho_2}\right)$$

(b) Final mud density

$$\rho_3 = \rho_1 - \frac{V_2}{V_3}(\rho_1 - \rho_2)$$

or

$$\rho_3 = \rho_1 - \left(\frac{V_2}{V_1 + V_2}\right)(\rho_1 - \rho_2)$$

TABLE 6.4 Equations for mud calculations: metric system

(1) To increase density from ρ_1 to ρ_3 for 10 m³ of initial mud volume

(a) Mass of barite required

$$M = \frac{42\,500(\rho_3 - \rho_1)}{(4250 - \rho_3)}\ \text{kg}$$

densities are in kg/m^3

(b) Number of sacks of barite

$$S = \frac{996.7(\rho_3 - \rho_1)}{(4250 - \rho_3)}$$

(c) Increase in pit volume

$$V = \frac{10(\rho_3 - \rho_1)}{(4250 - \rho_3)}\ \text{m}^3$$

(2) To reduce mud weight from ρ_1 to ρ_3

(a) Volume of liquid (water or oil) required:

$$V_2 = V_3\frac{(\rho_1 - \rho_3)}{(\rho_1 - \rho_2)}\ \text{m}^3$$

where V_3 is final mud volume or

$$V_2 = V_1\frac{(\rho_1 - \rho_3)}{(\rho_3 - \rho_2)}$$

where V_1 is the initial mud volume.

(b) Final mud density

$$\rho_3 = \rho_1 - \frac{V_2}{V_3}(\rho_1 - \rho_2)$$

$$\rho_3 = \rho_1 - \frac{V_2}{V_1 + V_2}(\rho_1 - \rho_2)$$

(b) Oil:

$$\rho_2 = 823\ \text{kg/m}^3\ (6.87\ \text{ppg})$$

$$V_{\text{oil}} = \frac{100(3000 - 2700)}{(2700 - 823)} = 15.98\ \text{m}^3$$

$$\%\ \text{oil in mud} = \frac{\text{volume of oil}}{\text{total volume of new mud}}$$

$$= \left(\frac{15.98}{100 + 15.98}\right) \times 100$$

$$= 13.8$$

In Imperial units:

(a) Water:

$$\rho_2 = 8.33\ \text{ppg}$$

$$\rho_3 = 22.6\ \text{ppg}$$

$$\rho_1 = 25.1\ \text{ppg}$$

$$V_1 = 629\ \text{bbl}$$

$$V_w = 629\left(\frac{25.1 - 22.6}{22.6 - 8.33}\right)$$

$$= 110\ \text{bbl}$$

(b) Oil:

$$V_{\text{oil}} = 629\left(\frac{25.1 - 22.6}{22.6 - 6.87}\right)$$

$$= 99.99\ \text{bbl}$$

$$\simeq 100\ \text{bbl}$$

$$\% \text{ oil in mud} = \frac{\text{volume of oil}}{\text{total volume of new mud}}$$
$$= \left(\frac{100}{629 + 100}\right) \times 100$$
$$= 13.7$$

(*Note:* 1 US bbl $= 0.159\ m^3$.)

MUD CONTAMINANTS

Drilling mud can be contaminated by different types of materials during drilling and cementing operations. The following is a summary of the most common types of contaminant.

Sodium chloride (common salt)

Sodium chloride enters the mud system during the drilling of salt domes, rock salt beds, evaporites or any bed containing salt water. Contamination of bentonite mud by salt leads to increased viscosity, yield point and gel strength, due mainly to the flocculation effect produced by the presence of high concentrations of sodium ions. A decrease in pH is also associated with increased salt content.

When freshwater mud comes into contact with a salt section (evaporites, etc.), it dissolves enough salt to reach saturation or near-saturation point. This process can lead to hole enlargement or even a complete washout if the salt section is near the surface. Hence, when drilling a salt section, it is customary to saturate the liquid phase with salt prior to drilling to prevent hole washout.

Routine measurements of mud chloride content are used to monitor the changes in mud salinity. A sudden increase in chloride content is a clear indication of salt water inflow and a possible kick. The properties of salt-contaminated mud are adjusted by the addition of thinners, to reduce viscosity and filtration rate, and caustic soda, to increase pH values.

Anhydrite and gypsum

Anhydrite and gypsum are basically forms of calcium sulphate ($CaSO_4$). Gypsum contains water and has the chemical formula $CaSO_4 \cdot 2H_2O$, while anhydrite contains no water and is described chemically as $CaSO_4$. Formations containing anhydrite and gypsum occur in nature as massive rocks or as thin interbedded sections within shale or limestone formations.

When anhydrite and gypsum come into contact with water they ionize as Ca^{2+} and SO_4^{2-}. The presence of high concentrations of calcium ions leads to clay flocculation and, in turn, increases in plastic viscosity, yield point, filtrate loss and gel strength. Flocculation is a result of replacement of Na^+ in the clay sheets by the more tightly bound calcium ions which reduce the degree of plate separation. Normally, caustic soda and chrome lignosulphonate are added to mud used to drill gypsum and anhydrite sections to reduce flocculation.

Anhydrite- or gypsum-contaminated mud may be treated by the addition of sodium carbonate (Na_2CO_3). This chemical removes the excess calcium ions, as shown by the following chemical equation:

$$Na_2CO_3 + CaSO_4 \rightarrow CaCO_3 + Na_2SO_4$$

The calcium carbonate is precipitated as an insoluble material.

Cement

Cement can enter the mud system from poor cement jobs or squeeze cementing. The effect of adding cement to mud is to increase viscosity, yield point, gel strength and pH.

It is normal to discard mud used to drill cement inside casing or when drilling the float collar and casing shoe. Mud contaminated by cement is normally treated by adding sodium bicarbonate ($NaHCO_3$), which precipitates calcium carbonate as an insoluble material, as shown below:

$$NaHCO_3 + Ca(OH)_2 \rightarrow CaCO_3 + NaOH + H_2O$$

The calcium content should always be kept to below 200 ppm so that the properties of mud can be maintained. As previously mentioned, the calcium cations (Ca^{2+}) prevent the separation of the clay plates, and promote flocculation and increased viscosity.

Other contaminants include soluble gases and drilled solids.

MUD CONDITIONING EQUIPMENT

Mud is prepared by mixing water (or oil) with bentonite (or attapulgite), barite and various other chemicals (see page 108) to prepare a fluid with suitable suspension and carrying properties. During the course of drilling, mud disperses and carries drill cuttings which change its original properties. Unless the drilled solids are removed, mud will lose much of its desired properties and can cause potential problems such as lost circulation.

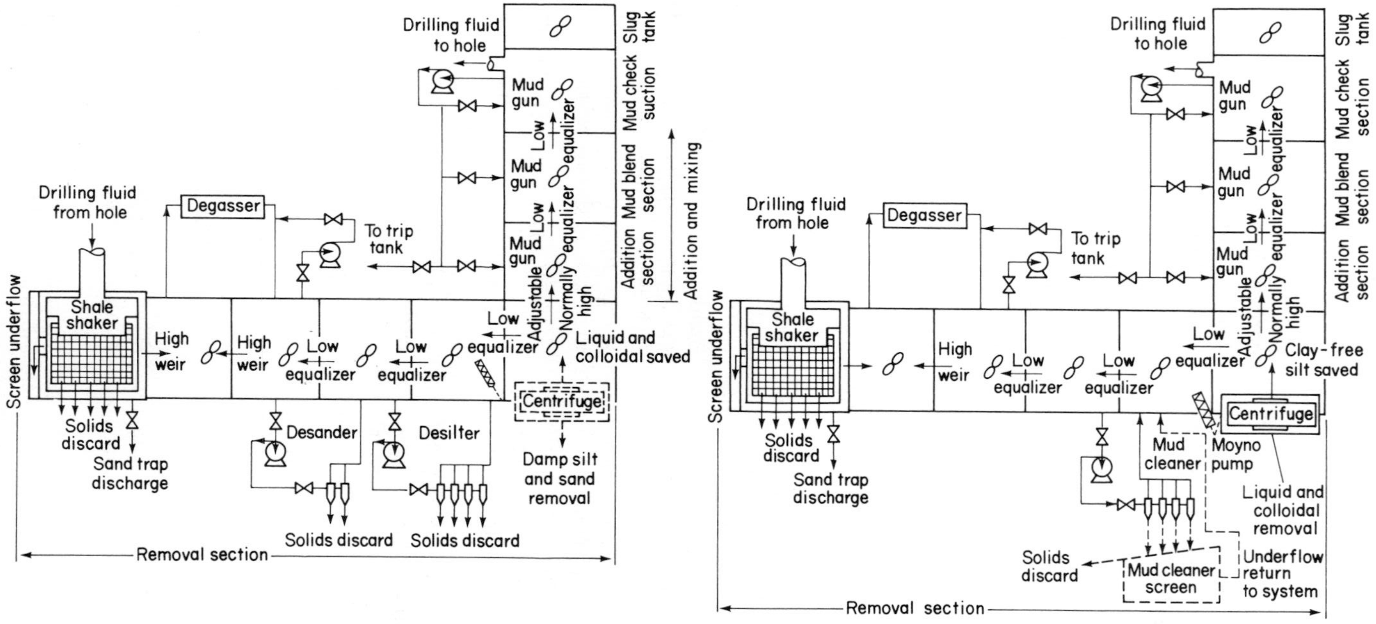

Fig. 6.10. Mud conditioning equipment. (a) Unweighted mud. Schematic of a complete removal system arranged for unweighted mud, showing relation of various components to one another. Other important sections shown include addition and suction sections. (b) Weighted mud. In a removal section arranged for weighted muds the degasser, mud cleaner and centrifuge are used as needed. (After Young & Robinson; courtesy of *World Oil*)[13]

The mud conditioning equipment aims at processing the mud returning from hole by removing unwanted solids and, when necessary, adding solids and chemicals. Figure 6.10 gives a schematic drawing of a complete set of mud conditioning equipment for both non-weighted and weighted (i.e. with barite) mud[13]. (See also Figure 7.1 for a schematic drawing of the circulating system.)

The components of a mud conditioning system can be conveniently divided into three sections: (1) the suction tank, which contains the conditioned mud which will be pumped down the hole; (2) the addition and mixing section; and (3) the removal section.

The addition and mixing section

The mixing operation involves the pouring of mud solids or chemicals through a hopper connected to a high-shear jet, as shown in Figure 6.11. The high-shear jet ensures that a homogeneous mixture is obtained. The resulting mud is, again, vigorously agitated, with a mud gun or mechanical agitator, as shown in Figure 6.12. The resulting mud is then directed to the suction tank, where it is first handled by centrifugal or charge pumps. The charge pumps give the mud a pressure of 80–90 psi (5.5–6.2 bar) before it is delivered to the main rig pumps. Practical experience has shown that the efficiency of the main pumps improves when they receive fluid under pressure.

From the rig pumps, the mud then flows through a standpipe, rotary hose, kelly, drillpipe and drill collars and, finally, to the bit. Drilling mud issuing from the bit nozzles will then carry the drill-cuttings up the hole and finally to the shale shakers.

The removal section

The removal section consists of: (a) shale shakers; (b) desanders and desilters (i.e. hydrocyclones); (c) mud cleaners; and (d) centrifuges.

The shale shaker

Mud returning from the hole passes over (a) shale shaker(s), where drill cuttings are separated from the mud. A shale shaker consists of vibrating (or rotating) sieves with openings that are large enough to pass the mud and its solids (e.g. bentonite and barite) but small enough to retain the cuttings. Figure 6.13 shows a single- and a double-decker shale shaker. The cuttings are collected in a dump pit, where they are discarded, or they may be used by the geologist for formation analysis (see Figure 6.14).

The mud passing through the shale shaker falls (Figure 6.14) into a sloping pit (the sand trap) just below the shale shaker, where remaining small drill cuttings are removed by gravity settlement. The sand trap has dumping valves which are opened periodically to clean out trapped solids.

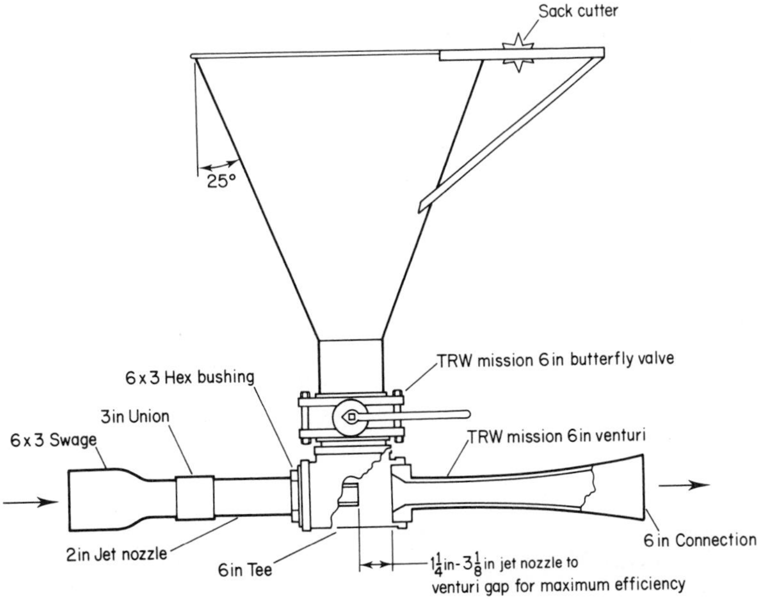

Fig. 6.11. Conventional mud mixer[14]. (Courtesy of Geolograph Pioneer)

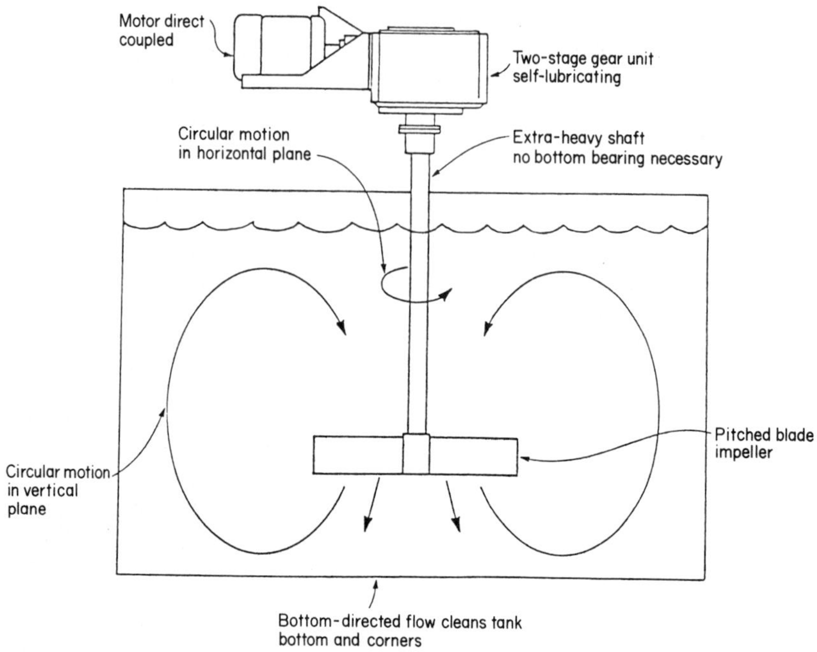

Fig. 6.12. Pit bull agitator[14]. (Courtesy of Geolograph Pioneer)

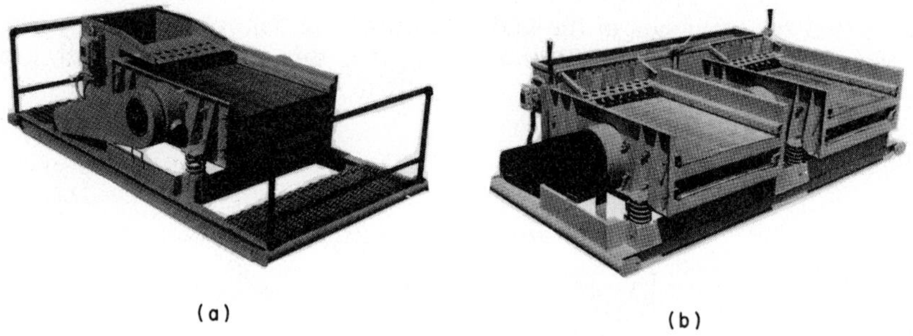

Fig. 6.13. Shale shakers[1]: (a) screening machine; (b) tandem screening machine. (Courtesy of Baroid/NL)

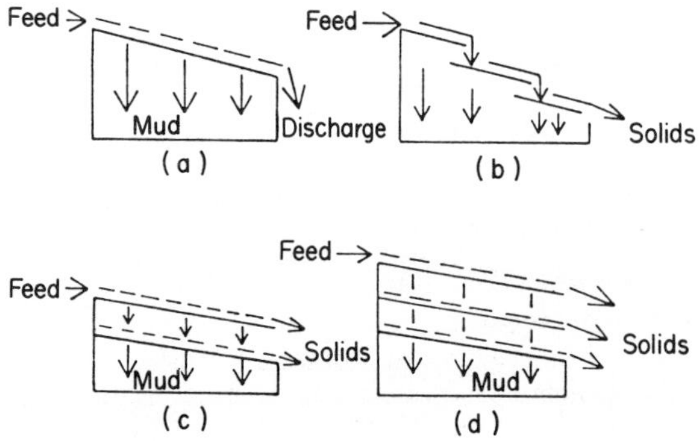

Fig. 6.14. Schematic of different types of shale shaker[11].

Desanders and desilters

Mud returning from the shale shaker is sent to a set of desanders and desilters by a small centrifugal pump. The desanders and desilters are hydrocyclones which operate on the principle that solids can be separated from fluid by utilising centrifugal forces (see Figure 6.15).

A hydrocyclone has no moving internal parts, but consists simply of:

(a) An upper cylindrical section with a tangential feed tube. The cylindrical section is fitted with a vortex finder pipe ending below the tangential feed tube. The

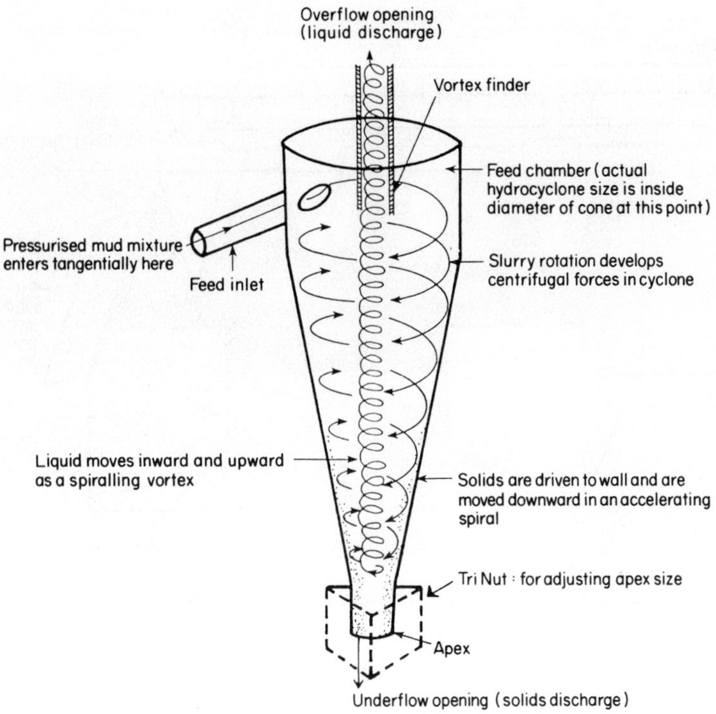

Fig. 6.15. Schematic of hydrocyclone[11]. (Courtesy of IMCO)

clean mud flows out from the vortex finder pipe.
(b) A conical section ending with an apex through which the solids (underflow) discharge.

Mud is injected tangentially to the hydrocyclone and the resulting centrifugal force drives the solids to the walls of the hydrocyclone and, finally, discharges them from the apex with a small volume of mud. The fluid portion of mud leaves the top of the cyclone as an overflow which is sent to the active pit to be pumped downhole again.

Hydrocyclones are arranged so that the large, sandy particles (i.e. above 74 μm) are first removed. This set of hydrocyclones is described as a desander. From the desander, the mud is directed to a desilter, where the small-sized units (2–74 μm) are removed. Since the barite size is below 74 μm, it follows that desilters cannot be used with weighted muds.

The operating size range for any hydrocyclone is dependent upon the internal diameter of the cone. Figure 6.16 gives the equivalent size of mud removed for various cyclone sizes: a desander would have a size of 6–12 in and a desilter would have a size of 4–5 in.

Mud cleaner

When weighted muds are used, desanders and desilters are replaced by a mud cleaner in order to save the barite (see Figure 6.10). Basically, a mud cleaner

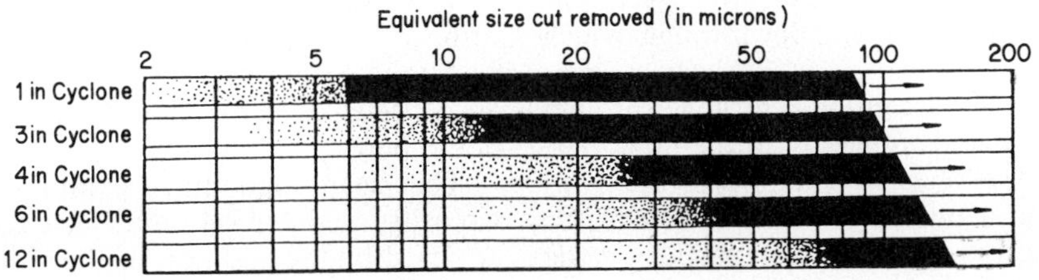

Fig. 6.16. Equivalent size cut removed for various cyclone sizes[11]. (Courtesy of IMCO)

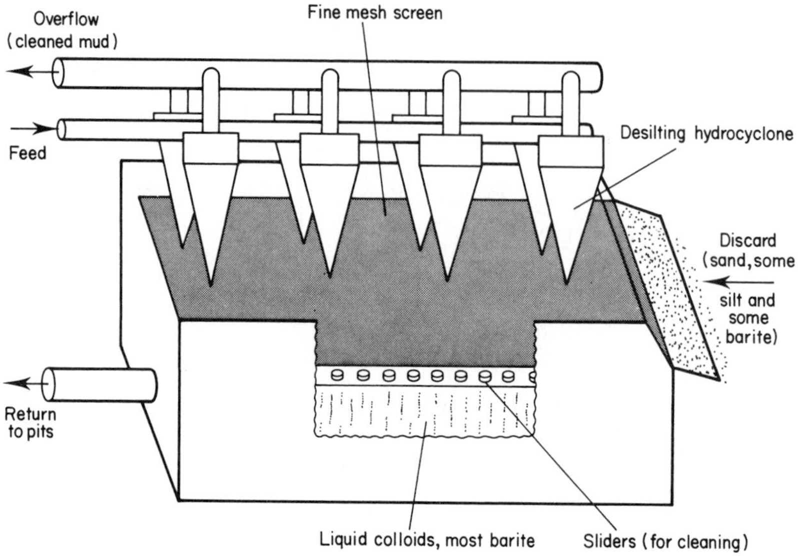

Fig. 6.17. Schematic of mud cleaner[11]. (Courtesy of IMCO)

consists[11] of a battery of hydro cyclones of 4 in ID placed above a high-energy vibrating screen with openings of size 40–125 μm (see Figure 6.17). In this arrangement, the hydrocyclone removes sandy and barite particles as an underflow which falls onto the vibrating screen. The barite will pass through the screen and is re-used, while the sandy particles are discarded.

This system has the advantage of saving the valuable components, such as barite, KCl, oil and mud[11]. Also, the discarded waste is much drier, thereby posing fewer disposal problems.

Centrifuges

Centrifuges also use centrifugal forces to separate heavy solids from the liquid and lighter components of the mud. A decanting centrifuge consists of a horizontal conical steel bowl, rotating at high speed (as shown in Figure 6.18). The bowl contains a double-screw-type conveyor which rotates in the same direction as the steel bowl, but at a slightly lower speed. The conveyor contains a hollow spindle through which mud is fed in.

Mud feed enters the decanting centrifuge through the hollow axle and is distributed[14] to the bowl. The centrifugal force developed by rotation of the bowl holds the slurry in a pond against the walls of the bowl. In this pond the silt and sandy particles settle against the walls and the conveyor blade scrapes or pushes the settled solids along towards the narrow end of the bowl, where they are collected as damp particles with no free liquid. The liquid and the clay particles (<2 μm) which cannot be separated by the centrifuge are collected as an overflow from ports at the large end of the bowl.

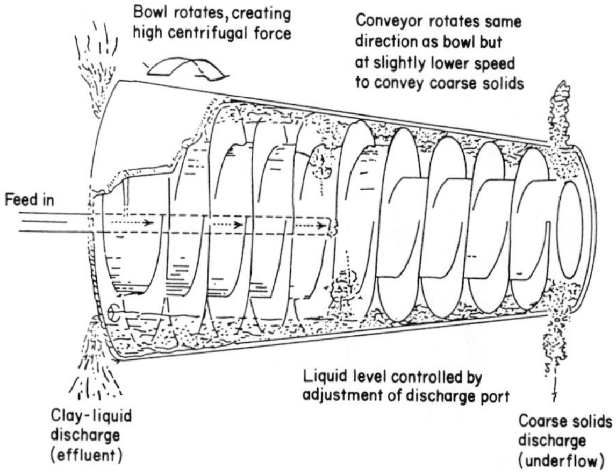

Fig. 6.18. Cross-section of a decanting centrifuge[11]. (Courtesy of IMCO)

The clean mud may require some treatment if viscosity, yield and gel values have changed, and this is achieved by the addition of chemicals, bentonite, etc. Conditioned mud is then picked up by the charge pumps and rig pumps to resume its journey downhole.

References

1. Baroid/NL Industries Inc. (1979). *Manual of Drilling Fluids Technology*. Baroid/NL Industries Publications.
2. Schlumberger (1972). *Log Interpretation, Volume 2: Principles*. Schlumberger Publications.
3. Rogers, F. (1963). *Composition and Properties of Oilwell Drilling Fluids*. Gulf Publishing Company, Houston.
4. Chillingarian, G. and Vorabutr, P. (1981). *Drilling and Drilling Fluids*. Elsevier Scientific, Amsterdam.
5. Van Olphen, H. (1963). *An Introduction to Colloid Chemistry*. Interscience, New York, London.
6. API RP 29 (1962). *Suggested Procedures for Laboratory Evaluation of Drilling Mud Properties*, API Production Department.
7. API Bull 13D (1980). *The Rheology of Oilwell Drilling Fluids*. API Production Department.
8. Borst, R. L. and Keller, W. D. (1969). Scanning electron microscope of API reference clay minerals and other selected samples. *Int. Clay Conf.*, Tokyo, Japan, pp. 871–901.
9. Deily, F., Holman, W., Lindblom, G. and Patton, J. (1967). New low-solids polymer mud designed to cut well costs. *World Oil*, July, 77–80.
10. Kelly, J. (1983). Drilling fluid selection, performance and quality control. *JPT*, May, 889–898.
11. IMCO (1979). *Applied Mud Technology* (Manual). IMCO Services Publications.
12. API RP13B (1982). *Standard Procedures for Testing Drilling Fluids*. API, Ninth Edition, May.
13. Young, G. and Robinson, L. (1982). How to design a mud system for optimum solids removal. *World Oil*, Sept., Oct., Nov.
14. Geolograph (1971). *Geolograph Pioneer Manual*.

Problems

1. Show that the mass of clay per bbl (m^3) of initial mud required to increase mud density from ρ_1 to ρ_3 is given by

$$M = 873.6\left(\frac{\rho_3 - \rho_1}{20.8 - \rho_3}\right)\text{lbm}$$

or

$$M = 2490\left(\frac{\rho_3 - \rho_1}{2490 - \rho_3}\right)\text{kg}$$

where ρ_2 = density of clay = 20.8 lbm/gal (2490 kg/m^3).

2. An 8% water–bentonite mud is to be weighted by adding barite so that the final mud weight is 75 pcf (1200 kg/m^3). Calculate the quantity of barite (in sacks) required to prepare 943 bbl (150 m^3) of this weighted mud.

Owing to lost circulation problems, a reduction in mud weight from 75 pcf to 56 pcf (1200 kg/m^3 to 900 kg/m^3) is required. Determine the volume of oil (specific gravity 0.8) required to bring about this reduction and the percentage of oil in mud.

Answer: 784 sacks; 2830 bbl; 75%.

3. Determine the volume of an 8% water–bentonite mud required to be mixed with barite to prepare 943 bbl of mud having a density of 1200 kg/m^3. How much bentonite was initially used?

Answer: 899 bbl; 272 sacks.

4. How many sacks of barite are required to raise the mud weight of 755 bbl from 77 pcf to 92 pcf? Calculate the new mud density when 126 bbl of oil is added to the new system. Determine the quantity of barite required to maintain a density of 92 pcf.

Specific gravity of oil = 0.8

Answer: 1038 sacks; 87 pcf; 346 sacks.

Chapter 7

Rig Hydraulics

INTRODUCTION

Proper utilisation of hydraulic energy at the drill bit and the determination of pressure losses at the various parts of the drill string are some of the topics which could be discussed under the heading, 'Rig Hydraulics'. Several models exist for the calculation of pressure losses in pipes and annuli. Each model is based on a set of assumptions which cannot be completely fulfilled in any drilling situation.

This chapter will cover the following topics:

Pressure losses
Surface connection losses
Pipe and annular losses
Pressure drop across bit
Optimisation of bit hydraulics
Surface pump pressure
Hydraulic criteria
Comparison of BHHP and IF criteria
Nozzle selection
Optimum flow rate
Field method of optimising bit hydraulics
A practical check on the efficiency of the bit hydraulics programme

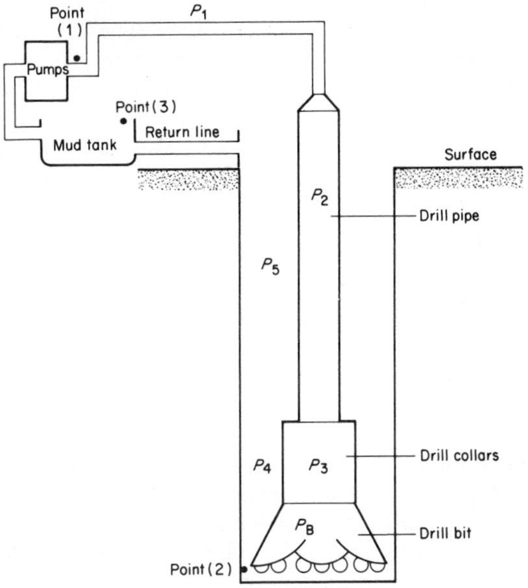

Fig. 7.1. Schematic drawing of the circulating system. Points (1) and (3) are assumed to be at the same level.

PRESSURE LOSSES

Consider the schematic drawing in Figure 7.1, where the various parts of the drill string, drill bit and surface connections are included. Owing to frictional forces, the circulating system will lose energy when fluid is pumped from point (1) to point (2) and back to point (3) in the mud tank. Therefore, the first objective of rig hydraulics is to calculate pressure losses resulting from frictional forces in each part of the circulating system.

Such pressure losses in the various parts of the circulating system can conveniently be discussed under four headings: (1) surface connection losses; (2) pipe losses; (3) annular losses; and (4) losses across bit. These pressure losses depend upon the type of fluid used and the type of flow in the circulating system.

SURFACE CONNECTION LOSSES (P_1)

Pressure losses in surface connections (P_1) are those taking place in standpipe, rotary hose, swivel and kelly. The task of estimating surface pressure losses is complicated by the fact that such losses are dependent on the dimensions and geometries of surface connections. These dimensions can vary with time, owing to continuous wear of surfaces by the drilling fluids. The following general equation may be used to evaluate pressure losses in surface connections:

$$P_1 = E\rho^{0.8}Q^{1.8}(\mathrm{PV})^{0.2} \text{ psi} \tag{7.1}$$

or

$$P_1 = E\rho^{0.8}Q^{1.8}(\mathrm{PV})^{0.2} \text{ bar} \tag{7.1a}$$

where ρ = mud weight (lbm/gal, or kg/l); Q = volume flow rate (gpm, or l/min); E = a constant depending on type of surface equipment used; and PV = plastic viscosity (cP).

In practice, there are only four types of surface equipment; each type is characterised by the dimensions of standpipe, kelly, rotary hose and swivel. Table 7.1 summarises the four types of surface equipment.

The values of the constant E in Equations (7.1) and (7.1a) are given in Table 7.2.

PIPE AND ANNULAR LOSSES

Pipe losses take place inside the drillpipe and drill collars, and are designated in Figure 7.1 as P_2 and P_3, respectively. Annular losses take place around the drill collars and drillpipe, and are designated as P_4 and P_5 in Figure 7.1. The magnitudes of P_2, P_3, P_4 and P_5 depend on: (a) dimensions of drillpipe (or drill collars), e.g. inside and outside diameter and length; (b) mud rheological properties, which include mud weight, plastic viscosity and yield point; and (c) type of flow, which may be laminar, plug or turbulent.

It should be noted that the actual behaviour of drilling fluids downhole is not accurately known and fluid properties measured at the surface usually assume different values at bottom hole conditions. Several models exist for the calculation of pressure losses, each producing different values for the same existing conditions.

Only two models will be used here: the Bingham plastic model and the power-law model. In Chapter 5 all the necessary equations were derived and Tables 7.3 and 7.4 summarise these equations.

TABLE 7.2 Values of the constant E

Surface equipment type	*Value of* E	
	Imperial units	*Metric units*
1	2.5×10^{-4}	8.8×10^{-6}
2	9.6×10^{-5}	3.3×10^{-6}
3	5.3×10^{-5}	1.8×10^{-6}
4	4.2×10^{-5}	1.4×10^{-6}

PRESSURE DROP ACROSS BIT

The objective of any hydraulics programme is to optimise pressure drop across the bit such that maximum cleaning of bottom hole is achieved.

For a given length of drill string (drillpipe and drill collars) and given mud properties, pressure losses P_1, P_2, P_3, P_4 and P_5 will remain constant. However, the pressure loss across the bit is greatly influenced by the sizes of nozzles used, and the latter determine the amount of hydraulic horsepower available at the bit. The smaller the nozzle the greater the pressure drop and the greater the nozzle velocity. In some situations where the rock is soft to medium in hardness, the main objective is to provide maximum cleaning and not maximum jetting action. In this case a high flow rate is required with bigger nozzles. These points are discussed later under 'Optimisation of Bit Hydraulics' (page 141).

TABLE 7.1 Four types of surface equipment

Surface equipment type	*Standpipe*		*Rotary hose*		*Swivel*		*Kelly*	
	Length (ft)	*ID (in)*	*Length (ft)*	*ID (in)*	*Length (ft)*	*ID (in)*	*Length (ft)*	*ID (in)*
1	40	3.0	40	2.0	4	2.0	40	2.25
2	40	3.5	55	2.5	5	2.5	40	3.25
3	45	4.0	55	3.0	5	2.5	40	3.25
4	45	4.0	55	3.0	6	3.0	40	4.00

TABLE 7.3 Summary of equations (Imperial units)

Bingham plastic model

Pipe flow

(1) Determine average velocity:

$$\bar{V} = \frac{24.5Q}{D^2}\ \text{ft/min}$$

(2)
$$V_c = \frac{97\,\text{PV} + 97\sqrt{(\text{PV})^2 + 8.2\rho D^2\,\text{YP}}}{\rho D}\ \text{ft/min}$$

(3) If $\bar{V} > V_c$, flow is turbulent; use

$$P = \frac{8.91\times10^{-5}\rho^{0.8}Q^{1.8}(\text{PV})^{0.2}L}{D^{4.8}}\ \text{psi}$$

If $\bar{V} < V_c$, flow is laminar; use

$$P = \frac{L}{300D}\left[\text{YP} + \frac{(\text{PV})\bar{V}}{5D}\right]\ \text{psi}$$

Annular flow

(1) Determine average velocity:

$$\bar{V} = \frac{24.5Q}{D_h^2 - OD_p^2}\ \text{ft/min}$$

(2) Determine V_c from

$$V_c = \frac{97\,\text{PV} + 97\sqrt{(\text{PV})^2 + 6.2\rho D_e^2\,\text{YP}}}{\rho D_e}\ \text{ft/min}$$

where $D_e = D_h - OD_{dp}$ (or OD_{dc})

(3a) If $\bar{V} > V_c$ flow is turbulent; use

$$P = \frac{8.91\times10^{-5}\rho^{0.8}Q^{1.8}(\text{PV})^{0.2}L}{(D_h - OD)^3(D_h + OD)^{1.8}}\ \text{psi}$$

where OD is the outside diameter of drill pipe or drill collars.

(3b) If $\bar{V} < V_c$, flow is laminar; use

$$P = \frac{L\,\text{PV}\,\bar{V}}{60\,000D_e^2} + \frac{L\,\text{YP}}{200D_e}\ \text{psi}$$

where D_e is the annular distance and $\bar{V}$ is the average velocity (ft/min).

Pressure loss across bit

(1) From previous calculations, determine pressure drop across the bit, using

$$P_{bit} = P_{standpipe} - (P_{dp} + P_{dc} + P_{adp} + P_{adc})$$

(2) Determine nozzle velocity (ft/s) from

$$V_n = 33.36\sqrt{\frac{P_{bit}}{\rho}}$$

(3) Determine total area of nozzles from

$$A = 0.32\frac{Q}{V_n}\ \text{in}^2$$

(4) Determine nozzle sizes in multiples of 32 from

$$d_N = \left(\sqrt{\frac{4A}{3\pi}}\right)\times 32$$

Power-law model

Pipe flow

(1) Determine n and K:

$$\theta_{600} = 2\,\text{PV} + \text{YP}\quad \theta_{300} = \text{PV} + Y$$

$$\text{PV} = \theta_{600} - \theta_{300}$$
$$\text{YP} = 2\theta_{300} - \theta_{600}$$

$$n = 3.32\log\left(\frac{\theta_{600}}{\theta_{300}}\right),\quad K = \frac{\theta_{300}}{(511)^n}$$

(2)
$$V_c = \left[\frac{5.82(10^4)K}{\rho}\right]^{1/(2-n)}\cdot\left[\frac{1.6}{D}\,\frac{3n+1}{4n}\right]^{n/(2-n)}\ \text{ft/min}$$

$$\bar{V} = \frac{24.5Q}{D^2}\ \text{ft/min}$$

(3) If $\bar{V} > V_c$ flow is turbulent; use:

$$P = \frac{8.91(10^{-5})\rho^{0.8}Q^{1.8}(\text{PV})^{0.2}L}{D^{4.8}}\ \text{psi}$$

If $\bar{V} < V_c$, flow is laminar; use:

$$P = \left[\frac{1.6\bar{V}}{D}\,\frac{(3n+1)}{4n}\right]^n\frac{KL}{300D}\ \text{psi}$$

Annular flow

(1) Determine n and K as above

(2)
$$V_c = \left[\frac{3.878(10^4)K}{\rho}\right]^{1/(2-n)}\times\left[\frac{2.4}{D_h - OD_p}\left(\frac{2n+1}{3n}\right)\right]^{n/(2-n)}\ \text{ft/min}$$

$$\bar{V} = \frac{24.5Q}{D_h^2 - OD_p^2}$$

(3) If $\bar{V} > V_c$, flow is turbulent; use

$$P = \frac{8.91(10^{-5})\rho^{0.8}Q^{1.8}(\text{PV})^{0.2}L}{(D_h - OD)^3(D_h + OD)^{1.8}}\ \text{psi}$$

If $\bar{V} < V_c$, flow is laminar; use

$$P = \left[\frac{2.4\bar{V}}{(D_h - OD)}\,\frac{(2n+1)}{3n}\right]^n\frac{KL}{300(D_h - OD)}\ \text{psi}$$

Pressure loss across bit

Equations are similar to those of the Bingham plastic model.

Field units in Imperial units:

OD = outside diameter (in)
D = inside diameter (in)
L = length (ft)
ρ = density lbm/gal (ppg)
V = velocity (ft/s)
PV = viscosity (cP)
YP = yield point (lbf/100 ft^2)

TABLE 7.4 Summary of hydraulics equations: metric units

Bingham plastic model

Pipe flow

(1) Determine average velocity, $\bar{V}$:

$$\bar{V} = 21.2(Q/D^2) \text{ m/s}$$

Q in l/min; D in mm.

(2) Determine critical velocity, V_c:

$$V_c = 1.5\left(\frac{\text{PV} + \sqrt{(\text{PV})^2 + 0.1064\rho D^2\,\text{YP}}}{\rho D}\right) \text{m/s}$$

(3a) If $\bar{V} > V_c$, flow is turbulent; use

$$P = \frac{55.5\rho^{0.8} Q^{1.8} (\text{PV})^{0.2} L}{D^{4.8}} \text{ bar}$$

(3b) If $V < V_c$, flow is laminar; use

$$P = \frac{32 \times 10^{-2} L\,\text{PV}\,\bar{V}}{D^2} + \frac{2.55\,\text{YP} \times 10^{-2}}{D} \text{ bars}$$

Annular flow

(1) Determine average velocity

$$\bar{V} = \frac{24.5}{D_h^2 - OD_p^2} \text{ m/s}$$

(2) Determine critical velocity, V_c:

$$V_c = 1.5\left(\frac{\text{PV} + \sqrt{(\text{PV})^2 + 0.079\rho D_e^2\,\text{YP}}}{\rho D_e}\right)$$

where $D_e = D_h = OD_p$

(3a) If $\bar{V} > V_c$, flow is turbulent; use

$$P = \frac{55.5\rho^{0.8} Q^{1.8} (\text{PV})^{0.2} L}{(D_h - OD_p)^3 (D_h + OD_p)^{1.8}} \text{ bars}$$

(3b) If $V_c < \bar{V}$, flow is laminar; use

$$P = \frac{48 \times 10^{-2} L\,\text{PV}\,\bar{V}}{(D_h - OD_p)^2} + \frac{2.87 \times 10^{-2} L\,\text{YP}}{(D_h - OD_p)} \text{ bars}$$

Pressure loss across bit

(1) From previous calculations, determine the available pressure drop across the bit as follows:

$$P_{\text{bit}} = P_{\text{standpipe}} - (P_{\text{dp}} + P_{\text{dc}} + P_{\text{adp}} + P_{\text{adc}})$$

(2) Determine nozzle velocity from

$$V_n = 13.44\sqrt{\frac{P_{\text{bit}}}{\rho}} \text{ m/s}$$

(3) Determine total area of nozzles from

$$A = \frac{100}{6}\frac{Q}{V_n} \text{ mm}^2$$

(4) Determine nozzle sizes in multiples of 32 from

$$d_N = 1.2598\sqrt{\frac{4A}{3\pi}}$$

(*Note:* μ = (PV)/3.2 for fully developed turbulent flow.)

Power-law model

Pipe flow

(1) Determine n and K from

$$n = 3.32 \log \frac{\theta_{600}}{\theta_{300}}, \quad K = \frac{\theta_{300}}{(511)^n}$$

(2) Determine average velocity, V, and critical velocity, V_c, from

$$\bar{V} = 21.2\left(\frac{Q}{d^2}\right) \text{m/s}$$

and

$$V_c = \left[\frac{0.179K}{\rho}\right]^{1/(2-n)} \left[\frac{2000}{D}\left(\frac{3n+1}{n}\right)\right]^{n/(1-n)} \text{m/s}$$

(3a) If $\bar{V} > V_c$, flow is turbulent; use

$$P = \frac{55.5\rho^{0.8} Q^{1.8} (\text{PV})^{0.2} L}{D^{4.8}} \text{ bar}$$

(3b) If $\bar{V} < V_c$, flow is laminar; use

$$P = \frac{19.16 \times 10^{-3} KL}{D}\left[\frac{2000\bar{V}}{D}\left(\frac{3n+1}{4n}\right)\right]^n \text{ bars}$$

Annular flow

(1) Determine n and K as above.

(2) Determine average velocity, $\bar{V}$, and critical velocity, V_c, from

$$\bar{V} = 21.2\left(\frac{Q}{D_h^2 - OD_p^2}\right) \text{m/s}$$

$$V_c = \left[\frac{0.119K}{\rho}\right]^{1/(2-n)} \left[\frac{4000}{(D_h - OD_p)}\left(\frac{2n+1}{n}\right)\right]^{n/(2-n)} \text{m/s}$$

(3a) If $\bar{V} > V_c$, flow is turbulent; use

$$P = \frac{55.5\rho^{0.8} Q^{1.8} (\text{PV})^{0.2} L}{(D_h - OD_p)^3 (D_h + OD_p)^{1.8}} \text{ bars}$$

(3b) If $\bar{V} < V_c$, flow is laminar; use

$$P = \frac{18.9 \times 10^{-3} KL}{(D_h - OD_p)}\left[\frac{4000\,\bar{V}}{(D_h - OD_p)}\frac{(2n+1)}{n}\right]^n \text{ bars}$$

Pressure loss across bit

Equations are similar to those of the Bingham plastic model.

Field units:

OD = outside diameter
D = inside diameter (mm)
L = length (m)
ρ = density (kg/l)
V = velocity (m/s)
PV = viscosity (cP)
YP = yield point $(\text{lbf}/100\ \text{ft})^2$ (as obtained from a viscometer)

To determine the pressure drop across the bit, add the total pressure drops across the system, i.e. $P_1 + P_2 + P_3 + P_4 + P_5$, to give a total value of P_c (described as the system pressure loss). Then determine the pressure rating of the pump used. If this pump is to be operated at, say, 80–90% of its rated value, then the pressure drop across the bit is simply pump pressure minus P_c.

The equations necessary for the calculation of jet velocity, pressure drop across bit and nozzle sizes are given in Tables 7.3 and 7.4.

Example 7.1

Using the Bingham plastic and power-law models, determine the various pressure drops, nozzle velocity and nozzle sizes for a section of 12.25 in (311 mm) hole. Two pumps are used to provide 700 gpm (2650 l/min).

Data:

plastic velocity	= 12 cP
yield point	= 12 lb/100 ft^2
	= (0.479 × 12) N/m^2
mud weight	= 8.824 lb/gal (1.057 kg/l)
drillpipe ID	= 4.276 in (108.6 mm)
OD	= 5 in (127 mm)
length	= 6480 ft (1975 m)
drill collars ID	= 2.875 in (73 mm)
OD	= 8 in (203 mm)
length	= 620 ft (189 m)

Last casing was $13\frac{3}{8}$ in (340 mm) with an ID of 12.565 in (319 mm). $13\frac{3}{8}$ in casing was set at 2550 ft (777 m). The two pumps are to be operated at a maximum standpipe pressure of 2200 psi (151.7 bar). Assume a surface equipment type of 4.

Solution

The solution to this example will be presented in Imperial units, with a summary of the Bingham plastic solution in metric units.

1. *Bingham plastic model*

Surface losses From Equation (7.1), pressure losses in surface equipment P_1 are given by

$$P_1 = E\rho^{0.8}Q^{1.8}(\text{PV})^{0.2}$$

From Table 7.2, the value of the constant E for type 4 is 4.2×10^{-5}; hence, Equation (7.1) becomes

$$P_1 = 4.2 \times 10^{-5}\rho^{0.8}Q^{1.8}(\text{PV})^{0.2}$$
$$= 4.2 \times 10^{-5}(8.824)^{0.8}(700)^{1.8}(12)^{0.2}$$
$$= 52 \text{ psi}$$

Pipe losses

Pressure losses inside drillpipe

$$\text{Average velocity } \bar{V} = \frac{24.5Q}{D^2} = \frac{24.5 \times 700}{(4.276)^2}$$
$$= 937.97 \text{ ft/min}$$

Critical velocity

$$V_c = \frac{97\ \text{PV} + 97\sqrt{(\text{PV})^2 + 8.2\rho D^2\ \text{YP}}}{\rho D} \text{ ft/min}$$

$$= \frac{97 \times 12 + 97\sqrt{(12)^2 + 8.2 \times 8.824 \times (4.276)^2 \times 12}}{8.824 \times 4.276}$$

$$= 356 \text{ ft/min}$$

Since $\bar{V} > V_c$, flow is turbulent and pressure drop inside drill pipe is calculated from:

$$P_2 = \frac{8.91 \times 10^{-5}\rho^{0.8}Q^{1.8}(\text{PV})^{0.2}L}{D^{4.8}}$$

$$= \frac{8.91 \times 10^{-5}(8.824)^{0.8}700^{1.8}12^{0.2} \times 6480}{(4.276)^{4.8}}$$

$$= 669.9 \text{ psi}$$
$$\approx 670 \text{ psi}$$

Pressure losses inside drill collars Following the same procedure as for drillpipe losses, we obtain

$$\bar{V} = \frac{24.5 \times 700}{(2.875)^2} = 2074.9 \text{ ft/min}$$

$$V_c = \frac{97 \times 12 + 97\sqrt{(12)^2 + 8.2 \times 8.824 \times (2.875)^2 \times 12}}{8.824 \times 2.875}$$

$$V_c = 373 \text{ ft/min}$$

Since $\bar{V} > V_c$, flow is turbulent and pressure loss inside drill collars P_3 is determined from

$$P_3 = \frac{8.91 \times 10^{-5}(8.824)^{0.8}(700)^{1.8}(12)^{0.2} \times 620}{(2.875)^{4.8}}$$

$$= 431 \text{ psi}$$

Annular pressure losses A rough sketch always helps in simplifying the problem. From Figure 7.2 it can be seen that part of the drillpipe is inside the casing and the rest is inside open hole. Hence, pressure loss calculations around the drillpipe must be split into (a) losses around the drillpipe inside the casing and (b) losses around the drillpipe in open hole.

Pressure losses around drillpipe

Cased hole section

$$\bar{V} = \frac{24.5Q}{(\text{ID}_c)^2 - (\text{OD}_{dp})^2} = \frac{24.5 \times 700}{(12.565)^2 - (5)^2}$$

$$= 129.1 \text{ ft/min}$$

where the subscripts 'c' and 'dp' refer to casing and drill pipe, respectively.

$$V_c = \frac{97 \times 12 + 97 \times \sqrt{(12)^2 + 6.2 \times (12.565 - 5)^2 \times 8.824 \times 12}}{8.824 \times (12.565 - 5)}$$

$$= 299.6 \text{ ft/min}$$

Since $\bar{V} < V_c$, flow is laminar and the pressure loss around the drillpipe in the cased hole section is determined from:

$$P_a = \frac{L(\text{PV})\bar{V}}{60\,000(\text{ID}_c - \text{OD}_{dp})^2} + \frac{L(\text{YP})}{200(\text{ID}_c - \text{OD}_{dp})}$$

where $L = 2550$ ft and $\bar{V}$ is in ft/min.

$$P_a = \frac{2550 \times 12 \times 129.01}{60\,000(12.565 - 5)^2} + \frac{2550 \times 12}{200(12.565 - 5)}$$

$$= 1.15 + 20.22 = 21.4$$

$$P_a \simeq 21 \text{ psi}$$

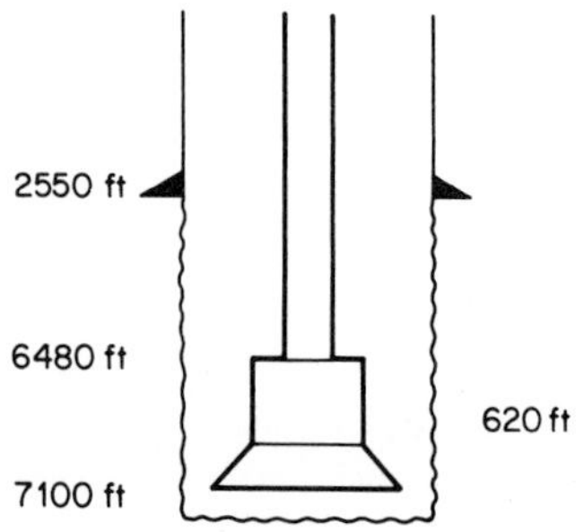

Fig. 7.2. Illustration of annular pressure loss calculation.

Open hole section

$$\bar{V} = \frac{24.5 \times 700}{(12.25)^2 - (5)^2}$$

$$= 137 \text{ ft/min}$$

$$V_c = 300.4 \text{ ft/min}$$

Since $\bar{V} < V_c$, flow is laminar and the pressure loss around the drillpipe in the open-hole section is determined from

$$P_b = \frac{3930 \times 12 \times 137}{60\,000(12.25 - 5)^2} + \frac{3930 \times 12}{200(12.25 - 5)}$$

$$= 35 \text{ psi}$$

(where $L = 6480 - 2550 = 3930$ ft, and L = length of drillpipe in the open-hole section).

Hence, total pressure drop around drillpipe is the sum of P_a and P_b. Thus,

$$P_5 = P_a + P_b = 21 + 35 = 56 \text{ psi}$$

Pressure losses around drill collars

$$\bar{V} = \frac{24.5 \times 700}{(12.25)^2 - (8)^2} = 199.3 \text{ ft/min}$$

$$V_c = \frac{97 \times 12 + 97 \times \sqrt{(12)^2 + 6.2 \times (8.824)(12.25 - 8)^2 \times 12}}{8.824 \times (12.25 - 8)}$$

$$= 314 \text{ ft/min}$$

Since $\bar{V} < V_c$, flow is laminar and pressure loss around drill collars is calculated from:

$$P_4 = \frac{620 \times 12 \times 199.3}{60\,000(12.25 - 8)^2} + \frac{620 \times 12}{200(12.25 - 8)}$$

$$= 1.37 + 8.75 = 10 \text{ psi}$$

Pressure drop across bit Total pressure loss in circulating system, except bit

$$= P_1 + P_2 + P_3 + P_4 + P_5$$

$$= 52 + 670 + 431 + 10 + 56$$

$$= 1219 \text{ psi}$$

Therefore, pressure drop available for bit (P_{bit})

$$= 2200 - 1219 = 981 \text{ psi}$$

$$\text{Nozzle velocity} = 33.36\sqrt{\frac{P_{bit}}{\rho}}$$

$$= 33.36\sqrt{\frac{981}{8.824}}$$

$$= 351.7 \text{ ft/s}$$

Total area of nozzles,

$$A_T = \frac{0.32\ Q}{V_n}$$

$$= \frac{0.32 \times 700}{351.7}$$

$$= 0.6369\ \text{in}^2$$

Nozzle size (in multiples of $\frac{1}{32}$)

$$= 32\sqrt{\frac{4A_T}{3\pi}}$$

$$= 16.64$$

Hence, select two nozzles of size 17 and one of size 16. The total area of these nozzles is 0.6397 in^2, which is slightly larger than the calculated area of 0.6368 in^2. (See also Example 5.3, page 103.)

Summary of Bingham plastic calculations in metric units

$P_1 = 3.49$ bar

$P_2 = 46.22$ bar

$P_3 = 29.77$ bar

$P_4 = 0.68$ bar

$P_5 = 3.74$ bar

$P_{bit} = 151.7 - (P_1 + P_2 + P_3 + P_4 + P_5) = 67.8$ bar

$A_T = 410.32\ \text{mm}^2$

Nozzle sizes: two 17s and one 16.

2. *Power-law model*

Surface losses

$$P_1 = 4.2 \times 10^{-5} \rho^{0.8} Q^{1.8} (\text{PV})^{0.2} = 52\ \text{psi}$$

Pipe losses

Pressure losses inside drillpipe (P_2)

$$\theta_{600} = 2(\text{PV}) + \text{YP} = 2(12) + 12 = 36$$

$$\theta_{300} = \text{PV} + \text{YP} = 12 + 12 = 24$$

$$n = 3.32 \log\left(\frac{\theta_{600}}{\theta_{300}}\right) = 0.5846 \simeq 0.585$$

$$K = \frac{300}{(511)}n = 0.6263 \simeq 0.626$$

$$V_c = \left[\frac{5.82(10^4)K}{\rho}\right]^{[1/(2-n)]}\left[\frac{1.6}{D}\left(\frac{3n+1}{4n}\right)\right]^{[n/(2-n)]}$$

where D = ID of drillpipe and ρ = density of mud.

$$V_c = \left[\frac{5.82 \times 10^4 \times 0.626}{8.824}\right]^{[1/(2-0.585)]}$$

$$\times \left[\frac{1.6 \times (3 \times 0.585 + 1)}{4.276 \times 4 \times 0.585}\right]^{[0.585/(2-0.585)]}$$

$$= (4128.88)^{0.707}(0.4405)^{0.413}$$

$$= (360.08) \times (0.713)$$

$$= 256.7\ \text{ft/min}$$

$$\bar{V} = \frac{24.5Q}{D^2} = \frac{24.5 \times 700}{(4.276)^2} = 937.97\ \text{ft/min}$$

Since $\bar{V} > V_c$, flow is turbulent and pressure loss inside drill pipe, P_2, is calculated by use of the turbulent flow equation given in Table 7.3:

$$P_2 = \frac{8.91(10^{-5})(8.824)^{0.8}(700)^{1.8}(12)^{0.2}(6480)}{(4.276)^{4.8}}$$

$$= 670\ \text{psi}$$

Pressure losses inside drill collars (P_3) As before, find V_c and $\bar{V}$:

$$V_c = 301\ \text{ft/min}$$

$$\bar{V} = \frac{24.5Q}{D^2} = \frac{24.5 \times 700}{(2.875)^2} = 2074.9\ \text{ft/min}$$

Since $\bar{V} > V_c$, flow is turbulent. Hence, pressure losses inside drill collars, P_3, are

$$P_3 = \frac{8.91(10^{-5})(8.824)^{0.8}(700)^{1.8}(12)^{0.2} \times (620)}{(2.875)^{4.8}}$$

$$P_3 = 431\ \text{psi}$$

Annular losses Because casing has been set, annular losses will have to be determined in the open hole section and cased hole section (see Figure 7.2).

Pressure losses around drillpipe

Cased hole section Values of n and K were calculated above:

$$n = 0.585 \text{ and } K = 0.626$$

$$L = \text{length of cased section} = 2550\ \text{ft}$$

Annular distance = ID$_c$ − OD = inside diameter of 13$\frac{3}{8}$ casing — outside diameter of drillpipe = 12.565 − 5 = 7.565 in

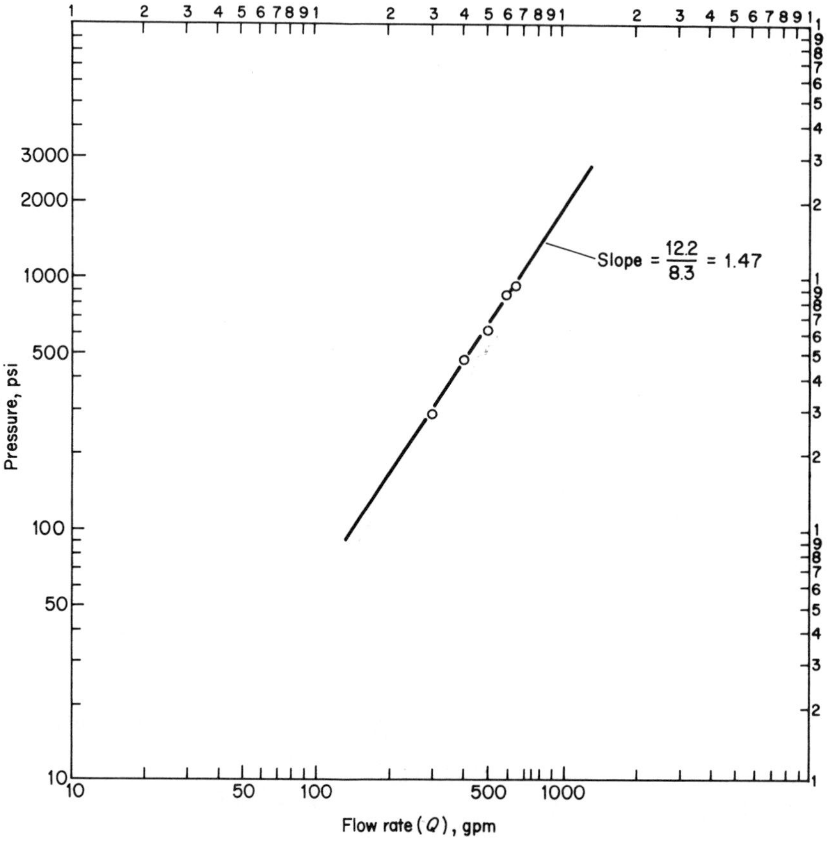

Fig. 7.3. Hydraulics data of Example 2. Maximum surface pressure (P_s) = 2500 psi.

$$V_c = \left[\frac{3.878 \times 10^4 K}{\rho}\right]^{[1/(2-n)]} \times \left[\frac{2.4}{(\mathrm{ID_c} - \mathrm{OD_{dp}})}\left(\frac{2n+1}{3n}\right)\right]^{[n/(2-n)]}$$

$$= 193 \text{ ft/min}$$

$$\bar{V} = \frac{24.5Q}{D_c^2 - (\mathrm{OD_{dp}})^2} = 129 \text{ fpm}$$

Since $\bar{V} < V_c$, flow is laminar and pressure losses around the drillpipe in the cased hole section are given by:

$$P_a = \left[\frac{2.4\bar{V}}{(\mathrm{ID_c} - \mathrm{OD_{dp}})}\left(\frac{2n+1}{3n}\right)\right]^n \frac{KL}{300(\mathrm{ID_c} - \mathrm{OD_{dp}})}$$

$$= \left[\frac{2.4 \times 129}{(12.565 - 5)}\frac{(1.17 + 1)}{3 \times 0.585}\right]^{0.585} \frac{0.626 \times 2550}{300 \times 7.565}$$

$$= 6.98 \text{ psi}$$

$$\approx 7 \text{ psi}$$

Open hole section Length of drillpipe in open hole section = 6480 − 2550 = 3930 ft

$$\text{Annular distance} = \text{hole diameter} - \text{OD of drillpipe}$$

$$= 12.25 - 5 = 7.25 \text{ in}$$

$$\bar{V} = \frac{24.5Q}{D^2 - (\mathrm{OD_{dp}})^2} = \frac{24.5 \times 700}{(12.25)^2 - (5)^2} = 137 \text{ fpm}$$

$$V_c = 196 \text{ fpm}$$

Since $\bar{V} < V_c$, flow is laminar and pressure loss around drillpipe in open hole is

$$P_b = \left[\frac{2.4 \times 137}{(12.25 - 5)}\frac{(1.17 + 1)}{3 \times 0.585}\right]^{0.585} \frac{0.626 \times 3930}{300 \times 7.25}$$

$$= 11.9 \text{ psi} \approx 12 \text{ psi}$$

Therefore, total pressure loss around drillpipe is

given by

$$P_5 = P_a + P_b$$
$$= 7 + 12 = 19 \text{ psi}$$

Pressure losses around drill collars

$$V_c = 231 \text{ ft/min}$$

$$\bar{V} = \frac{24.5 \times Q}{D^2 - \text{OD}^2_{\text{dc}}} = \frac{24.5 \times 700}{(12.25)^2 - (8)^2} = 199 \text{ fpm}$$

Since $\bar{V} < V_c$, flow is laminar:

$$P_4 = \left[\frac{2.4 \times 199}{(12.25 - 8)} \frac{(1.17 + 1)}{3 \times 0.585}\right]^{0.585} \frac{0.626 \times 620}{300(12.25 - 8)}$$
$$= 5.46 \text{ psi}$$
$$\approx 5 \text{ psi}$$

Pressure drop across bit Pressure losses inside and around the drillpipe and drill collars remain constant so long as: (a) mud properties remain unchanged; and (b) physical dimensions are unchanged. Therefore, total pressure losses through the system except bit

$$= P_1 + P_2 + P_3 + P_4 + P_5$$
$$= 52 + 670 + 431 + 19 + 5$$
$$= 1177 \text{ psi}$$

Therefore,

$$\text{pressure drop across bit} = \text{pump pressure} - 1178$$
$$= 2200 - 1177$$
$$= 1023 \text{ psi}$$

Using the same procedure as presented for the Bingham plastic solution, we obtain

$$\text{nozzle velocity} = V_n = 359 \text{ ft/s}$$

$$\text{total nozzle area, } A_T = 0.6239 \text{ in}^2$$

nozzle size (in multiples of $\frac{1}{32}$) = 16.47. Hence, select two 16s and one 17.

The total area of the nozzles is 0.6144 in^2, which is slightly less than 0.6239 in^2. If two 17s and one 16 were selected, then the total flow area would be 0.6397 in^2, which is slightly larger than the calculated area of 0.6239 in^2. In practice, one opts for the smaller size so that sufficient pressure drop is expended across the bit to give optimum hydraulics.

Calculations in metric units for the power-law model are left as an exercise for the reader.

3. *Comparison of the two models*

From the above results, it is obvious that the two models produce different nozzle sizes: the Bingham plastic model produced two 17s and one 16, whereas the power-law model produced two 16s and one 17. In practice, this difference is not considered serious, and if the mud pumps are capable of producing more than 2200 psi, then it is likely that three nozzles of size 16 will be chosen.

The reader should note also that the turbulent flow equations presented here use a turbulent viscosity term equal to (PV)/3.2 and not the plastic viscosity. If the plastic viscosity term is used instead, then pressure losses will be 26% higher than those calculated by our turbulent flow equation. It is the author's experience that the use of the turbulent viscosity term (i.e. (PV)/3.2) provides pressure loss values that are in agreement with field results.

OPTIMISATION OF BIT HYDRAULICS

All hydraulics programmes start by calculating pressure drops in the various parts of the circulating system. Pressure losses in surface connections, inside and around the drillpipe, inside and around drill collars, are calculated, and the total is taken as the pressure loss in the circulating system, excluding the bit. This pressure loss is normally given the symbol P_c.

Several hydraulics slide rules are available from bit manufacturers for calculating P_c. The slide rule is not suitable for calculating annular pressure losses, owing to: (a) the fact that annular pressure losses are normally small and may be beyond the scale of the slide rule; and (b) the fact that annular pressures are frequently laminar in nature and most slide rules use turbulent flow models.

A recent paper published in *World Oil*[1] offers a quick method of calculating system pressure losses. A 10% factor is incorporated into the drill string pressure losses to account for annular pressure losses. For shallow wells this method is quite satisfactory; however, for deep wells the annular pressure losses must be determined by using either the power-law model or the Bingham plastic model equations.

SURFACE PRESSURE

The system pressure losses, P_c, having been determined the question is how much pressure drop can be tolerated at the bit (P_{bit}). The value of P_{bit} is controlled entirely by the maximum allowable surface pump pressure.

Most rigs have limits on maximum surface pressure, especially when high volume rates — in excess of 500 gpm (1893 l/min) — are used. In this case, two pumps are used to provide this high quantity of flow. On land rigs typical limits on surface pressure are in the range 2500–3000 psi for well depths of around 12 000 ft. For deep wells, heavy-duty pumps are used which can have pressure ratings up to 5000 psi.

Hence, for most drilling operations, there is a limit on surface pump pressure, and the criteria for optimising bit hydraulics must incorporate this limitation.

HYDRAULIC CRITERIA

There exist two criteria for optimising bit hydraulics: (1) maximum bit hydraulic horsepower (BHHP); and (2) maximum impact force (IF). Each criterion yields different values of bit pressure drop and, in turn, different nozzle sizes. The engineer is faced with the task of deciding which criterion he is to choose. Moreover, in most drilling operations the circulation rate has already been fixed, to provide adequate annular velocity. This leaves only one variable to optimise: the pressure drop across the bit, P_{bit}. We shall examine the two criteria in detail and offer a quick method for optimising bit hydraulics.

Maximum bit hydraulic horsepower

The pressure loss across the bit is simply the difference between the standpipe pressure and P_c. However, for optimum hydraulics the bit pressure drop must be a certain fraction of the maximum available surface pressure. For a given volume flow rate optimum hydraulics is obtained when the bit hydraulic horsepower assumes a certain percentage of the available surface horsepower.

Surface hydraulic horsepower (HHP_s) is the sum of hydraulic horsepower at bit (BHHP) and hydraulic horsepower in the circulating system (HHP_c). Mathematically this can be expressed as:

$$HHP_s = BHHP + HHP_c$$

or

$$BHHP = HHP_s - HHP_c \tag{7.2}$$

In the case of limited surface pressure, the maximum pressure drop across the bit, as a function of available surface pressure, gives the maximum hydraulic horsepower at the bit for an optimum value of flow rate. In other words, the first term in equation (7.2) must be maximised for maximum BHHP. Equation (7.2) can be written as follows:

$$BHHP = \frac{P_s Q}{1714} - \frac{P_c Q}{1714} \tag{7.2a}$$

where P_s = maximum available surface pressure—also the standpipe pressure as read on the surface gauge (psi); P_c = pressure loss in the circulating system (psi); and Q = volume flow rate (gpm).

The pressure drop in the circulating system, P_c, can be related to Q as

$$P_c = KQ^n \tag{7.3}$$

where K = a constant; and n = index representing degree of turbulence in the circulating system.

Combining Equations (7.2a) and (7.3) gives

$$BHHP = \frac{P_s Q}{1714} - \frac{KQ^{n+1}}{1714} \tag{7.4}$$

Differentiating Equation (7.4) with respect to Q and setting $dBHHP/dQ = 0$ gives

$$P_s = (n+1)KQ^n$$

or

$$P_s = (n+1)P_c \tag{7.5}$$

Also

$$P_c = P_s - P_{bit} \tag{7.6}$$

Substituting Equation (7.6) in Equation (7.5) and simplifying gives

$$P_{bit} = \frac{n}{n+1} P_s \tag{7.7}$$

In the literature several values of n have been proposed, all of which fall in the range 1.8–1.86. Hence, when $n = 1.86$, Equation (7.7) gives $P_{bit} = 0.65\ P_s$. In other words, for optimum hydraulics, the pressure drop across the bit should be 65% of the total available surface pressure.

The actual value of n can be determined in the field by running the mud pump at several speeds and reading the resulting pressures. A graph of $P_c (= P_s - P_{bit})$ against Q is then drawn. The slope of this graph is taken as the index n.

Maximum impact force

In the case of limited surface pressure Robinson[2] showed that, for maximum impact force, the pressure drop across the bit (P_{bit1}) is given by

$$P_{bit1} = \frac{n}{n+2} P_s \tag{7.8}$$

where n = slope of P_c vs. Q; and P_s = maximum available surface pressure.

Derivation of Equation (7.8) is left as an exercise for the reader (see Problem 1 at the end of this chapter).

The bit impact force (IF) can be shown to be a function of Q and P_{bit} according to the following equation:

$$IF = \frac{Q\sqrt{\rho P_{bit}}}{58} \tag{7.9}$$

where ρ = mud weight (ppg).

Comparison of BHHP and IF criteria

The ratio R, of pressure drop across bit, as given by the BHHP criterion, and that as given by the IF criterion (Equations 7.7 and 7.8) is

$$R = \frac{n}{n+1}P_s \div \frac{n}{n+2}P_s$$

or

$$R = \frac{n+2}{n+1} \tag{7.10}$$

Equation (7.10) shows that the pressure drop across the bit, as determined by the maximum BHHP criterion, is always larger than that given by the maximum IF method. For laminar flow, $n = 1$ and R assumes its maximum value of 1.5. Actual field values of n are larger than 1. Results of Table 7.5, which has been constructed for values of n between 1 and 2, indicate that R decreases parabolically with increasing value of n, but can never assume unity. In other words, P_{bit} is always larger than P_{bit1}.

NOZZLE SELECTION

Smaller nozzle sizes are always obtained when the maximum BHHP method is used, as it gives larger values of P_{bit} than those given by the maximum IF method. The following equations may be used to determine total flow area and nozzle sizes:

$$A_T = 0.0096\, Q\sqrt{\frac{\rho}{P_{bit}}}\ \text{in}^2 \tag{7.11}$$

$$d_N = 32\sqrt{\frac{4A_T}{3}} \tag{7.12}$$

where A_T = total flow area (in^2); and d_N = nozzle size in multiples of $\frac{1}{32}$ in.

TABLE 7.5 Ratio $P_{bit}/P_{bit\,1}$ as a function of index n

n	$R = \frac{n+2}{n+1}$
1.0	1.50
1.2	1.45
1.5	1.40
1.8	1.36
2.0	1.33

OPTIMUM FLOW RATE

Optimum flow rate is obtained using the optimum value of P_c, n and maximum surface pressure, P_s. For example, using the maximum BHHP criterion, P_c is determined from

$$P_c = P_s - P_{bit} = P_s - \frac{n}{n+1}P_s$$

$$P_c = \frac{1}{n+1}P_s \tag{7.13}$$

The value of n is equal to the slope of the P_c–Q graph. The optimum value of flow rate, Q_{opt}, is obtained from the intersection of the P_c value (as determined by Equation 7.13) and the P_c–Q graph (see example 7.2).

FIELD METHOD OF OPTIMISING BIT HYDRAULICS

The index n can only be determined on site and is largely controlled by *downhole* conditions. The following method for determining n was suggested by Robinson[2] and is summarised here briefly.

(1) Prior to POH current bit for next bit change, run the pump at four or five different speeds and record the resulting standpipe pressures.
(2) From current nozzle sizes and mud weight determine pressure losses across the bit for each value of flow rate, using Equation (7.11) or a hydraulics slide rule.
(3) Subtract P_{bit} from standpipe pressure to obtain P_c.
(4) Plot a graph of P_c against Q on log–log graph paper and determine the slope of this graph, which is the index n in Equations (7.7), (7.8) and (7.13).
(5) For the next bit run, Equation (7.7) or (7.8) is used to determine P_{bit} that will produce maximum bit hydraulic horsepower. Nozzle sizes are then selected by use of this value of P_{bit}.

For a particular rig and field the index n will not vary widely if the same drilling parameters are used.

For standardisation purposes it is recommended that the above test be run at three depths for each bit run. The average value of n for each bit run can then be used for designing optimum hydraulics.

Example 7.2

Prior to changing of bit in a $12\frac{1}{4}$ in hole, the standpipe pressures listed in Table 7.6 were recorded at different flow rates with the present bit still on bottom. Present hole depth is 6528 ft and the next bit is expected to drill down to 8000 ft. Other relevant data are:

present nozzle sizes = three $\frac{16}{32}$
mud weight = 8.7 ppg (65 pcf)
current flow rate = 600 gpm
maximum allowable surface pressure = 2500 psi

Determine the optimum hydraulic parameters for the next bit run, using BHHP and IF criteria.

Solution

Maximum bit hydraulic horsepower criterion

From Equation (7.11), pressure drop across bit for the three given nozzle sizes (3 16s) is

$$P_{\text{bit}} = \frac{9.22 \times 10^{-5} Q^2 \rho}{A_T^2}$$

where

A_T = total area of the three nozzles

$$= 3 \times \frac{\pi}{4} \times \left(\frac{16}{32}\right)^2 = 0.589 \text{ in}^2$$

Hence,

$$P_{\text{bit}} = 231.179 \times 10^{-5} Q^2 \qquad (7.14)$$

Equation (7.14) can then be used to calculate the pressure drop across the bit for the given volume flow rates. Alternatively, an hydraulics slide rule may be used to calculate P_{bit}. Once P_{bit} is determined, P_c is then simply the difference between the standpipe pressure and P_{bit}. Table 7.7 summarises these results.

TABLE 7.6 Raw data for Example 7.2

Flow rate (gpm)	*Standpipe pressure (psi)*
300	500
400	850
500	1200
600	1700
650	1900

TABLE 7.7 Processed data of Example 7.2

Q *(gpm)*	*Standpipe pressure (psi)*	P_{bit} *(psi)*	P_c *(psi)*
300	500	208	292
400	850	370	480
500	1200	575	625
600	1700	830	870
650	1900	980	920

Figure 7.3 gives a plot of P_c against Q. The slope of this graph can be measured directly using a ruler or, more accurately, by a curve-fitting technique. The difference is usually small if the best straight line passing through the majority of points is drawn. Figure 7.3 gives a value of n equal to 1.47. The equations necessary for the calculations of the various hydraulics parameters are arranged conveniently as given below:

$$P_{\text{bit}} = \frac{n}{(n+1)} P_s \qquad (7.7)$$

$$A_n = 0.0096 \sqrt{\frac{\rho}{P_{\text{bit}}}} \qquad (7.11)$$

$$d = 32 \sqrt{\frac{4A_n}{3\pi}} \qquad (7.12)$$

$$\text{IF} = \frac{Q}{58} \sqrt{\rho P_{\text{bit}}} \qquad (7.9)$$

where A_n = total flow area of nozzles (in^2); ρ = mud weight (ppg); d = nozzle size as a fraction of 32; and IF = impact force (lb).

For the current flow rate of 600 gpm Using Equations (7.7), (7.11), (7.12) and (7.9), we obtain

$$P_{\text{bit}} = 1488 \text{ psi}$$

$$A_n = 0.4404 \text{ in}^2$$

$$d = 13.8 \text{ or one 13 and two 14s}$$

$$\text{IF} = 1179 \text{ lb}$$

Optimum circulation pressure, $P_c = 2500 - 1488 = 1012$ psi.

Optimum flow rate

The value of optimum flow rate, Q_{opt}, is obtained from the intersection of the line $P_c = 1012$ psi and the P_c vs. Q graph. Figure 7.3 gives a Q_{opt} of 660 gpm.

Using Equations (7.11), (7.12) and (7.9), we obtain

nozzle sizes = 14.3 or two 14s and one 15

impact force = 1295 lb

Maximum impact force criterion

From Figure 7.3, the slope of the graph is, again, 1.47.

$$P_{\text{bit}1} = \frac{n}{n+2} \times P_s$$

$$= \frac{1.47}{1.47+2} \times 2500$$

$$= 1059 \text{ psi}$$

Optimum circulation pressure, $P_c = 2500 - 1059 = 1441$ psi.

The intersection of $P_c = 1441$ and Figure 7.2 gives an optimum flow rate (Q_{opt}) of 840 gpm. Hence,

$$A_n = 0.0096\, Q \sqrt{\frac{\rho}{P_{\text{bit}1}}}$$

$$A_n = 0.7309 \text{ in}^2$$

$d = 17.8$ or two 18s and one 17

$$\text{impact force (IF)} = \frac{Q}{58}\sqrt{\rho P_{\text{bit}1}} = 1390 \text{ lb}$$

Comparison

The results of Example 7.2 show that the BHHP criterion gives better hydraulics in terms of small nozzle sizes and higher jet velocities. The IF criterion gives a slightly higher impact force than does the BHHP criterion.

A practical check on the efficiency of the bit hydraulics programme

(1) Determine pressure drop across bit, P_{bit}.
(2) Determine bit hydraulic horsepower (BHHP):

$$\text{BHHP} = \frac{P_{\text{bit}} \times Q}{1714} \text{ hp}$$

or

$$\text{BHHP} = P_{\text{bit}} \times Q \text{ kW}$$

(3) Divide BHHP obtained above by area of bit to determine K, where

$$K = \frac{\text{BHHP}}{\pi d^2/4}$$

(4) For maximum cleaning, K should be between 3 and 6 HHP/in^2 (i.e. 3.74–6.94 Watts/mm^2).

References

1. Brouse, M. (1982). Practical hydraulics: A key to efficient drilling. *World Oil*, Oct.
2. Robinson, L. (1982). Optimising bit hydraulics increases penetration rate. *World Oil*, July.

Problems

1. Using the maximum impact force criterion, prove that, for the case of limited surface pressure, the pressure drop across the bit, P_{bit}, is given by

$$P_{\text{bit}} = \left(\frac{n}{n+2}\right) \times P_s$$

where n = slope of circulation pressure vs. circulation rate; and P_s = standpipe pressure.

Hint: (a) Express the impact force, F, in terms of V, Q and ρ, to obtain

$$F = \rho Q V \tag{P.1}$$

(b) Pressure loss across the bit is given by

$$P_{\text{bit}} = \frac{K \rho Q^2}{A^2} \tag{P.2}$$

where K is a constant.

(c) From Equations (P.1) and (P.2) obtain the relationship

$$F = K_1 Q P_{\text{bit}}^{0.5}$$

where K_1 is a constant.

(d) Using $P_s = P_{bit} + P_c$ and $P = K_1 Q P_{bit}^{0.5}$, and differentiating F with respect to Q, obtain the following relationship:

$$P_{bit} = \left(\frac{n}{n+2}\right) P_s$$

2. Using the Bingham plastic model, calculate for the well described below: (a) the circulating pressure, P_c; (b) the nozzle sizes; (c) the bottom hole pressure while circulating; and (d) the equivalent circulating density.

depth	= 9500 ft (2896 m)
hole diameter	= 8.5 in (215.9 mm)
drillpipe	= 5 in/4.276 in (127 mm/108.6 mm) 9000 ft (2743 m)
drill collars	= 8 in/3 in (203.2 mm/76.2 mm) 500 ft (152.4 m)
mud weight	= 13 ppg (1.56 kg/l)
yield point	$= 30 \dfrac{\text{lb}}{100\ \text{ft}^2}$
viscosity	= 20 cP
circulation rate	= 350 gpm (1325 l/min)
maximum operating pressure of mud pumps	= 2500 psi (172 bar)
surface equipment type	= 4

Answer: ((a) 2469, (b) three 11s, (c) 6480, (d) 13.12 ppg)

3. Repeat the above example using the power-law model.

4. Assume that the circulation pressure, P_c, is related to flow rate according to the equation $P_c \propto Q^{1.86}$. Determine the optimum circulating pressure, circulation rate, pressure drop across bit and nozzle sizes, using data from Example 7.2 and the BHHP and IF criteria.

5. Prior to changing the bit in a $12\frac{1}{4}$ in hole, the following standpipe pressures were recorded at different flow rates with the present bit on bottom:

Flow rate (gpm)	*(l/min)*	*Standpipe pressure (psi)*	*(bar)*
252	954	390	26.9
336	1272	620	42.7
420	1590	920	63.4
504	1908	1240	85.5
630	2385	1840	126.9

Present nozzle sizes = three 16s
Present hole depth = 6572 ft (2403 m)
Next hole depth = 8300 ft (2530 m)
Mud weight = 8.3 ppg (0.995 kg*l)
Flow rate not to exceed 600 gpm, to limit hole erosion

Determine the optimum nozzle sizes for the given flow rate. Also determine the optimum flow rate that the BHHP criterion yields.

Answer: Slope = 1.68; nozzles: one 13 and two 14s; $Q_{opt} = 660$ gpm.

Chapter 8

Straight and Directional Hole Drilling

In this chapter the following topics will be discussed:

Straight hole drilling
- causes of hole deviation
- mechanical factors
- formation characteristics
- bottom hole assembly
- stabilisers and reamers

Directional drilling
- reasons for directional drilling
- geometry of a directional well
- types of directional well
- bottom hole assemblies for directional wells
- directional surveying instruments
- deflection tools
- orientation of deflection tools
- calculation methods
- dog-legs in directional wells

STRAIGHT HOLE DRILLING

Causes of hole deviation

In rotary drilling the basic components of a drill string comprise a drill bit, stabilisers, drill collars and drill pipe to surface.

Rock breakage is achieved by the application of weight onto the bit teeth (supplied by the drill collars) and rotation (supplied by a rotary table). The applied weight overcomes the compressive strength and crushes the rock surface, while rotation produces a tearing or shearing action. The two mechanisms combine to produce chips or cuttings of various sizes which are flushed to the surface by a stream of suitable fluid, such as mud or air. A new surface of rock will then be exposed to the bit teeth, allowing further hole to be made.

The ultimate direction of the new hole will be dictated by the drill string and formation characteristics. Figure 8.1 shows that the elastic drill string will bend under the applied forces and will contact the walls of the hole at some point described as the 'tangency point'. Thus, the forces in Figure 8.1 can be resolved as: (1) axial load — the total load applied on the bit (W in Figure 8.1); (2) pendulum force — the weight of drill collars below the tangency point; and (3) the formation reaction. Forces (1) and (2) are known as 'mechanical factors'.

Hole direction will be dictated by the forces 1, 2 and 3. Forces 1 and 2 can be quantified and predicted at any position within the hole. Formation reaction, however, varies with type of formation and is also highly variable within the same rock type, which makes it difficult to quantify and predict. The formation reaction depends also on the type and action of the bit.

Mechanical factors

The mechanical factors which contribute to hole deviation include the axial load (W) and the pendulum force (F), as shown in Figure 8.1. The axial load is the total weight-on bit and is compressive in nature. The drill string, being elastic, will buckle under the action of load W and, as a result, the axis along which W acts (i.e. axis of drill collars) is displaced from the centreline of the hole.

Near the bit, W can be resolved into two components: W_1 along the axis of the hole and W_2 per-

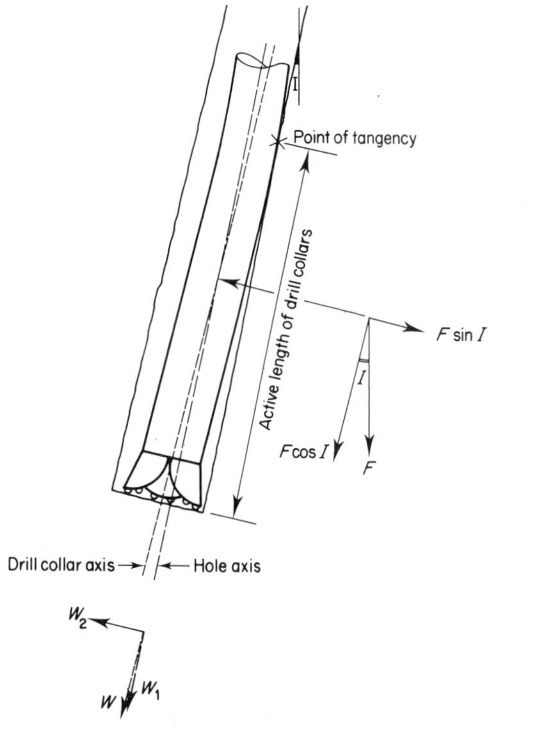

Fig. 8.1. Effects of mechanical factors on hole deviation. Hole inclination = I; weight-on-bit = W; pendulum force = F.

pendicular to the axis of the hole. The component W_2 is responsible for hole deviation, and its magnitude increases with increasing clearance between hole and drill collars and with increases in weight-on-bit. From Figure 8.1, the deviation caused by component W_2 is to the left.

The pendulum force arises due to gravity and hole inclination, and its magnitude is dependent on the active length of drill collars between the drill bit and their first point of tangency to the wall of the hole (see also Bottom Hole Assembly section). The force F can also be resolved into two components: $F \cos I$ along the centreline of the hole and $F \sin I$ perpendicular to hole axis (more correctly, the forces are resolved along and perpendicular to drill collar axis, however this assumption produces small errors). From Figure 8.1, deviation due to the component $F \sin I$ is to the right.

The magnitude and direction of the resultant hole deviation due to mechanical factors is dependent on the difference between W_2 and $F \sin I$.

Formation characteristics

The major source of natural hole deviation is formation characteristics. All hydrocarbon deposits (oil and gas) exist in sedimentary-type formations, having a laminated or bedded structure. Further, the sedimentary rock may be composed of alternating soft and hard bands. Soft bands are easily drilled and may be washed out by the drilling fluid, an oversize hole being produced, as shown in Figure 8.2. The drill collars will deflect the bit laterally within the washout[1] before the next hard band is approached. Continuous drilling of soft and hard bands and deflection of the bit within the washouts will ultimately produce a hole within the hard bands; that is, not in line with that in the soft bands (Figure 8.2). Thus, an unwanted deviation is produced with possibly a severe dog-leg.

Lamination in sedimentary rocks is another factor contributing to natural hole deviation.

In homogeneous, horizontally-bedded formations the bit will cut equal chips on both sides of the teeth, and the bit will drill a straight hole.

In dipping laminated formations the drill bit tends to cut unequal chips on each side of its teeth, which results in lateral movement of the bit and, ultimately, hole deviation. Practical experience has shown that the direction of bit deflection is related to the angle of dip. When the angle of dip is less than 45°, the bit tends to drill up-dip (Figure 8.3). When the angle of dip is greater than 45°, the bit tends to drill down-dip. Bit deflection is often described as 'bit walk'. Also, practical experience has shown that the angle of deviation is always less than the angle of dip.

Besides dip, other important formation character-

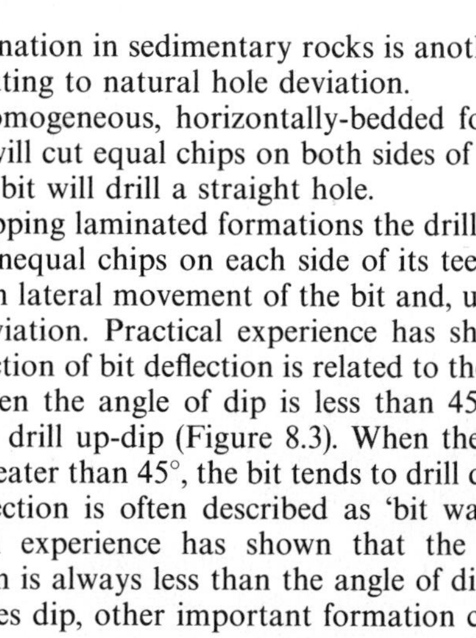

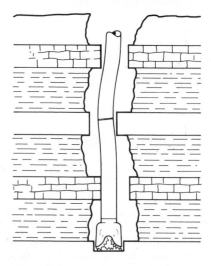

Fig. 8.2. Illustration of effect of hard and soft formations on hole deviation. Alternating hard and soft formations cause offset ledges, and washouts allow drill string to change direction. (After Wilson, 1976)[1]

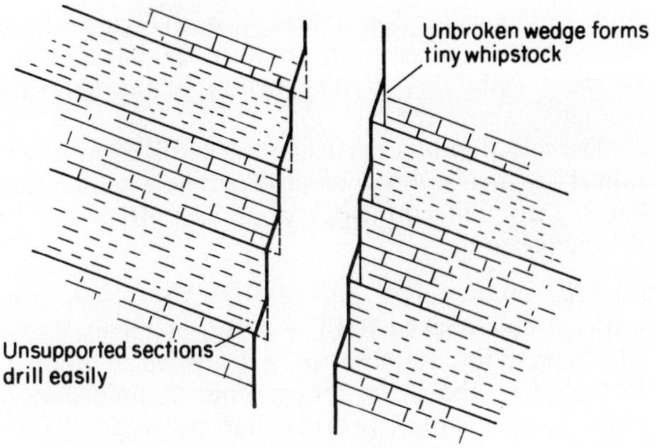

Fig. 8.3. When formations dipping less than 45° are drilled, often the bit will tend to drill updip, causing a deviated hole. In theory, small whipstocks formed at formation laminations cause the bit to drill off course. (After Wilson, 1976)[1]

istics which contribute to hole deviation include faulting, fracturing and fissuring, and degree of drillability.

In combination, the above-mentioned factors determine the formation's contribution to deviation. Formation tendency to deviation is normally described as 'crooked hole tendency'.

The degrees of crooked hole tendency are classified as mild, medium and severe. Mild crooked hole tendency produces little or no deviation and is normally associated with the drilling of hard and isotropic rocks. Medium and severe crooked hole tendencies are associated with medium and soft formations, respectively. These rock types normally show a great degree of dipping, fracturing and variation in strength.

The above classification for formation tendency to deviation can be used to select the most appropriate drilling assembly for maintaining or changing hole deviation.

Bottom hole assembly

In general, there are three types of bottom hole assembly (BHA): slick, pendulum and packed BHA.

Slick BHA

A slick BHA may be defined as one in which no stabiliser is used. The drill string simply consists of bit, drill collars and drillpipe. Thus, this type of BHA is suitable only for mild crooked hole formations. Slick BHAs are seldom used except to drill to kick-off point in directional wells.

Pendulum BHA

Bottom hole assemblies using the pendulum technique are primarily used to reduce hole deviation, as exemplified by the drop-off section of an S-type directional well (see page 153). Pendulum BHAs are also used to drill soft, unconsolidated formations in surface holes.

The pendulum technique relies on the principle that the force of gravity may be used to deflect the hole back to vertical. The force of gravity (F in Figure 8.1) is related to the length of drill collars between the drill bit and the point of tangency. This length is described as the 'active length' of drill collars. Thus, increasing the active length means that the vertical force is increased and the hole is quickly deflected back towards the vertical. The actual force which brings about the deflection towards vertical is the component, $F \sin I$, of the vertical force, F, which acts perpendicular to the well-bore axis. The along-hole component of the vertical force is $F \cos I$, which attempts to maintain the present hole direction. Hence, increasing the value of F causes the component $F \sin I$ to increase at a much faster rate than the along-hole force, $F \cos I$.

The pendulum assembly consists of a drill bit; several drill collars to provide the pendulum force; one or more stabilisers; drill collars; heavy-walled drillpipe; and drillpipe to surface. Maximum pendulum force can be obtained by placing the first string stabiliser as high as possible above the bit so that the active length of the drill collars will not touch the walls

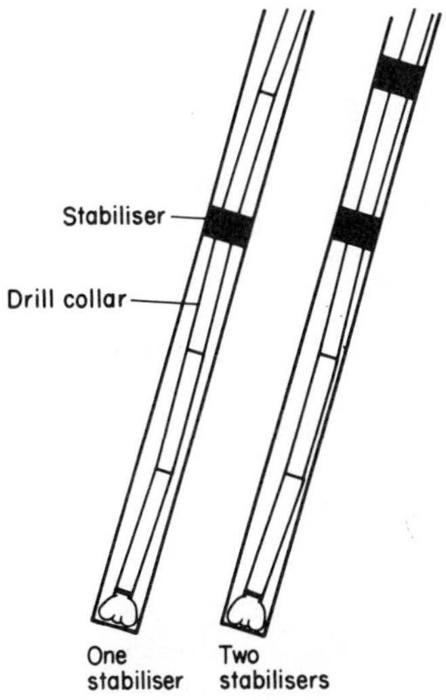

Fig. 8.4. Pendulum assemblies. (After Wilson, 1979)[2]

of the hole[2]. However, if the distance from the bit to the first stabiliser is made too large, there is a danger that the 'active length' of the drill collars will contact the low side of the hole, thus reducing the pendulum effect. Wilson[2] pointed out that a second stabiliser 30 ft above the first string stabiliser should be used to reduce the lateral force on the first stabiliser. Figure 8.4 shows a BHA utilising the pendulum technique.

The main disadvantage of the pendulum assembly is that the unstabilised bit can drill an undersized or misaligned hole[2] which makes it difficult for the casing to be run. A usable hole can be obtained with the pendulum assembly if the OD of drill collars is chosen such that[2]

$$\text{minimum OD of drill collars} = 2(\text{casing coupling OD}) - \text{bit OD} \quad (8.1)$$

By use of Equation (8.1) Table 8.1 can be constructed for minimum drill collar OD, which can be used to drill typical hole sizes with the pendulum BHA.

Packed hole assembly

The packed hole assembly relies on the principle that two points will contact and follow a sharp curve, while three points will follow a straight line (Figure 8.5). A three-point stabilisation is obtained by placing three or more stabilisers in the portion of the hole immediately above the bit.

The selection of a particular packed BHA is determined by the severity of crooked hole tendencies. Thus, three different types of packed BHA can be distinguished:

(1) *Mild crooked hole packed BHA.* A typical mild crooked hole packed BHA is shown in Figure 8.6(a). The three-point stabilisation is provided at Zone 1, immediately above the bit; at Zone 2, immediately above a short, large OD drill collar; and at Zone 3, on top of a standard drill collar.

If a vibration dampener is to be included, it should be placed at Zone 2 for maximum effectiveness, since this lacks the necessary stiffness[2]. A vibration dampener is used to absorb bit bounce and prevent vibrations from being transmitted to the surface.

(2) *Medium crooked hole packed BHA.* In this type of packed BHA a second stabiliser is included at Zone 1 in order to provide increased bit stabilisation against the deviation effects of the formation. Zones 2 and 3 are similar to those of Type (1) packed BHA. Figure 8.6(b) shows a medium crooked hole packed BHA.

(3) *Severe crooked hole packed BHA.* In this type of packed BHA three stabilisers are included at Zone 1. Zones 2 and 3 are similar to those of Type (1) packed BHA (see Figure 8.6c).

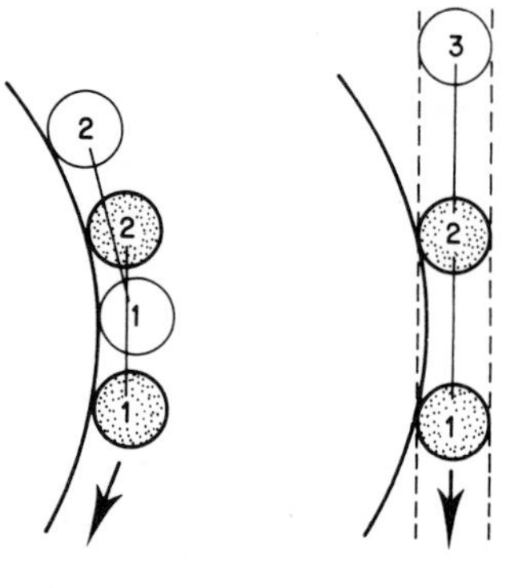

Fig. 8.5. The packed bottom hole assembly results from the basic idea that three points cannot contact and form a curved hole. (After Wilson, 1976)

TABLE 8.1 (after Reference 2)

Bit (or hole) size (in)	*Casing size OD (in)*	*Casing size Coupling (in)*	*Minimum drill collar OD (in)*
$8\frac{1}{2}$	7	7.656	6.562
$12\frac{1}{4}$	$9\frac{5}{8}$	10.625	9
$17\frac{1}{2}$	$13\frac{3}{8}$	14.375	11.250
24	$18\frac{5}{8}$	19.750	15.500

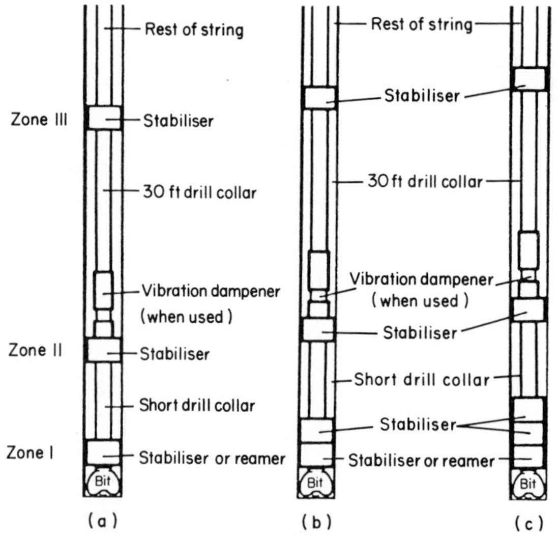

Fig. 8.6. Packed hole assemblies for mild (a), medium (b) and severe (c) crooked hole tendencies[2]. (After Wilson, 1979)

Stabilisers and reamers

The choice of stabilisers is dictated by the type of formation being drilled. For soft formations a stabiliser with a large contact area is used, for effective bit stabilisation. For hard rock, stabilisers with small wall contact areas are chosen[2]. For a discussion on different types of stabilisers, see Chapter 2.

Reamers are supplied with rotating tungsten carbide cones in order to ream out ridges left by the drill bit, thereby helping to produce a near-gauge hole. Reamers are, therefore, best placed immediately behind the bit and used particularly for drilling hard rocks.

Since the OD of stabilisers and reamers is very close to the diameter of the hole being drilled, the annular space between hole and stabiliser (or reamer) is very small. This leads to a much greater annular velocity around the stabilisers and reamers, and to possible hole erosion (or washout) in soft formations.

Measurement of hole verticality

Drift in vertical holes can be tracked with a mechanical pendulum unit housed in a special protective barrel (see Figure 8.7). The instrument measures deviation in degrees, and a record is made on a paper disc when punched by the pendulum-balanced stylus. The paper disc is divided into concentric circles, each circle representing 1° of deviation. The instrument is designed to produce two punches at 180° apart. The drift instrument is also equipped with a timing device which can be set to control the movement of the paper disc at the required depth.

A survey with a drift indicator is normally made every 500 ft or prior to bit change. The time required for the drift instrument to travel to hole bottom is set on the timing device, and the instrument is then dropped from the surface to land on a baffle plate on top of the drill bit. The disc housing the paper then moves up to meet the pendulum stylus and a record of deviation is made. On the way down, the disc rotates 180° and moves up again to record another punch. This second punch should read the same as the first punch, and is normally made to check that the instrument is functioning properly.

In straight holes the objective is to drill a well as nearly vertical as possible. In directional wells the hole is deliberately deflected to follow a given direction and the objective is to maintain hole deviation and direction in order to intersect the target zone. The reader should note that, in some directional wells, hole deviation is maintained to a certain depth and then reduced to reach the target area.

Fig. 8.7. Drift indicator[3]. (Courtesy of Eastman Whipstock)

Usually in vertical wells the objective is to keep the drift angle to a minimum in order to minimise the horizontal displacement. Deviation in straight hole is avoided for the following reasons: (a) to keep the well within a particular lease or pay zone; (b) to comply with governmental regulations; and (c) to avoid wear to the drill string and casing, resulting from sudden changes in hole angle or dog-legs. Dog-legs are instrumental in forming key-seats and causing pipe sticking (see Chapter 12). Dog-legs can also prevent a good cement job, as the cement cannot circulate between the walls of the hole and the casing at the dog-leg point[1].

DIRECTIONAL DRILLING

A directional well may be defined as one in which deviation is deliberately initiated in order to intersect (a) particular zone(s).

Reasons for directional drilling

There are certain situations in which a deviated well is the only practical way to reach a particular producing zone. The following is a summary of these situations (see also Figure 8.8).

(1) *Offshore development.* A whole offshore field may be developed by drilling the required number of wells from one platform (in deep waters) or from an artificial island (in very shallow waters) (see Figure 8.8a). The wells will have to be deviated at different angles from the vertical in order to reach the outer boundaries of the field.
(2) *Fault drilling.* Wells drilled through a fault plane are unstable, owing to the slip and shearing effects which the casing experiences due to fault movement. A well directed across or parallel to the fault plane will avoid such hazards, as shown in Figure 8.8(b).
(3) *Inaccessible and restricted areas.* When a reservoir is located below a mountain or in a highly populated area, deviation drilling is the only solution to the development of the oil field (Figure 8.8c).
(4) *Sidetracking.* In some cases part of the drill string is left in the hole; for example, owing to pipe sticking to the walls of the hole. The portion of steel which is left in the well is described as 'fish'. If the fish cannot be recovered, drilling can only proceed by changing the course of the well from a point just above the fish. This procedure is described as sidetracking. It involves deviating the well from its original direction (Figure 8.8d).
(5) *Salt dome drilling.* When an oil reservoir exists below a salt dome, practical experience has shown that it is necessary to drill a deviated well as shown in Figure 8.8(e) in order to avoid the salt dome. Casing strings set in salt domes are subjected to a collapsing effect resulting from stresses induced by the creep of the salt section. A deviated well may be planned to avoid the salt dome altogether and is deflected just above the oil zone.
(6) *Relief wells.* A directional well may be drilled to intersect a blowing well, so that heavy mud can be pumped in to kill the well.
(7) *Wildcatting.* The term 'wildcatting' refers to exploration drilling in areas containing potential hydrocarbon structures. The exploration statistics show that one well in nine results in a petroleum discovery. Thus, if the initial well is not successful, it is much cheaper to drill a deviated well from the existing well. This technique results in a considerable saving on the initial cost of drilling and setting of surface and intermediate casings. This technique is, therefore, analogous to sidetracking.

Geometry of a directional well

A directional well is drilled from the surface to reach a target area along the shortest possible path. Owing to changing rock properties, the hole path rarely follows a single plane but, instead, changes its inclination and direction continuously. Thus, the deviated well should be viewed in three dimensions, such that hole inclination and hole direction are specified at each position.

Figure 8.9(a) gives a three-dimensional view of a directional well, showing the vertical and horizontal plans of the well. The following parameters define a directional well (Figure 8.9b and c).

(1) *Inclination* is the angle between the vertical and a tangent to the well path at any point.
(2) *Azimuth* is the angle measured in a horizontal plane between the direction of north and a point on the well path. Thus, a point with an azimuth of 50° means that the direction of the point is at 50° from the north.

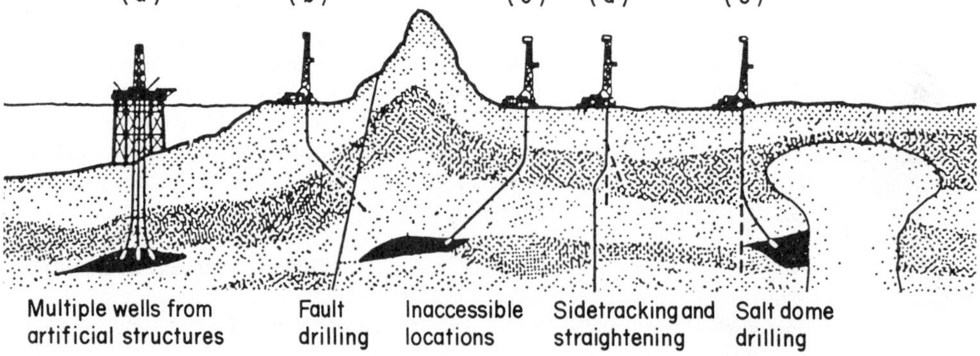

Fig. 8.8. Directional drilling applications[3]. (Courtesy of Eastman Whipstock)

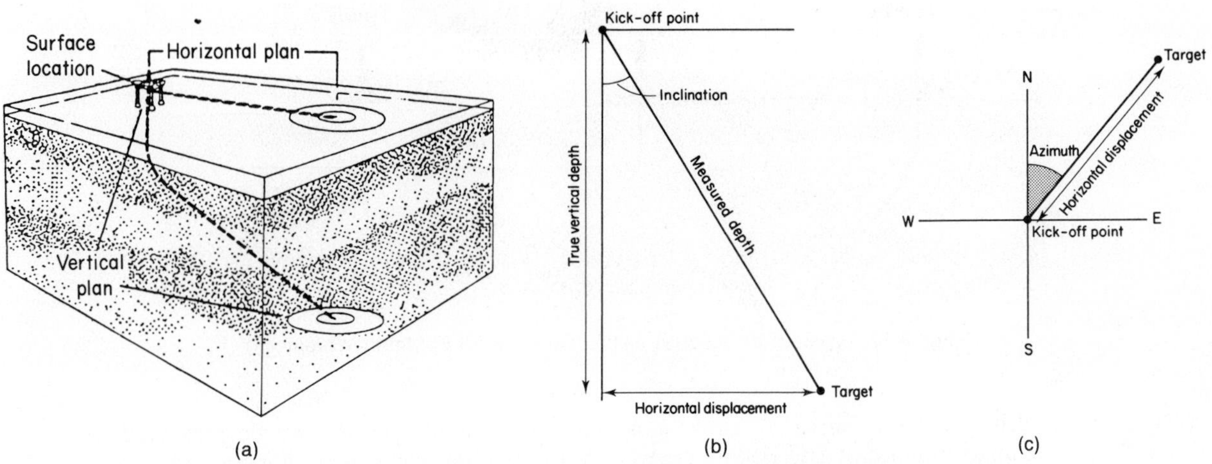

Fig. 8.9. A directional well: (a) three-dimensional view; (b) vertical section; (c) horizontal section.

In practice, two northern directions are recognised, geographic and magnetic norths. Geographic north refers to the north pole, while magnetic north specifies the northern direction of the earth's magnetic field. It is, in practice, normal to measure the magnetic north with a magnetic compass.

The two northern directions of the earth rarely coincide and a correction is normally applied to magnetic north to obtain the true geographic north. This correction is described as 'magnetic declination'.

(3) *Vertical depth* is the true vertical depth from surface to the target.

(4) *Horizontal displacement* is the horizontal distance of the target zone from the platform reference point.

The horizontal displacement and azimuth of the target at any point on the well path can be used to determine the northing and easting coordinates of the point in question (see page 163).

(5) *Dog-leg* is defined as the angular change between two points on the well path, and can result from changes in inclination, direction or both. The dog-leg over a specified depth interval (e.g. 100 ft) is termed 'dog-leg severity'.

(6) *Kick-off point* is the point at which the well is deflected from the vertical.

Types of directional wells

In general, there are three basic patterns of directional wells (Figure 8.10).

(1) *Type I.* A directional well of this type is deflected at a shallow depth and the inclination is locked in until the target zone is penetrated. Wells of Type I are used for[3]: (a) moderate depth drilling; (b) single-zone production; (c) wells requiring no intermediate casing; and (d) deep wells requiring large lateral displacement.

(2) *Type II.* This type is also called the S-type well, as its shape resembles the letter S (Figure 8.10). The well is deflected at a shallow depth until the maximum required inclination is achieved. The well path is then locked in and, finally, the inclination is reduced to a lower value or, in some cases, the well is returned back to vertical by gradually dropping off the angle. This type is mainly practised in multi-zone production and for relief wells, and requires very close supervision.

(3) *Type III.* This is similar to Type I, except that the well is deflected at a much deeper position in order to avoid, for example, a salt dome. Type III is also practised in sidetracking and wildcat drilling.

Bottom hole assemblies (BHAs) for directional wells

Depending on the section of the deviated well being drilled, a special BHA is required to maintain or drop deviation. From practical experience, the following BHAs were found suitable for the various parts of the deviated well:

(1) *Build-up section.* Bit/near-bit stabiliser/two or three non-magnetic drill collars/stabiliser/drill collar/stabiliser/3 drill collars/stabiliser/required number of drill collars/heavy-walled drillpipe/drillpipe.

(2) *Tangent (or locked-in) section.* Bit/near-bit sub/non-magnetic drill collar/stabiliser/drill collar/stabiliser/drill collar/stabiliser/3 drill collars/stabiliser/required number of drill collars/heavy-walled drillpipe/drillpipe.

(3) *Drop-off section.* Bit/non-magnetic drill collar/drill

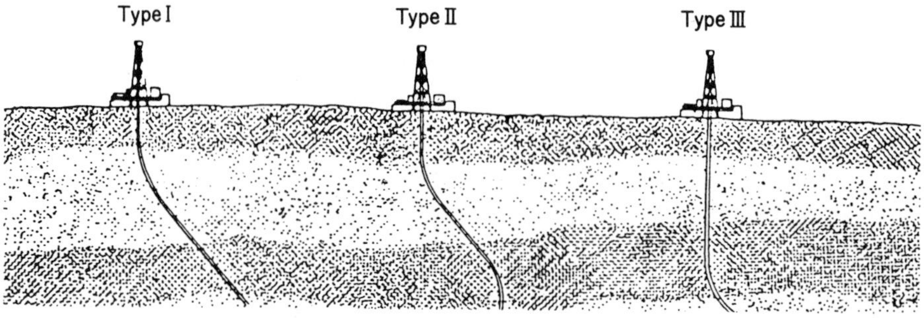

Fig. 8.10. Types of directional well[3]. (Courtesy of Eastman Whipstock)

collar/stabiliser/drill collar/stabiliser/3 drill collars/stabiliser/required number of drill collars/heavy-walled drillpipe/drillpipe.

Type (2) BHA is a packed-hole assembly, in which hole angle is maintained by packing the hole with stabilisers as discussed above (page 150).

Type (3) BHA is a pendulum assembly, in which the active length of the drill collars is used to return the hole angle back to vertical.

Directional surveying instruments

Vertical and directional wells are surveyed for the following reasons[4]:

(1) To monitor the progress of the well. Actual directional data can be used to plot the course of the well and can then be compared with the planned course. Any deviation can be corrected and the well brought back to the required direction.
(2) To prevent collision of the present well with nearby wells, as in platform drilling.
(3) To determine the amount of orientation required to position deflection tools in the appropriate direction.
(4) To determine the exact location of the bottom hole in terms of vertical depth, departure, azimuth, northing and easting. This information may be required in the event of a blowout, when a relief well is required to kill the well.
(5) To calculate the values of dog-leg severity.

Several types of surveying instruments are in use, including single- and multi-shot magnetic instruments and gyroscopes. Survey instruments can be dropped from the surface (except the gyro) or can be run on wire line to land in non-magnetic drill collar, commonly known as K-Monel.

Magnetic single shot

The magnetic single-shot instrument records, simultaneously, hole inclination and magnetic north direction of an uncased hole at a single measured depth (or station). The instrument consists of (a) an angle-indicating unit, (b) a camera section, (c) a timing device and (d) a battery pack. Figure 8.11 gives a schematic

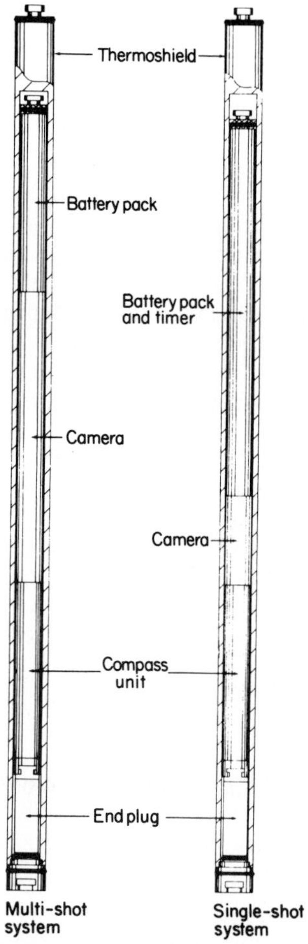

Fig. 8.11. Schematic drawing of magnetic survey instruments[4]. (Courtesy of Scientific Drilling Controls)

drawing of the magnetic single and multi-shot instruments.

The angle-indicating unit comprises a magnetic compass card and pendulum assembly. The compass card is pivoted on a single point and is divided into 360 degrees in five-degree increments[4]. The card is kept level by a suitable mechanism in order to eliminate the effects of the vertical component of the earth's magnetic field. The magnetic north recorded by the card is a measure of the deflection caused by the horizontal component of the earth's magnetic field.

The pendulum assembly consists of either a plumb bob or an inclinometer unit, and has a glass target between the pendulum and the compass[4]. The glass plate has a series of concentric circles in increments of one degree to represent drift or inclination angle.

The entire angle-indicating unit is built into one case and is fluid-filled to dampen movement, thereby protecting all the internal parts from shock damage.

A sectional drawing of the plumb-bob unit is given in Figure 8.12.

The camera section is placed above the angle unit and comprises a lamp assembly, lens and film holder. The camera has no shutter, and its exposure is controlled by the time for which the light is on.

The timing device can be either mechanical or electronic, and is used to perform two functions. First, it holds the action of the camera until the single-shot instrument is located in the non-magnetic collar above the bit. Second, it controls the time for which the lamps are switched on for proper film exposure.

The battery pack supplies electric current (power) for the timer and the camera.

The magnetic single-shot instrument is run on a wire line to hole bottom or dropped from the surface, where it sits on a baffle plate placed on top of the bit. When the instrument is seated on the baffle plate, the timing device activates the camera and a photographic record of the positions of the plumb bob and the magnetic compass card is made. The magnetic compass card will align itself with the north magnetic field and it is this position that is photographed. The instrument is then pulled out of hole and the film is developed and placed in a chart recorder. Hole inclination is determined as the number of concentric rings

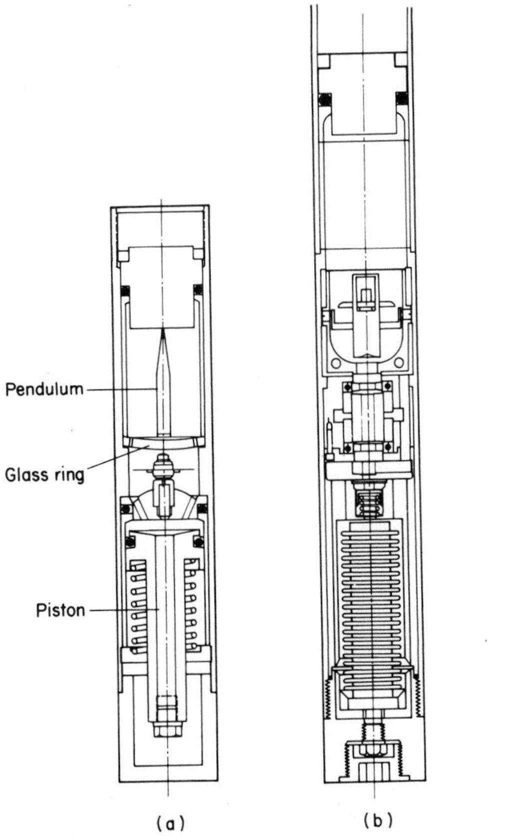

Fig. 8.12. Pendulum assembly: (a) plumb-bob angle unit; (b) drift arc inclinometer.

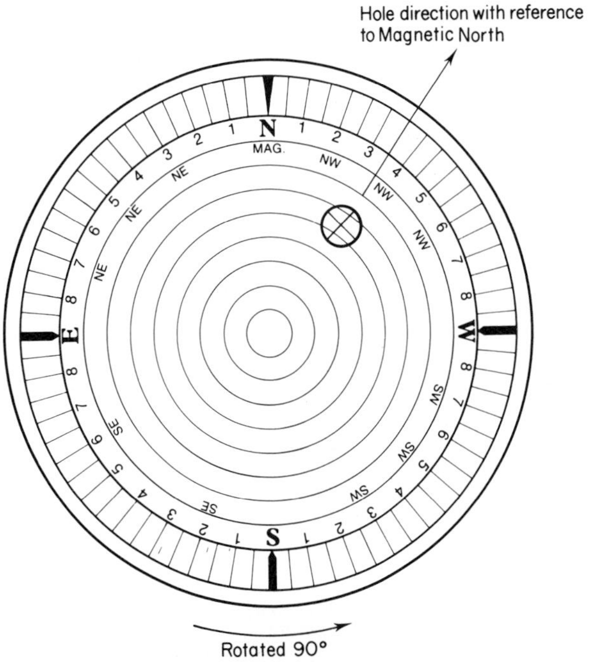

Fig. 8.13. Schematic drawing of magnetic single and multi-shot instruments.

(or concentric circles) from the centre, as shown in Figure 8.13. Each concentric circle is equivalent to 1°. Hole direction or azimuth is obtained by drawing a line from the centre dot through the intersection of the plumb bob hairs to the edge of the compass card, as shown in Figure 8.13. Hole deviation and direction in Figure 8.13 are $5\frac{1}{2}$° and N35°W, respectively.

It should be observed that the magnetic compass only records the direction of the horizontal component of the earth's magnetic field force at the well in question. This component gives the direction of magnetic north. The earth's magnetic north was found to be slightly displaced from the earth's geographic north. The angle between magnetic and geographic norths is described as the declination angle or magnetic declination. Magnetic north can be to the east (+ve) or west of the geographic north, as shown in Figure 8.14. When the magnetic north is to the east of the geographic north, the declination angle should be added to the observed bearing with respect to magnetic north to obtain the bearing with respect to geographic north. Bearing with respect to true north is described as 'true bearing'. West declination is subtracted from the observed magnetic bearing to obtain the true bearing.

It is, therefore, important to determine the magnetic declination of the field being developed and to add or subtract this value to or from the observed magnetic bearing in order to determine the true bearing of the well with respect to geographic north.

The following example shows the calculations required to convert magnetic bearings to true bearings and true azimuth.

Example 8.1

Determine the azimuth with respect to true north of the following wells:

Well	*Observed bearing with respect to magnetic north*	*Declination*
1	N45°E	3° west
2	N45°E	3° east
3	S80°W	5° west

Solution

True north = magnetic north ± (declination)

Well 1

$$\text{Bearing with respect to true north} = 45° + (-3°) = 42°$$

Azimuth = N42°E

(*Note:* Azimuth is the angle measured with respect to true north.)

Well 2

$$\text{Bearing with respect to true north} = 45° + (+3°) = 48°$$

Azimuth = N48°E

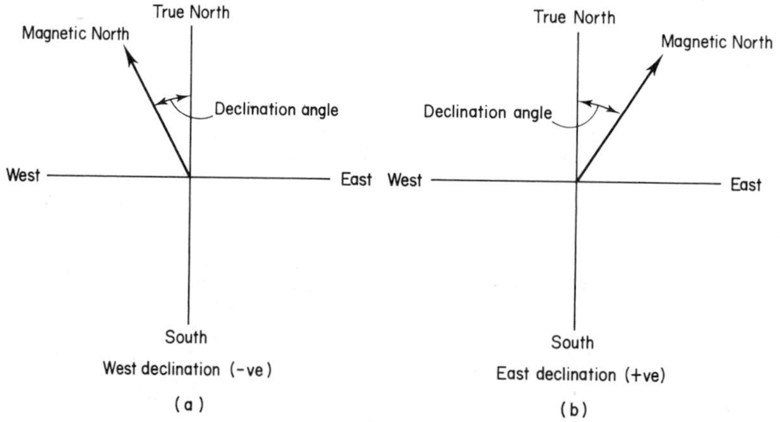

Fig. 8.14. Definition of declination angle: (a) west declination (−ve); (b) east declination (+ve). (*Note*: True north is geographic north)

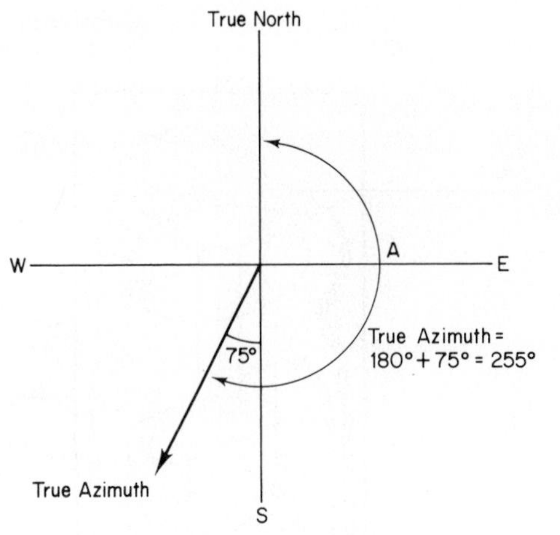

Fig. 8.15. Example 8.1.

Well 3 Referring to Figure 8.15,

$$\text{Bearing with respect to true north} = \text{S}80^\circ\text{W} + (-5^\circ)$$
$$= \text{S}75^\circ\text{W}$$
$$\text{Azimuth} = \text{angle A}$$
$$= 180^\circ + 75^\circ$$
$$= 255^\circ$$

Quality of readings depends on location with respect to earth's surface, hole inclination and azimuth.

The magnetic single-shot readings are liable to distortion by magnetic effects from the drill string, resulting from induction effects produced by the earth's magnetic field. For this reason, the drill string normally incorporates at least one non-magnetic drill collar (i.e. K-Monel) where the survey instrument is landed. The K-Monel will protect the magnetic compass from unwanted distortions.

Magnetic multi-shot

The magnetic multi-shot instrument is similar to the single-shot, except that a film magazine is used so that inclination and magnetic north are recorded at frequent intervals (every 15 to 20 seconds). Also, the camera is a modified movie camera.

The instrument can be run on a wire line to land in a non-magnetic collar or simply dropped from the surface. A continuous record of the inclination and azimuth of the open hole interval is obtained as the drill string is pulled out of hole. Surveys are taken each time a stand is 'broken off', when the remainder of the drill string is stationary.

Magnetic multi-shot instruments are suitable for wells free from magnetic materials or magnetic geological formations.

A new generation of multi-shot tools survey the well bore and store the information in digital form in downhole memory for later recovery.

Gyroscopes

The magnetic north (or azimuth) recorded by most magnetic survey instruments is liable to error, due to magnetic disturbances caused by casings of nearby wells.

To overcome these errors, gyroscopes are used because of their ability to maintain a preset direction, irrespective of the earth's magnetic field or factors disturbing it. Gyros are most widely used in orienting mud motors and bent sub assemblies in areas where magnetic instruments will be influenced by magnetic fields of nearby casings, as in platform drilling.

The principal element of the gyro is a rapidly spinning weighted wheel (Figure 8.16a), which maintains a preset direction by resisting changes in direction, owing to its inertia. Thus, if the wheel is made to rotate parallel to the earth's axis of rotation (i.e. N–S direction), then it will maintain this direction, thereby serving as a direction indicator.

The axis of the wheel is horizontal and is mounted inside two perpendicular gimbals to allow the gyro to maintain its direction irrespective of gyro casing movement[6] (see Figure 8.16b). The inner gimbal is made up of the spin motor and the housing. The motor spins at a rate between 18 000 and 42 000 rpm, depending on gyro size and type of spin motor used.

The outer gimbal is fitted with a compass card similar to that used in the magnetic single shot. The compass card is set to a known direction, normally to the north, to serve as a reference for survey directional data.

The pendulum assembly, a single-shot camera (for gyro single shot) and a timer are fitted above the compass card. The orientation of the gyro compass must be checked prior to running in hole, which is done by aligning the tool to a known reference direction by use of a telescope alignment kit. For single shot the gyro assembly is then run on wire line to hole bottom. A picture of the angle-indicating unit (i.e. plumb bob and gyro compass) is taken at the required depth. The survey tool is pulled out of hole to surface and the gyro film is retrieved. The film is developed and placed in a chart reader to determine hole direction and hole inclination.

For gyro multi-shot, the procedure is to lower the tool on wire line through a section of cased well bore. The tool is stopped at set depth intervals, e.g. every 100 ft, and a survey is taken by the multi-shot camera. Some check surveys are taken on the outrun, typically every 400 ft.

All gyros have a tendency to drift from their preset direction, and survey readings should be corrected for this effect. Drift is due entirely to friction in the

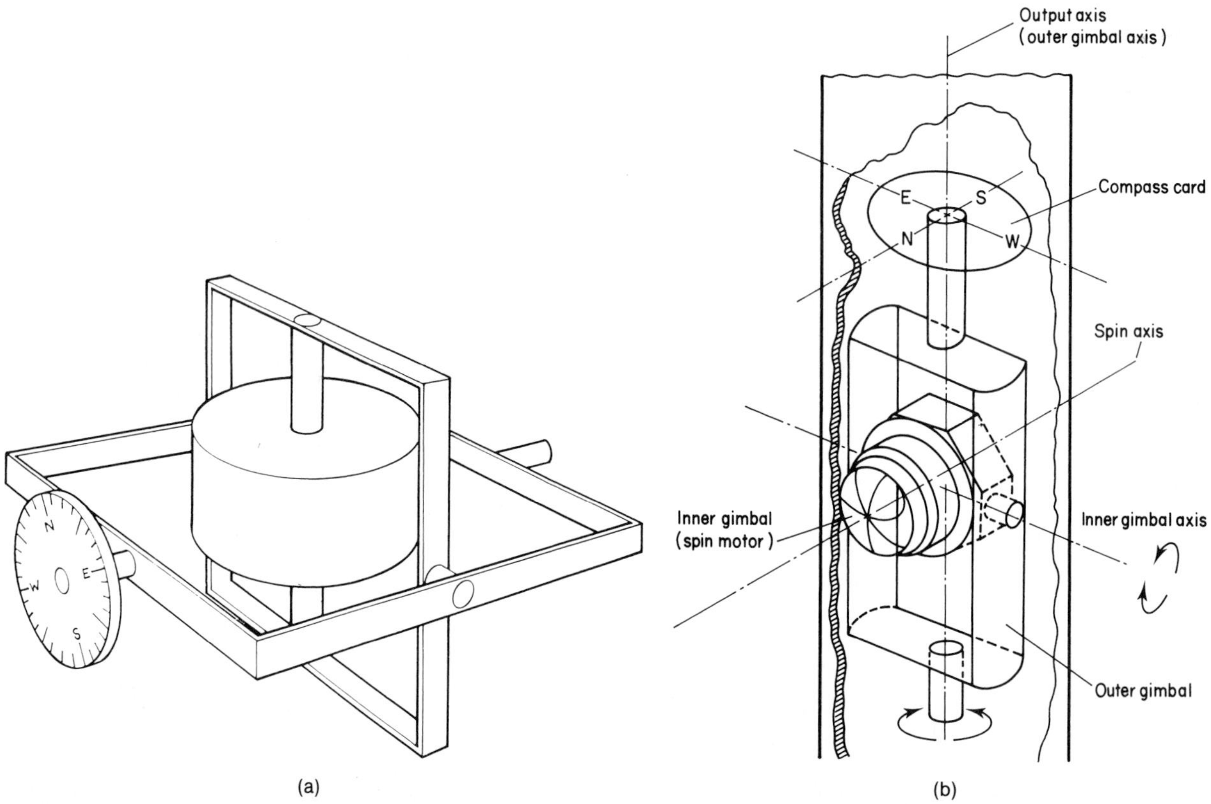

Fig. 8.16. (a) Principle of the gyroscope[4]. (Courtesy of Scientific Drilling Controls) (b) Gyroscope surveying downhole tool[5]. Gimbal configuration allows the gyroscope to maintain its orientation regardless of gyroscope casing movement. Spin motor and housing make up the inner gimbal, and the outer gimbal is attached to the compass card. The outer gimbal can rotate 360°. This permits the gyroscope casing to move around the gyroscope while the gyroscope compass card maintains its primary orientation. Likewise, the inner gimbal axis allows the casing to tilt without disturbing the gyroscope. (After Plight and Blount, 1980)[5]

bearings of the spinning wheel, gyro imbalance and motor speed instability[5]. During surveying, the tool is kept stationary every 15 min for 5 min, to monitor gyro drift.

Deflection tools

The required degree of deviation from the vertical is initiated by employing a special deflection tool. There exist a variety of deflection tools, each suitable for a particular formation type. The following is a summary of the most common types.

Bent sub

The bent sub is a small collar with a tool joint specially machined to provide an offset, as shown in Figure 8.17. The upper thread is cut concentric to the

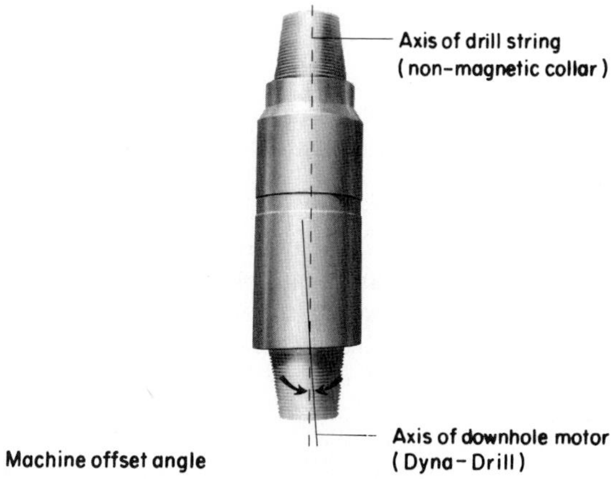

Fig. 8.17. Bent sub.

sub axis, while the lower thread is inclined at 1–3° to the axis of the upper thread.

The bent sub is used in conjunction with a downhole motor, and the direction in which the bent sub is faced determines the direction of deviation.

The offset angle creates a deflection in the bottom hole assembly (BHA), resulting in a smooth arc in the drilled section of the hole. The bent sub can also contain a muleshoe orienting sleeve and a key. The orienting sleeve is aligned with a scribe line on the outside of the bent sub. The orienting sleeve provides a means of determining the actual orientation of the BHA.

The drill string containing the bent sub is run to the hole bottom where deflection is required and a survey instrument (see page 163) is run inside the BHA. The survey instrument contains a camera which takes a photograph of the orientation of the bent sub sleeve. The photograph is analysed at the surface and the present orientation is determined. This is then compared with the planned orientation, and the drill string may be rotated to the left or right to position the BHA in the required direction.

Whipstock

The whipstock consists of a long, inverted steel wedge concave on the inside so that it can hold the BHA (Figure 8.18). The whipstock is normally attached to a spiral stabiliser by a shearing pin which can be broken by simply setting weight on the bit. The spiral stabiliser is placed directly above the bit so that the bit is forced to follow the curve of the whipstock. The whipstock also contains a chisel point, as shown in Figure 8.18, to anchor the whipstock to the hole bottom and prevent it from rotating during drilling.

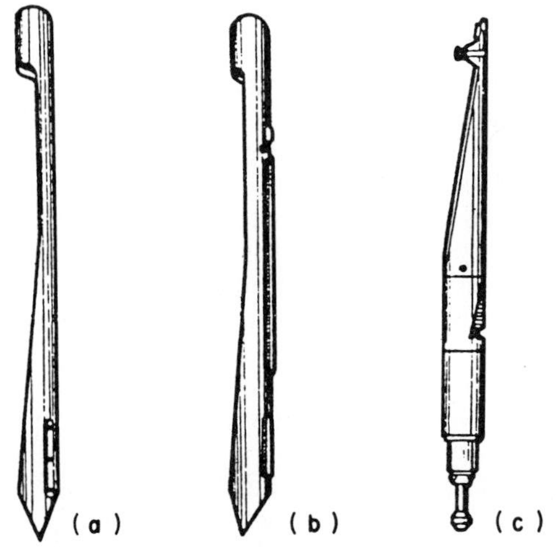

Fig. 8.18. Types of whipstock[3]. (Courtesy of Eastman Whipstock)

A drill string containing a whipstock is run to the hole bottom and weight is applied to break the shear pin to disengage the whipstock and set it firmly on the hole bottom. A small-sized bit is used to initiate a pilot hole with a diameter smaller than the final hole diameter. The pilot hole is drilled for a distance of 10–15 ft and then surveyed. If the hole is drilled in the required direction, it is then opened up using a normal drill bit.

There are three types of whipstock in common use:

(1) The standard whipstock (Figure 8.18a).

(2) The circulating whipstock (Figure 8.18b). This contains a circulating port intended to clean the bottom of the hole so that a firm seat is provided for the whipstock. This type is used in holes with problems of bottom hole fills.

(3) The permanent casing whipstock (Figure 8.18c). This is used to by-pass collapsed casing or fish in the cased section of the hole.

Whipstocks are now only run on land in some parts of the world and are largely replaced by downhole mud motors and bent subs.

Jet bit

A modified bit containing two small jets and one large nozzle is run to the bottom of the hole and oriented in the required direction. The size of the large nozzle is in the range 3/4–7/8 in[7]. The hole is then jetted, the maximum possible circulation rate being used until a pilot hole is drilled. The hole is surveyed and, if it is in the required direction, a normal drill bit with a proper BHA is run to drill the deviated section of the well.

Jetting is only applicable in soft rocks, where the rock matrix can be crushed by the compressive forces of the pumped mud (usually less than 3000 ft depth).

Downhole motors

Deflection of holes can be achieved by the use of downhole motors driven by the circulating mud. The downhole motor provides rotation for the bit, thereby eliminating the need to rotate the drill string from the surface. The downhole motor is normally placed above the bit, and attached to it directly is a bent sub having the appropriate offset angle. Figure 8.19 gives a schematic drawing of the equipment required with a downhole motor.

The main advantages of downhole motors are: (a) they drill a smoothly curved hole in both build-up and drop-off sections; (b) they give better control of dog-leg severity and better hole conditions; (c) they permit the use of surface read-out measurement-while-drilling (MWD) tools, which give continuous read-out

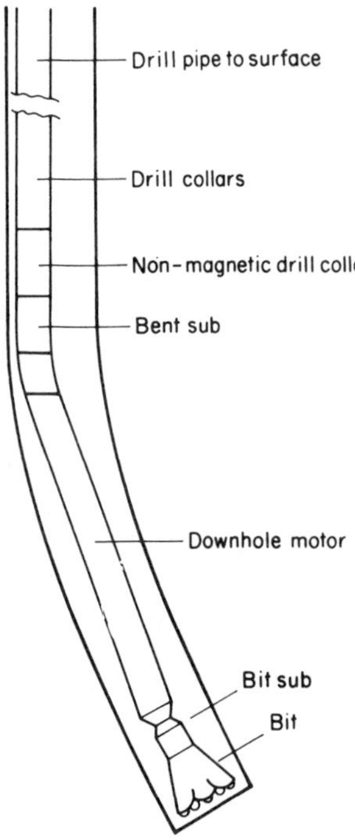

Fig. 8.19. Drilling assembly using a downhole motor and a bent sub.

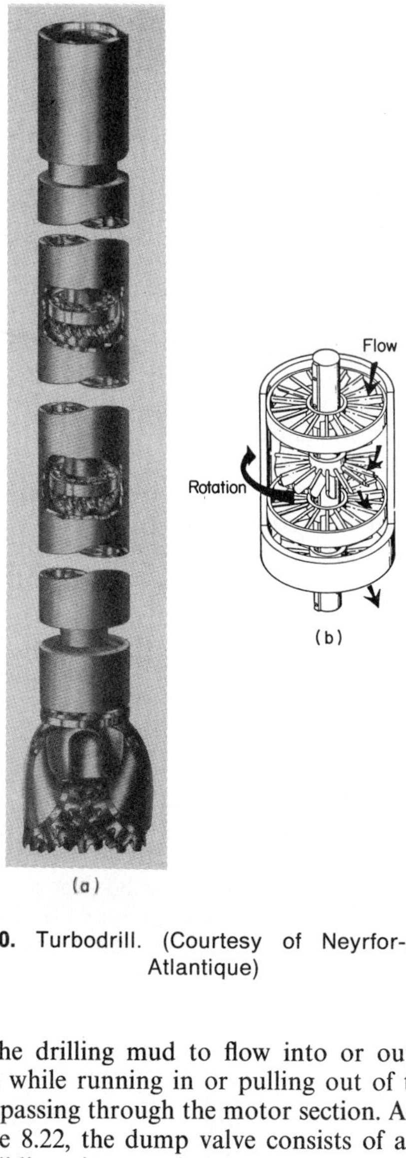

Fig. 8.20. Turbodrill. (Courtesy of Neyrfor-Alsthom-Atlantique)

of survey data, including tool face orientation.

There are two types of downhole motor: (1) turbine-type motor (turbodrill) and (2) positive displacement motor.

Turbine-type motor The turbine-type motor consists of multi-stage blade-type motor and stator sections, a thrust-bearing section and a drive shaft. The number of rotor/stator sections can vary from 25 to 250.

The stator (Figure 8.20) remains stationary and its main function is to deflect the mud to the rotor blades. The rotor blades are attached to the drive shaft, which, in turn, is attached to the bit. Mud under higher pressure is pumped down the drill string to the motor section, where it is deflected by the stator blades to the rotor blades. This, then, imparts rotation to the rotor and, in turn, to the drive shaft and drill bit.

Positive displacement motor The positive displacement motor is a type of pump used in reverse application such that high pressure is translated into rotation. A positive displacement motor (such as a Dyna-drill) consists of four main components: (1) the dump valve; (2) the motor assembly; (3) the connecting rod assembly; and (4) the bearing assembly.

Figure 8.21 shows a positive displacement motor with its main four components. The dump valve allows the drilling mud to flow into or out of the drillpipe while running in or pulling out of the hole without passing through the motor section. As shown in Figure 8.22, the dump valve consists of a spring-loaded sliding piston, a sleeve seat and external ports. Under low pressures (0–25 psi), the spring holds the piston off its seat, thereby allowing mud to vent out of the external port (Figure 8.22a). When mud pressure exceeds 25 psi, as during drilling, the piston is forced down, thereby closing the external ports and directing the high-pressured mud to the motor assembly (Figure 8.22b).

The motor assembly consists of a rotor and a stator. The stator is the outer tube and has a spherical spiral cavity lined with a resistant rubber-like material. The rotor is a solid steel spiral shaft with a circular cross-section, as shown in Figure 8.23.

As shown in Figure 8.23, the final profile of the internal motor is made up of small cavities between

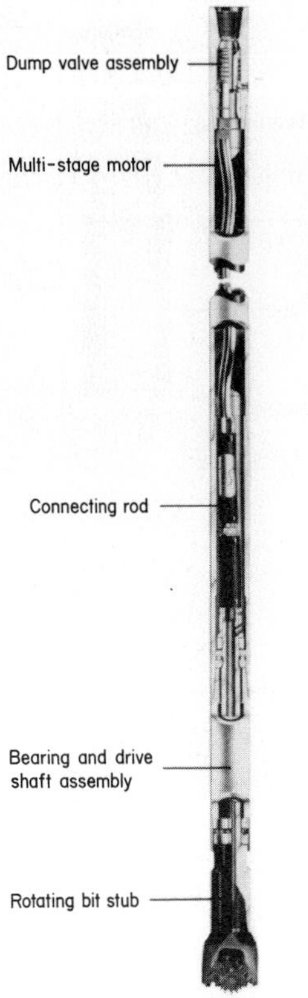

Fig. 8.21. The Dyna-Drill dump valve[6]. (Courtesy of Dyna-Drill)

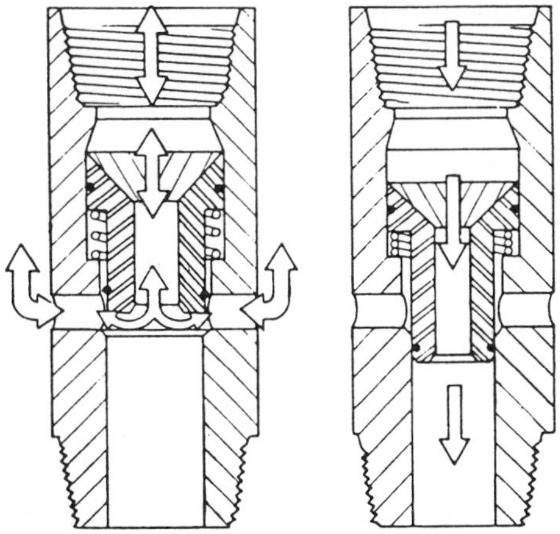

Fig. 8.22. Dump valve[6] (a) open: 0–25 psi (no pump); (b) closed: 25 psi or more (pump on). (Courtesy of Dyna-Drill)

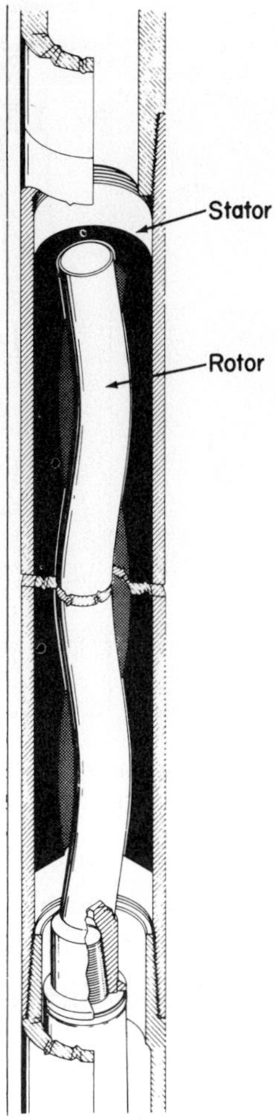

Fig. 8.23. Motor assembly of a Dyna-Drill[6]. (Courtesy of Dyna-Drill)

the rotor and stator. Rotation occurs when mud, pumped under high pressure is forced into the cavities between rotor and stator.[6]

The connecting rod assembly simply transmits the rotation to the drive shaft assembly, which is connected to the drill bit. The rod also converts the eccentric rotation of the rotor to the concentric (axial) rotation of the drive shaft (Figure 8.24).

The bearing section is shown in Figure 8.25. The upper thrust bearing protects the tool from piston loads when running off the bottom or without bit loads. The radial support bearing provides radial support to the drive shaft and also directs 5–10% of volume flow rate to the lower thrust bearing for cooling and lubrication.

The main advantages of the positive displacement

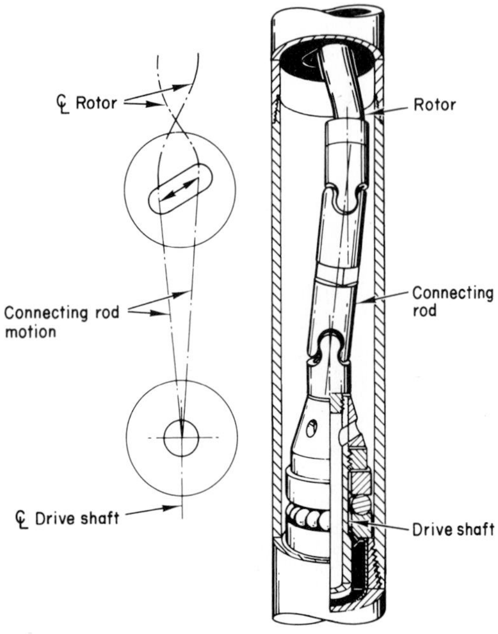

Fig. 8.24. Connecting rod assembly of a Dyna-Drill[6]. (Courtesy of Dyna-Drill)

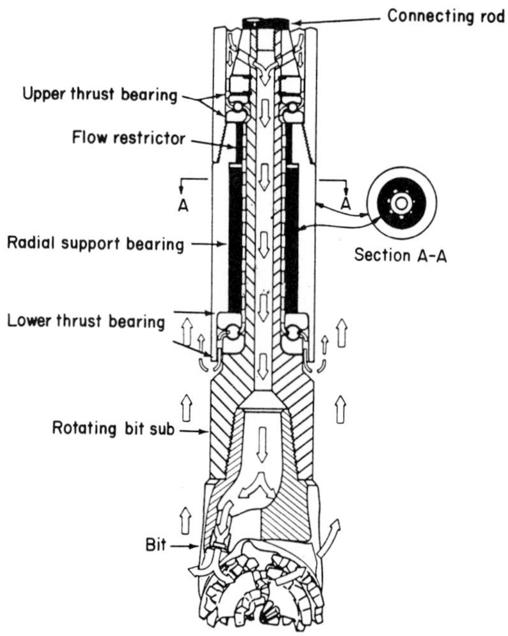

Fig. 8.25. Bearing assembly of a Dyna-Drill[6]. (Courtesy of Dyna-Drill)

motor are that it can be operated by mud, gas or air, and can tolerate up to 10% lost circulation material.

Orientation of deflection tools

Survey instruments can be used to determine the present orientation of the deflection tool at hole bottom. Very often the present orientation of the deflection tool does not coincide with the required hole direction and the bottom hole assembly must be rotated from the surface to reorient the deflection tool.

The amount of orientation required can be obtained by use of a graphical method known as the Ragland vector diagram. The Ragland diagram can be constructed from the following data: (a) initial hole deviation and direction and (b) maximum permissible dog-leg.

Example 8.2

Using a graphical method, determine the required orientation of the deflection tool to change the hole direction from N30°E to N40°E, assuming that hole deviation is 7° and maximum dog-leg severity is 2°/100 ft. Also determine the new hole inclination.

Solution (Figure 8.26)

(1) Draw a horizontal line of 7 units in length representing hole inclination. The direction of this line is N30°E. Also, mark the ends of this line as O at zero, and C at 7 units.

(2) Plot a circle of 2 units in radius with its centre at C. The 2 units represent the maximum dog-leg severity.

(3) From point O draw a line at 10° to OC. The 10° represents the difference in azimuth between the present hole position and final hole position, i.e. N40°E − N30°E = 10°. Mark this line as OM. Line OM gives the deviation and direction of the final hole.

(4) Angle θ between OC and radius CM gives the required orientation of the tool. From Figure 8.26, the value of angle θ ia 47°. Thus, to obtain a final hole direction of N40°E, the deflecting tool should be

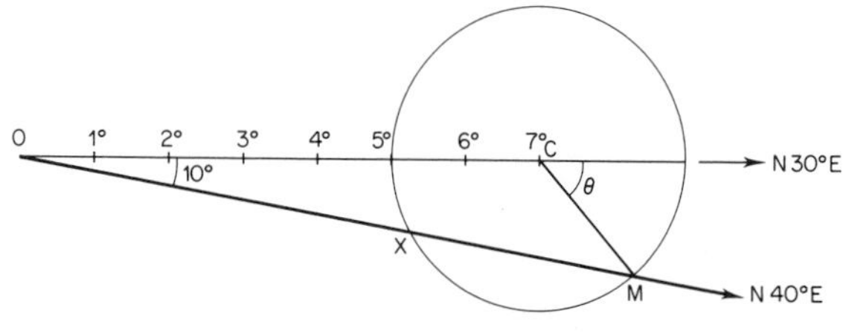

Fig. 8.26. Ragland vector diagram (Example 8.2): θ = orientation of deflection tool with respect to hole bottom; by measurement $\theta = 47°$.

oriented at 47° to the right of the present hole direction of N30°E. Hence, the required settting of the deflection tool is N(47 + 30)°E, or N77°E.

(5) The new hole inclination is represented by the length of line OM. From Figure 8.26, the new hole inclination is 8.5° at N40°E.

Example 8.3

Using the data of Example 8.2, determine the alternative tool face setting and the expected new hole inclination.

Solution

From Figure 8.27, the angle between lines CE (dotted line) and CX represents the alternative tool face setting. The length of line OX represents the resulting hole deviation.

From Figure 8.27,

$$\text{tool face setting} = 153^\circ$$

and

$$\text{hole inclination} = 5.3^\circ$$

Methods of orientation

In this section only two methods will be discussed — the muleshoe and the magnetic. A recent method for orientation utilises mud-pulse measurement-while-drilling (MWD) tools. The MWD method also provides information such as hole inclination and hole azimuth. The MWD method relies on the principle that downhole information can be transmitted through a flowing mud in the form of pressure pulses. These pulses are then processed at the surface by surface read-out equipment.

Muleshoe method The muleshoe method uses an orienting sub known as a 'muleshoe'. The muleshoe contains a key which can be aligned with the scribe line on the deflecting tool (e.g. bent sub or whipstock).

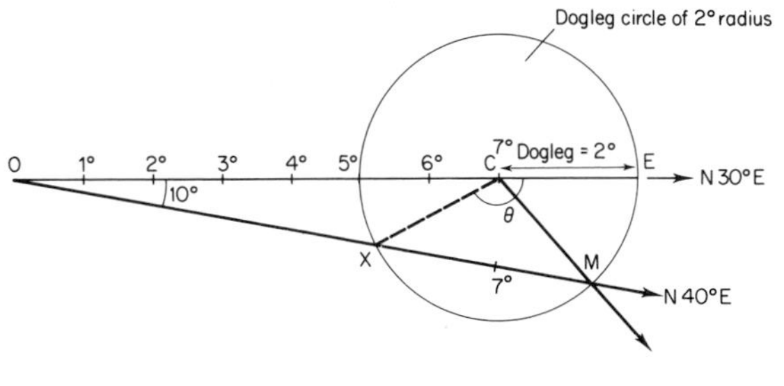

Fig. 8.27. Ragland vector diagram (Example 8.3): By measurement $\theta = 153^\circ$.

A non-magnetic collar is connected to the orienting sub and the complete drill string is run to hole bottom. A single-shot survey instrument is run to the orienting sub, where it slides into the orienting sub key through a special keyway cut in its surface. The reference line of the deflecting tool (i.e. the key) is then recorded in direct relation to the magnetic north as registered by the magnetic compass of the survey instrument.

On the surface the developed film will graphically show the relationship between tool face orientation, hole inclination and magnetic north. By use of simple arithmetic, the actual orientation of the tool is determined. If the tool is not facing in the required direction, the amount of rotation (in degrees) is then applied at the surface to position the tool in the appropriate direction.

Magnetic method The magnetic method is suitable for hole deflection created by jetting where an orienting sub is not required. It uses a non-magnetic collar containing six orienting magnets arranged in physical alignment with the deflecting tool face or with mathematical compensation. The survey (or orienting) instrument comprises a swinging pendulum, a regular compass and a 'needle-type' compass, which will be locked in place by the magnets in the collar[3].

The survey instrument is run on a wire line inside the drill pipe until it lands in the non-magnetic drill collar; the needle-type compass will be attracted to the orienting magnets; and a camera will photograph this position, superimposed on the regular compass. Thus, the relative position of the orienting magnets (and, in turn, the tool face direction) and direction of hole are determined. The hole deviation is indicated by the swinging pendulum.

Calculation methods

Directional surveys are normally run at intervals of 100 ft (30 m) to monitor the well progress while drilling. If the well is found to deviate from its planned direction, correction measures can be taken to bring the well back onto course. Hence, survey data provide a pictorial image of the path of the well and the data are normally plotted on the same graph as that of the planned course.

Various methods have been proposed for determining the well geometry from the measured survey data. The following are the most common types: (1) tangenttial method; (2) balanced tangential method; (3) angle averaging method; (4) radius of curvature method; (5) minimum curvature method; and (6) mercury method. Much of the following discussion is based on a paper by Craig and Randall[8].

The various calculation methods will be presented for an idealised case in which the well path is in a single plane, building up at a constant rate of inclination of (say) 3°/100 ft or (0.98°/10 m). Relevant data required for each method include inclination angle, I, azimuth, A, and measured hole depth, MD. The intention is to calculate the hole true vertical depth, VD, the horizontal displacement or departure, D, and the easting and northing co-ordinates of the new hole position.

Tangential method

The principle of the tangential method is to use the inclination and azimuth (direction) angles at the lower end of the course only for determining VD, D, easting and northing. Referring to Figure 8.28, assume that the measured inclination and azimuth at Station 2 are I_2 and A_2, respectively. Then

$$\Delta VD = \Delta MD \cos I_2 \quad (8.2)$$

$$\Delta D = \Delta MD \sin I_2 \quad (8.3)$$

$$\Delta E = \Delta D \sin A_2 = \Delta MD \sin I_2 \sin A_2 \quad (8.4)$$

$$\Delta N = \Delta D \cos A_2 = \Delta MD \sin I_2 \cos A_2 \quad (8.5)$$

where ΔMD = increment of course length; ΔVD = increment of true vertical depth; ΔD = increment of horizontal displacement or departure; ΔE = increment of easting co-ordinate; and ΔN = increment of northing co-ordinate.

This method has the advantage of being the most simple to use; however, calculations based on it result in greater lateral displacement and less vertical displacement. The errors are proportional to the magnitudes of measured angles and course length[8]. This method is now obsolete.

Balanced tangential method

The balanced tangential method uses the measured angles at the top and the bottom of the course, which results in tangentially balanced survey calculations and, in turn, a smoother well-bore profile.

Referring to Figure 8.29, inclination angles I_1 and I_2 and azimuth angles A_1 and A_2 are measured at two survey stations. The measured hole course, ΔMD, is divided into two equal increments of ΔMD/2. The top increment is used in conjunction with angles I_1 and A_1 to calculate horizontal displacement D_1, vertical depth increment, ΔVD_1, and increments of easting and northing, ΔE_1 and ΔN_1, respectively. Similarly, the bottom increment of ΔMD/2 is used to calculate ΔD_2, ΔVD_2, ΔE_2 and ΔN_2.

Thus,

$$\Delta D_1 = \left(\frac{\Delta MD}{2}\right) \sin I_1 \quad (8.5)$$

$$\Delta D_2 = \left(\frac{\Delta MD}{2}\right) \sin I_2 \quad (8.6)$$

total departure or horizontal displacement,

$$\Delta D = \Delta D_1 + \Delta D_2 = \frac{\Delta MD}{2}(\sin I_1 + \sin I_2) \quad (8.7)$$

$$\Delta VD_1 = \frac{\Delta MD}{2} \cos I_1$$

$$\Delta VD_2 = \frac{\Delta MD}{2} \cos I_2$$

The total increment of true vertical depth between station 1 and station 2 is

$$\Delta VD = \Delta VD_1 + \Delta VD_2$$

$$= \frac{\Delta MD}{2}(\cos I_1 + \cos I_2) \quad (8.8)$$

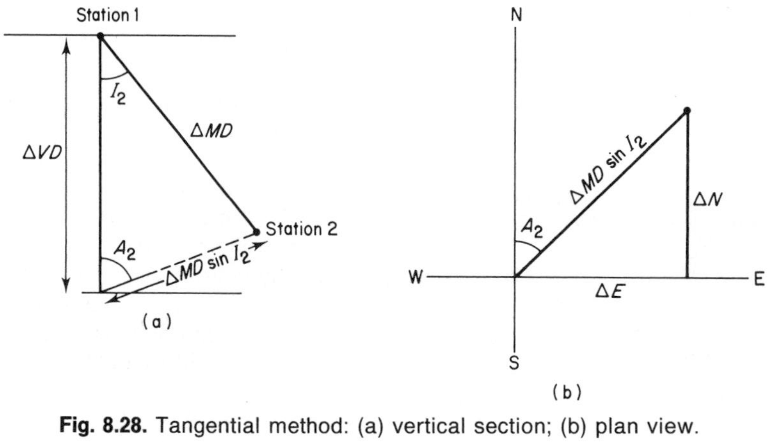

Fig. 8.28. Tangential method: (a) vertical section; (b) plan view.

Also,

$$\Delta N = \Delta N_1 + \Delta N_2 = \Delta D_1 \cos A_1 + \Delta D_2 \cos A_2$$

$$= \frac{\Delta MD}{2}(\sin I_1 \cos A_1 + \sin I_2 \cos A_2) \quad (8.9)$$

$$\Delta E = \Delta E_1 + \Delta E_2$$

$$\Delta E = \frac{\Delta MD}{2}(\sin I_1 \sin A_1 + \sin I_2 \sin A_2) \quad (8.10)$$

Angle averaging method

As the name implies, in the angle averaging method inclination angles I_1 and I_2 and direction angles A_1 and A_2 are averaged and used for the calculation of departure, total vertical depth, and easting and northing increments. This method is simple and provides accurate means of calculating wellbore surveys[8].

Referring to Figure 8.29, average values of inclination and direction angles are $(I_1 + I_2)/2$ and $(A_1 + A_2)/2$, respectively. Thus,

$$\Delta D = \Delta MD \sin\left(\frac{I_1 + I_2}{2}\right) \quad (8.11)$$

$$\Delta VD = \Delta MD \cos\left(\frac{I_1 + I_2}{2}\right) \quad (8.12)$$

$$\Delta E = \Delta MD \sin\left(\frac{I_1 + I_2}{2}\right) \sin\left(\frac{A_1 + A_2}{2}\right) \quad (8.13)$$

$$\Delta N = \Delta MD \sin\left(\frac{I_1 + I_2}{2}\right) \cos\left(\frac{A_1 + A_2}{2}\right) \quad (8.14)$$

Radius of curvature method

The radius of curvature method uses the top and bottom angles to generate a space curve having a spherical arc shape which passes through the two stations[8]. This effectively assumes that the well bore is smooth, which makes the method less sensitive to changes in position, especially if a severe dog-leg exists in the interval[8].

Assume that the well-bore section measured between two stations is represented by segment *ab*, as shown in Figure 8.30.

$$\text{Rate of change in inclination} = \frac{I_2 - I_1}{\Delta MD}$$

where I_2 and I_1 are in radians.

Radius of curvature of segment *ab*

$$= \frac{1}{\text{rate of change in inclination}}$$

or

$$R = \frac{1}{\dfrac{I_2 - I_1}{\Delta MD}} = \frac{\Delta MD}{I_2 - I_1}$$

Since I_1 and I_2 are measured in degrees, it is necessary to convert them to radians by multiplying by the factor $\frac{360}{2\pi}$. Thus,

$$R = \frac{\Delta MD}{(I_2 - I_1)} \times \frac{360}{2\pi}$$

$$= \frac{360\,\Delta MD}{2\pi(I_2 - I_1)}$$

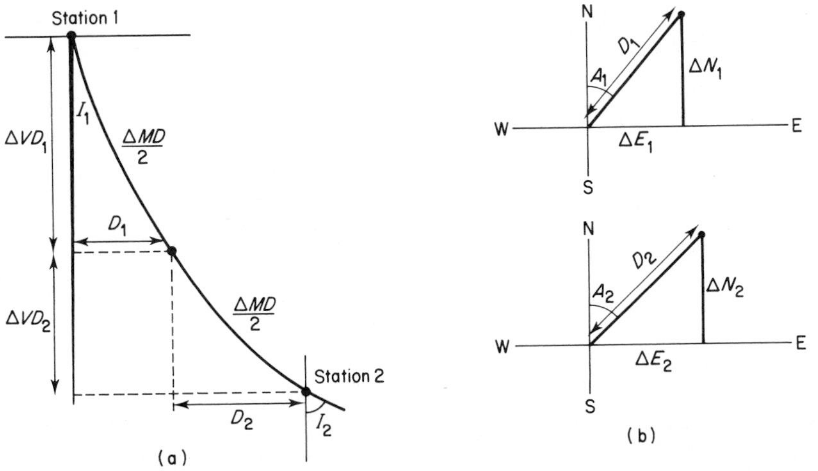

Fig. 8.29. Balanced tangential method: (a) vertical section; (b) plan view.

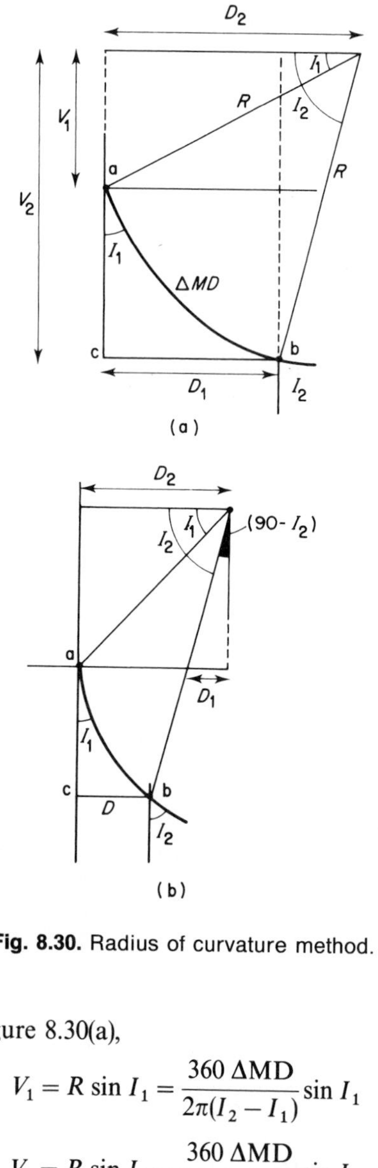

Fig. 8.30. Radius of curvature method.

From Figure 8.30(a),

$$V_1 = R \sin I_1 = \frac{360\,\Delta\text{MD}}{2\pi(I_2 - I_1)} \sin I_1$$

$$V_2 = R \sin I_2 = \frac{360\,\Delta\text{MD}}{2\pi(I_2 - I_1)} \sin I_2$$

Therefore,

$$\Delta\text{VD} = \text{V}_2 - \text{V}_1 = \frac{360\,\Delta\text{MD}}{2\pi(I_2 - I_1)}(\sin I_2 - \sin I_1) \quad (8.15)$$

Departure, $D = cb = D_2 - D_1$ (see Figure 8.30b)

$$D_2 = R \cos I_1$$

$$D_1 = R \sin(90 - I_2) = R \cos I_2$$

$$\Delta D = D_2 - D_1 = R(\cos I_1 - \cos I_2)$$

$$\Delta D = \frac{360\,\Delta\text{MD}}{2\pi(I_2 - I_1)}(\cos I_1 - \cos I_2) \quad (8.16)$$

Length bc, or departure, is assumed to be curved, and consequently the radius of curvature in the horizontal plane is given by

$$R_1 = \frac{1}{\dfrac{A_2 - A_1}{\Delta D}} = \frac{\Delta D}{A_2 - A_1} = \frac{360}{2\pi}\,\frac{\Delta\text{MD}(\cos I_1 - \cos I_2)}{(A_2 - A_1)(I_2 - I_1)}$$

where A_1 and A_2 are the azimuth angles measured at stations (a) and (b). Once again, angles I_1, I_2, A_1, A_2 are in degrees and must be converted to radians; hence, R_1 becomes

$$R_1 = \frac{(360)^2}{4\pi^2}\,\frac{\Delta\text{MD}(\cos I_1 - \cos I_2)}{(A_2 - A_1)(I_2 - I_1)}$$

Referring to the plan view of the well as given in Figure 8.31, the northing and easting co-ordinates are obtained as follows:

$$\Delta\text{N} = \text{N}_2 - \text{N}_1 = R_1 \sin A_2 - R_1 \sin A_1$$

$$= R_1(\sin A_2 - \sin A_1)$$

$$\Delta\text{N} = \frac{(360)^2\Delta\text{MD}(\cos I_1 - \cos I_2)(\sin A_2 - \sin A_1)}{4\pi^2(A_2 - A_1)(I_2 - I_1)} \quad (8.17)$$

Similarly,

$$\Delta\text{E} = \text{E}_2 - \text{E}_1$$

$$= R_1 \cos A_1 - R_1 \cos A_2$$

$$= R_1(\cos A_1 - \cos A_2)$$

$$\Delta\text{E} = \frac{(360)^2\Delta\text{MD}(\cos I_1 - \cos I_2)(\cos A_1 - \cos A_2)}{4\pi^2(A_2 - A_1)(I_2 - I_1)} \quad (8.18)$$

Minimum curvature method

The minimum curvature method is one of the most accurate calculation methods. It relies on the principle of minimising the total curvature within the physical constraints of the well bore[8]. This will then produce a smooth, circular arc. A ratio factor is used to smooth the well-bore path between the two stations. The ratio factor, RF, is defined by the curvature or dog-leg of the well-bore section, as follows:

$$\text{RF} = \left(\frac{2}{\text{DL}}\right)\text{(in radians)} \quad \tan\left(\frac{\text{DL}}{2}\right)\text{(in degrees)} \quad (8.19)$$

where

$$\text{DL} = \text{dog-leg angle; and}$$

$$\cos\text{DL} = \cos(I_2 - I_1) - \sin I_1 \times \sin I_2[1 - \cos(A_2 - A_1)] \quad (8.20)$$

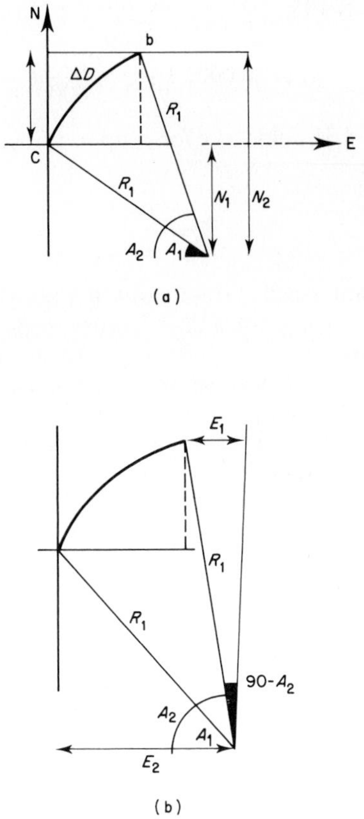

Fig. 8.31. Radius of curvature method.

The minimum curvature method equations are similar to those of the balanced tangential method, except that the survey data are multiplied by a ratio factor, RF, to produce the required smoothing effect. Thus,

$$\Delta VD = \frac{\Delta MD}{2}(\cos I_1 + \cos I_1)RF \quad (8.21)$$

$$\Delta N = \frac{\Delta MD}{2}(\sin I_1 \cos A_1 + \sin I_2 \cos A_2)RF \quad (8.22)$$

$$\Delta E = \frac{\Delta MD}{2}(\sin I_1 \sin A_1 + \sin I_2 \sin A_2)RF \quad (8.23)$$

Mercury method

This method is basically similar to the balanced tangential method with an allowance for the length of the survey instrument. Thus, the length of well bore between two survey stations is composed of a straight line representing the length of the instrument, and a line representing the distance between the survey stations[8].

The necessary equations are

$$\Delta VD = \left(\frac{\Delta MD - STL}{2}\right)(\cos A_2 + \cos A_1) + STL \cos A_2 \quad (8.24)$$

$$\Delta N = \left(\frac{\Delta MD - STL}{2}\right)(\sin I_1 \cos A_1 + \sin I_2 \cos A_2) + STL \sin I_2 \cos A_2 \quad (8.25)$$

$$\Delta E = \left(\frac{\Delta MD - STL}{2}\right)(\sin I_1 \sin A_1 + \sin I_1 \sin A_2) + STL \sin I_2 \sin A_2 \quad (8.26)$$

where STL is the length of the survey instrument.

Dog-leg in directional wells

A dog-leg can be defined as any sudden change in hole inclination and/or hole direction. The dog-leg over a specified interval (e.g. 100 ft or 10 m) is described as 'dog-leg severity'.

Excessive dog-legs result in fatigue failure of drill-pipe, drill collars and tool joints, and also results in keyseats, grooved casing, etc.[10] Fatigue failures of pipe results when the pipe is subjected to repeated stress reversals.

The bending of pipe near a dog-leg produces compression in the fibres of the joint at point A (Figure 8.32) and produces tension at point B. The total tensile stress at any point is determined as the sum of the tensile stress due to weight carried plus the induced stresses. Thus, at point A the resultant tensile stress (weight carried + induced compressive stress of compression) is less than that at point B (weight carried + induced tensile stress). Hence, as the pipe is rotated, the stress at the periphery of the pipe in the dog-leg areas varies between a maximum at B and a minimum at A. The repeated change in the magnitude of stress results in a type of failure described as fatigue failure.

The magnitude of dog-leg is dependent on changes in hole inclination, I, and/or direction, A, and can be determined from the following equation:

$$\cos DL = \cos(I_2 - I_1) - \sin I_1 \sin I_2 \times [1 - \cos(A_2 - A_1)] \quad (8.27)$$

where DL = dog-leg angle. Dog-leg severity is simply the ratio DL/100 ft.

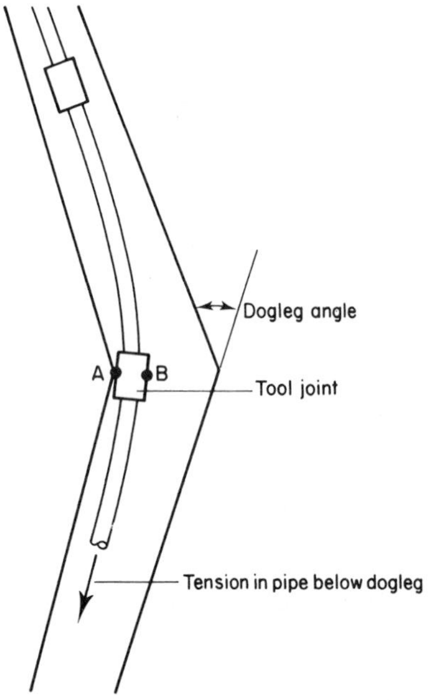

Fig. 8.32. Dog-leg.

Lubinski[9] has shown that the greater the tension to which drillpipe is subjected, the smaller the dog-leg severity (or hole curvature) which the pipe may withstand without fatigue failure occurring. The maximum permissible dog-leg severity in degrees per 100 ft, C, can be determined[10] from

$$C = \frac{432\,000}{\pi} \frac{\sigma_b}{ED} \frac{\tanh(KL)}{KL} \tag{8.28}$$

where

$$K = \sqrt{\frac{T}{EI}} \tag{8.29}$$

E = Young's modulus = 30×10^6 psi for steel; D = drillpipe OD (in); L = half the distance between tool joints (in) = 180 in for range 2 drillpipe; T = buoyant weight (including tool joints) suspended below the dog-leg (lb); σ_b = maximum permissible bending stress (psi); I = drillpipe moment of inertia with respect to its diameter (in^4) = $(\pi/64)\ (D^4 - d^4)$; and d = drillpipe ID (in).

The maximum bending stress, σ_b, is determined from

for Grade E:

$$\sigma_b = 19\,500 - \frac{10}{67}\sigma_t - \frac{0.6}{(670)^2}(\sigma_t - 33\,500)^2 \tag{8.30}$$

(This equation holds true for σ_t values up to 67 000 psi.)

for Grade S-135:

$$\sigma_b = 20\,000\left(1 - \frac{\sigma_t}{145\,000}\right) \tag{8.31}$$

(Equation 8.31 holds true for σ_t up to values of 133 400 psi) where

$$\sigma_t = \frac{T}{A} \tag{8.32}$$

σ_t = buoyant tensile stress (psi); T = buoyant weight below the dog-leg (lb); and A = cross-sectional area of drillpipe (in^2).

It has recently been shown that the use of steel or rubber drillpipe protectors greatly reduces the drillpipe fatigue and permits an increase in permissible dog-leg severity. Drillpipe protectors are simply short, cylindrical pieces of rubber or steel having the same OD as the tool joint. Drillpipe protectors act as supports and spread the bending stress over the entire length of drillpipe. As shown in Figure 8.33(a), a drillpipe without protectors would contact the walls of the hole in the dog-leg section in the manner shown. The degree of contact increases with the increased tensile force below the dog-leg. If a protector is added, the pipe contact is limited to the tool joints and the protector, with the bending stress being distributed over the length of the pipe as shown in Figure 8.33(b).

Figure 8.34 compares the maximum permissible dog-leg severity for pipes without and with protectors. Thus, at a tensile force of 200 000 lb, the maximum permissible dog-leg severity for a pipe without protectors is 2.7°/100 ft. The addition of one protector in the centre of the drillpipe increases the permissible dog-leg severity to 5.4°/100 ft. The addition of two

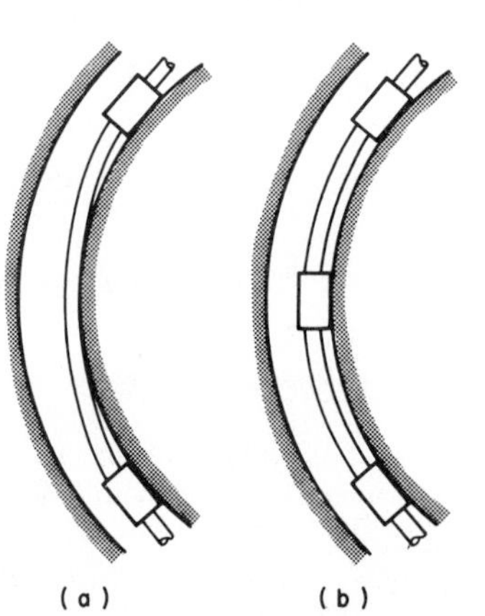

Fig. 8.33. Drillpipe-to-hole contact in a gradual dog-leg: (a) drillpipe without protectors; (b) drillpipe with protector.

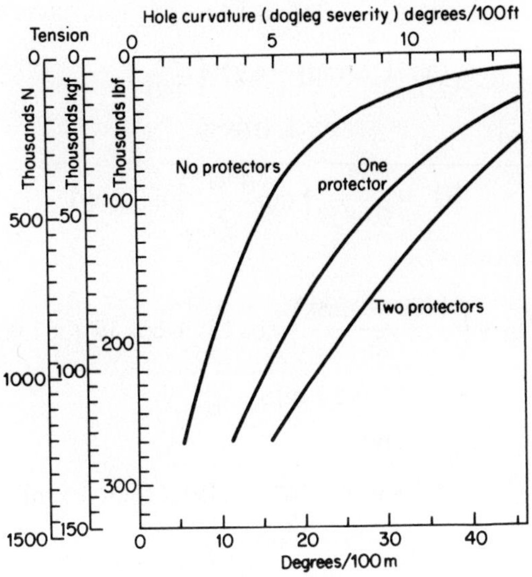

Fig. 8.34. Fatigue curves for gradual dog-legs ($4\frac{1}{2}$ in, 16.60 lbm/ft; 11.4 cm, 24.7 kg/m) grade E drillpipe. (After Reference 11)

protectors at a 10 ft spacing would increase the permissible dog-leg severity to 7.5°/100 ft.

Example 8.4

The following data refer to a directionally drilled well:

KOP = 1000 ft
northing co-ordinate of surface location = 2400 ft
easting co-ordinate of surface location = 200 ft

Assume that the KOP is situated directly below the surface location, so that they both have the same co-ordinates.

Survey data at two stations are as follows:

Station	*Measured depth* (ft)	*Inclination, I*	*Corrected azimuth, A*
1	1200	15	320
2	1400	19	310

Use the radius of curvature and minimum curvature methods to calculate the well path between stations 1 and 2 and dog-leg severity.

Solution

Radius of curvature

$$\Delta VD = \frac{360}{2\pi}\frac{MD}{(I_2 - I_1)}(\sin I_2 - \sin I_1)$$

$$= \frac{360(1400-1200)}{2\pi(19-15)}(\sin 19° - \sin 15°)$$

$$= 191.22 \text{ ft } (58.3 \text{ m})$$

True vertical depth at station 2 = KOP + ΔVD (between stations 1 and 2)

$$= 1000 + 191.22$$

$$= 1191.22 \text{ ft } (363.1 \text{ m})$$

Departure,

$$\Delta D = \frac{360(1400-1200)}{2\pi(I_2 - I_1)}(\cos I_1 - \cos I_2)$$

$$= 58.44 \text{ ft } (17.81 \text{ m})$$

$$\Delta N = \frac{360^2 \Delta MD(\cos I_1 - \cos I_2)(\sin A_2 - \sin A_1)}{4\pi^2 (A_2 - A_1)(I_2 - I_1)}$$

$$= \frac{360^2(1400-1200)(\cos 15° - \cos 19°) \times (\sin 310° - \sin 320°)}{4\pi^2(310-320)(19-15)}$$

$$= 41.29 \text{ ft } (12.6 \text{ m})$$

$$\Delta E = \frac{(360)^2 \Delta MD(\cos I_1 - \cos I_2)(\cos A_1 - \cos A_2)}{4\pi^2(A_2 - A_1)(I_2 - I_1)}$$

$$= \frac{(360)^2(1400-1200)(\cos 15° - \cos 19°) \times (\cos 320° - \cos 310°)}{(4\pi^2)(310-320)(19-15)}$$

$$= -41.29 \text{ ft } (-12.6 \text{ m})$$

$$\text{Azimuth} = 360 \pm \alpha$$

$$\tan \alpha = \frac{\Delta E}{\Delta N} = \frac{-41.29}{41.24}$$

$$\alpha = -45.03°$$

Azimuth between stations = 314.97°

Northing co-ordinate at station 2

$$= \text{northing of KOP} + \Delta N$$

$$= 2400 + 41.29$$

$$= 2441.29 \text{ ft } (744 \text{ m})$$

Easting co-ordinate at station 2

$$= \text{easting of KOP} + \Delta E$$
$$= 200 - 41.29$$
$$= 158.71 \text{ ft } (48.4 \text{ m})$$

Dog-leg severity

$$= \frac{\text{dog-leg angle}}{\text{length of interval between stations 1 and 2}}$$

Using

$$\cos DL = \cos(I_2 - I_1) - \sin I_1 \sin I_2 \times (1 - \cos(A_2 - A_1))$$
$$= \cos(19° - 15°) - \sin 15° \sin 19° \times (1 - \cos(310° - 320°))$$
$$= 0.9963$$
$$DL = 4.93°$$

$$\text{Dog-leg severity} = \frac{\text{dog-leg angle}}{\text{interval drilled}} \times 100 \text{ ft}$$
$$= \frac{4.92°}{(1400 - 1200)} \times 100$$
$$= 2.46(\text{deg}/100 \text{ ft})$$

Minimum curvature method

$$\Delta VD = \frac{\Delta MD}{2}(\cos I_1 + \cos I_2)RF$$

$$\Delta D = \frac{\Delta MD}{2}(\sin I_1 + \sin I_2)RF$$

$$\Delta N = \frac{\Delta MD}{2}(\sin I_1 \cos A_1 + \sin I_2 \cos A_2)RF$$

$$\Delta E = \frac{\Delta MD}{2}(\sin I_1 \sin A_1 + \sin I_2 \sin A_2)RF$$

$$RF = \frac{2}{DL} \tan\left(\frac{DL}{2}\right)$$

$$\cos DL = \cos(I_2 - I_1) - \sin I_1 \times \sin I_2 \times [1 - \cos(A_2 - A_1)]$$
$$= \cos(19° - 15°) - \sin 15° \sin 19° \times [1 - \cos(310° - 320°)]$$
$$= 0.9963$$

$$DL = 4.93°$$
$$DL(\text{rad}) = 4.92°\left(\frac{2\pi}{360}\right)$$
$$= 0.0859$$
$$RF = \frac{2}{0.0859}\left(\tan\frac{4.92°}{2}\right) = 1.0003$$

Therefore,

$$\Delta VD = \left(\frac{1400 - 1200}{2}\right)(\cos 15° + \cos 19°) \times 1.0003$$
$$= 191.20 \text{ ft } (58.3 \text{ m})$$

True vertical depth

$$= 1000 + 191.20 = 1191.2 \text{ ft } (363.1 \text{ m})$$

$$\Delta N = \left(\frac{1400 - 1200}{2}\right)(\sin 15° \cos 320° + \sin 19° \cos 310°) \times 1.0003$$
$$= \frac{200}{2}(0.1983 + 0.2093) \times 1.0003$$
$$= 40.77 \text{ ft } (12.4 \text{ m})$$

$$\Delta E = \left(\frac{1400 - 1200}{2}\right)(\sin 15° \sin 320° + \sin 19° \sin 310°) \times 1.0003$$
$$= -41.59 \text{ ft } (-12.7 \text{ m})$$

$$\Delta D = \frac{200}{2}(\sin 15° + \sin 19°) \times 1.0003$$
$$= 58.46 \text{ ft } (1.7082 \text{ m})$$

$$\text{Azimuth between stations} = 360 \pm \alpha$$

$$\tan \alpha = \frac{\Delta E}{\Delta N} = \frac{-41.59}{+40.77} = -1.0201$$
$$\alpha = -45.57°$$

(The negative sign means that the angle is in the second or fourth quadrant; in our case α is in the fourth quadrant, since E is negative.) Therefore,

$$\text{azimuth} = 360° - 45.57°$$
$$= 314.43°$$

$$\text{Northing at station } 2 = 2400 + 40.77 = 2440.8 \text{ ft}$$
$$\text{Easting at station } 2 = 200 - 41.59 = 158.4 \text{ ft}$$
$$\text{Dog-leg severity} = 2.46(\text{deg}/100 \text{ ft})$$

(as calculated in the first part).

References

1. Wilson, G. E. (1976). How to drill a usable hole. *World Oil*, September.
2. Wilson, G. E. (1979). How to select bottomhole drilling assemblies. *Petroleum Engineer International*, March, April.
3. Eastman Whipstock. *Introduction to Directional Drilling*. Eastman Whipstock Publications.
4. Scientific Drilling Controls. *Handbook on Surveying Instruments*. Scientific Drilling Controls Publications.
5. Plite, J. and Blount, S. (1980). How to avoid gyro misruns. *Oil and Gas Journal*.
6. Dyna-Drill. *Dyna-Drill Catalogue*. Dyna-Drill Publications.
7. Pickett, G. W. (1967). Techniques and deflection tools in high angle drilling: past, present and future. *Journal of Petroleum Engineering*.
8. Craig, J. T., Jr. and Randall, B. V. (1976). Directional survey calculation. *Petroleum Engineer*, March, 38–54.
9. Lubinski, A. (1961). Maximum permissible doglegs in rotary borehole. *Journal of Petroleum Technology*, February.
10. API RP 7G (1981). *Drillstem Design and Operating Limits*. API Publications.
11. Lubinski, A. and Williamson, J. S. (1984). Usefulness of steel or rubber drillpipe protector. *Journal of Petroleum Technology*, April, 628–636.

Problems

Problem 8.1

Given hole inclination/azimuth = 8°/90° magnetic; maximum dog-leg severity = 3°/100 ft. If the hole inclination is to be maintained constant at 8°, determine the azimuth change and the tool face setting.

Answer: 21.5°; 100° left or right.

Problem 8.2

Complete the following table using the radius of curvature method:

Measured depth	*Hole inclination*	*Azimuth*	*Co-ordinates*		*Vertical depth*	*Dog-leg severity*
			north	*east*		
15 000	57.2°	324.8	6241.6	−8099.3	9820.7	
15 100	58.5°	325.0				
15 200	58.6°	325.1				
15 300	58.1°	325.4				
15 400	58.4°	325.7				
15 500	58.5°	325.2	6591.2	−8341.7	10 083.5	0.5

Chapter 9

Fracture Gradient

In oil well drilling, the fracture gradient may be defined as the minimum total *in situ* stress divided by the depth.

Knowledge of fracture gradient is essential to the selection of proper casing seats, for the prevention of lost circulation and to the planning of hydraulic fracturing for the purpose of increasing the well productivity in zones of low permeability. Accurate knowledge of the fracture gradient is of paramount importance in areas where selective production and injection is practised. In such areas the adjacent reservoirs consist of several sequences of dense and porous zones such that, if a fracture is initiated (during drilling or stimulation), it can propagate, establishing communication between hydrocarbon reservoirs and can extend down to a water-bearing zone.

The fracture gradient is dependent upon several factors, including type of rock, degree of anisotropy, formation pore pressure, magnitude of overburden and degree of tectonics within the area. It follows that any analytical prediction method will have to incorporate all of the above factors in order to yield realistic values of the fracture gradient.

In this chapter the aim is to introduce the various methods currently used in the oil industry to determine or predict the fracture gradient of rock. Topics discussed include:

Definitions
Determination of fracture gradient
  direct method
  indirect methods
  effects of hole deviation
Selection of casing seats

DEFINITIONS

Before the mathematical equations necessary for the calculation of the fracture gradient are presented, the following terms must be clearly understood.

Overburden stress

Overburden stress (σ_v) is defined as the stress arising from the weight of rock overlying the zone under consideration. In geologically relaxed areas having little tectonic activity, the overburden gradient (=stress/depth) is taken as 1 psi/ft (0.2262 bar/m). In tectonically active areas, as in sedimentary basins which are still undergoing compaction or in highly faulted areas, the overburden gradient varies with depth, and an average value of 0.8 psi/ft is normally taken as being representative of the overburden gradient. Thus, a zone at 10 000 ft will have the following overburden stress:

in tectonically relaxed areas
overburden stress = 10 000 ft × 1 psi/ft
= 10 000 psi (689 bar)

in tectonically active areas
overburden stress = 10 000 ft × 0.8 psi/ft
= 8000 psi (552 bar)

In general, the overburden gradient varies from field to field[1] and increases with depth, owing to rock compaction. For a given field, accurate values of overburden gradient can be obtained by averaging density logs from several wells drilled in the area. A graph of bulk density against depth is then plotted as shown in Figure 9.1(a). The density–depth graph can be converted to an overburden gradient–depth graph (Figure 9.1b) by the use of the relation

$$\text{overburden stress} = \text{bulk density} \times \text{depth} \times \text{acceleration due to gravity}$$

In porous formations the overburden stress, σ_v, is supported jointly by the rock matrix stress, σ_s, and the formation pore pressure, P_f. Thus,

$$\sigma_v = \sigma_s + P_f$$

Formation pore pressure

Formation pore pressure is defined as the pressure exerted by the formation fluids on the walls of the rock pores. The pore pressure supports part of the weight of the overburden, while the other part is supported by the grains of the rock. (The terms pore pressure, formation pressure and fluid pressure are synonymous, referring to formation pore pressure.)

Formations are classified according to the magnitude of their pore pressure gradients. In general, two types of formation pressure are recognised:

(1) *Normal pore pressure* (or *hydropressure*). A formation is said to be normally pressured when its pore pressure is equal to the hydrostatic pressure of a full column of formation water[1]. Normal pore pressure is usually of the order of 0.465 psi/ft (0.105 bar/m). Hence, at 5000 ft a normally pressured zone will have a pore pressure of 5000 ft × 0.465 psi/ft = 2325 psi (160 bar).

(2) *Abnormal formation pressure* (or *geopressure*). This type exists in zones which are not in direct communication with its adjacent strata. The boundaries of the abnormally pressured zone are impermeable, preventing fluid communication and making the trapped fluid support a larger proportion of the overburden stress.

The maximum value of abnormal formation pressure is 1 psi/ft for tectonically relaxed areas and 0.8 psi/ft for active areas. Exceptions to these values were found in certain parts of Iran and Russia in which the abnormal formation pressure is in excess of the overburden gradient.

Formation pressures (normal and abnormal) can be detected by geophysical and logging methods. Geophysical methods provide prediction of formation pressure before the well is drilled, while logging methods provide information after the well or section of well has been drilled. Logging tools are run on a wire line inside the well. They include electrical, sonic, neutron, bulk density and lithology logs[2].

Determination of normal and abnormal pressures from well logs is beyond the scope of this book, and

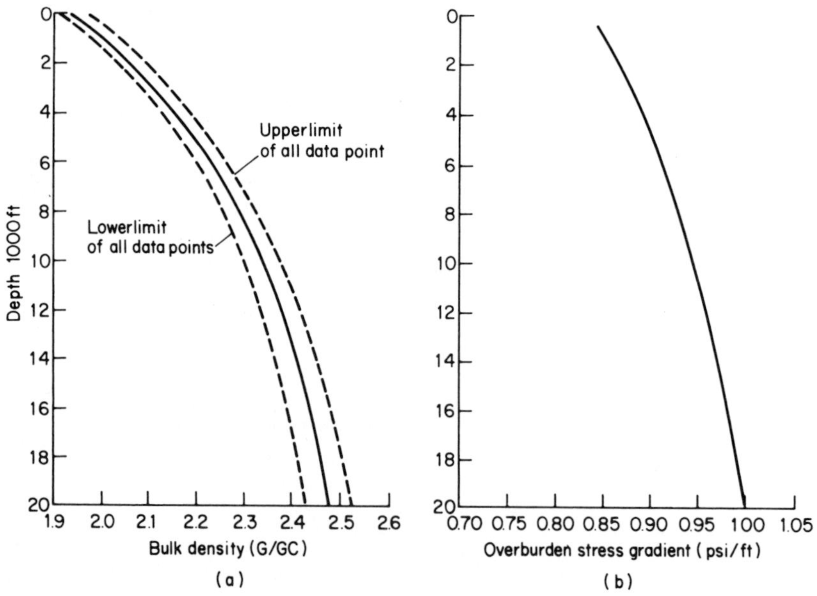

Fig. 9.1. (a) Composite bulk density curve from density log data for the Gulf Coast; (b) composite overburden stress gradient for all normally compacted Gulf Coast sediments. (After Eaton[8], 1969)

for further details the reader is advised to consult Reference 3.

Rock strength

Rock strength can be specified in terms of tensile strength, compressive strength, shear strength or impact strength. In the context of fracture gradient, only the tensile strength of rock is of importance.

The tensile strength of rock is defined as the pulling force required to rupture a rock sample divided by the sample's cross-sectional area. The tensile strength of rock is very small and is of the order of 0.1 of the compressive strength. Thus, a rock is more likely to fail in tension than in compression.

The *in situ* principal stresses

At any point below the earth's surface three mutually perpendicular stresses exist as shown in Figure 9.2. The maximum principal stress, σ_1, is normally vertical and is equal to the overburden stress in vertical holes. The value of the overburden stress is 1 psi/ft for relaxed areas and 0.8 psi/ft for tectonically active areas.

The intermediate and minimum total principal stresses (σ_2 and σ_3, respectively) are horizontal, and directly influence the fracturing of rock. In theory, the fluid pressure required to rupture a borehole should be greater than or equal to the minimum principal stress. However, the creation of a borehole within the earth's surface produces a magnification of stresses around the borehole walls such that the resulting stresses are several times larger than the least principal stress.

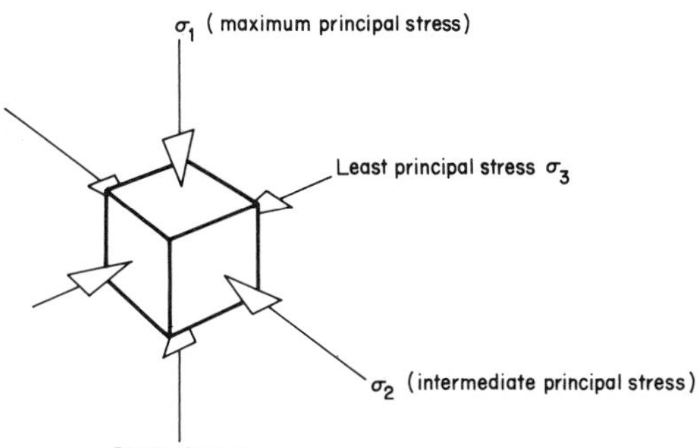

Fig. 9.2. *In situ* principle stresses of the earth.

Formation breakdown pressure

The formation breakdown pressure is the pressure required to overcome the well-bore stresses in order to fracture the formation in the immediate vicinity of the wellbore.

Fracture gradient

Fracture gradient is defined as the minimum total *in situ* stress divided by the depth.

Determination of fracture gradient

Two methods are used for determining fracture gradient: direct and indirect. The direct method relies on determining the pressure required to fracture the rock and the pressure required to propagate the resulting fracture. The indirect method uses stress analysis to predict fracture gradient.

Direct method

The direct method uses mud to pressurise the well until the formation fractures. The value of the surface pressure at fracture is noted and is added to the hydrostatic pressure of mud inside the hole to determine the total pressure required to fracture the formation. This pressure is described as the formation breakdown pressure.

The test can be made in the open hole section below surface or intermediate casing and uses the drill string shown in Figure 9.3. The hole is first filled with fresh mud and the annular preventer is closed. A surface pumping unit having accurate pressure gauges is used to pump small increments of mud, $\frac{1}{8}$–$\frac{1}{4}$ bbl. After each increment, the shut-in pressure is observed and plotted against time or volume of mud pumped in.

Figure 9.4 gives a simplified schematic drawing of pressure plotted against time during a fracturing test.

Up to point A the formation can withstand the total pressure imposed (surface pressure + hydrostatic pressure) without fracturing. The portion OA is analogous to the elastic portion of the stress–strain graph of metals. At point A the formation begins to take fluid, resulting in a non-linear relationship between pressure and time (or volume) as depicted by line AB.

At point B the applied pressure exceeds the wellbore stresses, which results in formation breakdown and a sudden drop in surface pressure, as depicted by line BC. The pressure at point B is described as the formation breakdown pressure. Continued pressurisation will then merely extend the fractures created by the breakdown pressure. The pressure required to

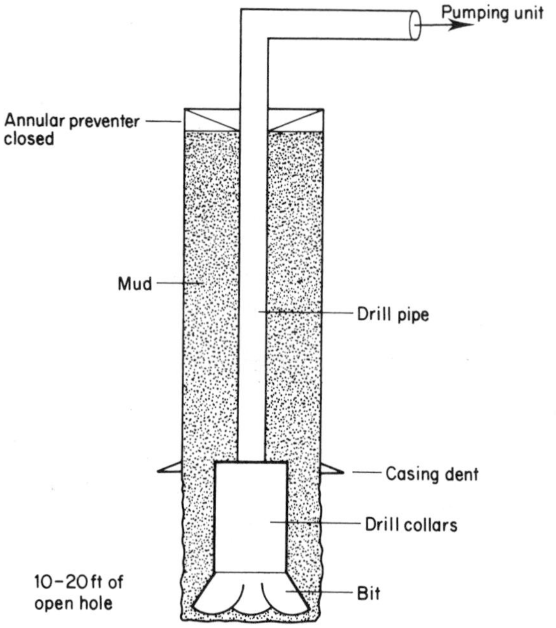

Fig. 9.3. Schematic drawing of a small fracturing, leak-off or casing seat test assembly.

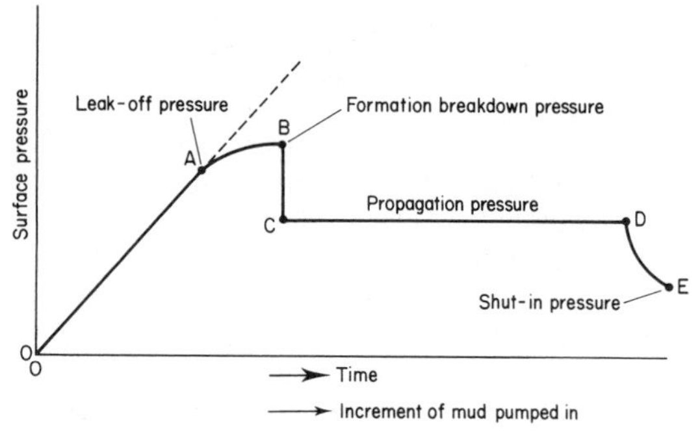

Fig. 9.4. Leak-off test.

extend such fractures is described as the fracture propagation pressure; its magnitude is much lower than the breakdown pressure. The fracture propagation pressure is normally taken to be equal to the minimum principal stress, σ_3, on the assumption that a fracture outside the vicinity of the hole can be made when the value of σ_3 is exceeded. Point E in Figure 9.4 represents the instantaneous shut-in pressure after the pump is stopped. This pressure is known as the fracture closure pressure, being the pressure required to keep the fracture from just closing.

A variation of the aforementioned test is the leak-off test, in which the test is stopped after the first sign of leak-off. Leak-off tests are normally run in the open hole section below surface or intermediate casing. The test is performed after 10–20 ft of open hole is drilled below the casing seat, in order to determine the maximum pressure that the casing seat can withstand before taking fluid while the next hole is being drilled. In other words, the test determines the maximum mud weight that can be used to drill the next hole without fracturing the casing seat. The leak-off pressure is dependent on the penetration properties of mud and the permeability of the formation. Thus, in a formation of low permeability and when a non-penetrating fluid is used, little or no leak-off will be observed before fracturing occurs. To ensure that leak-off is detected as soon as it occurs, pumping should be stopped at regular intervals and the shut-in pressure should be noted. The difference between the shut-in pressures of successive cycles should be noted. If an increase in this difference is observed, then this is a clear indication that fluid leak-off is taking place, and pumping should be stopped immediately. The leak-off data should also be plotted on a graph of pressure against time. This type of plot will produce line OA in Figure 9.4.

A third version of the pressure–time test is the casing seat test, in which the open hole is pressurised to below the leak-off pressure. The test is primarily used to check whether the casing seat can withstand the maximum mud weight that will be used to drill the next hole. If the current mud is ρ_{m1} and the final mud is ρ_{m2}, then the maximum surface pressure on a casing seat test is given by

$$\frac{(\rho_{m2} - \rho_{m1})}{144} \times D$$

where D = casing seat depth (ft); and ρ_{m1} and ρ_{m2} = mud weights (pcf).

For example, if the depth of a $9\frac{5}{8}$ in casing seat is 8900 ft, $\rho_{m1} = 65$ pcf, $\rho_{m2} = 72$ pcf, then the surface pressure which can be used to check the strength of the casing seat is $(72 - 65/144) \times 8900 = 433$ psi. The 433 psi will be applied in small increments, and, if the formation should break at any time below 433 psi, pumping must be stopped and the fracture pressure (surface pressure + hydrostatic mud pressure) must be noted. The next mud weight should be based on a value below the fracture pressure.

Indirect methods

Indirect methods rely on the use of stress analysis methods for predicting the fracture gradient. The following is a summary of the most widely used methods.

Hubbert and Willis method

The Hubbert and Willis method[4] is based on the premise that fracturing occurs when the applied fluid pressure exceeds the sum of the minimum effective stress and formation pressure. The fracture plane is assumed to be always perpendicular to the minimum principal stress.

The effective stress is defined as the difference between the total stress and the pore pressure. Thus, the minimum effective stress, σ'_3, is given by

$$\sigma'_3 = \sigma_3 - P_f$$

Also,

$$\sigma'_2 = \sigma_2 - P_f$$

and

$$\sigma'_1 = \sigma_1 - P_f$$

For porous media, the failure of the material is controlled by the magnitude of the effective stress only and not the total stress.

An expression for the minimum effective principal stress in terms of the overburden stress can be derived by use of Hooke's law of elasticity. It can be shown[5] that the strains in the three principal directions are given by

$$\varepsilon_1 = \frac{\sigma'_1}{E} - \frac{v}{E}(\sigma'_2 + \sigma'_3) \tag{9.1}$$

$$\varepsilon_2 = \frac{\sigma'_2}{E} - \frac{v}{E}(\sigma'_1 + \sigma'_3) \tag{9.2}$$

$$\varepsilon_3 = \frac{\sigma'_3}{E} - \frac{v}{E}(\sigma'_2 + \sigma'_1) \tag{9.3}$$

where σ'_1, σ'_2 and σ'_3 are the maximum, intermediate and minimum effective principal stresses, respectively; ε_1, ε_2 and ε_3 are the principal strains; E = Young's modulus of rock; and v = Poisson's ratio.

In tectonically relaxed areas, $\sigma'_2 = \sigma'_3$ and the strains ε_2 and ε_3 are equal. Also, during hydraulic pressurising, the near bore-hole horizontal strains, ε_2 and ε_3, are virtually unchanged and are equal to zero, owing to the lateral restraints by adjacent strata. Hence, putting $\varepsilon_2 = 0$ in Equation (9.2) or $\varepsilon_3 = 0$ in Equation (9.3) gives

$$0 = \frac{\sigma'_3}{E} - \frac{v}{E}(\sigma'_2 + \sigma'_1)$$

But

$$\sigma'_2 = \sigma'_3$$

Therefore,

$$\sigma'_3 = \left(\frac{v}{1 - v}\right)\sigma'_1 \tag{9.4}$$

Most rocks have a Poisson's ratio of approximately 0.25. Substituting this value of v in Equation (9.4) gives

$$\sigma'_3\,(\text{or } \sigma'_2) = \left(\frac{0.250}{1 - 0.25}\right)\sigma'_1$$

or

$$\sigma'_3\,(= \sigma'_2) = \frac{1}{3}\sigma'_1 \tag{9.5}$$

According to the Hubbert and Willis method, the total injection (or fracturing) pressure, FP, required to keep open and extend a fracture is given by:

$$\text{FP} = \sigma'_3 + P_f \tag{9.6}$$

where FP = fracturing pressure; and P_f = formation pressure.

Combining Equations (9.5) and (9.6) gives

$$\text{FP} = \frac{1}{3}\sigma'_1 + P_f \tag{9.7}$$

Also,

$$\sigma'_1 = \sigma_1 - P_f$$

and

$$\sigma_1 = \sigma_v = \text{the overburden stress}$$

Hence,

$$\sigma'_1 = \sigma_v - P_f \tag{9.8}$$

Substituting Equation (9.8) in Equation (9.7) yields

$$\text{FP} = \frac{1}{3}(\sigma_v - P_f) + P_f$$

$$= \frac{1}{3}\sigma_v + \frac{2}{3}P_f \tag{9.9}$$

Dividing Equation (9.9) by depth D gives the fracture gradient, FG:

$$\text{FG} = \left(\frac{\text{FP}}{D}\right) = \frac{1}{3}\frac{\sigma_v}{D} + \frac{2}{3}\frac{P_f}{D} \tag{9.10}$$

The main disadvantage of the Hubbert and Willis method is that it predicts a higher fracture gradient in abnormal pressure and a lower fracture gradient in subnormal pressure formations.

Example 9.1

Given that the formation pressure at 5000 ft is 2400 psi and the overburden stress is 1 psi/ft (determined from bulk density logs), estimate the formation fracture gradient at 5000 ft.

From Equation (9.10)

$$FG = \frac{1}{3}\left(\frac{\sigma_v}{D}\right) + \frac{2}{3}\frac{P_f}{D}$$

$$= \frac{1}{3}\left(1\,\frac{\text{psi}}{\text{ft}}\right) + \frac{2}{3}\left(\frac{2400\text{ psi}}{5000\text{ ft}}\right)$$

$$= 0.653\text{ psi/ft}$$

Matthews and Kelly method

The Hubbert and Willis method was found not to apply in soft rock country such as the Gulf of Mexico and the northern North Sea. Matthews and Kelly[6] modified the Hubbert and Willis method by changing Equation (9.7) to

$$FP = K_i(\sigma'_1) + P_f$$

$$= K_i(\sigma_v - P_f) + P_f \quad (9.11)$$

where K_i is the matrix stress coefficient (dimensionless) for the depth at which the value of σ'_1 would be the normal matrix stress. The value of K_i should be determined from actual fracture data of nearby wells. Matthews and Kelly refer to the effective stress, σ'_1, as the matrix stress.

The use of the matrix stress coefficient implicitly eliminates the assumption that $\sigma_2 = \sigma_3$, as used in the Hubbert and Willis method. However, Equation (9.11) can only be used if fracture data of nearby wells are available for which a graph of K_i against depth can be established (see Figure 9.5).

The Matthews and Kelly method is used as follows:

(1) Assume a normal compaction in which the formation pore pressure gradient is 0.465 psi/ft and the overburden stress is 1 psi/ft. The maximum effective principal stress is given by

$$\sigma'_1 = \sigma_v - P_f = 1 - 0.465 = 0.535\text{ psi/ft}$$

(2) Determine the equivalent depth, D_i, corresponding to the assumed normal compaction, i.e.

$$\sigma'_1 = 0.535 \times D_i$$

or

$$D_i = \frac{\sigma'_1\text{ (psi)}}{0.535\text{ (psi/ft)}}\text{ ft} \quad (9.12)$$

(3) Using a graph of K_i against depth for the field under consideration, determine the value of K_i corresponding to D_i.

(4) Use Equation (9.11) to determine the fracture propagation pressure or the fracture gradient.

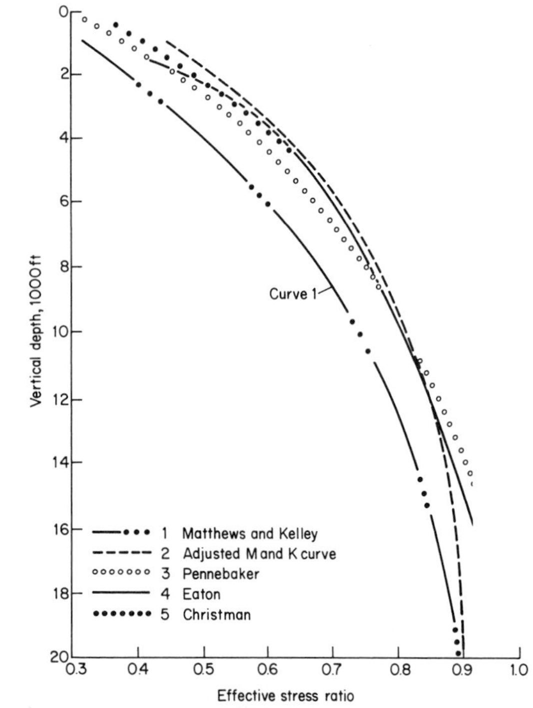

Fig. 9.5. Effective stress ratio curves from various authors. (After Pilkington, 1978[7])

Example 9.2

Use the data of Example 9.1 and Figure 9.5 (curve 1) to determine the fracture pressure and fracture gradient according to the Matthews and Kelly method.

Solution

Equivalent depth for normal compaction

$$= D_i = \frac{\sigma'_1}{0.535}$$

$$= \frac{\sigma_v - P_f}{0.535}$$

$$= \frac{\left(5000\text{ ft} \times \dfrac{1\text{ psi}}{\text{ft}}\right) - 2400}{0.535}$$

$$= 4860\text{ ft}$$

From Figure 9.5 (curve 1), K_i corresponding to a depth of 4860 ft is 0.55. Hence, Equation (9.11) becomes

$$\begin{aligned}FP &= K_i(\sigma_v - P_f) + P_f \\ &= 0.55(5000 \times 1 - 2400) + 2400 \\ &= 3830 \text{ psi}\end{aligned}$$

$$\begin{aligned}\text{Fracture gradient} &= \frac{FP}{5000} = \frac{3830}{5000} \\ &= 0.766 \text{ psi/ft}\end{aligned}$$

Eaton method

The Eaton method[8] is the most widely used in the oil industry. It is basically a modification of the Hubbert and Willis method, in which both overburden stress and Poisson's ratio are assumed to be variable. Poisson's ratio is a rock property which describes the effect produced in one direction as stress is applied in a perpendicular direction. Thus, for a two-dimensional case, if σ_y is a stress applied in the y direction and σ_x is the resulting stress in the x direction, then Poisson's ratio $(\nu) = \frac{\varepsilon_x}{\varepsilon_y}$, where ε_x is the strain in the x-direction and ε_y is the strain in the y-direction. Most rocks tested under laboratory conditions produce a Poisson's ratio of 0.25–0.3. Under field conditions, however, the rock is subjected to a much greater degree of confinement and Poisson's ratio can vary from 0.25 (or less) to a maximum value of 0.5.

The Eaton equation can be derived by combining Equations (9.4) and (9.6) to give

$$FP = \left(\frac{\nu}{1-\nu}\right)\sigma'_1 + P_f$$

or

$$FG = \left(\frac{\nu}{1-\nu}\right)\frac{\sigma'_1}{D} + \frac{P_f}{D}$$

Replacing σ'_1 by $(\sigma_v - P_f)$ yields

$$FG = \left(\frac{\nu}{1-\nu}\right)\left(\frac{\sigma_v - P_f}{D}\right) + \frac{P_f}{D} \tag{9.13}$$

Eaton argued that Poisson's ratio for a given field should be fairly constant and may be determined from previous data obtained from wells in the same field or area. Then, by rearranging Equation (9.13) as

$$\frac{\nu}{1-\nu} = \frac{FG - \dfrac{P_f}{D}}{\left(\dfrac{\sigma_v - P_f}{D}\right)}$$

Poisson's ratio, ν, can be determined. A plot of Poisson's ratio against depth can then be established as shown in Figure 9.6.

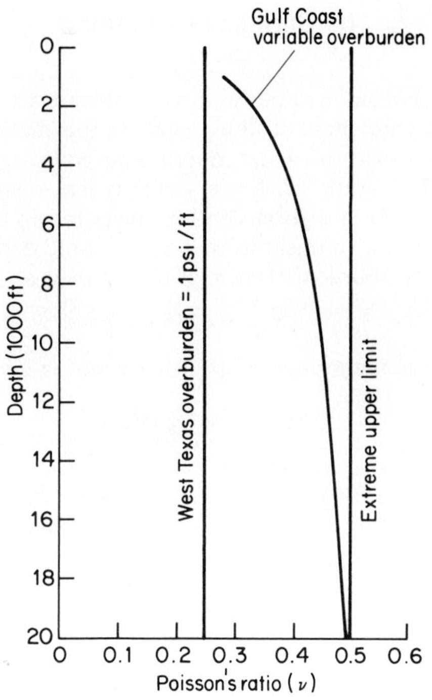

Fig. 9.6. Variation of Poisson's ratio with depth. (After Eaton, 1969[8])

By use of the Eaton method, the fracture gradient is determined as follows:

(1) Establish a pore pressure–depth graph for the field under consideration.
(2) Establish a density–depth graph and convert to an overburden stress (σ_v)–depth graph by multiplying the density values by 0.4335 to convert to psi/ft.
(3) Establish a Poisson's ratio–depth graph.
(4) Use Equation (9.13) to predict the fracture gradient for future wells.

Example 9.3

Using the data of Example 9.1 and Figure 9.6, determine the fracture gradient at 5000 ft by the Eaton method.

Solution

As we have no pore pressure–depth and overburden stress–depth graphs, we shall assume a normal compaction in which $\sigma_v = 1$ psi/ft, or $\sigma_v = 5000$ psi at 5000 ft. From Figure 9.6 (curve 1), $\nu = 0.4$ at 5000 ft. Hence, Equation (9.13) gives

$$FG = \left(\frac{0.4}{1-0.4}\right)\left(\frac{5000-2400}{5000}\right) + \frac{2400}{5000}$$

(where $P_f = 2400$ psi). Therefore,

$$FG = 0.827 \text{ psi/ft}$$

The Christman's method

The Christman's method[9] is essentially a modification of the Eaton method and is designed to predict fracture gradient in offshore fields. In this method, the depth consists of water depth and hole depth, as shown in Figure 9.7. Since water is less dense than rock, the FG at a given depth is lower for an offshore well than for an onshore well at the same depth.

The overburden stress σ_v is determined as follows:

$$\sigma_v = 0.4335\bar{\rho} \tag{9.14}$$

where $\bar{\rho}$ is the average bulk density and is given by

$$\bar{\rho} = \frac{\rho_w D_w + \rho_R D_h}{D} \tag{9.15}$$

where ρ_w = density of sea-water = 1.02 g/cm^3; D_w = depth of water (ft); ρ_R = average density of rock; and D = total depth = $D_w + D_h$ (D_h = depth below mudline, in ft). Combining Equations (9.14) and (9.15) gives

$$\sigma_v = \frac{0.4335}{D}(\rho_w D_w + \rho_R D_h) \tag{9.16}$$

and the fracture gradient, FG, is given by

$$\mathrm{FG} = F\left(\frac{\sigma_v - P_f}{D}\right) + \frac{P_f}{D} \tag{9.17}$$

where F = the stress ratio factor and must be calculated from fracture data of previous wells.

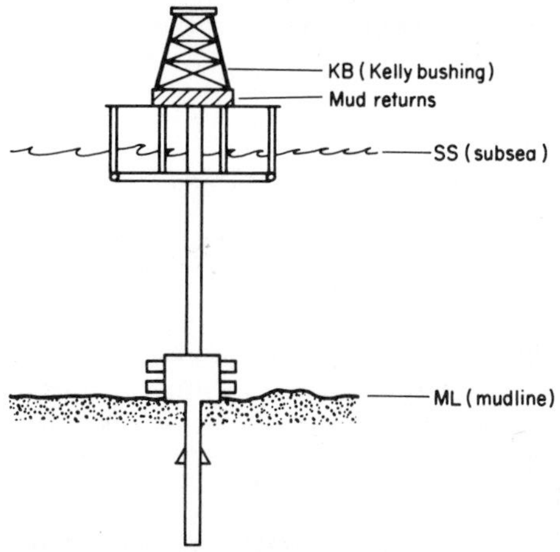

Fig. 9.7. Offshore reference depths. (After Christman, 1973[9])

Formation breakdown gradient

Knowledge of the formation breakdown gradient (FBG) is of paramount importance during kick situations when the casing shut-in pressure, CSIP, is being monitored, so that the sum of the hydrostatic pressure of fluids in the annulus and CSIP is always kept below the FBG at the casing shoe. It may be argued that, for added safety, control should be based on the fracture propagation pressure ($=\sigma_3$) rather than the FBG. However, in deviated wells and in areas where σ_3 is considerably different from σ_2, the calculated FBG can be lower than σ_3.

With the assumption that an oil well is a hole in an infinite plate, it can be shown that the formation breakdown gradient (FBG) is given by

$$\mathrm{FBG} = 3\sigma_3 - \sigma_2 + T + P_f \tag{9.18}$$

where σ_3 = smaller horizontal principal tectonic effective stress; σ_2 = larger horizontal principal tectonic effective stress; T = tensile strength of rock; and P_f = pore pressure.

Equation (9.18) is only valid when no fluid invades or penetrates the formation. In porous and permeable rocks the drilling mud normally penetrates the formation, thereby changing the magnitude of the stress concentrations around the borehole. The effect of fluid penetration is to create a force radially outward which reduces the stress concentrations at the walls of the hole, thereby making it more easy to fracture. Haimson and Fairhurst[10] modified Equation (9.18) to take into account the effects of fluid penetration, to obtain

formation breakdown pressure

$$= \frac{3\sigma_3 - \sigma_2 + T}{2 - \alpha\left(\dfrac{1-2v}{1-v}\right)} + P_f \tag{9.19}$$

where $\alpha = 1 - (C_r/C_b)$; C_r = the rock matrix compressibility; C_b = the rock bulk compressibility; and $0.5 \leqslant \alpha \leqslant 1$ when $0 \leqslant v \leqslant 0.5$

and, finally,

$$0 \leqslant \alpha\left(\frac{1-2v}{1-v}\right) \leqslant 1$$

Haimson and Fairhurst[10] refer to the formation breakdown pressure as fracture initiation pressure. Hence, when $\alpha\{(1-2v)/(1-v)\} = 1$, Equation (9.19) reduces to Equation (9.18) for the case of no fluid penetration. For the extreme case where there is complete fluid penetration, $\alpha\dfrac{(1-2v)}{(1-v)} = 0$, and

Equation (9.19) reduces to

formation breakdown pressure (FBP)

$$= \frac{1}{2}(3\sigma_3 - \sigma_2 + T) + P_f \qquad (9.20)$$

Thus, the effect of fluid penetration is to greatly reduce the magnitude of the formation breakdown pressure.

Assuming $\sigma_2 = \sigma_3$, Equation (9.19) can be modified to give

formation breakdown pressure

$$= \frac{2\sigma_3 + T}{2 - \alpha\left(\frac{1-2v}{1-v}\right)} + P_f \qquad (9.21)$$

Also

$$\sigma_3 = \left(\frac{v}{1-v}\right)(\sigma_v - P_f) \qquad (9.22)$$

Substituting Equation (9.22) in Equation (9.18) gives

formation breakdown gradient for no fluid penetration

$$= 2\left(\frac{v}{1-v}\right)(\sigma_v - P_f) + T + P_f \qquad (9.23)$$

Similarly, for the case of fluid penetration, when $\sigma_3 = \sigma_2$, Equation (9.19) reduces to

formation breakdown pressure

$$= \frac{2\left(\frac{v}{1-v}\right)(\sigma_v - P_f) + T}{2 - \alpha\left(\frac{1-2v}{1-v}\right)} + P_f \qquad (9.24)$$

Effects of hole deviation

All the previous equations were derived for the case of vertical holes in which the vertical principal stress had no influence on the magnitude of fracture pressure. However, as the hole starts deviating from vertical, the overburden stress starts contributing to the fracture pressure, thereby reducing the magnitude of the fracture gradient.

It can be shown (11) that the fracture pressure for a directional well is given by

$$\text{FBP} = 3\sigma_3 - \sigma_2 \cos^2\theta - \sigma_1 \sin\theta + T + P_f \qquad (9.25)$$

where θ = angle of deviation from vertical. And, when $\sigma_3 = \sigma_2$, equation (9.25) reduces to

$$\text{FBP} = \sigma_3(3 - \cos^2\theta) - \sigma_1 \sin^2\theta + T + P_f \qquad (9.26)$$

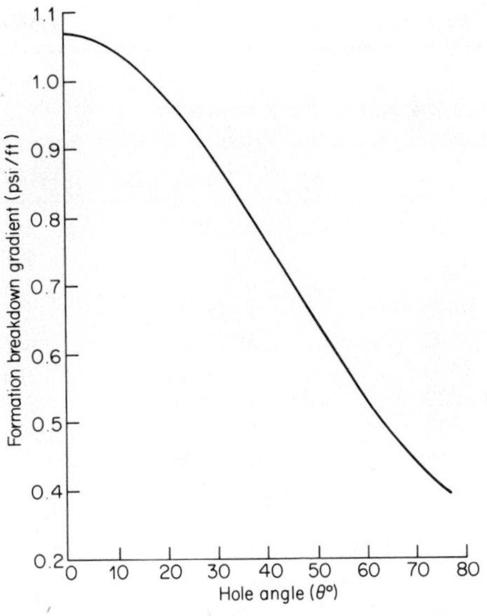

Fig. 9.8. Effect of hole angle on formation breakdown gradient.

Figure 9.8 shows a plot of formation breakdown gradient (=FBP/depth) against depth for hole angles ranging from 0 to 80°.

CASING SEAT SELECTION

Accurate knowledge of pore pressure and fracture gradient plays a major role in the selection of proper casing seats which would allow the drilling of the next hole without fracturing. Pore pressure, mud weight and fracture gradient are used collectively to select proper casing seats.

The following example illustrates a method for the selection of casing seats.

Example 9.4

Using the data in columns 1 and 2 of Table 9.1, calculate the fracture gradient at the various depths using Equation (9.13), and select appropriate casing seats. Assume $v = 0.4$.

The following steps may be used to complete columns 3–6 of Table 9.1:

(1) The required hydrostatic pressure of mud is taken as equal to pore pressure + 200 psi, where 200 psi is the magnitude of overbalance. Any reasonable value of overbalance may be used.

(2) Calculate the pore pressure and mud pressure

TABLE 9.1 Example 9.4

(1) *Depth below sea level (ft)*	*(2)* *Pore pressure (psi)*	*(3)* *Mud pressure = pore pressure + 200 psi (psi)*	*(4)* *Pore pressure gradient (psi/ft)*	*(5)* *Mud pressure gradient (psi/ft)*	*(6)* *Fracture gradient (psi/ft)*
3000	1320	1520	0.44	0.51	0.81
5000	2450	2650	0.49	0.53	0.83
8300	4067	4267	0.49	0.51	0.83
8500	4504	4704	0.53	0.55	0.84
9000	5984	6184	0.66	0.69	0.89
9500	6810	7010	0.72	0.74	0.91
10 000	7800	8000	0.78	0.80	0.93
11 000	10 171	10 371	0.92	0.94	0.97

gradients by simply dividing pore pressure and mud pressure by depth to obtain the gradient in psi/ft.
(3) Calculate the fracture gradient by use of Equation (9.13):

$$FG = \left(\frac{v}{1-v}\right)\frac{(\sigma_v - P_f)}{D} + \frac{P_f}{D}$$

By use of $v = 0.4$, this equation becomes

$$FG = \frac{2}{3}\left(\frac{\sigma_v - P_f}{D}\right) + \frac{P_f}{D}$$

or

$$FG = \frac{2}{3}\left(\frac{\sigma_v}{D} - \frac{P_f}{D}\right) + \frac{P_f}{D}$$

where P_f/D is the pore pressure gradient. Taking $\sigma_v/D = 1$ psi/ft, and using Table 9.1, the fracture gradient at 5000 ft is calculated as follows:

$$FG = \frac{2}{3}(1 - 0.49) + 0.49 = 0.83 \text{ psi/ft}$$

The same equation can be used to calculate FG at other depths (see Table 9.1).

Casing seat selection
(1) On the same graph paper, plot the pore pressure gradient, the mud pressure gradient and the fracture gradient against depth as shown in Figure 9.9.
(2) Starting at hole TD (11 000 ft), draw a vertical line through the mud gradient until it intersects the frac-

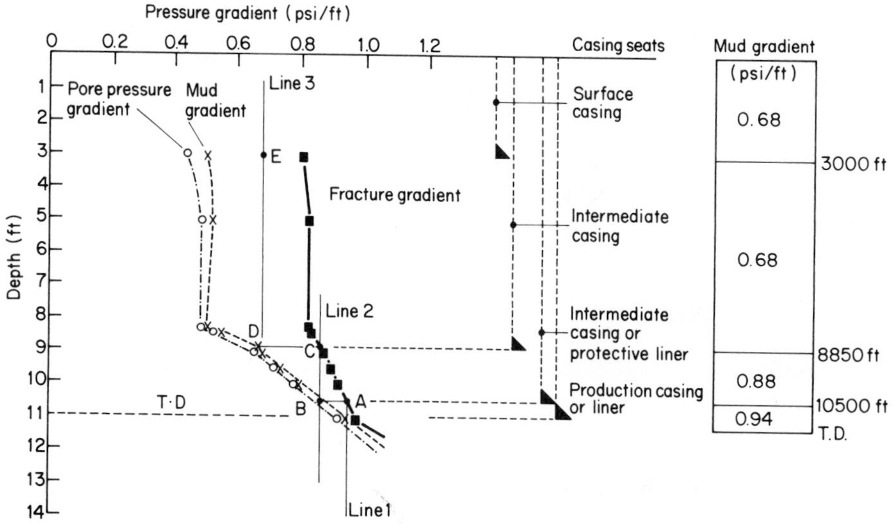

Fig. 9.9. Casing seat selection (Example 9.4).

ture gradient line. In our example the mud gradient at TD is 0.94 psi/ft and a vertical line through it (line 1 in Figure 9.9) intersects the fracture gradient line at 10 500 ft (point A in Figure 9.9). Above 10 500 ft, the mud gradient, 0.94 psi/ft, will exceed the fracture gradient, and a reduction in mud weight is required. Hence, between 10 500 ft and 11 000 ft the open hole should be cased with either a production liner or a production casing.

(3) Above 10 500 ft the hole must be drilled with a mud weight less than 0.94 psi/ft. The new mud gradient is obtained by drawing a horizontal line from point A to the mud gradient line. Point B in Figure 9.9 gives the new mud gradient as 0.88 psi/ft. Move vertically from point B until the fracture gradient line is intersected at 8850 ft at point C. Point C establishes the maximum depth that can be drilled before changing to the new mud gradient of 0.88 psi/ft. Hence, between points B and C a protective liner should be used. Alternatively, an intermediate casing can be set at point B.

(4) From point C move horizontally to the mud gradient line to point D, where the mud gradient is 0.68 psi/ft. A vertical line from point D shows that a hole can be drilled with a mud gradient of 0.68 psi/ft to surface without fracturing the formation.

Suggested casing types at the various setting depths are given in Figure 9.9.

References

1. Snyder, R. and Suman, G. (1978, 1979). *World Oil's Handbook of High Pressure Well Completions. World Oil.*
2. Schlumberger (1972). *Log Interpretation Principles*, Vol. 1. Schlumberger Publications.
3. Fertl, W. H. (1976). *Abnormal Formation Pressures.* Elsevier, Amsterdam.
4. Hubbert, M. and Willis, D. (1957). Mechanics of hydraulic fracturing.
5. Benham, P. and Warnock, F. (1973). *Mechanics of Solids and Structures.* Pitman, London.
6. Matthews, W. and Kelly, J. (1967). How to predict formation pressure and fracture gradient. *Oil and Gas Journal*, February, 92–106.
7. Pilkington, P. E. (1978). Fracture gradient estimates in tertiary basins. *Petroleum Engineer*, May, 138–148.
8. Eaton, B. A. (1969). Fracture gradient prediction and interpretation of hydraulic fracture treatments. *Journal of Petroleum Technology*, October, 21–29.
9. Christman, S. A. (1973). Offshore fracture gradients. *Journal of Petroleum Technology*, August, 910–914.
10. Haimson, B. and Fairhurst, C. (1967). Initiation and extension of hydraulic fractures in rocks. *Society of Petroleum Engineers Journal*, September, 310–311.
11. Bradley, W. B. (1979). Failure of inclined boreholes. *Transactions of the ASME*, December, **101**, 232–239.

Chapter 10

Casing Design

In this chapter the following topics will be discussed:

Functions of casing
Casing types
Strength properties
Casing specification
Casing design
Buoyancy effects
Bending forces
Shock loads
Casing design example
Compression in conductor pipe
Maximum load casing design

FUNCTIONS OF CASING

The functions of casing may be summarised as follows.

(1) To keep the hole open and to provide a support for weak, or fractured formations. In the latter case, if the hole is left uncased, the formation may cave in and redrilling of the hole will then become necessary.
(2) To isolate porous media with different fluid/pressure regimes from contaminating the pay zone. This is basically achieved through the combined presence of cement and casing. Therefore, production from a specific zone can be made.
(3) To prevent contamination of near-surface fresh-water zones.
(4) To provide a passage for hydrocarbon fluids; most production operations are carried out through special tubings which are run inside the casing.
(5) To provide a suitable connection for the wellhead equipment (e.g. the christmas tree). The casing also serves to connect the blowout prevention equipment (BOP) which is used to control the well while drilling.
(6) To provide a hole of known diameter and depth to facilitate the running of testing and completion equipment.

CASING TYPES

In practice, it would be much cheaper to drill a single-size hole to total depth (TD), probably with a small-diameter drill bit, and then to case the hole from the surface to the TD. However, the presence of high-pressured zones at different depths along the well bore, and the presence of weak, unconsolidated formations or sloughing shaly zones, necessitates running casing to seal off these troublesome zones and to allow of drilling to TD. Different sizes of casing are therefore run to case off the various sections of hole; a large-size casing is run at the surface followed by one or several intermediate casings and finally a small size casing for production purposes. Many different size combinations are run in different parts of the world. In this chapter sizes used most commonly in the North Sea and the Middle East have been quoted.

The types of casing currently in use are as follows.

Stove pipe

Stove pipe (or marine conductor or foundation pile for offshore drilling) is run to prevent washout of near-surface unconsolidated formations, to provide a circulation system for the drilling mud and to ensure the stability of the ground surface upon which the rig is sited. This pipe does not carry any wellhead equipment and can be driven into the ground with a pile driver. The normal size for a stove pipe ranges from 26 in to 42 in.

Conductor pipe

Conductor pipe is run from the surface to some shallow depth to protect near-surface unconsolidated formations, seal off shallow-water zones, provide protection against shallow gas flows and provide a circuit for the drilling mud to protect the foundations of the platform (offshore). One or more BOPs may be mounted on this casing or a diverter system if the setting depth of the conductor pipe is shallow. In the Middle East a typical size for a conductor pipe is either $18\frac{5}{8}$ in (473 mm) or 20 in (508 mm). In the North Sea exploration wells the size of the conductor pipe is usually 30 in (762 mm). Conductor pipe is always cemented to the surface and is either used to support subsequent casing strings and wellhead equipment or simply cut at the surface after setting the surface casing.

Surface casing

Surface casing is run to prevent caving of weak formations that are encountered at shallow depths. This casing should be set in competent rocks such as hard limestone. This will ensure that the formations at the casing shoe will not fracture at high hydrostatic pressures which may be used later. The surface casing also serves to provide protection against shallow blowouts as drilling progresses, hence BOPs are connected to the top of this string. The setting depth of this string of casing is chosen such that troublesome formations, thief zones, water sands, shallow hydrocarbon zones and build-up sections of deviated wells may be protected. A typical size of this casing is $13\frac{3}{8}$ in (340 mm) in the Middle East and $18\frac{5}{8}$ in or 20 in in North Sea operations.

Intermediate casing

Intermediate casing is usually set in the transition zone below or above an over-pressured zone or is run to seal off a severe-loss zone or to protect against problem formations, such as salt sections or caving shales. Good cementation of this casing must be ensured, to prevent communication behind the casing between the lower hydrocarbon zones and upper water formations. Multistage cementing may be used to cement long strings of intermediate casing. The most common size of this casing is $9\frac{5}{8}$ in (244.5 mm).

Production casing

Production string represents the last casing string. It is run to isolate producing zones, to provide reservoir fluid control, and to permit selective production in multizone production. This is the string through which the well will be completed. The normal size is 7 in (177.8 mm).

Liner casing

A liner is a string of casing that does not reach the surface. Liners are hung on the intermediate casing, by use of a suitable arrangement of a packer and slips called a liner hanger. In liner completions both the liner and the intermediate casing act as the production string. Because a liner is set at the bottom and hung from the intermediate casing, the major design criterion for a liner is the ability to withstand the maximum collapse pressure; see section on 'Casing Design' for further details.

The advantages of a liner are: (a) total costs of the production string are reduced, and running and cementing times are reduced; (b) the length of reduced diameter is reduced, which allows completion with adequate sizes of production tubings.

The disadvantages of a liner are: (a) possible leak across a liner hanger; and (b) difficulty in obtaining a good primary cementation due to the narrow annulus between the liner and the hole.

Types of liner

Drilling liners are used to isolate lost circulation or abnormally pressured zones to permit deeper drilling.

Production liners are run instead of a full casing to provide isolation across the producing or injection zones.

The *tie-back liner* is a section of casing extending upwards from the top of an existing liner to the surface or well head.

The *scab liner* is a section of casing that does not reach the surface. It is used to repair existing damaged casing. It is normally sealed with packers at top and bottom and in some cases is also cemented.

The *scab tie-back liner* is a section of casing extending from the top of an existing liner but does not reach the surface. The scab tie-back liner is usually cemented in place.

STRENGTH PROPERTIES

Casing strength properties are normally specified as: (1) yield strength for (a) pipe body and (b) coupling; (2) collapse strength; and (3) burst strength for (a) plain pipe and (b) coupling.

Yield strength

When a specimen of any metal such as steel is slowly loaded by compression or tension, a gradual decrease or increase in its length is observed. If the increments of load are plotted against extension (or contraction), a graph similar to that shown in Figure 10.1 is obtained. Up to a certain load, any increase in load will be accompanied by a proportional increase in length in accordance with Hooke's law:

$$\sigma = E\varepsilon \tag{10.1}$$

where σ = applied stress = load ÷ cross-sectional area; and ε = deformation (strain) = elongation/original length.

Hooke's law is only applicable along the straight portion of the graph referred to as the 'elastic range', line OA in Figure 10.1. The point on the load–elongation graph (or σ–ε graph) where Hooke's law is no longer applicable marks a change from elastic to plastic behaviour. Along the plastic portion the metal yields, resulting in permanent deformation and loss of strength. Point B in Figure 10.1 defines the yield strength of the material. Up to point A, loading results in no damage to the internal structure, and removal of the load will result in the specimen regaining its original length and shape. Loading beyond the yield point results in a change in the internal structure of the material and in a loss of strength. The original shape and length of the specimen are not regained upon removal of the load. Hence, it is important not to exceed the yield strength of casing during running, cementing and production operations, to prevent possible failure of the casing string. The ratio between stress and strain along the portion OA defines Young's modulus, E. Other important properties on the load-deformation graph are shown in Figure 10.1.

When quoting the strength of the casing, it is customary to use the yield strength of the casing, normally measured in lb or kN.

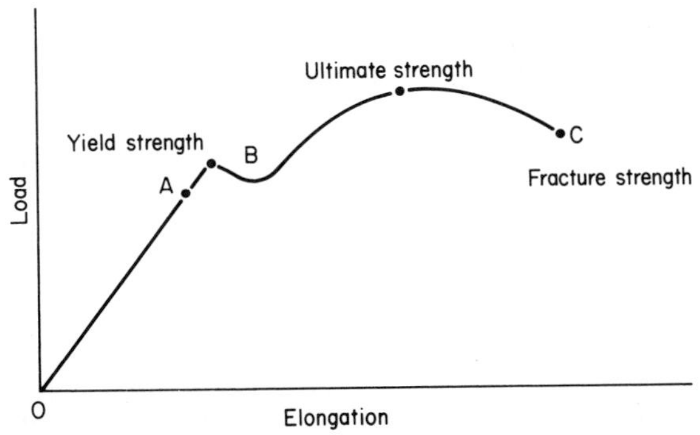

Fig. 10.1. Graph of load versus extension for metals.

API defines the yield strength as the tensile stress required to produce a total elongation of 0.5% of the gauge length, as determined by an extensometer or by multiplying dividers[1].

The most common types of casing joint are threaded on both ends and are fitted with a threaded coupling at one end only. The coupling is the box end of the casing joint. The strength of the coupling may be higher or lower than the yield strength of the main body of the casing joint. Hence, manufacturers supply data on both body and coupling strengths to be used in casing design calculations, as will be shown later. There are also integral casings available (i.e. without couplings) in which the threads are cut from internal–external upset pipe ends.

Collapse strength

Collapse strength is defined as the maximum external pressure required to collapse a specimen of casing. The procedure for determining the collapse strength of casing is detailed in API *Bulletin* 5C3 (1980).

In practice, two types of collapse are observed: elastic and plastic. In elastic collapse the specimen fails before it deforms, while in plastic collapse a certain deformation takes place prior to failure of the specimen.

Elastic collapse

The elastic collapse pressure, P_c, may be determined from the following formula:

$$P_c = \frac{2E}{1-\nu^2} \times \frac{1}{\frac{D}{t}\left(\frac{D}{t}-1\right)^2} \tag{10.2}$$

where E = Young's modulus of steel; ν = Poisson's ratio; t = casing thickness; and D = the outside diameter of the casing.

In Imperial units $E = 30 \times 10^6$ psi and $\nu = 0.3$; hence, Equation (10.2) simplifies to

$$P_c = \frac{46.95 \times 10^6}{\frac{D}{t}\left(\frac{D}{t}-1\right)^2} \text{ psi} \tag{10.3}$$

In metric units, Equation (10.2) becomes

$$P_c = \frac{2.198 \times 10^6}{\frac{D}{t}\left(\frac{D}{t}-1\right)^2} \text{ bar} \tag{10.4}$$

Equations (10.3) and (10.4) are applicable to the range of D/t values given in Table 10.1.

TABLE 10.1 *D/t* range for elastic collapse. (Courtesy of API[2])

1 Grade†	2 *D/t* range
H-40	42.70 and greater
-50*	38.83 ,, ,,
J-K-55 & D	37.20 ,, ,,
-60*	35.73 ,, ,,
-70*	33.17 ,, ,,
C-75 & E	32.05 ,, ,,
L-80 & N-80	31.05 ,, ,,
-90*	29.18 ,, ,,
C-95	28.25 ,, ,,
-100*	27.60 ,, ,,
P-105	26.88 ,, ,,
P-110	26.20 ,, ,,
-120*	25.01 ,, ,,
-125	24.53 ,, ,,
-130*	23.94 ,, ,,
-135	23.42 ,, ,,
-140	23.00 ,, ,,
-150	22.12 ,, ,,
-155*	21.70 ,, ,,
-160*	21.32 ,, ,,
-170	20.59 ,, ,,
-180*	19.93 ,, ,,

† Grades indicated without letter designation are not API grades but are grades in use or grades being considered for use and are shown for information purposes.

TABLE 10.2 Formula factors and *D/t* ranges for plastic collapse. (Courtesy of API[2])

1	2	3	4	5
Grade†	Formula factor			*D/t* range
	A	*B*	*C*	
H-40	2.950	0.0463	755	16.44–26.62
-50*	2.976	0.0515	1056	15.24–25.63
J-K-55 & D	2.990	0.0541	1205	14.80–24.99
-60*	3.005	0.0566	1356	14.44–24.42
-70*	3.037	0.0617	1656	13.85–23.38
C-75 & E	3.060	0.0642	1805	13.67–23.09
L-80 & N-80	3.070	0.0667	1955	13.38–22.46
-90*	3.106	0.0718	2254	13.01–21.69
C-95	3.125	0.0745	2405	12.83–21.21
-100*	3.143	0.0768	2553	12.70–21.00
P-105	3.162	0.0795	2700	12.56–20.66
P-110	3.180	0.0820	2855	12.42–20.29
-120*	3.219	0.0870	3151	12.21–19.88
-125	3.240	0.0895	3300	12.12–19.65
-130*	3.258	0.0920	3451	12.02–19.40
-135	3.280	0.0945	3600	11.90–19.14
-140	3.295	0.0970	3750	11.83–18.95
-150	3.335	0.1020	4055	11.67–18.57
-155*	3.356	0.1047	4204	11.59–18.37
-160*	3.375	0.1072	4356	11.52–18.19
-170	3.413	0.1123	4660	11.37–18.45
-180*	3.449	0.1173	4966	11.23–17.47

† Grades indicated without letter designation are not API grades but are grades in use or grades being considered for use and are shown for information purposes.

Plastic collapse

The minimum collapse pressure (P_p) in the plastic range may be calculated from the following equation:

$$P_p = Y\left(\frac{A}{D/t} - B\right) - C \qquad (10.5)$$

where A, B and C are constants depending on the grade of steel used and Y is yield strength.

Equation (10.5) is applicable for the range of D/t values given in Table 10.2. The ratio D/t should first be determined, and if it falls in the range given in Table 10.2, then Equation (10.5) is applicable and the values of A, B and C are read directly from this table.

Transition collapse pressure

The collapse behaviour, P_T, of steel in the transition zone between elastic and plastic failure is described by the following formula:

$$P_T = Y\left(\frac{F}{D/t} - G\right) \text{psi}$$

where F and G are constants, given by

$$F = \frac{46.95 \times 10^6 \left(\dfrac{3B/A}{2 + B/A}\right)^3}{Y\left[\dfrac{3B/A}{2+(B/A)} - (B/A)\right]\left[1 - \dfrac{3B/A}{2+(B/A)}\right]^2} \qquad (10.6)$$

$$G = \frac{FB}{A}$$

The range of D/t values applicable to Equation (10.6) is given in Table 10.3 together with F and G values.

TABLE 10.3 Formula factors and D/t range for transition collapse. (Courtesy of API[2])

1	2	3	4
Grade†	*Formula factors*		*D/t range*
	F	*G*	
H-40	2.047	0.031 25	26.62–42.70
-50*	2.003	0.0347	25.63–38.83
J-K-55 & D	1.990	0.0360	24.99–37.20
-60*	1.983	0.0373	24.42–35.73
-70*	1.984	0.0403	23.38–33.17
C-75 & E	1.985	0.0417	23.09–32.05
L-80 & N-80	1.998	0.0434	22.46–31.05
-90*	2.017	0.0466	21.69–29.18
C-95	2.047	0.0490	21.21–28.25
-100*	2.040	0.0499	21.00–27.60
P-105	2.052	0.0515	20.66–26.88
P-110	2.075	0.0535	20.29–26.20
-120*	2.092	0.0565	19.88–25.01
-125	2.102	0.0580	19.65–24.53
-130*	2.119	0.0599	19.40–23.94
-135	2.129	0.0613	19.14–23.42
-140	2.142	0.0630	18.95–23.00
-150	2.170	0.0663	18.57–22.12
-155*	2.188	0.068 25	18.37–21.70
-160*	2.202	0.0700	18.19–21.32
-170	2.123	0.0698	18.45–20.59
-180*	2.261	0.0769	17.47–19.93

† Grades indicated without letter designation are not API grades but are grades in use or being considered for use and are shown for information purposes.

Burst (or internal yield) strength

Burst strength is defined as the maximum value of internal pressure required to cause the steel to yield. The minimum burst pressure for a casing is calculated by use of Barlow's formula:

$$P = 0.875\left(\frac{2\,Y\,t}{D}\right) \qquad (10.7)$$

where t = thickness of casing (in); D = OD of casing (in); and Y = minimum yield strength (psi).

Equation (10.7) gives burst resistance for a minimum yield of 87.5% of the pipe wall, allowing for a 12.5% variation of wall thickness due to manufacturing defects.

CASING SPECIFICATION

A casing is specified by the following parameters: (a) outside diameter and wall thickness; (b) weight per unit length, normally weight per foot or metre; (c) type of coupling; (d) length of joint; and (e) grade of steel.

Outside diameter and wall thickness

As mentioned previously, different casing sizes are run at different parts of the hole to allow the drilling of the hole to its total depth with minimum risk. Since pressures vary along any section of hole, it is possible to run a casing string having the same outside diameter but with different thicknesses or strength properties (i.e. grade). Thus, a heavy or high grade casing can be run only along the portions of hole containing high pressures or is run near the surface, where tensile stresses are high. This arrangement provides the most economical way of selecting a given casing string, as will be shown later.

Weight per unit length

API defines three types of casing weight: (1) nominal weight; (2) plain end weight; and (3) threaded and coupled weight.

Nominal weight

The term 'nominal weight' is used primarily for the purpose of identification of casing types during ordering. It is expressed in lb/ft or kg/m. Nominal weights are not exact weights and are normally based on the calculated, theoretical weight per foot for a 20 ft length of threaded and coupled casing joint.

Nominal weight, W_n, is calculated from the following formula:

$$W_n = 10.68(D - t)t + 0.0722D^2 \text{ lb/ft} \qquad (10.8)$$

where D = outside diameter (in) and t = wall thickness (in).

Casing weights required for design purposes are reported as nominal weights; see Table 10.4.

Plain end weight

The plain end weight is the weight of the casing joint without the inclusion of threads and couplings. The plain end weight can be calculated by use of the following formula, taken from API Standards, *Bulletin* 5C3:

$$W_{pe} = 10.68(D - t)t \text{ lb/ft} \qquad (10.9)$$

or

$$W_{pe} = 0.02466(D - t)t \text{ kg/m} \qquad (10.10)$$

where W_{pe} = plain end weight (lb/ft or kg/m); and D and t are as defined before.

TABLE 10.4 Properties of 20″, $21\frac{1}{2}$″ and $24\frac{1}{2}$″ casings. (Courtesy of Mannesmann Rohrenwerke[3])

1	2	3	4	5	6	7	8	9	10	11	12
Size: outside diameter D **in** mm	Nominal weight, threads and coupling **lb/ft** kg/m	Grade	Pipe		Threaded and coupled			Extreme-line		Collapse resistance P_c **psi** bar	Pipe body yield strength P_y **1000 lb** kN
			Wall thickness t	Inside diameter d	Drift diameter	Regular	Special clearance	Drift diameter	Outside diameter W		
						Outside diameter					
						W	W_c				
			in mm								
20 508.0	**94.00** 140	**J-55**	**0.438** 11.13	**19.124** 485.7	**18.936** 480.97	**21.000** 533.4				**520** 36	**1480** 6583
	106.50 159	**J-55**	**0.500** 12.70	**19.000** 482.6	**18.812** 477.82	**21.000** 533.4				**770** 53	**1685** 7495
	133.00 198	**J-55**	**0.635** 16.13	**18.730** 475.7	**18.542** 470.97	**21.000** 533.4				**1500** 103	**2125** 9452
	94.00 140	**K-55**	**0.438** 11.13	**19.124** 485.7	**18.936** 480.97	**21.000** 533.4				**520** 36	**1480** 6583
	106.50 159	**K-55**	**0.500** 12.70	**19.000** 482.6	**18.812** 477.82	**21.000** 533.4				**770** 53	**1685** 7495
	133.00 198	**K-55**	**0.635** 16.13	**18.730** 475.7	**18.542** 470.97	**21.000** 533.4				**1500** 103	**2125** 9452
	133.00* 198	**C-75**	**0.635** 16.13	**18.730** 475.7	**18.542** 470.97	**21.000** 533.4				**1600** 110	**2897** 12 886
	169.00* 252	**C-75**	**0.812** 20.62	**18.376** 466.7	**18.188** 461.98	**21.000** 533.4				**2920** 201	**3671** 16 329

	133.00*	**L-80**	**0.635**	**18.730**	**18.542**	**21.000**	**1600**	**3091**
	198		16.13	475.7	470.97	533.4	110	13 749
	169.00*	**L-80**	**0.812**	**18.376**	**18.188**	**21.000**	**3020**	**3916**
	252		20.62	466.7	461.98	533.4	208	17 419
	106.50*	**N-80**	**0.500**	**19.000**	**18.812**	**21.000**	**770**	**2450**
	159		12.70	482.6	477.82	533.4	53	10 898
	133.00*	**N-80**	**0.635**	**18.730**	**18.542**	**21.000**	**1600**	**3091**
	198		16.13	475.7	470.97	533.4	110	13 749
	169.00*	**N-80**	**0.812**	**18.376**	**18.188**	**21.000**	**3020**	**3916**
	252		20.62	466.7	461.98	533.4	208	17 419
	133.00*	**C-95**	**0.635**	**18.730**	**18.542**	**21.000**	**1600**	**3670**
	198		16.13	475.7	470.97	533.4	110	16 325
	169.00*	**C-95**	**0.812**	**18.376**	**18.188**	**21.000**	**3240**	**4650**
	252		20.62	466.7	461.98	533.4	223	20 684
21½	**115.00***	**J-55**	**0.500**	**20.500**	**20.312**	**22.500**	**620**	**1814**
546.1	171		12.70	520.7	515.92	571.5	43	8069
	141.00*	**J-55**	**0.625**	**20.250**	**20.062**	**22.500**	**1200**	**2254**
	210		15.88	514.3	509.57	571.5	83	10 026
	115.00*	**K-55**	**0.500**	**20.500**	**20.312**	**22.500**	**620**	**1814**
	171		12.70	520.7	515.92	571.5	43	8069
	141.00*	**K-55**	**0.625**	**20.250**	**20.062**	**22.500**	**1200**	**2254**
	210		15.88	514.3	509.57	571.5	83	10 026
	115.00*	**N-80**	**0.500**	**20.500**	**20.312**	**22.500**	**620**	**2639**
	171		12.70	520.7	515.92	571.5	43	11 739
	141.00*	**N-80**	**0.625**	**20.250**	**20.062**	**22.500**	**1220**	**3279**
	210		15.88	514.3	509.57	571.5	84	14 586
24½**	**140.00***	**J-55**	**0.531**	**23.438**	**23.141****	**25.500**	**500**	**2199**
622.3	209		13.50	595.3	587.78	647.7	34	9782
	140.00*	**K-55**	**0.531**	**23.438**	**23.141****	**25.500**	**500**	**2199**
	209		13.50	595.3	587.78	647.7	34	9782
	140.00*	**N-80**	**0.531**	**23.438**	**23.250**	**25.500**	**500**	**3199**
	209		13.50	595.3	590.55	647.7	34	14 230

* Non-API.

TABLE 10.4 *Continued*

13	14	15	16	17	18	19	20	21	22	23	24	25	26	27
	Internal yield pressure at minimum yield						*Joint strength*							
Plain-end or extreme-line	*Round thread*		*Buttress/BDS/Omega*				*Threaded and coupled*						*Extreme-line*	
	Short	*Long*	*Regular coupling*		*Special clearance coupling*		*Round thread*		*Buttress/BDS/Omega*				*BDS/Omega*	
							Short	*Long*	*Regular coupling*		*Special clearance coupling*			
			Same grade	*Higher grade*	*Same grade*	*Higher grade*			*Same grade*	*Higher grade*	*Same grade*	*Higher grade*	*Standard joint*	*Optional joint*
			P_i **psi** *bar*								*P_j* ***1000 lb*** *kN*			
2110	**2110**	**2110**	**2110**	**2110**			**784**	**907**	**1402**	**1402**				
145	145	145	145	145			3487	4035	6236	6236				
2410	**2410**	**2410**	**2410**	**2410**			**913**	**1057**	**1596**	**1596**				
166	166	166	166	166			4061	4702	7099	7099				
3060	**3060**	**3060**	**3060**	**3060**			**1192**	**1380**	**2012**	**2012**				
211	211	211	211	211			5302	6139	8950	8950				
2110	**2110**	**2110**	**2110**	**2110**			**824**	**955**	**1479**	**1479**				
145	145	145	145	145			3665	4248	6579	6579				
2410	**2410**	**2410**	**2410**	**2410**			**960**	**1113**	**1683**	**1683**				
166	166	166	166	166			4270	4951	7486	7486				
3060	**3060**	**3060**	**3060**	**3060**			**1253**	**1453**	**2123**	**2123**				
211	211	211	211	211			5574	6463	9444	9444				
4170	**4170**	**4170**	**4170**	**4170**			**1604**	**1857**	**2704**	**2704**				
288	288	288	288	288			7135	8260	12 028	12 028				

5330	**440**	**4780**	**4610**	**4610**	**2088**	**2418**	**3426**	**3426**
367	303	330	318	318	9288	10 756	15 240	15 240
4450	**4450**	**4450**	**4450**	**4450**	**1692**	**1958**	**2849**	**2849**
307	307	307	307	307	7526	8710	12 673	12 673
5680	**4690**	**4990**	**4920**	**5680**	**2202**	**2549**	**3610**	**3610**
392	323	344	339	392	9795	11 339	16 058	16 058
3500	**3500**	**3500**	**3500**	**3500**	**1307**	**1514**	**2281**	**2281**
241	241	241	241	241	5814	6734	10 146	10 146
4450	**4450**	**4450**	**4450**	**4450**	**1707**	**1976**	**2877**	**2877**
307	307	307	307	307	7593	8790	12 798	12 798
5680	**4690**	**4990**	**4920**	**5680**	**2222**	**2573**	**3645**	**3645**
392	323	344	339	392	9884	11 445	16 214	16 214
5280	**5280**	**5280**	**5280**	**5280**	**1985**	**2296**	**3340**	**3340**
364	364	364	364	364	8830	10 213	14 857	14 857
6750	**5570**	**5930**	**5840**	**6750**	**2584**	**2990**	**4120**	**4232**
465	384	409	403	465	11 494	13 300	18 327	18 825
2240	**2240**	**2240**			**941**	**1095**		
154	154	154			4186	4871		
2800	**2800**	**2800**			**1208**	**1406**		
193	193	193			5373	6254		
2240	**2240**	**2240**			**987**	**1151**		
154	154	154			4390	5120		
2800	**2800**	**2800**			**1267**	**1477**		
193	193	193			5636	6570		
3260	**3260**	**3260**			**1348**	**1567**		
225	225	225			5996	6970		
4070	**4070**	**4070**			**1731**	**2012**		
281	281	281			7700	8950		
2090	**2090**	**2090**			**1062**	**1247**		
144	144	144			4724	5547		
2090	**2090**	**2090**	**2090**		**1110**	**1306**	**1873**	
144	144	144	144		4938	5809	8330	
3030	**3030**	**3030**	**3030**		**1523**	**1787**	**2786**	
209	209	209	209		6775	7949	12 393	

Threaded and coupled weight

The threaded and coupled weight is the average weight of a joint including the threads at both ends and a coupling at one end when power-tight. This weight is calculated from the following formula:

$$W = \frac{\left(20 - \frac{N_L + 2J}{24}\right) W_{pe} + \text{weight of coupling} - \text{weight removed in threading two pipe ends}}{20} \tag{10.11}$$

where W = threaded and coupled weight (lb/ft); N_L = coupling length (in); J = distance from end of pipe to centre of coupling in the power-tight position (in); and W_{pe} = plain end weight, as calculated in Equations (10.9).

Types of coupling and thread

A coupling is a short section of casing and is used to connect two casing joints. A casing joint is externally threaded at both ends. The most common type of coupling is internally threaded from each end. API specifies that a coupling should be of the same grade as the pipe body.

In general, the casing and coupling are specified by the type of threads (or connection) cut in the pipe or coupling. API[4] defines three principal elements of thread (see Figures 10.2–10.5): (1) thread height or depth, defined as the distance between the thread crest and the thread root measured normal to the axis of the thread; (2) lead, defined as the distance from one point on a thread to a corresponding point on the adjacent thread, as measured parallel to the thread axis; (3) taper, defined as the change in diameter of a thread expressed in inches per foot of thread length; and (4) thread form — most casing threads are squared or V-shaped.

The following are the most widely used connections.

(a) API 8 round thread: This thread is a V-type thread with an included angle of 60° between the flanks and with 8 threads per inch on a 0.0625 in per inch of taper, as shown in Figure 10.2. The thread crest and roots are truncated with a radius. The API 8 round casing is threaded on both ends of a non-upset pipe, and individual pipes are joined together by means of an internally threaded coupling. API round thread couplings are of two types: (1) short thread coupling (STC); and (2) long thread coupling (LTC). The STC and LTC are weaker than the pipe body, the LTC being capable of transmitting higher axial loads.

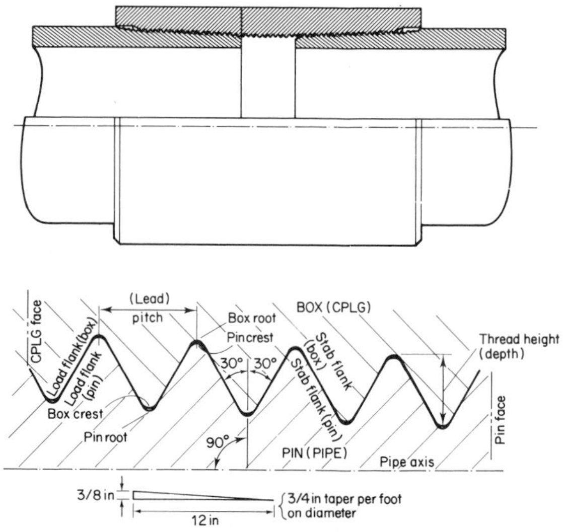

Fig. 10.2. Round thread casing and tubing thread configuration. (Courtesy of API)

(b) Buttress thread. Figure 10.3 shows a cross-section of a buttress connection and a thread profile. The buttress thread is capable of transmitting higher axial loads than the API 8 round thread and can be distinguished by the conical crests and roots for $13\frac{3}{8}$ in OD and smaller casings and by the flat crests and roots for 16 in OD or larger casings. A proper thread compound is used to increase leak resistance. A buttress connection has five threads per inch.

(c) VAM thread: This is a modified version of the buttress thread, providing a double metal-to-metal seal at the pin end. Figure 10.4 shows a VAM-type coupling and thread profile. The VAM thread profile has five threads per inch. The VAM thread takes its name from the manufacturer Vamonec and it is one example of the many proprietary threads made by specific manufacturers.

(d) Extreme line threaded coupling: API extreme line casing is externally and internally threaded on internal–external upset ends. The upset ends are specially machined to have increased wall thickness, to compensate for the loss of metal due to threading. The thread profile is trapezoidal, providing a metal-to-metal seal at both the pin end and the external shoulder. This makes the extreme line couplings suitable for use in elevated temperatures and pressures. For casing sizes 5–$7\frac{5}{8}$ in the threads are of 6 in pitch (six threads per inch). For casing sizes $8\frac{5}{8}$–$10\frac{3}{4}$ in the threads are of 5 in pitch. Figure 10.5 shows a section of an extreme line connection and a thread profile.

(e) Buttress double seal (BDS) thread: This is a modification of the buttress thread, having five

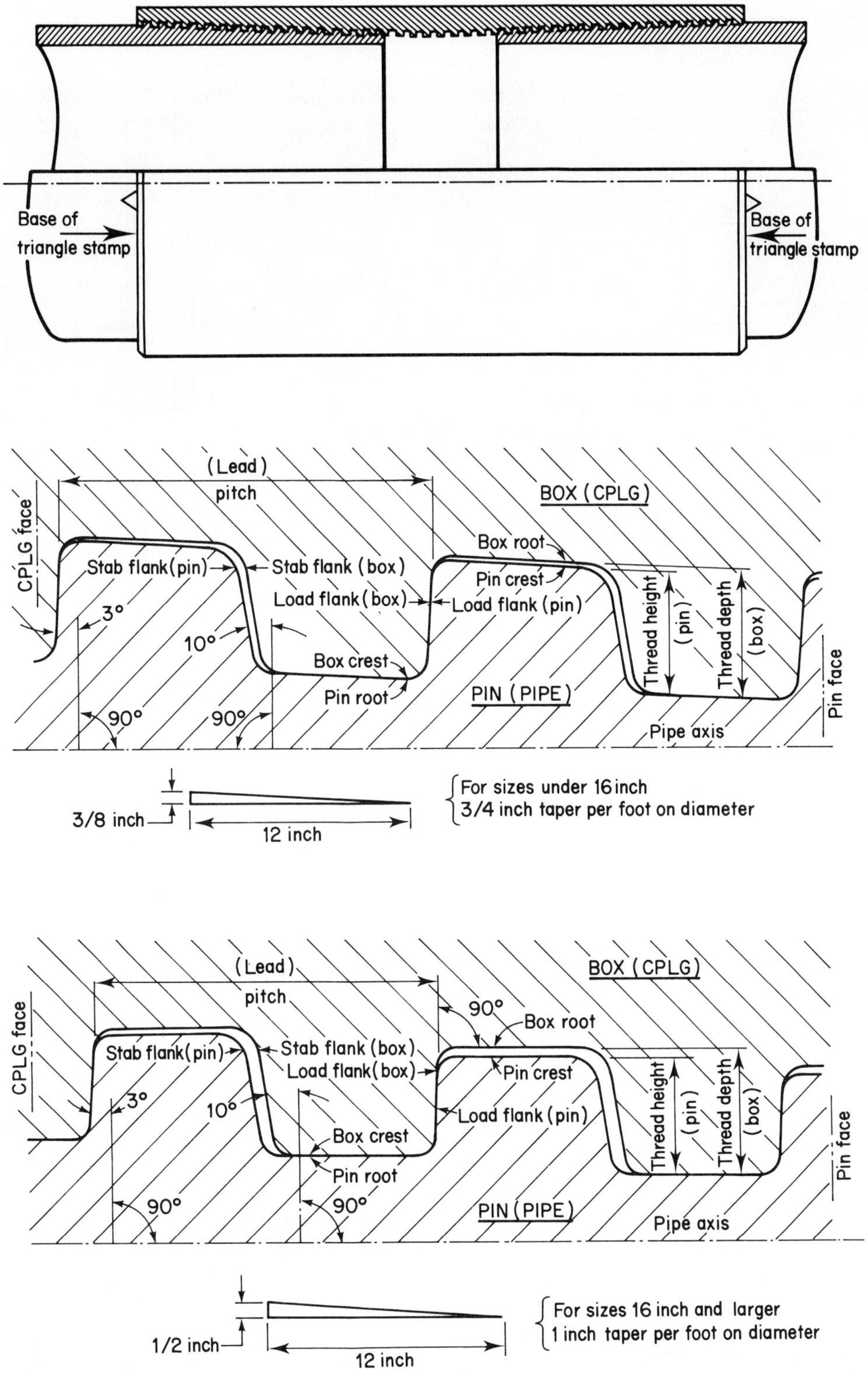

Fig. 10.3. Buttress thread configuration: (a) $13\frac{3}{8}$-in outside diameter and small casing; (b) 16-in outside diameter and larger casing. (Courtesy of API)

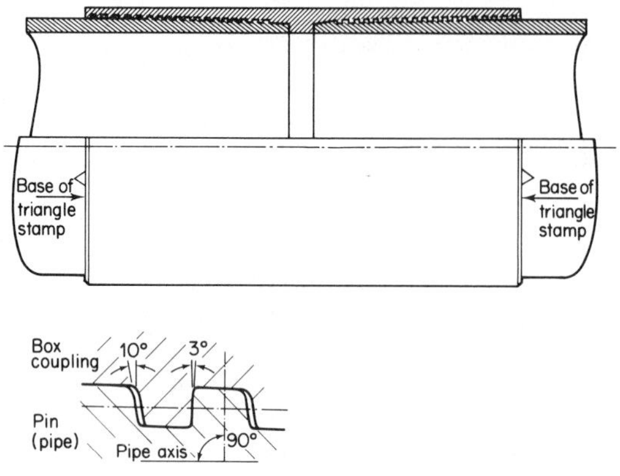

Fig. 10.4. Vam thread configuration.

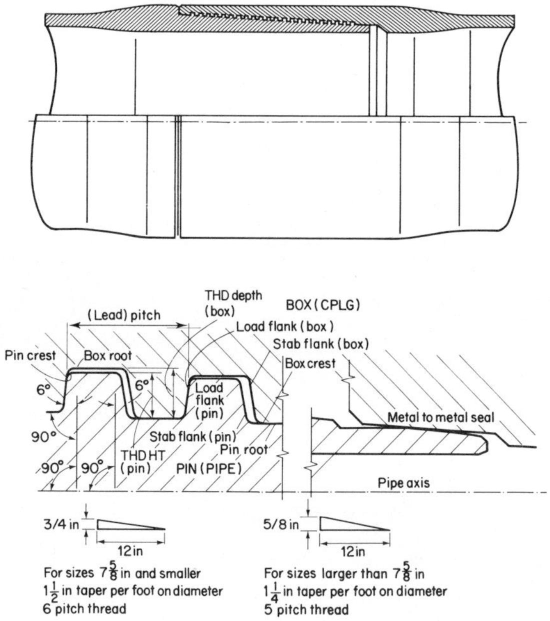

Fig. 10.5. Extreme-line casing thread configuration. (Courtesy of API)

threads per inch and a taper of 1:16. Sealing by this thread is provided in two ways: (1) a metal-to-metal seal between the cylindrical mandrel of the box and the conical shape of the pin; and (2) sealing through the internal bevel of the shoulder. Figure 10.6 shows a BDS connection and a thread profile.

There are also other proprietary thread brands, including NL Atlas Bradford, Hydril and Armco. Full details of these types can be obtained from the manufacturers.

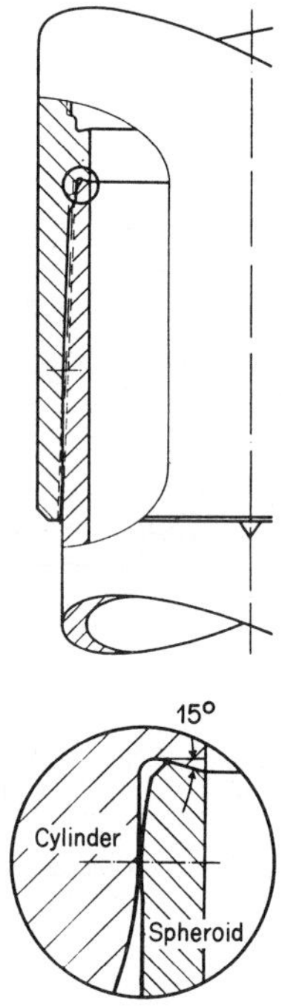

Fig. 10.6. Buttress double seal[3] (5 threads per inch; sizes, $4\frac{1}{2}$–$13\frac{3}{8}$ in). (Courtesy of Mannesmann)

Length of joint

API[1] has specified three ranges in which a pipe length must lie. These are as follows:

Range	*Length* (ft)	*Average length* (ft)
1	16–25	22
2	25–34	31
3	over 34	42

Grade of steel

The raw steel material used for the manufacturing of casing has no definite microstructure. The microstructure of steel can be greatly changed by the addition of special alloys and by heat treatment. Thus, different grades of casing can be manufactured to suit different drilling situations. API lists eight different grades of casing, as follows:

Grade	*Minimum yield strength* (psi)	*Minimum tensile strength* (psi)
H40	40 000	60 000
J55	55 000	70 000–95 000
K55	55 000	70 000–95 000
C-75	75 000	95 000
N-80	80 000	100 000
L-80	80 000	100 000
C-95	95 000	105 000
P110	110 000	125 000

Tables 10.4–10.7 give the dimensions and strength properties of 20 in, $13\frac{3}{8}$ in, $9\frac{5}{8}$ in and 7 in casing, respectively.

CASING DESIGN

A casing design exercise involves the determination of factors that influence the failure of casing and the selection of the most suitable casing grade for a specific operation, both safely and economically. The casing programme should also reflect the completion and production requirements.

Solution of casing design problems depends heavily on stress analysis techniques and rock mechanics. The end product of such a design is a 'pressure vessel' capable of withstanding the expected internal and external pressures and stresses arising from the casing own weight. A safety margin is also included in the design, to allow for future deterioration of the casing's and for other unknown forces that may be encountered, such as corrosion, wear and thermal effects.

Casing design is also influenced by: (a) the loading conditions during drilling and production; (b) the strength properties of the casing seat (i.e. formation strength at casing shoe) and of available casing; (c) the degree of deterioration to which the pipe will be subjected during the entire life of the well; and (d) the availability of casing.

Design criteria

The criteria for casing design are as follows.

Tensile force

Tensile forces in casings originate from casing-own-weight, bending forces, and shock loading. In casing design the uppermost joint of the string is considered the weakest in tension, as it has to carry the total weight of the casing string. Selection is normally based on a safety factor of 1.6–1.8 for the top joint.

Collapse pressure

Collapse pressure originates from the column of mud used to drill the hole, and acts on the outside of the casing. Since the hydrostatic pressure of a column of mud increases with depth, collapse pressure is highest at the bottom and zero at the top.

Therefore,

$$\text{collapse pressure } (C) = \text{mud density} \times \text{depth} \times \text{acceleration due to gravity}$$

$$= \rho_m g h$$

$$C = \frac{\rho_m h}{144} \text{ psi} \qquad (10.12)$$

where ρ_m is in lbm/ft^3 and h is in ft, or

$$C = 0.052 \rho h \text{ psi} \qquad (10.13)$$

where ρ is in ppg.

In metric units collapse pressure, C, is given by

$$C = 0.0981\ \rho h \text{ bar} \qquad (10.14)$$

where ρ is in kg/l and h is in m.

The designer should ensure that the collapse pressure, C, never exceeds the collapse resistance of the casing. For this purpose, the casing collapse resistance is taken as the load at which the internal diameter of the casing yields. In designing for collapse, the casing is assumed empty for surface and production casing and partially empty for intermediate casing.

Burst pressure

The burst criterion in casing design is normally based on the maximum formation pressure that can be encountered during the drilling of next hole section. Also, it is assumed that in the event of a kick, the influx fluid(s) will displace the entire drilling mud, thereby subjecting the entire casing to the bursting effects of formation pressure.

At the top of the hole the external pressure due to the hydrostatic head of mud is zero and the internal pressure must be supported entirely by the casing body. Therefore, burst pressure is highest at the top and least at the casing shoe. As will be shown later, in production casing the burst pressure at the shoe can be higher than the burst pressure at the surface in situations when the production tubing leaks gas to the casing.

TABLE 10.5 Properties of $13\frac{3}{8}''$ casings. (Courtesy of Mannesmann Rohrenwerke[3])

1	2	3	4	5	6	7	8	9	10	11	12
Size: outside diameter D **in** mm	Nominal weight, threads and coupling **lb/ft** kg/m	Grade	Pipe		Threaded and coupled			Extreme-line		Collapse resistance P_c **psi** bar	Pipe body yield strength P_y **1000 lb** kN
			Wall thickness t	Inside diameter d	Drift diameter	Regular	Special clearance	Drift diameter	Outside diameter W		
						Outside diameter W	Outside diameter W_c				
			in mm								
$13\frac{3}{8}$ 339.7	**54.50** 81.2	**J-55**	**0.380** 9.65	**12.615** 320.4	**12.459** 316.46	**14.375** 365.1				**1130** 78	**853** 3794
	61.00 90.9	**J-55**	**0.430** 10.92	**12.515** 317.9	**12.359** 313.92	**14.375** 365.1				**1540** 106	**962** 4279
	68.00 101	**J-55**	**0.480** 12.19	**12.415** 315.3	**12.259** 311.38	**14.375** 365.1				**1950** 134	**1069** 4755
	72.00* 107	**J-55**	**0.514** 13.06	**12.347** 313.6	**12.191** 309.65	**14.375** 365.1				**2230** 154	**1142** 5080
	54.50 81.2	**K-55**	**0.380** 9.65	**12.615** 320.4	**12.459** 316.46	**14.375** 365.1				**1130** 78	**853** 3794
	61.00 90.9	**K-55**	**0.430** 10.92	**12.515** 317.9	**12.359** 313.92	**14.375** 365.1				**1540** 106	**962** 4279
	68.00 101	**K-55**	**0.480** 12.19	**12.415** 315.3	**12.259** 311.38	**14.375** 365.1				**1950** 134	**1069** 4755
	72.00* 107	**K-55**	**0.514** 13.06	**12.347** 313.6	**12.191** 309.65	**14.375** 365.1				**2230** 154	**1142** 5080
	61.00* 90.9	**C-75**	**0.430** 10.92	**12.515** 317.9	**12.359** 313.92	**14.375** 365.1				**1660** 114	**1312** 5836
	68.00 101	**C-75**	**0.480** 12.19	**12.415** 315.3	**12.259** 311.38	**14.375** 365.1				**2220** 153	**1458** 6486
	72.00 107	**C-75**	**0.514** 13.06	**12.347** 313.6	**12.191** 309.65	**14.375** 365.1				**2590** 179	**1558** 6930
	61.00* 90.9	**L-80**	**0.430** 10.92	**12.515** 317.9	**12.359** 313.92	**14.375** 365.1				**1670** 115	**1399** 6223
	68.00 101	**L-80**	**0.480** 12.19	**12.415** 315.3	**12.259** 311.38	**14.375** 365.1				**2260** 156	**1556** 6921
	72.00 107	**L-80**	**0.514** 13.06	**12.347** 313.6	**12.191** 309.65	**14.375** 365.1				**2670** 184	**1661** 7388
	61.00* 90.9	**N-80**	**0.430** 10.92	**12.515** 317.9	**12.359** 313.92	**14.375** 365.1				**1670** 115	**1399** 6223

68.00	**N-80**	**0.480**	**12.415**	**12.259**	**14.375**	**2260**	**1556**
101		12.19	315.3	311.38	365.1	156	6921
72.00	**N-80**	**0.514**	**12.347**	**12.191**	**14.375**	**2670**	**1661**
107		13.06	313.6	309.65	365.1	184	7388
61.00*	**MW-C-90**	**0.430**	**12.515**	**12.359**	**14.375**	**1670**	**1574**
90.9		10.92	317.9	313.92	365.1	115	7002
68.00*	**MW-C-90**	**0.480**	**12.415**	**12.259**	**14.375**	**2320**	**1750**
101		12.19	315.3	311.38	365.1	160	7784
72.00*	**MW-C-90**	**0.514**	**12.347**	**12.191**	**14.375**	**2780**	**1869**
107		13.06	313.6	309.65	365.1	192	8314
61.00*	**C-95**	**0.430**	**12.515**	**12.359**	**14.375**	**1670**	**1661**
90.9		10.92	317.9	313.92	365.1	115	7388
68.00	**C-95**	**0.480**	**12.415**	**12.259**	**14.375**	**2320**	**1847**
101		12.19	315.3	311.38	365.1	160	8216
72.00	**C-95**	**0.514**	**12.347**	**12.191**	**14.375**	**2820**	**1973**
107		13.06	313.6	309.65	365.1	194	8776
61.00*	**MW-C-95**	**0.430**	**12.515**	**12.359**	**14.375**	**1670**	**1661**
90.9		10.92	317.9	313.92	365.1	115	7388
68.00*	**MW-C-95**	**0.480**	**12.415**	**12.259**	**14.375**	**2320**	**1847**
101		12.19	315.3	311.38	365.1	160	8216
72.00*	**MW-C-95**	**0.514**	**12.347**	**12.191**	**14.375**	**3470**	**1973**
107		13.06	313.6	309.65	365.1	239	8776
61.00*	**P-110**	**0.430**	**12.515**	**12.359**	**14.375**	**1670**	**1924**
90.9		10.92	317.9	313.92	365.1	115	8558
68.00*	**P-110**	**0.480**	**12.415**	**12.259**	**14.375**	**2330**	**2139**
101		12.19	315.3	311.38	365.1	161	9515
72.00*	**P-110**	**0.514**	**12.347**	**12.191**	**14.375**	**2890**	**2284**
107		13.06	313.6	309.65	365.1	199	10 160
68.00*	**MW-125**	**0.480**	**12.415**	**12.259**	**14.375**	**2330**	**2431**
101		12.19	315.3	311.38	365.1	161	10 814
72.00*	**MW-125**	**0.514**	**12.347**	**12.191**	**14.375**	**2880**	**2596**
107		13.06	313.6	309.65	365.1	199	11 548
68.00*	**MW-140**	**0.480**	**12.415**	**12.259**	**14.375**	**2330**	**2722**
101		12.19	315.3	311.38	365.1	161	12 108
72.00*	**MW-140**	**0.514**	**12.347**	**12.191**	**14.375**	**2880**	**2907**
107		13.06	313.6	309.65	365.1	199	12 931
68.00*	**V-150**	**0.480**	**12.415**	**12.259**	**14.375**	**2330**	**2917**
101		12.19	315.3	311.38	365.1	161	12 975
72.00*	**V-150**	**0.514**	**12.347**	**12.191**	**14.375**	**2880**	**3115**
107		13.06	313.6	309.65	365.1	199	13 856
68.00*	**MW-155**	**0.480**	**12.415**	**12.259**	**14.375**	**2330**	**3014**
101		12.19	315.3	311.38	365.1	161	13 407
72.00*	**MW-155**	**0.514**	**12.347**	**12.191**	**14.375**	**2880**	**3219**
107		13.06	313.6	309.65	365.1	199	14 319

* Non-API.

TABLE 10.5 *Continued*

13	14	15	16	17	18	19	20	21	22	23	24	25	26	27
Internal yield pressure at minimum yield							*Joint strength*							
Plain-end or extreme-line	*Round thread*		*Buttress/BDS/Omega*				*Threaded and coupled*						*Extreme-line*	
	Short	*Long**	*Regular coupling*		*Special clearance coupling*		*Round thread*		*Buttress/BDS/Omega*					
							Short	*Long*	*Regular coupling*		*Special clearance coupling*			
			Same grade	*Higher grade*	*Same grade*	*Higher grade*			*Same grade*	*Higher grade*	*Same grade*	*Higher grade*	*Standard joint*	*Optional joint*
			P_i **psi** *bar*								P_j ***1000 lb*** *kN*			
2730	**2730**	**2730**	**2730**	**2730**			**514**	**596**	**909**	**909**				
188	188	188	188	188			2286	2651	4043	4043				
3090	**3090**	**3090**	**3090**	**3090**			**595**	**690**	**1025**	**1025**				
213	213	213	213	213			2647	3069	4559	4559				
3450	**3450**	**3450**	**3450**	**3450**			**675**	**783**	**1140**	**1140**				
238	238	238	238	238			3003	3483	5071	5071				
3700	**3700**		**3700**	**3700**			**730**		**1218**	**1218**				
255	255		255	255			3247		5418	5418				
2730	**2730**	**2730**	**2730**	**2730**			**547**	**636**	**1038**	**1038**				
188	188	188	188	188			2433	2829	4617	4617				
3090	**3090**	**3090**	**3090**	**3090**			**633**	**736**	**1169**	**1169**				
213	213	213	213	213			2816	3274	5200	5200				
3450	**3450**	**3450**	**3450**	**3450**			**718**	**835**	**1300**	**1300**				
238	238	238	238	238			3194	3714	5783	5783				
3700	**3700**		**3700**	**3700**			**776**		**1389**	**1389**				
255	255		255	255			3452		6179	6179				
4220	**4220**		**4220**	**4220**			**798**		**1345**	**1345**				
291	291		291	291			3550		5983	5983				
4710	**4710**		**4710**	**4710**			**906**		**1496**	**1496**				
325	325		325	325			4030		6655	6655				
5040	**5040**	**5040**	**5040**	**5040**			**978**	**1133**	**1598**	**1598**				
347	347	347	347	347			4350	5040	7108	7108				
4500	**4500**		**4500**	**4500**			**839**		**1389**	**1389**				
310	310		310	310			3732		6179	6179				
5020	**5020**		**5020**	**5020**			**952**		**1545**	**1545**				
346	346		346	346			4235		6873	6873				
5380	**5380**	**5380**	**5380**	**5380**			**1029**	**1191**	**1650**	**1650**				
371	371	371	371	371			4577	5298	7340	7340				

4500	**4500**		**4500**	**4500**	**849**		**1426**	**1426**
310	310		310	310	3777		6343	6343
5020	**5020**		**5020**	**5020**	**963**		**1585**	**1585**
346	346		346	346	4284		7050	7050
5380	**5380**	**5380**	**5380**	**5380**	**1040**	**1205**	**1693**	**1693**
371	371	371	371	371	4626	5360	7531	7531
5060	**5060**		**5060**	**5060**	**931**		**1514**	**1514**
349	349		349	349	4141		6735	6735
5650	**5650**		**5650**	**5650**	**1057**		**1683**	**1683**
390	390		390	390	4702		7486	7486
6050	**6050**		**6050**	**6050**	**1142**		**1798**	**1798**
417	417		417	417	5080		7998	7998
5340	**5340**		**5340**	**5340**	**982**		**1594**	**1594**
368	368		368	368	4368		7090	7090
5970	**5970**		**5970**	**5970**	**1114**		**1772**	**1772**
412	412		412	412	4955		7882	7882
6390	**6390**	**6390**	**6390**	**6390**	**1204**	**1392**	**1893**	**1893**
441	441	441	441	441	5356	6192	8420	8420
5340	**5340**		**5340**	**5340**	**991**		**1630**	**1630**
368	368		368	368	4408		7251	7251
5970	**5970**		**5970**	**5970**	**1125**		**1812**	**1812**
412	412		412	412	5004		8060	8060
6390	**6390**		**6390**	**6390**	**1215**		**1935**	**1935**
441	441		441	441	5405		8607	8607
6190	**6190**		**6190**	**6190**	**1143**		**1870**	**1870**
427	427		427	427	5084		8318	8318
6910	**6910**		**6910**	**6910**	**1297**		**2079**	**2079**
476	476		476	476	5769		9248	9248
7400	**7400**		**7400**	**7400**	**1402**		**2221**	**2221**
510	510		510	510	6236		9880	9880
7850	**7850**		**7850**	**7850**	**1459**		**2306**	**2306**
541	541		541	541	6490		10 258	10 258
8410	**8410**		**8410**	**8410**	**1577**		**2463**	**2463**
580	580		580	580	7015		10 956	10 956
8790	**8790**		**8790**	**8790**	**1632**		**2573**	**2573**
606	606		606	606	7259		11 445	11 445
9420	**9420**		**9420**	**9420**	**1763**		**2749**	**2749**
649	649		649	649	7842		12 228	12 228
9420	**9420**		**9420**	**9420**	**1747**		**2752**	
649	649		649	649	7771		12 242	
10 090	**10 090**		**10 090**	**10 090**	**1887**		**2939**	
696	696		696	696	8394		13 073	
9730	**9730**		**9730**	**9730**	**1804**		**2841**	
671	671		671	671	8025		12 637	
10 420	**10 420**		**10 420**	**10 420**	**1949**		**3034**	
718	718		718	718	8670		13 496	

TABLE 10.6 Properties of $9\frac{5}{8}''$ casings. (Courtesy of Mannesmann Rohrenwerke[3])

1	2	3	4	5	6	7	8	9	10	11	12
Size: outside diameter D **in** mm	Nominal weight, threads and coupling **lb/ft** kg/m	Grade	Pipe		Threaded and coupled			Extreme-line		Collapse resistance P_c **psi** bar	Pipe body yield strength P_y **1000 lb** kN
			Wall thickness t	Inside diameter d	Drift diameter	Regular	Special clearance	Drift diameter	Outside diameter W^{**}		
						Outside diameter W	Outside diameter W_c				
			in mm								
$9\frac{5}{8}$	**36.00**	J-55	**0.352**	**8.921**	**8.765**	**10.625**	**10.125**			**2020**	**564**
244.5	53.6		8.94	226.6	222.63	269.9	257.2			139	2509
	40.00	J-55	**0.395**	**8.835**	**8.679**	**10.625**	**10.125**	**8.599**	**10.100**	**2570**	**630**
	59.6		10.03	224.4	220.45	269.9	257.2	218.41	256.5	177	2802
	36.00	K-55	**0.352**	**8.921**	**8.765**	**10.625**	**10.125**			**2020**	**564**
	53.6		8.94	226.6	222.63	269.9	257.2			139	2509
	40.00	K-55	**0.395**	**8.835**	**8.679**	**10.625**	**10.125**	**8.599**	**10.100**	**2570**	**630**
	59.6		10.03	224.4	220.45	269.9	257.2	218.41	256.5	177	2802
	40.00	C-75	**0.395**	**8.835**	**8.679**	**10.625**	**10.125**	**8.599**	**10.100**	**2980**	**859**
	59.6		10.03	224.4	220.45	269.9	257.2	218.41	256.5	205	3821
	43.50	C-75	**0.435**	**8.755**	**8.599**	**10.625**	**10.125**	**8.599**	**10.100**	**3750**	**942**
	64.8		11.05	222.4	218.41	269.9	257.2	218.41	256.5	259	4190
	47.00	C-75	**0.472**	**8.681**	**8.525**	**10.625**	**10.125**	**8.525**	**10.100**	**4630**	**1018**
	70.0		11.99	220.5	216.54	269.9	257.2	216.54	256.5	319	4528
	53.50	C-75	**0.545**	**8.535**	**8.379**	**10.625**	**10.125**	**8.379**	**10.100**	**6380**	**1166**
	79.7		13.84	216.8	212.83	269.9	257.2	212.83	256.5	440	5187
	58.40*	C-75	**0.595**	**8.435**	**8.279**	**10.625**	**10.125**			**7570**	**1266**
	87.0		15.11	214.2	210.30	269.9	257.2			522	5631
	40.00	L-80	**0.395**	**8.835**	**8.679**	**10.625**	**10.125**	**8.599**	**10.100**	**3090**	**916**
	59.6		10.03	224.4	220.45	269.9	257.2	218.41	256.5	213	4075
	43.50	L-80	**0.435**	**8.755**	**8.599**	**10.625**	**10.125**	**8.599**	**10.100**	**3810**	**1005**
	64.8		11.05	222.4	218.41	269.9	257.2	218.41	256.5	263	4470
	47.00	L-80	**0.472**	**8.681**	**8.525**	**10.625**	**10.125**	**8.525**	**10.100**	**4750**	**1086**
	70.0		11.99	220.5	216.54	269.9	257.2	216.54	256.5	328	4831
	53.50	L-80	**0.545**	**8.535**	**8.379**	**10.625**	**10.125**	**8.379**	**10.100**	**6620**	**1244**
	79.7		13.84	216.8	212.83	269.9	257.2	212.83	256.5	456	5534
	58.40*	L-80	**0.595**	**8.435**	**8.279**	**10.625**	**10.125**			**7890**	**1350**
	87.0		15.11	214.2	210.30	269.9	257.2			544	6005
	40.00	N-80	**0.395**	**8.835**	**8.679**	**10.625**	**10.125**	**8.599**	**10.100**	**3090**	**916**
	59.6		10.03	224.4	220.45	269.9	257.2	218.41	256.5	213	4075

43.50	**N-80**	**0.435**	**8.755**	**8.599**	**10.625**	**10.125**	**8.599**	**10.100**	**3810**	**1005**
64.8		11.05	222.4	218.41	269.9	257.2	218.41	256.5	263	4470
47.00	**N-80**	**0.472**	**8.681**	**8.525**	**10.625**	**10.125**	**8.525**	**10.100**	**4750**	**1086**
70.0		11.99	220.5	216.54	269.9	257.2	216.54	256.5	328	4831
53.50	**N-80**	**0.545**	**8.535**	**8.379**	**10.625**	**10.125**	**8.379**	**10.100**	**6620**	**1244**
79.7		13.84	216.8	212.83	269.9	257.2	212.83	256.5	456	5534
58.40*	**N-80**	**0.595**	**8.435**	**8.279**	**10.625**	**10.125**			**7890**	**1350**
87.0		15.11	214.2	210.30	269.9	257.2			544	6005
40.00*	**MW-C-90**	**0.395**	**8.835**	**8.679**	**10.625**	**10.125**	**8.599**	**10.100**	**3260**	**1031**
59.6		10.03	224.4	220.45	269.9	257.2	218.41	256.5	225	4586
43.50*	**MW-C-90**	**0.435**	**8.755**	**8.599**	**10.625**	**10.125**	**8.599**	**10.100**	**4010**	**1130**
64.8		11.05	222.4	218.41	269.9	257.2	218.41	256.5	276	5026
47.00*	**MW-C-90**	**0.472**	**8.681**	**8.525**	**10.625**	**10.125**	**8.525**	**10.100**	**4990**	**1222**
70.0		11.99	220.5	216.54	269.9	257.2	216.54	256.5	344	5436
53.50*	**MW-C-90**	**0.545**	**8.535**	**8.379**	**10.625**	**10.125**	**8.379**	**10.100**	**7110**	**1399**
79.7		13.84	216.8	212.83	269.9	257.2	212.83	256.5	490	6223
58.40*	**MW-C-90**	**0.595**	**8.435**	**8.279**	**10.625**	**10.125**			**8560**	**1519**
87.0		15.11	214.2	210.30	269.9	257.2			590	6757
40.00*	**C-95**	**0.395**	**8.835**	**8.679**	**10.625**	**10.125**	**8.599**	**10.100**	**3330**	**1088**
59.6		10.03	224.4	220.45	269.9	257.2	218.41	256.5	230	4840
43.50*	**C-95**	**0.435**	**8.755**	**8.599**	**10.625**	**10.125**	**8.599**	**10.100**	**4130**	**1193**
64.8		11.05	222.4	218.41	269.9	257.2	218.41	256.5	285	5307
47.00*	**C-95**	**0.472**	**8.681**	**8.525**	**10.625**	**10.125**	**8.525**	**10.100**	**5080**	**1289**
70.0		11.99	220.5	216.54	269.9	257.2	216.54	256.5	350	5734
53.50*	**C-95**	**0.545**	**8.535**	**8.379**	**10.625**	**10.125**	**8.379**	**10.100**	**7330**	**1477**
79.7		13.84	216.8	212.83	269.9	257.2	212.83	256.5	505	6570
58.40*	**C-95**	**0.595**	**8.435**	**8.279**	**10.625**	**10.125**			**8870**	**1604**
87.0		15.11	214.2	210.30	269.9	257.2			612	7135
40.00*	**MW-C-95**	**0.395**	**8.835**	**8.679**	**10.625**	**10.125**	**8.599**	**10.100**	**4230**	**1088**
59.6		10.03	224.4	220.45	269.9	257.2	218.41	256.5	292	4840
43.50*	**MW-C-95**	**0.435**	**8.755**	**8.599**	**10.625**	**10.125**	**8.599**	**10.100**	**5600**	**1193**
64.8		11.05	222.4	218.41	269.9	257.2	218.41	256.5	386	5307
47.00*	**MW-C-95**	**0.472**	**8.681**	**8.525**	**10.625**	**10.125**	**8.525**	**10.100**	**7100**	**1289**
70.0		11.99	220.5	216.54	269.9	257.2	216.54	256.5	489	5734
53.50*	**MW-C-95**	**0.545**	**8.535**	**8.379**	**10.625**	**10.125**	**8.379**	**10.100**	**8850**	**1477**
79.7		13.84	216.8	212.83	269.9	257.2	212.83	256.5	610	6570
58.40*	**MW-C-95**	**0.595**	**8.435**	**8.279**	**10.625**	**10.125**			**9950**	**1604**
87.0		15.11	214.2	210.30	269.9	257.2			686	7135
40.00	**P-110**	**0.395**	**8.835**	**8.679**	**10.625**	**10.125**	**8.599**	**10.100**	**3480**	**1260**
59.6		10.03	224.4	220.45	269.9	257.2	218.41	256.5	240	5605
43.50	**P-110**	**0.435**	**8.755**	**8.599**	**10.625**	**10.125**	**8.599**	**10.100**	**4430**	**1381**
64.8		11.05	222.4	218.41	269.9	257.2	218.41	256.5	305	6143
47.00	**P-110**	**0.472**	**8.681**	**8.525**	**10.625**	**10.125**	**8.525**	**10.100**	**5310**	**1493**
70.0		11.99	220.5	216.54	269.9	257.2	216.54	256.5	366	6641
53.50	**P-110**	**0.545**	**8.535**	**8.379**	**10.625**	**10.125**	**8.379**	**10.100**	**7930**	**1710**
79.7		13.84	216.8	212.83	269.9	257.2	212.83	256.5	547	7606

TABLE 10.6 *Continued*

1	2	3	4	5	6	7	8	9	10	11	12
Size: outside diameter D **in** *mm*	*Nominal weight, threads and coupling* **lb/ft** *kg/m*	*Grade*	*Pipe*		*Threaded and coupled*			*Extreme-line*		*Collapse resistance* P_c **psi** *bar*	*Pipe body yield strength* P_y **1000 lb** *kN*
			Wall thickness t	*Inside diameter* d	*Drift diameter*	*Regular*	*Special clearance*	*Drift diameter*	*Outside diameter* W^{**}		
						Outside diameter					
						W	W_c				
						in *mm*					
9⅝	**58.40***	**P-110**	**0.595**	**8.435**	**8.279**	**10.625**	**10.125**			**9750**	**1857**
244.5	87.0		15.11	214.2	210.30	269.9	257.2			672	8260
	40.00*	**MW-125**	**0.395**	**8.835**	**8.679**	**10.625**	**10.125**	**8.599**	**10.100**	**3530**	**1432**
	59.6		10.03	224.4	220.45	269.9	257.2	218.41	256.5	243	6370
	43.50*	**MW-125**	**0.435**	**8.755**	**8.599**	**10.625**	**10.125**	**8.599**	**10.100**	**4620**	**1570**
	64.8		11.05	222.4	218.41	269.9	257.2	218.41	256.5	319	6984
	47.00*	**MW-125**	**0.472**	**8.681**	**8.525**	**10.625**	**10.125**	**8.525**	**10.100**	**5630**	**1697**
	70.0		11.99	220.5	216.54	269.9	257.2	216.54	256.5	388	7549
	58.50*	**MW-125**	**0.545**	**8.535**	**8.379**	**10.625**	**10.125**	**8.379**	**10.100**	**8440**	**1943**
	79.7		13.84	216.8	212.83	269.9	257.2	212.83	256.5	582	8643
	58.40*	**MW-125**	**0.595**	**8.435**	**8.279**	**10.625**	**10.125**			**10 550**	**2110**
	87.0		15.11	214.2	210.30	269.9	257.2			727	9386
	40.00*	**MW-140**	**0.395**	**8.835**	**8.679**	**10.625**	**10.125**	**8.599**	**10.100**	**3530**	**1604**
	59.6		10.03	224.4	220.45	269.9	257.2	218.41	256.5	243	7135

43.50*	**MW-140**	**0.435**	**8.755**	**8.599**	**10.625**	**10.125**	**8.599**	**10.100**	**4730**	**1758**
64.8		11.05	222.4	218.41	269.9	257.2	218.41	256.5	326	7820
47.00*	**MW-140**	**0.472**	**8.681**	**8.525**	**10.625**	**10.125**	**8.525**	**10.100**	**5890**	**1900**
70.0		11.99	220.5	216.54	269.9	257.2	216.54	256.5	406	8452
53.50*	**MW-140**	**0.545**	**8.535**	**8.379**	**10.625**	**10.125**	**8.379**	**10.100**	**8790**	**2177**
79.7		13.84	216.8	212.83	269.9	257.2	212.83	256.5	606	9684
58.40*	**MW-140**	**0.595**	**8.435**	**8.279**	**10.625**	**10.125**			**11 190**	**2363**
87.0		15.11	214.2	210.30	269.9	257.2			772	10 511
40.00*	**V-150**	**0.395**	**8.835**	**8.679**	**10.625**	**10.125**	**8.599**	**10.100**	**3530**	**1718**
59.6		10.03	224.4	220.45	269.9	257.2	218.41	256.5	243	7642
43.50*	**V-150**	**0.435**	**8.755**	**8.599**	**10.625**	**10.125**	**8.599**	**10.100**	**4750**	**1884**
64.8		11.05	222.4	218.41	269.9	257.2	218.41	256.5	328	8380
47.00*	**V-150**	**0.472**	**8.681**	**8.525**	**10.625**	**10.125**	**8.525**	**10.100**	**6020**	**2036**
70.0		11.99	220.5	216.54	269.9	257.2	216.54	256.5	415	9057
53.50*	**V-150**	**0.545**	**8.535**	**8.379**	**10.625**	**10.125**	**8.379**	**10.100**	**8970**	**2332**
79.7		13.84	216.8	212.83	269.9	257.2	212.83	256.5	618	10 373
58.40*	**V-150**	**0.595**	**8.435**	**8.279**	**10.625**	**10.125**			**11 570**	**2532**
87.0		15.11	214.2	210.30	269.9	257.2			798	11 263
40.00*	**MW-155**	**0.395**	**8.835**	**8.679**	**10.625**	**10.125**	**8.599**	**10.100**	**3530**	**1775**
59.6		10.03	224.4	220.45	269.9	257.2	218.41	256.5	243	7895
43.50*	**MW-155**	**0.435**	**8.755**	**8.599**	**10.625**	**10.125**	**8.599**	**10.100**	**4750**	**1947**
64.8		11.05	222.4	218.41	269.9	257.2	218.41	256.5	328	8661
47.00*	**MW-155**	**0.472**	**8.681**	**8.525**	**10.625**	**10.125**	**8.525**	**10.100**	**6050**	**2104**
70.0		11.99	220.5	216.54	269.9	257.2	216.54	256.5	417	9359
53.50*	**MW-155**	**0.545**	**8.535**	**8.379**	**10.625**	**10.125**	**8.379**	**10.100**	**9020**	**2410**
79.7		13.84	216.8	212.83	269.9	257.2	212.83	256.5	622	10 720
58.40*	**MW-155**	**0.595**	**8.435**	**8.279**	**10.625**	**10.125**			**11 720**	**2616**
87.0		15.11	214.2	210.30	269.9	257.2			808	11 637

* Non-API.

TABLE 10.6 *Continued*

13	14	15	16	17	18	19	20	21	22	23	24	25	26	27
Internal yield pressure at minimum yield							Joint strength							
Plain-end or extreme-line	Round thread		Buttress/BSD/Omega				Threaded and coupled						Extreme-line	
	Short	Long	Regular coupling		Special clearance coupling		Round thread		Buttress/BSD/Omega					
							Short	Long	Regular coupling		Special clearance coupling			
			Same grade	Higher grade	Same grade	Higher grade			Same grade	Higher grade	Same grade	Higher grade	Standard joint	Optional joint
			P_i **psi** bar								P_j **1000 lb** kN			
3520	**3520**	**3520**	**3520**	**3520**	**3520**	**3520**	**394**	**453**	**639**	**639**	**639**	**639**		
243	243	243	243	243	243	243	1753	2015	2842	2842	2842	2842		
3950	**3950**	**3950**	**3950**	**3950**	**3660**	**3950**	**452**	**520**	**714**	**714**	**714**	**714**	**770**	**770**
272	272	272	272	272	252	272	2011	2313	3176	3176	3176	3176	3425	3425
3520	**3520**	**3520**	**3520**	**3520**	**3520**	**3520**	**423**	**489**	**755**	**755**	**755**	**755**		
243	243	243	243	243	243	243	1822	2175	3358	3358	3358	3358		
3950	**3950**	**3950**	**3950**	**3950**	**3660**	**3950**	**486**	**561**	**843**	**843**	**843**	**843**	**975**	**975**
272	272	272	272	272	252	272	2162	2495	3750	3750	3750	3750	4337	4337
5390		**5390**	**5390**	**5390**	**4990**	**5390**		**694**	**926**	**926**	**926**	**926**	**975**	**975**
372		372	372	372	344	372		3087	4119	4119	4119	4119	4337	4337
5930		**5930**	**5930**	**5930**	**4990**	**5930**		**776**	**1016**	**1016**	**934**	**1016**	**975**	**975**
409		409	409	409	344	409		3452	4519	4519	4155	4519	4337	4337
6440		**6440**	**6440**	**6440**	**4990**	**6440**		**852**	**1098**	**1098**	**934**	**1098**	**1032**	**1032**
444		444	444	444	344	444		3790	4884	4884	4155	4884	4591	4591
7430		**7430**	**7430**	**7430**	**4990**	**7310**		**999**	**1257**	**1257**	**934**	**1229**	**1173**	**1053**
512		512	512	512	344	504		4444	5591	5591	4155	5467	5218	4684
8110		**8110**	**8110**	**8110**	**4990**	**7310**		**1098**	**1365**	**1365**	**934**	**1229**		
559		559	559	559	344	504		4884	6072	6072	4155	5467		
5750		**5750**	**5750**	**5750**	**5320**	**5750**		**727**	**947**	**947**	**934**	**947**	**975**	**975**
396		396	396	396	367	396		3234	4212	4212	4155	4212	4337	4337
6330		**6330**	**6330**	**6330**	**5320**	**6330**		**813**	**1038**	**1038**	**934**	**1038**	**975**	**975**
436		436	436	436	367	436		3616	4617	4617	4155	4617	4337	4337
6870		**6870**	**6870**	**6870**	**5320**	**6870**		**893**	**1122**	**1122**	**934**	**1122**	**1032**	**1032**
474		474	474	474	367	474		3972	4991	4991	4155	4991	4591	4591
7930		**7930**	**7930**	**7930**	**5320**	**7310**		**1047**	**1286**	**1286**	**934**	**1229**	**1173**	**1053**
547		547	547	547	367	504		4657	5720	5720	4155	5467	5218	4684
8650		**8650**	**8650**	**8650**	**5320**	**7310**		**1151**	**1396**	**1396**	**934**	**1229**		
596		596	596	596	367	504		5120	6210	6210	4155	5467		

5750	**5750**	**5750**	**5750**	**5320**	**5750**	**737**	**979**	**979**	**979**	**979**	**1027**	**1027**
396	396	396	396	367	396	3278	4355	4355	4355	4355	4568	4568
6330	**6330**	**6330**	**6330**	**5320**	**6330**	**825**	**1074**	**1074**	**983**	**1074**	**1027**	**1027**
436	436	436	436	367	436	3670	4777	4777	4373	4777	4568	4568
6870	**6870**	**6870**	**6870**	**5320**	**6870**	**905**	**1161**	**1161**	**983**	**1161**	**1086**	**1086**
474	474	474	474	367	474	4026	5164	5164	4373	5164	4831	4831
7930	**7930**	**7930**	**7930**	**5320**	**7310**	**1062**	**1329**	**1329**	**983**	**1229**	**1235**	**1109**
547	547	547	547	367	504	4724	5912	5912	4373	5467	5494	4933
8650	**8650**	**8650**	**8650**	**5320**	**7310**	**1167**	**1443**	**1443**	**983**	**1229**		
596	596	596	596	367	504	5191	6419	6419	4373	5467		
6460	**6460**	**6460**	**6460**	**5980**	**6460**	**804**	**1021**	**1021**	**983**	**1021**	**1027**	**1027**
445	445	445	445	412	445	3576	4542	4542	4373	4542	4568	4568
7120	**7120**	**7120**	**7120**	**5980**	**7120**	**899**	**1119**	**1119**	**983**	**1119**	**1027**	**1027**
491	491	491	491	412	491	3999	4978	4978	4373	4978	4568	4568
7720	**7720**	**7720**	**7720**	**5980**	**7310**	**987**	**1210**	**1210**	**983**	**1210**	**1087**	**1087**
532	532	532	532	412	504	4390	5382	5382	4373	5382	4835	4835
8920	**8920**	**8920**	**8920**	**5980**	**7310**	**1157**	**1386**	**1386**	**983**	**1229**	**1235**	**1109**
615	615	615	615	412	504	5147	6165	6165	4373	5467	5494	4933
9740	**9740**	**9740**	**9740**	**5980**	**7310**	**1272**	**1504**	**1504**	**983**	**1229**		
672	672	672	672	412	504	5658	6690	6690	4373	5467		
6820	**6820**	**6820**	**6820**	**6310**	**6820**	**847**	**1074**	**1074**	**1032**	**1074**	**1078**	**1078**
470	470	470	470	435	470	3768	4777	4777	4591	4777	4795	4795
7510	**7510**	**7510**	**7510**	**6310**	**7310**	**948**	**1178**	**1178**	**1032**	**1178**	**1078**	**1078**
518	518	518	518	435	504	4217	5240	5240	4591	5240	4795	4795
8150	**8150**	**8150**	**8150**	**6310**	**7310**	**1040**	**1273**	**1273**	**1032**	**1229**	**1141**	**1141**
562	562	562	562	435	504	4626	5663	5663	4951	5467	5075	5075
9410	**9410**	**9410**	**9410**	**6310**	**7310**	**1220**	**1458**	**1458**	**1032**	**1229**	**1297**	**1164**
649	649	649	649	435	504	5427	6486	6486	4591	5467	5769	5178
10 280	**10 280**	**10 280**	**10 280**	**6310**	**7310**	**1341**	**1583**	**1583**	**1032**	**1229**		
709	709	709	709	435	504	5965	7042	7042	4591	5467		
6820	**6820**	**6820**	**6820**	**6310**	**6820**	**858**	**1106**	**1106**	**1081**	**1106**	**1131**	**1131**
470	470	470	470	435	470	3817	4920	4920	4809	4920	5031	5031
7510	**7510**	**7510**	**7510**	**6310**	**7310**	**959**	**1213**	**1213**	**1081**	**1213**	**1131**	**1131**
518	518	518	518	435	504	4266	5396	5396	4809	5396	5031	5031
8150	**8150**	**8150**	**8150**	**6310**	**7310**	**1053**	**1311**	**1311**	**1081**	**1229**	**1197**	**1197**
562	562	562	562	435	504	4684	5832	5832	4809	5467	5325	5325
9410	**9410**	**9410**	**9410**	**9310**	**7310**	**1235**	**1502**	**1502**	**1081**	**1229**	**1359**	**1219**
649	649	649	649	435	504	5494	6681	6681	4809	5467	6045	5422
10 280	**10 280**	**10 280**	**10 280**	**6310**	**7310**	**1357**	**1630**	**1630**	**1081**	**1229**		
709	709	709	709	435	504	6036	7251	7251	4809	5467		
7900	**7900**	**7900**	**7900**	**7310**	**7900**	**988**	**1266**	**1266**	**1229**	**1266**	**1283**	**1283**
545	545	545	545	504	545	4395	5631	5631	5467	5631	5707	5707
8700	**8700**	**8700**	**8700**	**7310**	**8700**	**1106**	**1388**	**1388**	**1229**	**1388**	**1283**	**1283**
600	600	600	600	504	600	4920	6174	6174	5467	6174	5707	5707
9440	**9440**	**9440**	**9440**	**7310**	**9440**	**1213**	**1500**	**1500**	**1229**	**1500**	**1358**	**1358**
651	651	651	651	504	651	5396	6672	6672	5467	6672	6041	6041
10 900	**10 900**	**10 900**	**10 900**	**7310**	**9970**	**1422**	**1718**	**1718**	**1229**	**1573**	**1544**	**1386**
752	752	752	752	504	687	6325	7642	7642	5467	6997	6868	6165

TABLE 10.6 *Continued*

13	14	15	16	17	18	19	20	21	22	23	24	25	26	27
Internal yield pressure at minimum yield							*Joint strength*							
Plain-end or ex-treme-line	*Round thread*		*Buttress/BSD/Omega*				*Threaded and coupled*						*Extreme-line*	
	Short	*Long*	*Regular coupling*		*Special clearance coupling*		*Round thread*		*Buttress/BSD/Omega*					
							Short	*Long*	*Regular coupling*		*Special clearance coupling*			
			Same grade	*Higher grade*	*Same grade*	*Higher grade*			*Same grade*	*Higher grade*	*Same grade*	*Higher grade*	*Standard joint*	*Optional joint*
			P_i **psi** *bar*								P_j ***1000 lb*** *kN*			
11 900		**11 900**	**11 900**	**11 900**	**7310**	**9970**		**1564**	**1865**	**1865**	**1229**	**1573**		
820		820	820	820	504	687		6957	8296	8296	5467	6997		
8980		**8980**	**8980**	**8980**	**8310**	**8980**		**1108**	**1393**	**1393**	**1327**	**1393**	**1388**	**1388**
619		619	619	619	573	619		4929	6196	6196	5903	6196	6174	6174
9890		**9890**	**9890**	**9890**	**8310**	**9890**		**1240**	**1527**	**1527**	**1327**	**1527**	**1388**	**1388**
682		682	682	682	573	682		5516	6792	6792	5903	6792	6174	6174
10 730		**10 730**	**10 730**	**10 730**	**8310**	**9970**		**1361**	**1650**	**1650**	**1327**	**1573**	**1469**	**1469**
740		740	740	740	573	687		6054	7340	7340	5903	6997	6534	6534
12 390		**12 390**	**12 390**	**12 390**	**8310**	**9970**		**1595**	**1890**	**1890**	**1327**	**1573**	**1667**	**1497**
854		854	854	854	573	687		7095	8407	8407	5903	6997	7415	6659
13 520		**13 520**	**13 520**	**13 520**	**8310**	**9970**		**1754**	**2052**	**2052**	**1327**	**1573**		
932		932	932	932	573	687		7802	9128	9128	5903	6997		

10 050	**10 050**	**10 050**	**10 050**	**9310**	**9970**	**1239**	**1552**	**1552**	**1475**	**1552**	**1542**	**1542**
693	693	693	693	642	687	5511	6904	6904	6561	6904	6859	6859
11 070	**11 070**	**11 070**	**11 070**	**9310**	**9970**	**1386**	**1702**	**1702**	**1475**	**1573**	**1542**	**1542**
763	763	763	763	642	687	6165	7571	7571	6561	6997	6859	6859
12 010	**12 010**	**12 010**	**12 010**	**9310**	**9970**	**1521**	**1839**	**1839**	**1475**	**1573**	**1632**	**1632**
828	828	828	828	642	687	6766	8180	8180	6561	6997	7259	7259
13 870	**13 870**	**13 870**	**13 870**	**9310**	**9970**	**1783**	**2107**	**2107**	**1475**	**1573**	**1853**	**1663**
956	956	956	956	642	687	7931	9372	9372	6561	6997	8243	7397
15 150	**15 150**	**15 150**	**15 150**	**9310**	**9970**	**1961**	**2287**	**2287**	**1475**	**1573**		
1045	1045	1045	1045	642	687	8723	10 173	10 173	6561	6997		
10 770	**10 770**	**10 770**		**9970**		**1326**	**1658**		**1573**		**1645**	**1645**
743	743	743		687		5898	7375		6997		7317	7317
11 860	**11 860**	**11 860**		**9970**		**1483**	**1818**		**1573**		**1645**	**1645**
818	818	818		687		6597	8087		6997		7317	7317
12 870	**12 870**	**12 870**		**9970**		**1628**	**1965**		**1573**		**1740**	**1740**
887	887	887		687		7242	8741		6997		7740	7740
14 860	**14 860**	**14 860**		**9970**		**1909**	**2238**		**1573**		**1976**	**1774**
1025	1025	1025		687		8492	9955		6997		8790	7891
16 230	**16 230**	**16 230**		**9970**		**2098**	**2444**		**1573**			
1119	1119	1119		687		9332	10 871		6997			
11 130	**11 130**	**11 130**		**10 300**		**1369**	**1712**		**1622**		**1696**	**1696**
767	767	767		710		6090	7615		7215		7544	7544
12 260	**12 260**	**12 260**		**10 300**		**1532**	**1877**		**1622**		**1696**	**1696**
845	845	845		710		6815	8349		7215		7544	7544
13 300	**13 300**	**13 300**		**10 300**		**1681**	**2028**		**1622**		**1795**	**1795**
917	917	917		710		7477	9021		7215		7985	7985
15 360	**15 360**	**15 360**		**10 300**		**1971**	**2323**		**1622**		**2038**	**1829**
1059	1059	1059		710		8767	10 333		7215		9065	8136
16 770	**16 770**	**16 770**		**10 300**		**2167**	**2522**		**1622**			
1156	1156	1156		710		9639	11 218		7215			

TABLE 10.7 Properties of 7″ casing. (Courtesy of Mannesmann Rohrenwerke[3])

1	2	3	4	5	6	7	8	9	10	11	12
Size: outside diameter D **in** mm	Nominal weight, threads and coupling **lb/ft** kg/m	Grade	Pipe		Threaded and coupled			Extreme-line		Collapse resistance P_c **psi** bars	Pipe body yield strength P_y **1000 lb** kN
			Wall thickness t	Inside diameter d	Drift diameter	Regular	Special clearance	Drift diameter	Outside diameter W		
						Outside diameter W	Outside diameter W_c				
			in mm								
7	**20.00**	**J-55**	**0.272**	**6.456**	**6.331**	**7.656**				**2270**	**316**
177.8	29.8		6.91	164.0	160.81	194.5				157	1406
	23.00	**J-55**	**0.317**	**6.366**	**6.241**	**7.656**	**7.375**	**6.151**	**7.390**	**3270**	**366**
	34.3		8.05	161.7	158.52	194.5	187.3	156.24	187.7	225	1628
	26.00	**J-55**	**0.362**	**6.276**	**6.151**	**7.656**	**7.375**	**6.151**	**7.390**	**4320**	**415**
	38.7		9.19	159.4	156.24	194.5	187.3	156.24	187.7	298	1846
	20.00	**K-55**	**0.272**	**6.456**	**6.331**	**7.656**				**2270**	**316**
	29.8		6.91	164.0	160.81	194.5				157	1406
	23.00	**K-55**	**0.317**	**6.366**	**6.241**	**7.656**	**7.375**	**6.151**	**7.390**	**3270**	**366**
	34.3		8.05	161.7	158.52	194.5	187.3	156.24	187.7	225	1628
	26.00	**K-55**	**0.362**	**6.276**	**6.151**	**7.656**	**7.375**	**6.151**	**7.390**	**4320**	**415**
	38.7		9.19	159.4	156.24	194.5	187.3	156.24	187.7	298	1846
	23.00	**C-75**	**0.317**	**6.366**	**6.241**	**7.656**	**7.375**	**6.151**	**7.390**	**3770**	**499**
	34.3		8.05	161.7	158.52	194.5	187.3	156.24	187.7	260	2220
	26.00	**C-75**	**0.362**	**6.276**	**6.151**	**7.656**	**7.375**	**6.151**	**7.390**	**5250**	**566**
	38.7		9.19	159.4	156.24	194.5	187.3	156.24	187.7	362	2518
	29.00	**C-75**	**0.408**	**6.184**	**6.059**	**7.656**	**7.375**	**6.059**	**7.390**	**6760**	**634**
	43.2		10.36	157.1	153.90	194.5	187.3	153.90	187.7	466	2820
	32.00	**C-75**	**0.453**	**6.094**	**5.969**	**7.656**	**7.375**	**5.969**	**7.390**	**8230**	**699**
	47.7		11.51	154.8	151.61	194.5	187.3	151.61	187.7	567	3109
	35.00	**C-75**	**0.498**	**6.004**	**5.879**	**7.656**	**7.375**	**5.879**	**7.530**	**9710**	**763**
	52.1		12.65	152.5	149.33	194.5	187.3	149.33	191.3	669	3394
	38.00	**C-75**	**0.540**	**5.920**	**5.795**	**7.656**	**7.375**	**5.795**	**7.530**	**10 680**	**822**
	56.6		13.72	150.4	147.19	194.5	187.3	147.19	191.3	736	3656
	23.00	**L-80**	**0.317**	**6.366**	**6.241**	**7.656**	**7.375**	**6.151**	**7.390**	**3830**	**532**
	34.3		8.05	161.7	158.52	194.5	187.3	156.24	187.7	264	2366
	26.00	**L-80**	**0.362**	**6.276**	**6.151**	**7.656**	**7.375**	**6.151**	**7.390**	**5410**	**604**
	38.7		9.19	159.4	156.24	194.5	187.3	156.24	187.7	373	2687
	29.00	**L-80**	**0.408**	**6.184**	**6.059**	**7.656**	**7.375**	**6.059**	**7.390**	**7020**	**676**
	43.2		10.36	157.1	153.90	194.5	187.3	153.90	187.7	484	3007

	32.00 47.7	**L-80**	**0.453** 11.51	**6.094** 154.8	**5.969** 151.61	**7.656** 194.5	**7.375** 187.3	**5.969** 151.61	**7.390** 187.7	**8600** 593	**745** 3314
	35.00 52.1	**L-80**	**0.498** 12.65	**6.004** 152.5	**5.879** 149.33	**7.656** 194.5	**7.375** 187.3	**5.879** 149.33	**7.530** 191.3	**10 180** 702	**814** 3621
	38.00 56.6	**L-80**	**0.540** 13.72	**5.920** 150.4	**5.795** 147.19	**7.656** 194.5	**7.375** 187.3	**5.795** 147.19	**7.530** 191.3	**11 390** 785	**877** 3901
	23.00 34.3	**N-80**	**0.317** 8.05	**6.366** 161.7	**6.241** 158.52	**7.656** 194.5	**7.375** 187.3	**6.151** 156.24	**7.390** 187.7	**3830** 264	**532** 2366
	26.00 38.7	**N-80**	**0.362** 9.19	**6.276** 159.4	**6.151** 156.24	**7.656** 194.5	**7.375** 187.3	**6.151** 156.24	**7.390** 187.7	**5410** 373	**604** 2687
	29.00 43.2	**N-80**	**0.408** 10.36	**6.184** 157.1	**6.059** 153.90	**7.656** 194.5	**7.375** 187.3	**6.059** 153.90	**7.390** 187.7	**7020** 484	**676** 3007
	32.00 47.7	**N-80**	**0.453** 11.51	**6.094** 154.8	**5.969** 151.61	**7.656** 194.5	**7.375** 187.3	**5.969** 151.61	**7.390** 187.7	**8600** 593	**745** 3314
	35.00 52.1	**N-80**	**0.498** 12.65	**6.004** 152.5	**5.879** 149.33	**7.656** 194.5	**7.375** 187.3	**5.879** 149.33	**7.530** 191.3	**10 180** 702	**814** 3621
	38.00 56.6	**N-80**	**0.540** 13.72	**5.920** 150.4	**5.795** 147.19	**7.656** 194.5	**7.375** 187.3	**5.795** 147.19	**7.530** 191.3	**11 390** 785	**877** 3901
7 177.8	**23.00*** 34.3	**MW-C-90**	**0.317** 8.05	**6.366** 161.7	**6.241** 158.52	**7.656** 194.5	**7.375** 187.3	**6.151** 156.24	**7.390** 187.7	**4030** 278	**599** 2665
	26.00* 38.7	**MW-C-90**	**0.362** 9.19	**6.276** 159.4	**6.151** 156.24	**7.656** 194.5	**7.375** 187.3	**6.151** 156.24	**7.390** 187.7	**5740** 396	**679** 3020
	29.00* 43.2	**MW-C-90**	**0.408** 10.36	**6.184** 157.1	**6.059** 153.90	**7.656** 194.5	**7.375** 187.3	**6.059** 153.90	**7.390** 187.7	**7580** 523	**760** 3381
	32.00* 47.7	**MW-C-90**	**0.453** 11.51	**6.094** 154.8	**5.969** 151.61	**7.656** 194.5	**7.375** 187.3	**5.969** 151.61	**7.390** 187.7	**9370** 646	**839** 3732
	35.00* 52.1	**MW-C-90**	**0.498** 12.65	**6.004** 152.5	**5.879** 149.33	**7.656** 194.5	**7.375** 187.3	**5.879** 149.33	**7.530** 191.3	**11 170** 770	**916** 4075
	38.00* 56.6	**MW-C-90**	**0.540** 13.72	**5.920** 150.4	**5.795** 147.19	**7.656** 194.5	**7.375** 187.3	**5.795** 147.19	**7.530** 191.3	**12 810** 883	**986** 4386
	23.00 34.3	**C-95**	**0.317** 8.05	**6.366** 161.7	**6.241** 158.52	**7.656** 194.5	**7.375** 187.3	**6.151** 156.24	**7.390** 187.7	**4150** 286	**632** 2811
	26.00 38.7	**C-95**	**0.362** 9.19	**6.276** 159.4	**6.151** 156.24	**7.656** 194.5	**7.375** 187.3	**6.151** 156.24	**7.390** 187.7	**5870** 405	**717** 3189
	29.00 43.2	**C-95**	**0.408** 10.36	**6.184** 157.1	**6.059** 153.90	**7.656** 194.5	**7.375** 187.3	**6.059** 153.90	**7.390** 187.7	**7820** 539	**803** 3572
	32.00 47.7	**C-95**	**0.453** 11.51	**6.094** 154.8	**5.969** 151.61	**7.656** 194.5	**7.375** 187.3	**5.969** 151.61	**7.390** 187.7	**9730** 671	**885** 3937
	35.00 52.1	**C-95**	**0.498** 12.65	**6.004** 152.5	**5.879** 149.33	**7.656** 194.5	**7.375** 187.3	**5.879** 149.33	**7.530** 191.3	**11 640** 803	**966** 4297
	38.00 56.6	**C-95**	**0.540** 13.72	**5.920** 150.4	**5.795** 147.19	**7.656** 194.5	**7.375** 187.3	**5.795** 147.19	**7.530** 191.3	**13 420** 925	**1041** 4631
	23.00* 34.3	**MW-C-95**	**0.317** 8.05	**6.366** 161.7	**6.241** 158.52	**7.656** 194.5	**7.375** 187.3	**6.151** 156.24	**7.390** 187.7	**5650** 390	**632** 2811
	26.00* 38.7	**MW-C-95**	**0.362** 9.19	**6.276** 159.4	**6.151** 156.24	**7.656** 194.5	**7.375** 187.3	**6.151** 156.24	**7.390** 187.7	**7800** 538	**717** 3189

TABLE 10.7 *Continued*

1	2	3	4	5	6	7	8	9	10	11	12
Size: outside diameter D **in** mm	Nominal weight, threads and coupling **lb/ft** kg/m	Grade	Pipe		Threaded and coupled			Extreme-line		Collapse resistance P_c **psi** bars	Pipe body yield strength P_y **1000 lb** kN
			Wall thickness t	Inside diameter d	Drift diameter	Regular	Special clearance	Drift diameter	Outside diameter W		
						Outside diameter					
						W	W_c				
			in mm								
7 177.8	**29.00***	**MW-C-95**	**0.408**	**6.184**	**6.059**	**7.656**	**7.375**	**6.059**	**7.390**	**9200**	**803**
	43.2		10.36	157.1	153.90	194.5	187.3	153.90	187.7	634	3572
	32.00*	**MW-C-95**	**0.453**	**6.094**	**5.969**	**7.656**	**7.375**	**5.969**	**7.390**	**10 400**	**885**
	47.7		11.51	154.8	151.61	194.5	187.3	151.61	187.7	719	3937
	35.00*	**MW-C-95**	**0.498**	**6.004**	**5.879**	**7.656**	**7.375**	**5.879**	**7.530**	**11 640**	**966**
	52.1		12.65	152.5	149.33	194.5	187.3	149.33	191.3	803	4297
	38.00*	**MW-C-95**	**0.540**	**5.920**	**5.795**	**7.656**	**7.375**	**5.795**	**7.530**	**13 420**	**1041**
	56.6		13.72	150.4	147.19	194.5	187.3	147.19	191.3	925	4631
	23.00*	**P-110**	**0.317**	**6.366**	**6.241**	**7.656**	**7.375**	**6.151**	**7.390**	**4450**	**732**
	34.3		8.05	161.7	158.52	194.5	187.3	156.24	187.7	307	3256
	26.00	**P-110**	**0.362**	**6.276**	**6.151**	**7.656**	**7.375**	**6.151**	**7.390**	**6210**	**830**
	38.7		9.19	159.4	156.24	194.5	187.3	156.24	187.7	428	3692
	29.00	**P-110**	**0.408**	**6.184**	**6.059**	**7.656**	**7.375**	**6.059**	**7.390**	**8510**	**929**
	43.2		10.36	157.1	153.90	194.5	187.3	153.90	187.7	587	4132
	32.00	**P-110**	**0.453**	**6.094**	**5.969**	**7.656**	**7.375**	**5.969**	**7.390**	**10 760**	**1025**
	47.7		11.51	154.8	151.61	194.5	187.3	151.61	187.7	742	4559
	35.00	**P-110**	**0.498**	**6.004**	**5.879**	**7.656**	**7.375**	**5.879**	**7.530**	**13 010**	**1119**
	52.1		12.65	152.5	149.33	194.5	187.3	149.33	191.3	897	4978
	38.00	**P-110**	**0.540**	**5.920**	**5.795**	**7.656**	**7.375**	**5.795**	**7.530**	**15 110**	**1205**
	56.6		13.72	150.4	147.19	194.5	187.3	147.19	191.3	1042	5360
	23.00*	**MW-125**	**0.317**	**6.366**	**6.241**	**7.656**	**7.375**	**6.151**	**7.390**	**4650**	**832**
	34.3		8.05	161.7	158.52	194.5	187.3	156.24	187.7	321	3701
	26.00*	**MW-125**	**0.362**	**6.276**	**6.151**	**7.656**	**7.375**	**6.151**	**7.390**	**6460**	**944**
	38.7		9.19	159.4	156.24	194.5	187.3	156.24	187.7	445	4199
	29.00*	**MW-125**	**0.408**	**6.184**	**6.059**	**7.656**	**7.375**	**6.059**	**7.390**	**9120**	**1056**
	43.2		10.36	157.1	153.90	194.5	187.3	153.90	187.7	629	4697

32.00*	MW-125	0.453	6.094	5.969	7.656	7.375	5.969	7.390	11 720	1165
47.7		11.51	154.8	151.61	194.5	187.3	151.61	187.7	808	5182
35.00*	MW-125	0.498	6.004	5.879	7.656	7.375	5.879	7.530	14 330	1272
52.1		12.65	152.5	149.33	194.5	187.3	149.33	191.3	988	5658
38.00*	MW-125	0.540	5.920	5.795	7.656	7.375	5.795	7.530	16 760	1370
56.6		13.72	150.4	147.19	194.5	187.3	147.19	191.3	1156	6094
23.00*	MW-140	0.317	6.366	6.241	7.656	7.375	6.151	7.390	4760	932
34.3		8.05	161.7	158.52	194.5	187.3	156.24	187.7	328	4146
26.00*	MW-140	0.362	6.276	6.151	7.656	7.375	6.151	7.390	6690	1057
38.7		9.19	159.4	156.24	194.5	187.3	156.24	187.7	461	4702
29.00*	MW-140	0.408	6.184	6.059	7.656	7.375	6.059	7.390	9560	1183
43.2		10.36	157.1	153.90	194.5	187.3	153.90	187.7	659	5262
32.00*	MW-140	0.453	6.094	5.969	7.656	7.375	5.969	7.390	12 520	1304
47.7		11.51	154.8	151.61	194.5	187.3	151.61	187.7	863	5800
35.00*	MW-140	0.498	6.004	5.879	7.656	7.375	5.879	7.530	15 490	1424
52.1		12.65	152.5	149.33	194.5	187.3	149.33	191.3	1068	6334
38.00*	MW-140	0.540	5.920	5.795	7.656	7.375	5.795	7.530	18 260	1534
56.6		13.72	150.4	147.19	194.5	187.3	147.19	191.3	1259	6824
23.00*	V-150	0.317	6.366	6.241	7.656	7.375	6.151	7.390	4800	998
34.3		8.05	161.7	158.52	194.5	187.3	156.24	187.7	331	4439
26.00*	V-150	0.362	6.276	6.151	7.656	7.375	6.151	7.390	6890	1132
38.7		9.19	159.4	156.24	194.5	187.3	156.24	187.7	475	5035
29.00*	V-150	0.408	6.184	6.059	7.656	7.375	6.059	7.390	9800	1267
43.2		10.36	157.1	153.90	194.5	187.3	153.90	187.7	674	5636
32.00*	V-150	0.453	6.094	5.969	7.656	7.375	5.969	7.390	13 020	1398
47.7		11.51	154.8	151.61	194.5	187.3	151.61	187.7	898	6219
35.00*	V-150	0.498	6.004	5.879	7.656	7.375	5.879	7.530	16 230	1526
52.1		12.65	152.5	149.33	194.5	187.3	149.33	191.3	1119	6788
38.00*	V-150	0.540	5.920	5.795	7.656	7.375	5.795	7.530	19 240	1644
56.6		13.72	150.4	147.19	194.5	187.3	147.19	191.3	1327	7313
23.00*	MW-155	0.317	6.366	6.241	7.656	7.375	6.151	7.390	4780	1032
34.3		8.05	161.7	158.52	194.5	187.3	156.24	187.7	330	4591
26.00*	MW-155	0.362	6.276	6.151	7.656	7.375	6.151	7.390	6960	1170
38.7		9.19	159.4	156.24	194.5	187.3	156.24	187.7	480	5204
29.00*	MW-155	0.408	6.184	6.059	7.656	7.375	6.059	7.390	9890	1310
43.2		10.36	157.1	153.90	194.5	187.3	153.90	187.7	682	5827
32.00*	MW-155	0.453	6.094	5.969	7.656	7.375	5.969	7.390	13 230	1444
47.7		11.51	154.8	151.61	194.5	187.3	151.61	187.7	912	6423
35.00*	MW-155	0.498	6.004	5.879	7.656	7.375	5.879	7.530	16 570	1577
52.1		12.65	152.5	149.33	194.5	187.3	149.33	191.3	1142	7015
38.00*	MW-155	0.540	5.920	5.795	7.656	7.375	5.795	7.530	19 700	1697
56.6		13.72	150.4	147.19	194.5	187.3	147.19	191.3	1358	7549

* Non-API.

TABLE 10.7 *Continued*

13	14	15	16	17	18	19	20	21	22	23	24	25	26	27
	Internal yield pressure at minimum yield						*Joint strength*							
Plain-end or ex-treme-line	*Round thread*		*Buttress/BSD/Omega*				*Threaded and coupled*						*Extreme-line*	
			Regular coupling		*Special clearance coupling*		*Round thread*		*Buttress/BSD/Omega*					
	Short	*Long*					*Short*	*Long*	*Regular coupling*		*Special clearance coupling*			
			Same grade	*Higher grade*	*Same grade*	*Higher grade*			*Same grade*	*Higher grade*	*Same grade*	*Higher grade*	*Standard joint*	*Optional joint*
			P_i **psi** *bar*								P_j **1000 lb** *kN*			
3740	**3740**						**234**							
258	258						1041							
4360	**4360**	**4360**	**4360**	**4360**	**3950**	**4360**	**284**	**313**	**432**	**432**	**421**	**432**	**499**	**499**
301	301	301	301	301	272	301	1263	1392	1922	1922	1873	1922	2220	2220
4980	**4980**	**4980**	**4980**	**4980**	**3950**	**4980**	**334**	**367**	**490**	**490**	**421**	**490**	**506**	**506**
343	343	343	343	343	272	343	1486	1632	2180	2180	1873	2180	2251	2251
3740	**3740**						**254**							
258	258						1130							
4360	**4360**	**4360**	**4360**	**4360**	**3950**	**4360**	**309**	**341**	**522**	**522**	**522**	**522**	**632**	**632**
301	301	301	301	301	272	301	1375	1517	2322	2322	2322	2322	2811	2811
4980	**4980**	**4980**	**4980**	**4980**	**3950**	**4980**	**364**	**401**	**592**	**592**	**533**	**561**	**641**	**641**
343	343	343	343	343	272	343	1619	1784	2633	2633	2371	2495	2851	2851
5940		**5940**	**5940**		**5380**			**416**	**557**		**533**		**632**	**632**
410		410	410		371			1850	2478		2371		2811	2811
6790		**6790**	**6790**		**5380**			**489**	**631**		**533**		**641**	**641**
468		468	468		371			2175	2807		2371		2851	2851
7650		**7650**	**7650**		**5380**			**562**	**707**		**533**		**685**	**674**
527		527	527		371			2500	3145		2371		3047	2998
8490		**8490**	**7930**		**5380**			**633**	**779**		**533**		**761**	**674**
585		585	547		371			2816	3465		2371		3385	2998
9340		**8660**	**7930**		**5380**			**703**	**833**		**533**		**850**	**761**
644		597	547		371			3127	3705		2371		3781	3385
10 120		**8660**	**7930**		**5380**			**767**	**833**		**533**		**917**	**761**
698		597	547		371			3412	3705		2371		4079	3385
6340		**6340**	**6340**	**6340**	**5740**	**6340**		**435**	**565**		**533**		**632**	**632**
437		437	437	437	396	437		1935	2513		2371		2811	2811
7240		**7240**	**7240**	**7240**	**5740**	**7240**		**511**	**641**		**533**		**641**	**641**
499		499	499	499	396	499		2273	2851		2371		2851	2851

8160	8160	8160	8160	5740	7890	587	718		533		685	674
563	563	563	563	396	544	2611	3194		2371		3047	2998
9060	9060	8460	9060	5740	7890	661	791		533		2371	3385
10 800	9240	8460	10 800	5740	7890	801	833		533		917	761
745	637	583	745	396	544	3563	3705		2371		4079	3385
6340	6340	6340	6340	5740	6340	442	588	588	561	588	666	666
437	437	437	437	396	437	1966	2616	2616	2495	2616	2963	2963
7240	7240	7240	7240	5740	7240	519	667	667	561	667	675	675
499	499	499	499	396	499	2309	2967	2967	2495	2967	3003	3003
8160	8160	8160	8160	5740	7890	597	746	746	561	702	721	709
563	563	563	563	396	544	2656	3318	3318	2495	3123	3207	3154
9060	9060	8460	9060	5740	7890	672	823	823	561	702	801	709
625	625	583	625	396	544	2989	3661	3661	2495	3123	3563	3154
9960	9240	8460	9960	5740	7890	746	876	898	561	702	895	801
687	637	583	687	396	544	3318	3897	3995	2495	3123	3981	3563
10 800	9240	8460	10 800	5740	7890	814	876	968	561	702	965	801
745	637	583	745	396	544	3621	3897	4306	2495	3123	4293	3563
7130	7130	7130	7130	6460	7130	479	605	605	561	589	666	666
492	492	492	492	445	492	2131	2691	2691	2495	2620	2963	2963
8150	8150	8150	8150	6460	7890	563	687	687	561	589	675	675
562	562	562	562	445	544	2504	3056	3056	2495	2620	3003	3003
9180	9180	9180	9180	6460	7890	648	768	768	561	589	721	709
633	633	633	633	445	544	2882	3416	3416	2495	2620	3207	3154
10 190	10 190	9520	10 190	6460	7890	729	847	847	561	589	801	709
703	703	656	703	445	544	3243	3768	3768	2495	2620	3563	3154
11 210	10 390	9520	11 210	6460	7890	809	876	920	561	589	895	801
773	716	656	773	445	544	3599	3897	4092	2495	2620	3981	3563
12 150	10 390	9520	11 640	6460	7890	883	876	920	561	589	965	801
838	716	656	803	445	544	3928	3897	4092	2495	2620	4293	3563
7530	7530	7530		6810		505	636		589		699	699
519	519	519		470		2246	2829		2620		3109	3109
8600	8600	8600		6810		593	722		589		709	709
593	593	593		470		2638	3212		2620		3154	3154
9690	9690	9690		6810		683	808		589		757	744
668	668	668		470		3038	3594		2620		3367	3309
10 760	10 760	10 050		6810		768	891		589		841	744
742	742	693		470		3416	3963		2620		3741	3309
11 830	10 970	10 050		6810		853	920		589		940	841
816	756	693		470		3794	4092		2620		4181	3741
12 820	10 970	10 050		6810		931	920		589		1013	841
884	756	693		470		4141	4092		2620		4506	3741
7530	7530	7530	7530	6810	7530	512	659	659	617	659	732	732
519	519	519	519	470	519	2277	2931	2931	2745	2931	3256	3256
8600	8600	8600	8600	6810	7890	602	747	747	617	702	742	742
593	593	593	593	470	544	2678	3323	3323	2745	3123	3301	3301
9690	9690	9690	9690	6810	7890	692	836	836	617	702	793	779
668	668	668	668	470	544	3078	3719	3719	2745	3123	3527	3465

TABLE 10.7 *Continued*

13	14	15	16	17	18	19	20	21	22	23	24	25	26	27
	Internal yield pressure at minimum yield						*Joint strength*							
Plain-end or extreme-line	*Round thread*		*Buttress/BSD/Omega*				*Threaded and coupled*						*Extreme-line*	
	Short	*Long*	*Regular coupling*		*Special clearance coupling*		*Round thread*		*Buttress/BSD/Omega*					
							Short	*Long*	*Regular coupling*		*Special clearance coupling*			
			Same grade	*Higher grade*	*Same grade*	*Higher grade*			*Same grade*	*Higher grade*	*Same grade*	*Higher grade*	*Standard joint*	*Optional joint*
			P_i **psi** *bar*								P_j ***1000 lb*** *kN*			
10 760		**10 760**	**10 050**	**10 760**	**6810**	**7890**		**779**	**922**	**922**	**617**	**702**	**882**	**780**
742		742	693	742	470	544		3465	4101	4101	2745	3123	3923	3470
11 830		**10 970**	**10 050**	**11 640**	**6810**	**7890**		**865**	**964**	**1007**	**617**	**702**	**984**	**881**
816		756	693	803	470	544		3848	4288	4479	2745	3123	4377	3919
12 820		**10 970**	**10 050**	**11 640**	**6810**	**7890**		**944**	**964**	**1085**	**617**	**702**	**1061**	**881**
884		756	693	803	470	544		4199	4288	4826	2745	3123	4720	3919
8720		**8720**	**8720**	**8720**	**7890**	**8720**		**590**	**752**	**752**	**702**	**752**	**832**	**832**
601		601	601	601	544	601		2624	3345	3345	3123	3345	3701	3701
9960		**9960**	**9960**	**9960**	**7890**	**9960**		**693**	**853**	**853**	**702**	**853**	**844**	**844**
687		687	687	687	544	687		3083	3794	3794	3123	3794	3754	3754
11 220		**11 220**	**11 220**	**11 220**	**7890**	**10 760**		**797**	**955**	**955**	**702**	**898**	**902**	**886**
774		774	774	774	544	742		3545	4248	4248	3123	3995	4012	3941
12 460		**12 460**	**11 640**	**12 460**	**7890**	**10 760**		**897**	**1053**	**1053**	**702**	**898**	**1002**	**886**
859		859	803	859	544	742		3990	4684	4684	3123	3995	4457	3941
13 700		**12 700**	**11 640**	**13 700**	**7890**	**10 760**		**996**	**1096**	**1150**	**702**	**898**	**1118**	**1002**
945		876	803	945	544	742		4430	4875	5115	3123	3995	4973	4457
14 850		**12 700**	**11 640**	**14 850**	**7890**	**10 760**		**1087**	**1096**	**1239**	**702**	**898**	**1207**	**1002**
1024		876	803	1024	544	742		4835	4875	5511	3123	3995	5369	4457
9910		**9910**	**9910**	**9910**	**8970**	**9910**		**655**	**823**	**823**	**758**	**823**	**899**	**899**
683		683	683	683	618	683		2914	3661	3661	3372	3661	3999	3999
11 310		**11 310**	**11 310**	**11 310**	**8970**	**10 760**		**769**	**934**	**934**	**758**	**842**	**911**	**911**
780		780	780	780	618	742		3421	4155	4155	3372	3745	4052	4052
12 750		**12 750**	**12 750**	**12 750**	**8970**	**10 760**		**885**	**1045**	**1045**	**758**	**842**	**973**	**956**
879		879	879	879	618	742		3937	4648	4648	3372	3745	4284	4253

14 160	**14 160**	**13 220**	**14 160**	**8970**	**10 760**	**996**	**1152**	**1152**	**758**	**842**	**1082**	**957**
976	976	911	976	618	742	4430	5124	5124	3372	3745	4813	4257
15 560	**14 430**	**13 220**	**15 560**	**8970**	**10 760**	**1106**	**1183**	**1258**	**758**	**842**	**1207**	**1081**
1073	995	911	1073	618	742	4920	5262	5596	3372	3745	5369	4809
16 880	**14 430**	**13 220**	**15 870**	**8970**	**10 760**	**1207**	**1183**	**1315**	**758**	**842**	**1303**	**1081**
1164	995	911	1094	618	742	5369	5262	5849	3372	3745	5796	4809
11 090	**11 090**	**11 090**	**11 090**	**10 040**	**10 760**	**724**	**917**	**917**	**842**	**898**	**998**	**998**
765	765	765	765	692	742	3221	4079	4079	3745	3995	4439	4439
12 670	**12 670**	**12 670**	**12 670**	**10 040**	**10 760**	**855**	**1040**	**1040**	**842**	**898**	**1012**	**1012**
874	874	874	874	692	742	3803	4626	4626	3745	3995	4502	4502
14 280	**14 280**	**14 280**	**14 280**	**10 040**	**10 760**	**983**	**1164**	**1164**	**842**	**898**	**1081**	**1063**
985	985	985	985	692	742	4373	5178	5178	3745	3995	4809	4729
15 860	**15 860**	**14 810**	**15 870**	**10 040**	**10 760**	**1107**	**1283**	**1283**	**842**	**898**	**1202**	**1064**
1094	1094	1021	1094	692	742	4924	5707	5707	3745	3995	5347	4733
17 430	**16 170**	**14 810**	**15 870**	**10 040**	**10 760**	**1229**	**1315**	**1401**	**842**	**898**	**1341**	**1201**
1202	1115	1021	1094	692	742	5467	5849	6232	3745	3995	5965	5342
18 900	**16 170**	**14 810**	**15 870**	**10 040**	**10 760**	**1341**	**1315**	**1402**	**842**	**898**	**1447**	**1201**
1303	1115	1021	1094	692	742	5965	5849	6236	3745	3995	6437	5342
11 890	**11 890**	**11 890**		**10 760**		**776**	**979**		**898**		**1065**	**1065**
820	820	820		742		3452	4355		3995		4737	4737
13 570	**13 570**	**13 570**		**10 760**		**912**	**1110**		**898**		**1079**	**1079**
936	936	936		742		4057	4938		3995		4800	4800
15 300	**15 300**	**15 300**		**10 760**		**1049**	**1243**		**898**		**1153**	**1134**
1055	1055	1055		742		4666	5529		3995		5129	5044
16 990	**16 990**	**15 870**		**10 760**		**1180**	**1370**		**898**		**1282**	**1134**
1171	1171	1094		742		5240	6094		3995		5703	5044
18 680	**17 320**	**15 870**		**10 760**		**1310**	**1402**		**898**		**1431**	**1281**
1288	1194	1094		742		5827	6236		3995		6365	5698
20 250	**17 320**	**15 870**		**10 760**		**1430**	**1402**		**898**		**1544**	**1281**
1396	1194	1094		742		6261	6236		3995		6868	5698
12 280	**12 280**	**12 280**		**11 120**		**800**	**1010**		**926**		**1098**	**1098**
847	847	847		767		3559	4493		4119		4884	4884
14 030	**14 030**	**14 030**		**11120**		**940**	**1146**		**926**		**1113**	**1113**
967	967	967		767		4181	5098		4119		4951	4951
15 810	**15 810**	**15 810**		**11 120**		**1081**	**1282**		**926**		**1189**	**1169**
1090	1090	1090		767		4809	5703		4119		5289	5200
17 550	**17 550**	**16 400**		**11 120**		**1217**	**1414**		**926**		**1322**	**1170**
1210	1210	1131		767		5413	6290		4119		5881	5204
19 300	**17 900**	**16 400**		**11 120**		**1351**	**1446**		**926**		**1475**	**1321**
1331	1234	1131		767		6010	6432		4119		6561	5876
20 930	**17 900**	**16 400**		**11 120**		**1475**	**1446**		**926**		**1592**	**1321**
1443	1234	1131		767		6561	6432		4119		7082	5876

Compression load

A compression load arises in casings that carry inner strings. Production strings do not develop any compression load, since they do not carry inner strings.

Other loadings

Other loadings that may develop in the casing include: (a) bending with tongs during make-up; (b) pull-out of the joint and slip crushing; (c) corrosion and fatigue failure, both of the body and of the threads; (d) pipe wear due to running wire line tools and drill string assembly, which can be very detrimental to casing in deviated and dog-legged holes; and (e) additional loadings arising from treatment operations such as squeeze-cementing, acidising and hydraulic fracturing.

Only tensile force, collapse pressure, burst pressure and compression load will be considered in the casing design. Other loadings, with the exception of (e), cannot be determined by direct application of mathematical equations and will be accounted for through the use of 'safety factors'.

Safety factors

It is evident from the discussion of casing design criteria that exact values of loadings are difficult to determine. For example, if mud of 0.5 psi/ft (72 pcf) is assumed to exist on the outside of the casing during running of the casing, this value cannot be expected to remain constant throughout the life of the well. Deterioration of mud with time will reduce this value to, say, a salt-water value of 0.465 psi/ft. Hence, calculations of burst values assuming a column of mud of 0.5 psi/ft on the outside of the casing are not applicable throughout the life of the well. If the initial design of the casing is marginal, then any change in loading conditions may lead to bursting of the casing in the event of a gas leak from tubing during production.

Therefore, casing design is not an exact technique, because of the uncertainties in determining the actual loadings and also because of the change in casing properties with time, resulting from corrosion and wear. A safety factor is used to allow for such uncertainties in the casing design and to ensure that the rated performance of the casing is always greater than any expected loading. In other words, the casing strength is downrated by a chosen safety factor value.

Each operating company uses its own safety factors for specific situations. These values have been developed through many years of drilling and production experience. Usual safety factors are:

collapse: 0.85–1.125
burst: 1–1.1
tension: 1.6–1.8

The safety factor is determined as the ratio between the body resistance and the magnitude of the applied pressure. Thus the safety factor (SF) in burst is given by:

$$\text{SF} = \frac{\text{burst resistance of casing}}{\text{burst pressure}}$$

Combination strings

In a casing string, maximum tension occurs at the top and the design criterion requires high grade or heavy casing at the top. Burst pressures are most severe at the top and, again, casing must be strong enough on top to resist failure in burst. In collapse calculations, however, the worst conditions occur at the bottom and heavy casing must, therefore, be chosen for the bottom part to resist collapsing pressure.

Hence, the requirements for burst and tension criteria are different from the requirement for collapse and a compromise must be reached when designing for casing. This compromise is achieved in the form of a combination string. In other words, casings of various grades or differing weights are used at different depths of hole, each grade of casing being capable of withstanding the imposed loading conditions at that depth. Strong and heavy casing is used at the surface, light yet strong casing is used in the middle section, and heavy casing may be required at the bottom to withstand the high collapsing pressure.

The combination string method represents the most economical way of selecting casing consistent with safety. Although as many grades as possible could be used for a string of casing, practical experience has shown that the logistics of using more than two different grades of casing create problems for rig crews.

Biaxial effects

The combination of stresses due to the weight of the casing and external pressures are referred to as 'biaxial stresses'. Biaxial stresses reduce the collapse resistance of the casing in the plastic failure mode and must be accounted for in designing for deep wells or combination strings. The collapse resistance, P_{cc}, under tensile load is given by

$$P_{cc} = \frac{WP_c}{2AS_0}\left[\sqrt{\left\{4\left(\frac{AS_0}{W}\right)^2 - 3\right\}} - 1\right] \quad (10.15)$$

where W = weight supported by casing (lb); P_c = collapse resistance with no tensile load (psi); A = cross-sectional area of casing (in^2) $= \pi t(d_0 - t)$; d_0 = outside diameter of casing (in); t = thickness of casing (in); and S_0 = average yield stress of steel (psi) with zero load.

Setting $K = 2AS_0$ gives

$$\frac{P_{cc}}{P_c} = \frac{W}{K}\left[\sqrt{\left\{4\left(\frac{K}{2W}\right)^2 - 3\right\}} - 1\right]$$

or

$$\frac{P_{cc}}{P_c} = \frac{1}{K}[\sqrt{(K^2 - 3W^2)} - W] \qquad (10.16)$$

Biaxial loading generates forces within the surfaces of the casing which reduce the casing collapse resistance but increase its burst resistance.

A graphical solution of Equation (10.16) was presented in Figure 2.18 (page 41). Equation (10.16) can also be presented in a tabulated form, showing the percentage reduction in collapse resistance for a given unit weight carried by the casing (see Table 10.8). To use this table, determine the ratio between the weight to be carried by the top joint of the weakest casing and the yield strength of the casing. Then from the table determine the corresponding reduction in collapse strength. This reduction in collapse resistance applies to the top joint only and the casing effectively becomes stronger down the hole, since the weakest grade will be carrying less weight with increasing depth. Hence, it is only required to calculate the hydrostatic head due to mud at the top joint and compare this with the net collapse resistance of the top joint. A minimum SF of, say, 0.85 in collapse should be obtained; otherwise replace this joint by a heavier grade. This is explained in detail in Examples 10.1 and 10.4.

Graphical method for casing design

Nowadays many oil companies use a graphical technique for solving casing design problems. The method was first described in 1965 in a series of articles by Goins *et al.*[5] In this book, the graphical method is extended to include the effects of shock loading, maximum load conditions and gas leaks during production. In this method a graph of pressure against depth is first constructed. It is marked such that the top of the graph starts at zero to coincide with zero depth and zero pressure, as shown in Figure 10.8. Collapse, burst and sometimes fracture gradient lines are drawn on this graph. Strength values of available casing grades in collapse and burst are then plotted as vertical lines on this graph. Selection is made such that the casing chosen has strength properties which are higher than the maximum existing collapse and burst pressures.

TABLE 10.8 Reduction in collapse resistance due to biaxial effects

Tensile ratio = *weight carried* / *yield strength*	*Remaining collapse resistance (%)*
0	100
0.01	99.5
0.05	97.3
0.1	94.5
0.15	91.8
0.2	88.5
0.25	85.0
0.30	81.3
0.35	77.7
0.40	76.0
0.45	69.5
0.5	65.0
0.55	60.2
0.60	55.8
0.65	50.0
0.70	44.5
0.75	38.5
0.80	32.0
0.85	25.0
0.90	17.8
0.95	9.0
1.00 (or 100%)	0

Collapse line

The collapse line is determined as follows: (a) calculate the external load due to the mud column, H; (b) calculate the internal load also due to the mud inside the casing, H_1; (c) calculate the collapse pressure, C, as the difference between H and H_1,

$$C = H - H_1$$

At the surface, the external and internal pressures are both zero and the value of the collapse pressure, C, is zero. At the casing shoe, the collapse pressure, C, has its maximum value. On a pressure–depth graph, join the zero co-ordinates with the value of C at the casing shoe depth to obtain the collapse line.

Collapse calculations are, in some cases, based on 100% evacuation such that the internal pressure, H_1, is taken as zero. The 100% evacuation condition can only occur when (a) the casing is run empty; (b) there is complete loss of fluid into a thief zone; and (c) there is complete loss of fluid due to a gas blowout which subsequently subsides, e.g. shallow gas kick through a conductor pipe. Neither of these conditions should be allowed to occur in practice, and they are, in fact, extremely rare.

During lost circulation, the mud level in the well

drops to a height such that the remaining hydrostatic pressure of mud is equal to the formation pressure of the thief zone.

In shallow surface casing it is possible to empty out the casing of a large volume of mud if a loss of circulation is encountered. Some designers assume the surface casing to be completely empty when designing for collapse, irrespective of its setting depth, to provide an in-built safety factor in the design. This overdesign can be significantly reduced when the pressure of the reduced level of mud inside the casing is subtracted from the external collapsing pressure to give the effective collapse pressure.

Both of the above-mentioned methods are discussed in this chapter.

In intermediate casing complete evacuation is never achieved during loss of circulation, while in production casing the assumption of complete evacuation is only justified in artificial lift operations. In such operations gas is injected from the surface to reduce the hydrostatic column of liquid against the formation to help production. If the well pressure were bled to zero at surface, a situation of complete evacuation could exist. Another case of complete evacuation in a production casing may occur in a gas well when perforations plugging may allow the surface pressure to bleed to zero and, hence, give little pressure inside the casing.

Therefore, it is seen that 100% evacuation is an extreme case and a 40% evacuation level is normally used. In this case H_1 is calculated from a 60% filled casing. This will be discussed in detail below in the last section of this chapter.

Burst line

The burst line is determined as follows.

(1) Calculate the external load due to an assumed mud column of 0.465 psi/ft. This value is equal to the gradient of salt-saturated water. (Here it is assumed that the mud in which the casing is run will deteriorate with time, largely because of settlement of mud solids, resulting in a fluid with a much reduced pressure gradient.)
(2) Calculate the internal load due to the formation pressure.
(3) Calculate the burst pressure as the difference between (1) and (2).

The formation pressure in (2) is the formation pressure that can be encountered during the drilling of the next hole. In other words, an intermediate casing is designed for the maximum formation pressure that may result from a kick during drilling of the open hole of the production string.

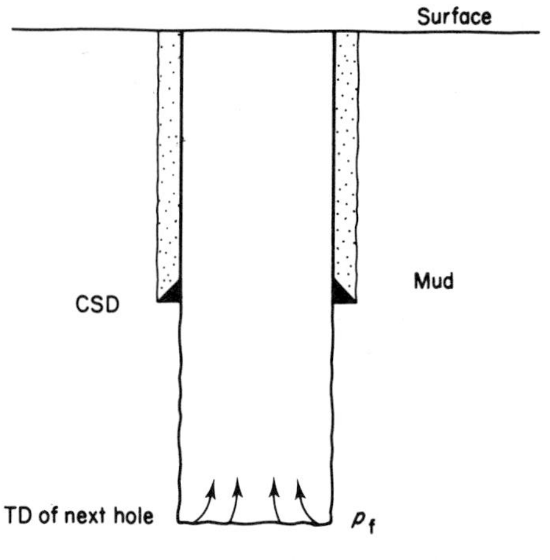

Fig. 10.7. Burst criterion in casing design (CSD = casing setting depth; TD = total depth; P_f = formation pressure at TD of next hole).

Referring to Figure 10.7, burst pressures at the casing shoe and at the surface are calculated by use of the following relationship:

burst pressure = internal pressure − external pressure

In a burst at the shoe

$$\text{external pressure} = \text{CSD} \times G_m \qquad (10.17)$$

$$\text{internal pressure} = P_f - (\text{TD} - \text{CSD})G \qquad (10.18)$$

The term (TD − CSD) × G represents the hydrostatic head of formation fluid between TD and casing setting depth. Formation pressure, P_f, will lose an amount of pressure equal to (TD − CSD) × G in travelling from TD to CSD; see Figure 10.7, in which P_f = formation pressure from next TD; G = gradient of formation fluid; CSD = casing setting depth; TD = total depth of next section of hole; and G_m = mud gradient.

Burst at shoe = Equation (10.18)–Equation (10.17)

$$= P_f - (\text{TD} - \text{CSD})G - \text{CSD} \times G_m \qquad (10.19)$$

It is assumed that the gradient of mud, G_m, deteriorates to that of salt water, such that $G_m = 0.465$ psi/ft (0.1052 bar/m).

In a burst at the surface

$$\text{external pressure} = 0$$

$$\text{internal pressure} = P_f - \text{TD} \times G$$

Therefore,

$$\text{burst at surface} = P_f - \text{TD} \times G \qquad (10.20)$$

In conventional casing design it is customary to assume a gas kick, thereby anticipating the worst possible type of kick. The gas gradient is of the order of 0.1 psi/ft (0.0266 bar/m), therefore causing a very small decrease in formation pressure as the gas rises up the well. When a gas kick is assumed, two points must be considered:

(1) The casing seat should be selected such that the gas pressure at the casing shoe is less than the formation breakdown pressure at the shoe.
(2) Gas pressure data must be available from reservoirs in the open hole section. In exploration wells where reservoir pressures are not known, the formation pressure at TD of the next open hole section is calculated from the maximum anticipated mud weight at that depth. A gas pressure equal to this value is used for the calculations of internal pressures.

In development areas reservoir pressures are normally determined by use of wire line logs, drill stem testing or production testing. These pressure values should be used in casing design.

Tensile forces

Tensile forces are determined as follows: (a) calculate the weight of casing in air (positive value); (b) calculate the buoyancy force (negative value); (c) calculate the bending force in deviated wells (positive value); and (d) calculate shock loads due to arresting casing.

Forces (a)–(c) exist always, whether the pipe is static or in motion, while forces in (d) only exist when the pipe is arrested in the rotary table.

Tensile loads (sometimes referred to as installation loads) must be determined accurately so that the yield strength of the casing chosen is never exceeded. Also, the installation loads must always be less than the rated derrick load capacity so that the casing can be run or pulled without causing damage to the derrick.

In the initial selection of casing, it is important to check that the casing can carry its own weight in mud, and when the casing is finally chosen, to calculate total tensile loads and compare them with the joint or pipe body yield values. A safety factor (= coupling or pipe body yield strength divided by total tensile loads) in tension of 1.6–1.8 should be obtained.

Example 10.1

The following three grades of $13\frac{3}{8}$ in (340 mm) casings are available in a company store. It is required to run a combination string based on collapse and tension only. The casing is run in 67 pcf (1.0734 kg/l) mud to 6200 ft (1890 m). Safety factors are 1.8 for tension and a minimum of 0.85 for collapse. (*Note:* Figures in parentheses are in metric units.)

Grade	*Weight* lbm/ft (kg/m)	*Collapse* psi (bar)
K55	54.5 (81.2)	1130 (78)
K55	68 (101)	1950 (134)
L80	72 (107)	2670 (184)

Grade	*Yield strength* × 1000 lb (kN) *Body*	*Coupling*
K55	853 (3794)	636 (2829)
K55	1069 (4755)	1300 (5783)
L80	1661 (7388)	1693 (7340)

Joint type: LTC for K55, 54.5 lb/ft and BTS for remaining grades.

Solution

(1) *Collapse*

$$\text{Collapse pressure} = \frac{67 \times 6200}{144} = 2884.7 \text{ psi}$$

$$\approx 2885 \text{ psi}$$

or

$$C = 0.0981\rho h = 199 \text{ bar}$$

(*Note:* Mud gradient is 67/144 = 0.4653 psi/ft or 0.1053 bar/m.)

On a graph of depth against pressure draw a collapse pressure line between zero at surface and 2885 psi (199 bar) at 6200 ft (1890 m). Draw the collapse resistances of the three grades as vertical lines, as shown in Figure 10.8.

From Figure 10.8, selection based on collapse is as shown at the top of page 222. (*Note*: Minimum safety factor in collapse = collapse resistance of casing divided by collapse pressure of mud column.)

Note that the last grade was only suitable down to a depth of 5400 ft (1740 m) for a safety factor of 1. However, since a minimum safety factor of 0.85 is to be used, this grade is suitable down to 6200 ft (1890 m), with the lowest safety factor being 0.93 at TD. Above 6200 ft (1890 m) the safety factor value in collapse increases and assumes a maximum value (at 4200 ft) of

$$\frac{1950}{\left(\frac{2500 \times 67}{144}\right)} = 1.7 \quad \left(\frac{134 \text{ bar}}{740 \text{ m} \times 0.1053 \text{ bar/m}} = 1.7\right)$$

Depth	*Grade and weight*	*Length of section*	*Minimum safety factor*
0–2500 ft (0–740 m)	K55, 54.5 lbm/ft (K55, 81.2 kg/m)	2500 ft (740 m)	1
2500–4200 ft (740–1270 m)	K55, 68 lbm/ft (K55, 101 kg/m)	1700 ft (530 m)	1
4200–6200 ft (1270–TD m)	L-80, 72 lb/ft (L-80, 107 kg/m)	2000 ft (620 m)	$\frac{184}{1840 \times 0.1053} = 0.93$

(2) *Tension* Casing-carrying capacity must be checked from the bottom joint to the surface. Two values of yield strength are given in the table of strength properties. One specifies the yield strength of pipe body and the other the yield strength of the coupling. The lower of these two values is used for the calculation of the safety factor in tension. Therefore, starting from the bottom, see table below.

Since a minimum safety factor of 1.8 is to be used in tension, the K55, 54.5 lbm/ft (81.2 kg/m) may be used if it is designed to carry a maximum weight, W, given by:

Imperial units

$$1.8 = \frac{636 \times 1000}{W}$$

$$W = 353\,333 \text{ lb}$$

Hence, useable weight of section of 54.5# = (Total weight which can be carried) – (weight of lower casing grades)

$$\text{weight of section of } 54.5\#^* = 353\,333 - 259\,600$$
$$= 93\,733 \text{ lb}$$

and length of usable section of K55, 54.5#

$$= \frac{93\,733 \text{ lb}}{54.5 \text{ lbm/ft}} = 1720 \text{ ft}$$

Metric units

$$1.8 = \frac{2829 \text{ kN}}{W}$$

$$W = 1571.667 \text{ kN}$$

*The symbol # indicates lb/ft.

Grade and weight	*Weight of section* lb (kN)	*Cumulative weight carried by top joint of section* × 1000 lb (kN)	*Safety factor* = $\frac{\textit{yield strength}}{\textit{weight carried}}$
L80, 72 lb/ft	2000 × 72 = 144 000	144	$\frac{1661}{144} = 11.5$
(L80, 107 kg/m)	$(620 \times 107 \times 10^{-3} = 650.8)$	(650.8)	$\left(\frac{7340}{650.8} = 11.3\right)$
K55, 68 lbm/ft	1700 × 68 = 115 600	259.6	$\frac{1069}{259.6} = 4.1$
(K55, 101 kg/m)	$(530 \times 101 \times g \times 10^{-3} = 525.13)$	(1175.93)	$\left(\frac{4755}{1175.93} = 4.04\right)$
K55, 54.5 lbm/ft	1500 × 54.5 = 136 250	395.85	$\frac{636}{396.85} = 1.6 < 1.8$
(K55, 81.2 kg/m)	$(740 \times 81.2 \times g \times 10^{-3} = 58\,946)$	(1765.4)	$\left(\frac{2829}{1765.4} = 1.6 < 1.8\right)$

Here g = acceleration due to gravity (9.81 m/s^2), 10^{-3} = factor to convert N to kN)

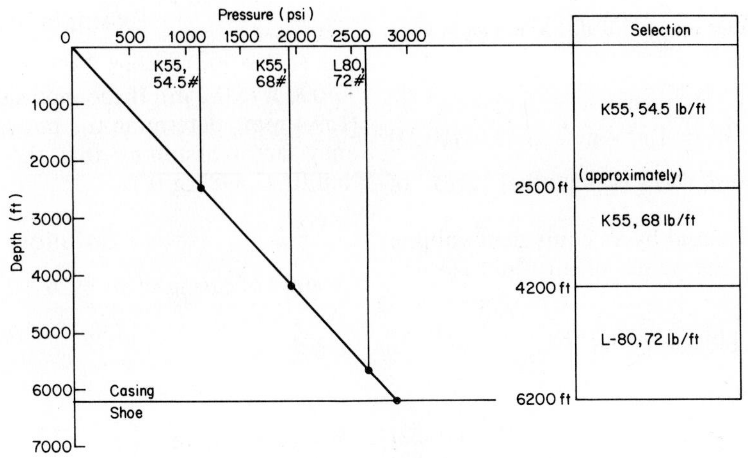

Fig. 10.8. Casing design (Example 1).

Useable weight of section of K55, 81.2 kg/m

$$= 1571.667 - 1175.93$$

$$= 395.737 \text{ kN}$$

and

length of usable section of K55, 81.2 kg/m

$$= \frac{395.737 \times 1000}{81.2 \times g}$$

$$= 497 \text{ m}$$

Remaining top length $= 2500 - 1720 = 780$ ft

$(740 - 497 = 243 \text{ m})$

A heavy casing must be used for the top 780 ft. Try K55, 68# (101 kg/m — i.e. next heavy casing).

Total weight that can be carried by the top joint of K55 is:

$$= 353\,333 + 780 \times 68 = 406\,373 \text{ lb}$$

$$(= 1571.667 + 243 \times 101 \times g \times 10^{-3} = 1812.434 \text{ kN})$$

SF in tension for K55, 68# at top joint

$$= \frac{1069 \times 10^3}{406\,373} = 2.6 \left(\frac{4755}{1812.434} = 2.6 \right)$$

Hence, the final casing selection, based on collapse and tension, is as follows:

Depth		*Grade and weight*
0– 780 ft	(0– 243 m)	K55, 68# (101 kg/m)
780–2500 ft	(243– 740 m)	K55, 54.5# (81.2 kg/m)
2500–4200 ft	(740–1270 m)	K55, 68# (101 kg/m)
4200–6200 ft	(1270–1890 m)	L80, 72# (107 kg/m)

In exploration wells the designer often discards grades which give a marginal safety factor. In fact, the above selection could well be simplified further to obtain added safety factors and to eliminate the risk of using the wrong joint in a critical section of the well. In this example grade K55, 54.5# (81.2 kg/m) is the weakest grade and can therefore be eliminated from our selection. Hence, final selection can be made as follows:

Depth	*Grade and weight*
0–4200 ft	K55, 68 lb/ft
(0–1270 m)	(K55, 101 kg/m)
4200–6200 ft	L80 72 lb/ft
(1270–1890 m)	(L80 107 kg/m)

BUOYANCY

Archimedes' principle states that all immersed bodies suffer from a buoyancy force which is equal to the weight of fluid displaced by the immersed body. In casing design a buoyancy force is required in the calculation of the installation load or effective tension force at the top joint. The buoyancy force acts on the bottom joint of the casing and results in a reduction in the hanging weight of the casing.

Consider a cylinder of 1 m (or 1 ft) in length, of density ρ_s, which is fully immersed in a fluid of density ρ_m. The cylinder is assumed to have an outside diameter of d_o and an inside diameter of d_i.

Weight = mass × acceleration due to gravity

$$\text{Air weight of cylinder} = \frac{\pi}{4}(d_o^2 - d_i^2) \times 1 \times \rho_s \times g$$

or

$$W_a = A_s \rho_s \times g \qquad (10.21)$$

where $A_s = (\pi/4)(d_o^2 - d_i^2)$ = cross-sectional area of cylinder.

If this cylinder is replaced by an equivalent volume of fluid of density ρ_m, the weight of this fluid is

$$W_m = \frac{\pi}{4}(d_o^2 - d_i^2)\rho_m \times g$$

or

$$W_m = A_s \rho_m \times g \qquad (10.22)$$

If the steel cylinder is now immersed in this fluid of density ρ_m, it displaces a volume of fluid equal to its own volume and will be subjected to a buoyancy force equal to W_m.

Hence, the effective or buoyant weight of the casing, W_B, is

$$W_B = W_a - W_m$$

$$W_B = (A_s \rho_s - A_s \rho_m) g$$

Multiplying and dividing the right-hand side of this expression by W_a gives

$$W_B = \frac{W_a}{W_a}(A_s \rho_s - A_s \rho_m) g$$

$$= W_a \left(\frac{A_s \rho_s}{W_a} - \frac{A_s \rho_m}{W_a} \right) g$$

$$W_B = W_a \left(1 - \frac{\rho_m}{\rho_s} \right)$$

or

$$W_B = W_a \times BF \qquad (10.23)$$

where

$$BF = \text{buoyancy factor} = \left(1 - \frac{\rho_m}{\rho_s} \right) \qquad (10.24)$$

(*Note:* The value of BF is always less than 1.)

Buoyancy force may be calculated as the difference between air weight and buoyant weight, or

$$W_m = \text{buoyancy force} = W_a - W_B$$

$$= W_a - W_a(BF)$$

$$W_m = W_a(1 - BF) \qquad (10.25)$$

Buoyancy force acts on the whole immersed body and its value is the same throughout the casing.

Example 10.2

7 in (177.8 mm) casing, 26# (38.7 kg/m), is to be set at 17 000 ft (5182 m). If the internal diameter is 6.276 in (159.4 mm), determine the buoyancy force and buoyancy factor assuming that the mud density is 93.5 lbm/ft^3 (1.498 kg/l).

Solution

Weight of casing in air = 26 × 17 000 = 442 000 lb

(38.7 × 5182 × g = 1 965 325 N

= 1965.33 kN)

$$\text{Buoyancy factor} = \left(1 - \frac{\rho_m}{\rho_s} \right) = \left(1 - \frac{93.5}{489.5} \right) = 0.809$$

or

$$BF = \left(1 - \frac{1.498}{7.85} \right) = 0.809$$

where density of steel = 489.5 lb/ft^3 (7.85 kg/l).

Buoyant weight of casing

= 0.809 × 442 000 = 357 578 lb

(0.809 × 1965.33 = 1589.95 kN)

Buoyancy force = 442 000 − 357 578 = 84 422 lb

= (1965.33 − 1589.95 = 375.38 kN)

BENDING FORCE

Bending forces arise if casing is run in highly deviated wells or in wells with severe dog-leg problems. An equation for the bending force can be derived by considering a beam subjected to pure bending, as shown in Figure 10.9. In the following derivation it is assumed that plane transverse sections of a beam will remain plane after bending and that the radius of curvature of the beam is large in comparison with the transverse dimensions. Also, Young's modulus of the beam has the same value in tension as in compression.

During pure bending, the upper surface of the beam stretches and therefore is in tension, while the lower surface shortens and therefore is in compression. Between the compressed and stretched surfaces there must exist a surface in which longitudinal deformation is zero. This surface is described as the neutral plane and a line parallel to this surface is known as the neutral axis, line NA in Figure 10.9.

Consider a longitudinal surface HJ at a distance y from NA which has the same length as the fibre KL at

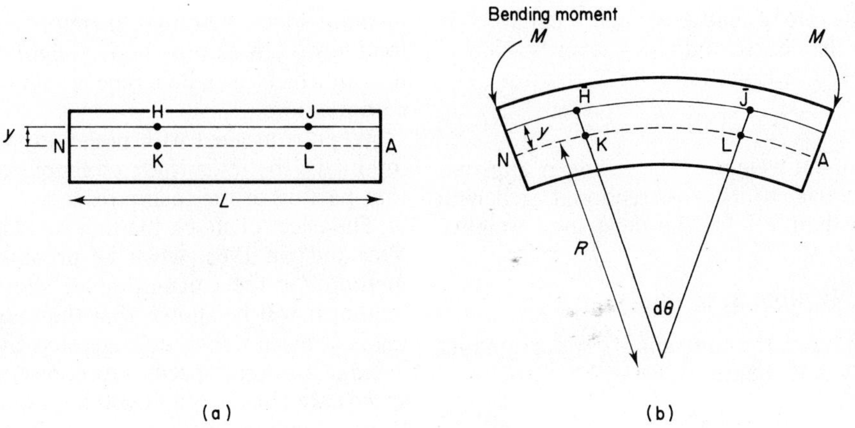

Fig. 10.9. Pure bending of a beam (NA = neutral axis).

the neutral axis (Figure 10.9a). After bending, the surface HJ deforms to an arc ($\bar{HJ}$ in Figure 10.9b) of radius R and included angle $d\theta$. The surface KL, being at the neutral surface, retains its original length.

Thus, the longitudinal strain, e, in the fibre $\bar{HJ}$ is

$$e = \frac{\bar{HJ} - HJ}{HJ} = \frac{(R + y)d\theta - Rd\theta}{Rd\theta}$$

where HJ = KL = $Rd\theta$. Therefore,

$$e = \frac{y}{R} \tag{10.26}$$

where y = distance of arc $\bar{HJ}$ from the neutral axis and R = radius of curvature of the deformed beam.

Assuming that the beam remains elastic after bending, then Hook's law, $E = \sigma/e$, is applicable (where σ is the longitudinal stress). Therefore,

$$\sigma = Ee = E\frac{y}{R} \tag{10.27}$$

If the original length of the beam is L and the total deformation angle is θ, then

$$NA = R\theta$$

But

$$NA = L$$

Therefore,

$$L = R\theta \tag{10.28}$$

Substituting Equation (10.28) in Equation (10.27) gives

$$\sigma = \frac{Ey}{L/\theta} = \frac{E\theta y}{L} \tag{10.29}$$

The maximum tensile stress (σ) occurs at the upper extreme end of the beam at $y = D/2$, where D is the diameter of the beam. Thus,

$$\sigma = \frac{E\theta D}{2L} \tag{10.30}$$

Also,

$$\text{bending force (FB)} = \sigma \times A$$

where A = cross-sectional area. Hence,

$$FB = \frac{ED}{2L}A\theta$$

Angle θ is normally expressed in degrees, while the above formula requires the angle in radians; hence,

$$FB = \frac{EDA}{2L}\theta\left(\frac{\pi}{180}\right) \tag{10.31}$$

where θ = change in angle of deviation (in degrees).

Equation (10.31) in field units

Imperial units

E = modulus of elasticity of steel: 30×10^6 psi

D = in; A = in^2; L = ft; θ = degrees

Therefore,

$$FB = \frac{30 \times 10^6(\text{lbf/in}^2) \times D(\text{in}) \times A(\text{in}^2) \times \theta}{2L(\text{ft}) \times \left(\frac{12\text{ in}}{\text{ft}}\right)}\left(\frac{\pi}{180}\right)$$

$$= 218.17 \times 10^2\frac{DA}{L}\theta \tag{10.32}$$

In practice, the rate of change of angle θ per 100 ft is used to indicate the degree of dog-leg severity. Hence, replacing L by 100 in Equation (10.32) gives

$$\text{FB} = 218DA\theta \qquad (10.33)$$

Also, the nominal weight of the casing is approximately proportional[5] to its cross-sectional area with an error of less than 2% for the most used weights. Mathematically,

$$W_N\,(\text{lbm/ft}) \propto A(\text{in}^2)$$

For most casing types, the constant of proportionality is approximately 3.46; hence,

$$W_N = 3.46A \qquad (10.34)$$

Substituting Equation (10.34) in Equation (10.33) gives

$$\text{FB} = 218D \times \frac{W_N}{3.46}\theta$$

$$\text{FB} = 63DW_N\theta \text{ lb} \qquad (10.35)$$

Metric units

$$E = 210 \times 10^9 \text{ N/m}^2$$

$$\text{FB} = 210 \times 10^9 \text{ N/m}^2 \times D\left(\text{mm} \times \frac{\text{m}}{1000 \text{ mm}}\right) \times A\left(\text{mm}^2 \times \frac{\text{m}^2}{10^6 \text{ mm}^2}\right) \times \left(\frac{\theta \times \frac{\pi}{180}}{2L(\text{m})}\right)$$

Dog-leg severity is normally expressed in degrees per 10 m; hence, replacing L by 10 results in

$$\text{FB} = 0.1833DA\theta \qquad (10.36)$$

The weight of casing in kg/m is related to the cross-sectional area in mm^2 by the following relationship:

$$W_N = 7.9 \times 10^{-3}A \qquad (10.36\text{a})$$

Combining Equations (10.36) and (10.36a) gives

$$\text{FB} = 23.2DW_N\theta \text{ kN} \qquad (10.37)$$

SHOCK LOADS

The running of casing strings (or drill pipe) requires that the pipe be decelerated during the last 5 ft before the slips are set in the rotary table to arrest the pipe. The process of arresting the casing sets up shock (or dynamic) stresses through the body of the pipe. These stresses (or loadings) are active for a very short period of time, affecting only one part of the pipe at any instant. Shock loadings, therefore, differ from static loadings (such as pipe body weight or bending loading), in which the whole pipe is subjected to tension or compression.

When combined with pipe body weight and bending loads, the magnitude of shock loads can be such that parting of the casing results.

The effect of shock loading was first recognised by Vreeland[6] in 1961, when he presented an analytical method for the calculation of shock loads. In this section it will be shown that the magnitude of shock loads is twice the value suggested by Vreeland when average running speeds are considered. When peak speeds are considered, it will be shown that the actual shock loading is four times that suggested by Vreeland[6].

Transmission of shock waves

When casing or drillpipe is suddenly arrested by slips, particles of the pipe near the slips are given a forward velocity V and a compressive stress σ. These particles will impinge on neighbouring particles, imparting to them both stress and velocity. The impacting process continues generating a compressive stress wave through the casing body (Figure 10.10). The compressive stress wave travels from the slips area to the casing shoe, where it will experience reflection.

Assume that the velocity of the stress wave is C_0; then, after time t, the wave will occupy a distance of C_0t along the casing, as shown in Figure 10.10(b).

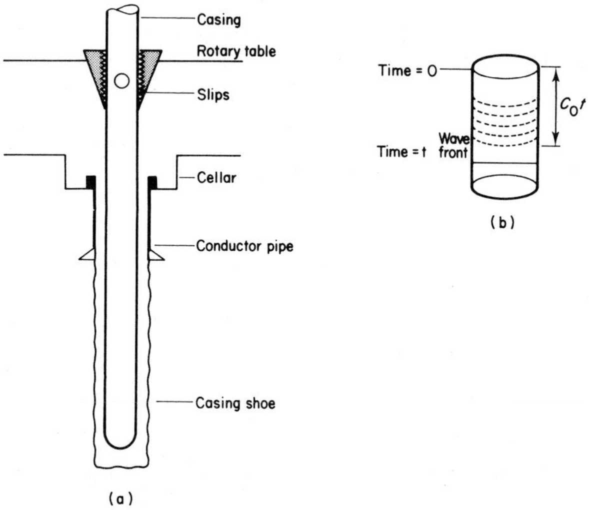

Fig. 10.10. (a) Shock loading; (b) element of casing under the influence of a shock stress wave.

During the same time t the casing particles move through a distance of Vt.

Using the law of conservation of momentum,

change of momentum of length C_0t

$$= \text{impulsive force} \times \text{time} \quad (10.38)$$

$$\text{impulsive force} = \text{stress} \times \text{cross-sectional area}$$

$$= \sigma A \quad (10.39)$$

Substituting Equation (10.39) in Equation (10.38) gives

$$m \times V = (\sigma A)t \quad (10.40)$$

where m = mass of element of casing of length C_0t $= (AC_0t)\rho$; and V = velocity of particles in the impacted area. Substituting the value of m in Equation (10.40) yields

$$(AC_0t)\rho \times V = (\sigma A)t$$

Therefore,

$$\sigma = \rho C_0 V \quad (10.41)$$

Thus, a compressive stress σ of magnitude given by Equation (10.41) propagates to the casing shoe.

The above analysis assumed a one-dimensional theory in which the casing is considered as a long thin bar. This assumption produces a negligible error in the final result[7].

Reflection of the generated stress wave

When the compressive stress wave arrives at the shoe, it encounters a free end in which particle movements are not resisted. At this end three changes occur. The wave is reflected, it changes its nature from compressive to tensile and it propagates up the casing string. At the casing shoe the reflected tensile stress cancels the compressive wave, which results in zero stress at this end.

The reflected tensile wave travels up the casing shoe to arrive at the surface, where it encounters a fixed end (since the casing is held in place by the slips). Particle movements are inhibited at a fixed end and the tensile wave reflects as a tensile wave. Thus, at a free end reflection of waves is accompanied by a change in sign, whereas at fixed ends reflection does not alter the sign of the original incident wave. Further details of the mechanics of wave reflection can be found in Reference 7.

Reflection of the stress wave at the surface results in a total stress of 2σ at the slips area. It is this stress which contributes to total loading at the slips. The stress wave is of such short duration that it dies away when the next casing joint (or drillpipe joint) is stabbed in.

For example, for a 10 000 ft well the time, t, required for the wave to travel from surface to shoe and back to surface is

$$t = \frac{2L}{C_0} = \frac{2 \times 10\,000 \text{ ft}}{17\,028 \text{ ft/s}} = 1.2 \text{ s}$$

where C_0 = one-dimensional wave velocity in steel = 17 028 ft/s. Thus, after a time 1.2 s no shock loading is experienced at the surface.

Since force is the product of stress and cross-sectional area, total shock (dynamic) forces at the surface are given by F_s:

$$F_s = (2\sigma) \times A \quad (10.42)$$

Combining Equations (10.42) and (10.41) gives

$$F_s = 2\rho C_0 V \times A \quad (10.43)$$

The shock force calculated from Equation (10.43) is twice the value given in Vreeland's paper[6].

Field version of Equation (10.43)

Equation (10.43) can be simplified by replacing the term ρC_0 by a numerical value and the cross-sectional area by the nominal weight per foot. For steel, the term ρC_0 is simplified as follows:

$$\rho C_0 = 489.5 \frac{\text{lbm}}{\text{ft}^3} \times 17\,028 \frac{\text{ft}}{\text{s}} = 8\,335\,206 \frac{\text{lbm}}{\text{ft}^2\text{s}} \quad (10.44)$$

Also, for most casing sizes, the cross-sectional area is related to the nominal weight per foot, with negligible error[3] through the following formula:

$$A = \frac{W_N}{3.46} \text{ in}^2 \quad (10.45)$$

Substituting Equations (10.44) and (10.45) in Equation (10.43) and simplifying yields

$$F_s = 1.04 \times 10^3 W_N \times V \text{ lb} \quad (10.46)$$

The velocity, V, may be taken as equal to the average running speed of 13 s per 40 ft of casing joint. Thus

$$F_s = 3200 W_N \text{ lb} \quad (10.47)$$

In metric units Equation (10.47) is

$$F_s = 9.4 W_N \text{ kN} \quad (10.48)$$

Peak velocity

The magnitude of the shock load is greatly increased when the peak velocity is considered. Assuming that the initial velocity of the casing is zero, then the peak velocity, V_s, is twice the average running speed of 13 s per 40 ft. Thus, Equation (10.46) becomes

$$F_s(\text{peak}) = 6400 W_N \qquad (10.49)$$

Example 10.3

Determine the magnitude of shock loading during the running of L80, 43.5 lbm/ft, $9\frac{5}{8}$ in casing, using: (a) an average running speed of 13 s per 40 ft; (b) a peak velocity of twice the average speed.

Solution

(a) From Equation (10.47)

$$F_s = 3200 \times W_N = 3200 \times 43.5$$

$$= 139\,200 \text{ lb}$$

(b) Equation (10.49) is applicable; hence,

$$F_s(\text{peak}) = 2 \times 139\,200$$

$$= 278\,400 \text{ lb}$$

The result of part (b) clearly shows that, for marginal casing design, the casing running speed is critical and should not exceed a maximum of 13 s per 40 ft. Shock loading can also be decreased by reducing the average running speed to below 13 s per 40 ft.

CASING DESIGN EXAMPLE (EXAMPLE 10.4)

An exploration well is to be drilled to a total depth of 13 900 ft (4327 m). Relevant data are as follows.

Drilling programme:

0–350 ft (107 m), 26 in (660.4 mm) hole
350–6200 ft (1890 m), $17\frac{1}{2}$ in (444.5 mm) hole
6200–10 400 ft (3170 m), $12\frac{1}{4}$ in (311.2 mm) hole
10 400–13 900 ft (4237 m), $8\frac{1}{2}$ in (215.9 mm) hole

Casing programme:

20 in (508 mm) casing to be set at 350 ft (107 m)
$13\frac{3}{8}$ in (339.7 mm) casing to be set at 6200 ft (1890 m)
$9\frac{5}{8}$ in (244.5 mm) casing to be set at 10 400 ft (3170 m)
7 in (177.8 mm) casing to be set at 13 900 ft (4237 m)

The casing head housing will be installed on the 20 in casing. The 7 in casing will be run to the surface.

Mud programme:

Down to 350 ft (107 m), mud weight is 65 pcf (1.041 kg/l)
Down to 6200 ft (1890 m), mud weight is 67 pcf (1.073 kg/l)
Down to 10 400 ft (3170 m), mud weight is 73 pcf (1.169 kg/l)
Down to 13 900 ft (4237 m), mud weight is 87 pcf (1.394 kg/l)

Safety factors:

Burst = 1.1
Collapse = 0.85
Tension = 1.8

Formation fluid gradient:

0–6200 ft (1890 m), P_f = 0.465 psi/ft (0.105 bar/m)
6200–10 400 ft (3170 m), P_f = 0.48 psi/ft (0.1086 bar/m)
10 400–13 900 ft (4237 m), P_f = 0.57 psi/ft (0.1289 bar/m)

The $12\frac{1}{4}$ in hole experiences a maximum dog-leg severity of 3°/100 ft. Other sections of the well experience negligible deviation. Shock loads are to be included in the design of $9\frac{5}{8}$ in and 7 in casing strings.

For collapse, burst and yield strength values refer to Tables 10.4–10.7.

Design suitable casing strings for the given hole sizes, taking into consideration the available casing grades and the maximum expected pressures.

Solution

1. Conductor pipe (20 in casing)

This pipe is set at 350 ft (107 m) and will be subjected to formation pressure from the next hole drilled to a depth of 6200 ft (1890 m). It will be assumed that no gas exists at this shallow depth and kick calculations will be based on a water kick situation in which formation gradient is 0.465 psi/ft (0.105 bar/m). Note that if gas is known to exist at shallow depths, it must be included in the calculations.

Collapse

Collapse pressure at surface = 0

$$\text{Collapse pressure at 350 ft} = \frac{\text{mud weight} \times \text{depth}}{144}$$

where mud weight is lbm/ft³. Therefore,

$$\text{collapse pressure at 350 ft} = \frac{65 \times 350}{144}$$

$$= 158 \text{ psi (11 bar)}$$

This pressure acts on the outside of the casing and for the worst possible situation assume that the casing is 100% evacuated (as is the case in a complete-loss circulation situation).

Burst

$$\text{Burst pressure} = \text{internal pressure} - \text{external pressure}$$

(a) Burst at shoe
From Figure 10.11,

$$\text{formation pressure at next TD} = 6200 \times 0.465$$

or

$$P_f = 2883 \text{ psi (199 bar)}$$

$$\text{Internal pressure} = P_f - (\text{TD} - \text{CSD}) \times G$$
$$= 2883 - (6200 - 350) \times 0.465$$
$$= 163 \text{ psi (11 bar)}$$

where G = gradient of invading fluid = 0.465 psi/ft.

$$\text{External pressure} = \text{casing setting depth} \times \text{mud gradient}$$

$$\text{external pressure} = (350 \times 65)/144$$
$$= 158 \text{ psi (11 bar)}$$

$$\text{Burst at shoe} = \text{internal pressure} - \text{external pressure}$$
$$= 163 - 158 = 5 \text{ psi (0.4 bar)}$$

(b) Burst at surface

$$\text{Burst at surface} = P_f - \text{TD} \times G$$
$$= 2883 - 6200 \times 0.465 = 0$$

It should be noted that the zero values were obtained as a result of the fact that a salt-water kick is considered. If instead a gas kick is considered, the burst pressure values at the shoe and surface will be 2135 psi and 2140 psi, respectively.

Selection A graph is not normally required and selection is determined by comparing the strength properties of available casing with existing pressures.

From Table 10.4 it can be seen that all the available grades have collapse and burst values above those calculated above. Hence, select grade K555, 94#, having collapse pressure = 520 psi (36 bar), burst pressure = 2110 psi (145 bar) and yield strength = 1 479 000 lb (6579 kN). It should be noted that grade K55, 94# is the lightest and the cheapest of the three available grades.

Since the casing head housing is installed on the 20 in casing, the latter will be subjected to compression forces resulting from the weights of subsequent casing strings.

This casing will be checked later to determine whether it is capable of carrying other casing strings.

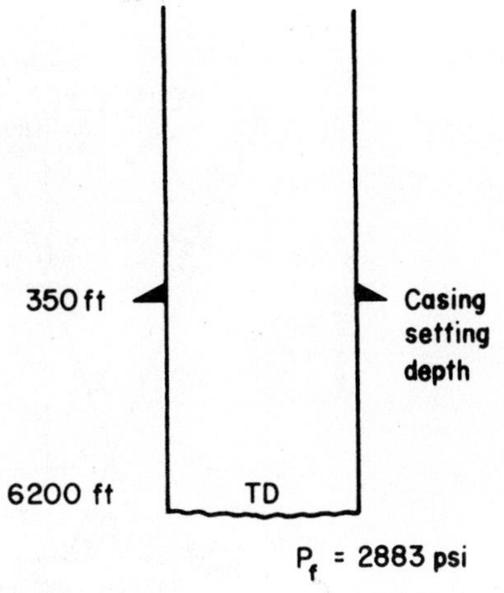

Fig. 10.11. Solution to Example 10.4 (20 in casing design).

2. $13\frac{3}{8}$ in casing

This string is set at 6200 ft and will be subjected, in the event of a kick, to formation pressures from the next hole drilled to a TD of 10 400 ft.

Collapse

$$\text{Collapse pressure at surface} = 0$$

$$\text{Collapse pressure at 6200 ft (1890 m)} = \frac{67 \times 6200}{144}$$
$$= 2885 \text{ psi (199 bar)}$$

The collapse line is drawn between 0 at the surface and 2885 psi at 6200 ft, as shown in Figure 10.12.

From Table 10.5 the collapse resistances of the available grades as adjusted for a safety factor of 0.85 are as follows:

Grade	*Weight* (lbm/ft)	*Coupling*	*Collapse resistance* *SF = 1*	*SF = 0.85*
K55	54.5	LTC	1130	$\frac{1130}{0.85} = 1329$
K55	68.0	BTS	1950	$\frac{1950}{0.85} = 2294$
L80	72.0	BTS	2670	$\frac{2670}{0.85} = 3141$

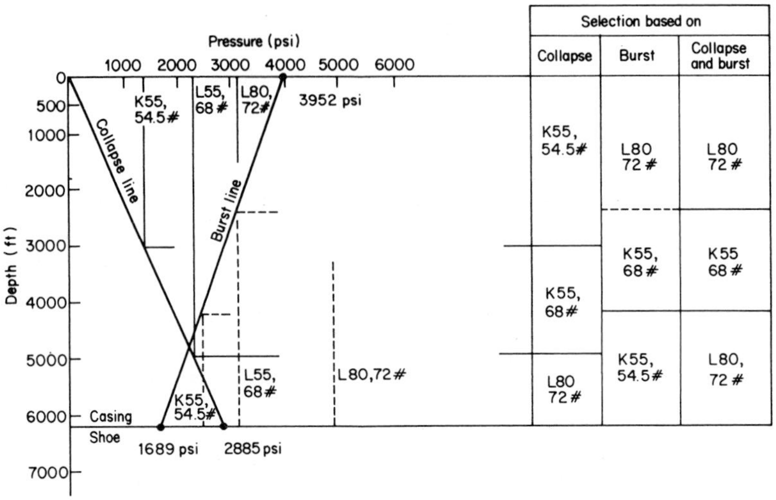

Fig. 10.12. $13\frac{3}{8}$-in casing design: (——) collapse lines; (-------) burst lines.

The collapse resistance values are plotted as vertical lines, as shown in Figure 10.12.

Burst

Formation pressure from next TD

$$= 10\,400 \times 0.48$$

$$= 4992 \text{ psi (344 bar)}$$

(see Figure 10.13).

$$\text{Burst at shoe} = \text{internal pressure} - \text{external pressure}$$

$$\text{Internal pressure} = P_f - (\text{TD} - \text{CSD}) \times G$$

$$= 4992 - (10\,400 - 6200) \times 0.1$$

$$= 4572 \text{ psi (315 bar)}$$

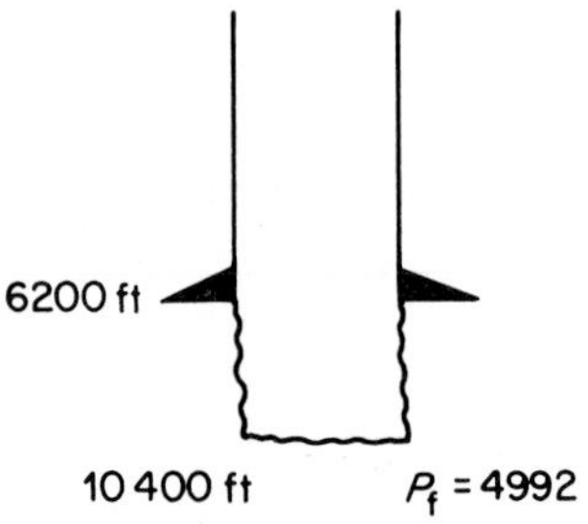

Fig. 10.13. Solution to Example 10.4 ($13\frac{3}{8}$ in casing design).

(where G = gradient of invading fluid, assumed to be gas having a 0.1 psi/ft gradient).

$$\text{External pressure} = \text{CSD} \times 0.465$$

where 0.465 psi/ft is the gradient of mud outside the casing. Therefore,

$$\text{external pressure} = 6200 \times 0.465$$

$$= 2883 \text{ psi (199 bar)}$$

Thus,

$$\text{burst at shoe} = 4572 - 2883$$

$$= 1689 \text{ psi (116 bar)}$$

$$\text{Burst at surface} = \text{internal pressure} - \text{external pressure}$$

$$\text{External pressure} = 0$$

$$\text{Internal pressure} = P_f - (\text{TD}) \times G$$

Therefore,

$$\text{burst at surface} = P_f - (\text{TD}) \times G$$

$$= 4992 - 10\,400 \times 0.1$$

$$= 3952 \text{ psi (273 bar)}$$

The burst line can now be drawn between 1689 psi at the shoe and 3952 psi at the surface; see Figure 10.12.

From Table 10.5, of casing properties, the burst resistances of the available grades are given below, together with adjustment for SF = 1.1.

Grade	*Weight* (lbm/ft)	*Coupling*	*Burst resistance* (psi) *SF = 1*	*SF = 1.1*
K55	54.5	LTC	2730	$\frac{2730}{1.1} = 2482$
K55	68.0	BTS	3450	$\frac{3450}{1.1} = 3136$
L80	72	BTS	5380	$\frac{5380}{1.1} = 4891$

The burst resistance values are drawn as vertical lines, as shown in Figure 10.12.

Selection Selection should consider the lightest weights first, as these grades are the cheapest. On the basis of collapse only, Figure 10.12 indicates that the given grades are suitable for the following depths:

0–3050 ft K55, 54.5#

3050–4950 ft K55, 68#

4950–6200 ft L80, 72#

On the basis of burst only, Figure 10.12 gives the following selection:

0–2400 ft L80, 72#

2400–4200 ft K55, 68#

4200–6200 ft K55, 54.5#

When selection is based on both collapse and burst, Figure 10.12 indicates that grade K55, 54.5# does not satisfy the burst requirement from 0 to 4200 ft. Also, grade K55, 68# does not satisfy burst from 0 to 2400 ft. Hence, selection from 0 to 2400 ft is limited to grade L80, 72#.

Below 2400 ft, grade K55, 68# is suitable for collapse from 0 to 4950 ft and for burst from 2400 ft to 4200 ft. Hence, the middle section consists of K55, 68# from 2400 to 4200 ft.

The last section of hole can only be satisfied by grade L80, 72# in both collapse and burst; see Figure 10.12. Hence, selection based on collapse and burst is (see table below):

Note that grade K55, 54.5# has been rejected, since it does not satisfy both collapse and burst at once along any section of the hole.

Tension If bending and shock forces are ignored, the suitability of selected grades in tension can be checked by comparing the weight in air carried by each section with its yield strength. For the $9\frac{5}{8}$ in and 7 in casing, effects of bending and shock loading will be included and buoyant weight will be considered to reduce the possibility of over-designing. Hence, starting from the bottom, see table at the top of page 232.

Note that yield strength values are obtained from Table 10.5 as the lowest value of either the body or coupling yield strength.

The safety factor must, at least, be equal to the required value of 1.8 if any of the selected grades is to satisfy the criterion of tension. The table overleaf produces values of SF of greater than 1.8, which indicates that the grades satisfy collapse, burst and tension.

Pressure testing After the casing is landed and cemented, it is the practice to test the casing prior to drilling the casing shoe. The testing pressure employed by some operating companies is 60% of the burst rating of the weakest grade of casing in the string. Hence,

testing pressure of $13\frac{3}{8}$ in

= 60% × burst pressure of K55, 68#

= 60% × 3450

= 2070 psi (143 bars)

During pressure testing an extra tensile force is exerted on the casing and the SF should, again, be >1.8 for the top joint (or the joint of weakest grade). Hence,

total tensile force during pressure testing at top joint

= buoyant weight of casing

+ tensile force due to pressure testing

$= \text{weight in air} \times \text{BF} + \frac{\pi}{4}(\text{ID})^2$

$\times$ testing pressure

Depth	*Grade and weight*	*Weight in air* (× 1000 lb)
0–2400 ft (732 m)	L80, 72#	2400 × 72 = 172.8
2400–4200 ft (1280 m)	K55, 68#	(4200–2400) × 68 = 122.4
4200–6200 ft (1890 m)	L80, 72#	(6200–4200) × 72 = 144.0
Total weight in air		= 439.2

Weight of section (× 1000 lb) (kN)	*Grade and weight*	*Cumulative weight* (× 1000 lb) (kN)	*Safety factor* = *yield strength* ÷ *cumulative weight*
144.0 (641)	L80, 72#	144.0 (641)	$\frac{1650}{144} = 11.5$
122.4 (544)	K55, 68#	266.4 (1185)	$\frac{835}{266.4} = 3.13$
172.8 (769)	L80, 72#	439.2 (1954)	$\frac{1650}{439.2} = 3.8$

$$\text{BF} = \left(1 - \frac{\rho_m}{\rho_s}\right)$$

$$\text{BF} = \left(1 - \frac{67}{489.5}\right) = 0.863$$

Table 10.5 gives the inside diameter of L80, 72# as 12.347 in (313.6 mm). Therefore,

$$\text{total tensile force} = (439.2 \times 0.863) \times 1000 + \frac{\pi}{4}(12.347)^2 \times 2070$$

$$= 379\,030 + 247\,847$$

$$= 626\,877 \text{ lb}$$

$$\text{SF in tension for top joint} = \frac{1\,661\,000}{626\,877} = 2.65$$

Biaxial effects Check the weakest grade of selected casing for biaxial effects as follows.

$$\text{Tensile ratio} = \frac{\text{weight carried by weakest joint}}{\text{yield strength of body (or coupling)}}$$

Weakest grade selected is the K55, 68#, having a body yield strength of 1 069 000 lb and a coupling strength (LTC) of 835 000 lb. Hence,

$$\text{tensile ratio} = \frac{266.4 \times 1000}{835\,000} = 0.319$$

For a tensile ratio of 0.319, Table 10.8 shows that the collapse resistance of the casing is reduced to approximately 80% of its original (under zero load) value. Hence,

$$\text{collapse resistance of K55, 68\# under biaxial loading} = 0.8 \times 1950 = 1560 \text{ psi (108 bars)}$$

Collapse pressure due to mud at 2400 ft (i.e. top joint of grade of the K55, 68#)

$$= \frac{67 \times 2400}{144} = 1117 \text{ psi (77 bars)}$$

Therefore,

SF in collapse for top joint of K55, 68#

$$= \frac{\text{collapse resistance}}{\text{collapse pressure}}$$

$$= \frac{1560}{1117} = 1.40$$

Final selection

Depth	*Grade and weight*
0–2400 ft (732 m)	L80, 72# (107 kg/m)
2400–4200 ft (1280 m)	K55, 68# (101 kg/m)
4200–6200 ft (1890 m)	L80, 72# (107 kg/m)

3. $9\frac{5}{8}$ in casing

The $9\frac{5}{8}$ in casing is set at 10 400 ft and will be subjected, in the event of a kick, to formation pressures from the next hole drilled to a TD of 13 900 ft.

Collapse

At surface

$$\text{collapse pressure} = 0$$

At shoe

$$\text{collapse pressure} = \frac{73 \times 10\,400}{144}$$

$$= 5272 \text{ psi (363.5 bars)}$$

Draw a line between 0 and 5272 psi as shown in Figure 10.14. From Table 10.6 collapse properties of available casing are as follows:

Grade	*Weight* (lbm/ft)	*Collapse pressure* *SF = 1*	*SF = 0.85*
C75	43.5	3750	$\frac{3750}{0.85} = 4412$
L80	47.0	4750	$\frac{4750}{0.85} = 5888$
C95	53.5	7330	$\frac{7330}{0.85} = 8624$

The above collapse resistances can be drawn as vertical lines, as shown in Figure 10.14.

Burst

The $9\frac{5}{8}$ in casing will be subjected in the event of a kick, to a formation pressure of:

$$0.57 \text{ psi/ft} \times 13\,900 \text{ ft} = 7923 \text{ psi (546 bar)}$$

$$\text{Burst at shoe} = \text{internal pressure} - \text{external pressure}$$

$$\text{Burst at shoe} = [P_f - (\text{TD} - \text{CSD}) \times G] - \text{CSD} \times 0.465$$

A gas kick is considered for this string; thus, $G = 0.1$ psi/ft. Therefore,

$$\text{burst at shoe} = 7923 - (13\,900 - 10\,400) \times 0.1 - 10\,400 \times 0.465 = 2737 \text{ psi (189 bars)}$$

(where TD = next hole depth = 13 900 ft).

$$\text{Burst at surface} = P_f - \text{TD} \times G$$

Therefore,

$$\text{burst at surface} = 7923 - 13\,900 \times 0.1 = 6533 \text{ psi (450.4 bar)}$$

The burst line can now be plotted between 6533 psi at the surface (i.e. at zero depth) and 2737 psi at 10 400 ft, as shown in Figure 10.14.

From Table 10.6 burst pressures of available grades of $9\frac{5}{8}$ in casing as adjusted for an SF = 1.1 are:

Grade	*Weight* (lbm/ft)	*Burst pressure* (psi) *SF = 1*	*SF = 1.1*
C75	43.50	5930	$\frac{5930}{1.1} = 5391$
L80	47.0	6870	$\frac{6870}{1.1} = 6245$
C95	53.50	9410	$\frac{9410}{1.1} = 8555$

Burst resistance lines are plotted, as shown in Figure 10.14.

Selection based on collapse and burst From Figure 10.14, selection based on collapse and burst is as shown at the top of page 234.

$$\text{Buoyant weight of casing} = 474.75 \times \text{BF}$$

$$\text{BF} = \left(1 - \frac{73}{489.5}\right) = 0.851$$

$$\text{Buoyant weight of casing} = 474.75 \times 0.851 = 404.012 \times 1000 \text{ lb}$$

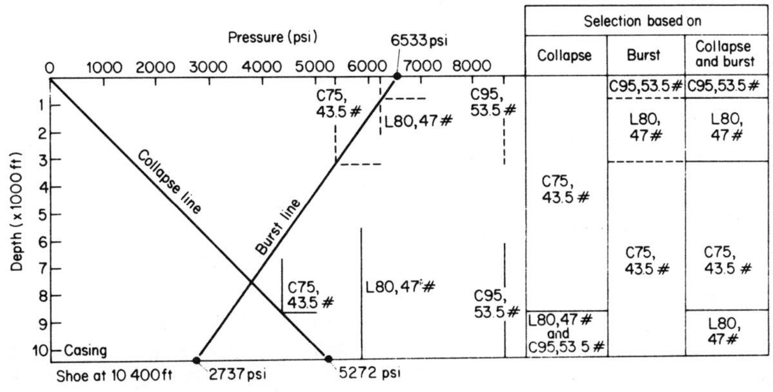

Fig. 10.14. $9\frac{5}{8}$ in casing design: (———) collapse lines; (-------) burst lines.

Depth	Grade and weight	Weight of section in air (× 1000 lb)
0–800 ft (244 m)	C95, 53.5#	42.80
800–3200 ft (975 m)	L80, 47#	112.80
3200–8700 ft (2652 m)	C75, 43.5#	239.25
8700–10 400 ft (3170 m)	L80, 47#	79.90
	Total weight in air	=474.75

Tension The suitability of the selected grades in tension will be investigated by considering the total tensile forces resulting from casing buoyant weight, bending force and shock load. Starting from the bottom, the weight carried by each section can be calculated, as follows:

Depth (ft)	Weight of each section (× 1000 lb)	Weight in air carried by top joint of each section
10 400–8700	79.90	79.90
8700–3200	239.25	79.90 + 239.25 = 319.15
3200–800	112.80	319.15 + 112.8 = 431.95
800–0	42.80	431.95 + 42.8 = 474.75

By use of the equations

$$\text{bending force} = 63\theta \times D \times W_N$$

$$\text{drag force} = 3200 \times W_N$$

where W_N is the weight per unit length, Table 10.10 can be constructed. Table 10.10 shows that all the selected grades satisfy the tension requirement.

Pressure testing

Testing pressure

= 60% of burst pressure of lowest grade (C75, 43.5#)

= 0.6 × 5930

= 3558 psi (245 bar)

During pressure testing, an extra tensile force is generated and selected grades with marginal SF should be checked. At 800 ft grade L80, 47# has the lowest SF of 1.8 (see Table 10.10); hence, this grade should be checked.

During pressure testing,

total tensile force

= buoyant load

+ tensile force due to pressure testing

From table 10.10,

buoyant force at 800 ft = 361.21(× 1000 lb)

$$\text{Total tensile load at 800 ft} = 361.21 \times 1000 + \frac{\pi}{4}(8.681)^2 \times 3558$$

$$= 580.623 \times 1000 \text{ lb}$$

$$\text{SF in tension} = \frac{1086}{580.623} = 1.87$$

Biaxial effects Check the weakest grade selected. Grade C75, 43.5# is the weakest grade, carrying a total buoyant load of 248.41 × 1000 lb, as shown in Table 10.10.

$$\text{Tensile ratio} = \frac{\text{weight carried}}{\text{yield strength}} = \frac{248.41}{942} = 0.264$$

From Table 10.8 it can be seen that, for a tensile ratio of 0.264, the collapse resistance reduces to 84% of its original value. Hence,

collapse resistance of C75, 43.5# under biaxial loading

= 0.84 × 3750

= 3150 psi (217.6 bar)

SF in collapse

$$= \frac{\text{collapse resistance under biaxial loading}}{\text{collapse pressure at 3200 ft}}$$

$$= \frac{3150}{\left(\dfrac{73 \times 3200}{144}\right)} = 1.94$$

Final selection

Depth	Grade and weight
0–800 ft (244 m)	C95, 53.5# (79.7 kg/m)
800–3200 ft (976 m)	L80, 47# (70 kg/m)
3200–9700 ft (2957 m)	C75, 43.5# (64.8 kg/m)
9700–10 400 ft (3170 m)	L80, 47# (70 kg/m)

TABLE 10.9 Available casing grades

Size, in (mm)	*Grade*	*Inside diameter, in (mm)*	*Weight, lb/ft (kg/m)*	*Coupling type*
20 (508)	K55	19.124 (485.7)	94.0 (140)	BTS
	K55	19.00 (482.6)	106.5 (159)	LTC
	K55	18.73 (475.7)	133.0 (198)	LTC
$13\frac{3}{8}$ (339.7)	K55	12.615 (320.4)	54.5 (81.2)	LTC
	K55	12.415 (315.3)	68.0 (101)	LTC
	L-80	12.347 (313.6)	72.0 (107)	BTS
$9\frac{5}{8}$ (244.5)	C-75	8.599 (218.4)	43.5 (64.8)	BTS
	L-80	8.681 (220.5)	47.0 (70)	BTS
	C-95	8.535 (216.8)	53.5 (79.7)	BDS
7 (177.8)	K55	6.276 (159.4)	26.0 (38.7)	BTS
	L-80	6.184 (157.1)	29.0 (43.2)	BTS
	C-95	6.184 (157.1)	29.0 (4.32)	BDS

BTS = buttress thread
BDS = buttress double seal thread
LTC = long (round) thread coupling
STC = short (round) thread coupling

4. 7 in casing

This string is set at 13 900 ft (4237 m).

Collapse

Collapse pressure at surface = 0

$$\text{Collapse pressure at 13 900 ft} = \frac{87 \times 13\,900}{144}$$

$$= 8398 \text{ psi (579 bar)}$$

This pressure acts on the outside of the casing, and for the worst possible situation assume that there is zero pressure inside the casing. Draw the collapse pressure line, as shown in Figure 10.15, between 0 psi at the surface and 8398 psi at 13 900 ft.

Collapse resistances, from Table 10.7, are as follows:

Grade	*Weight* (lbm/ft)	*Collapse resistance* *SF = 1*	*SF = 0.85*
K55	26.0	4320	$\frac{4320}{0.85} = 5082$
L80	29.0	7020	$\frac{7020}{0.85} = 8259$
C95	29.0	7820	$\frac{7820}{0.85} = 9200$

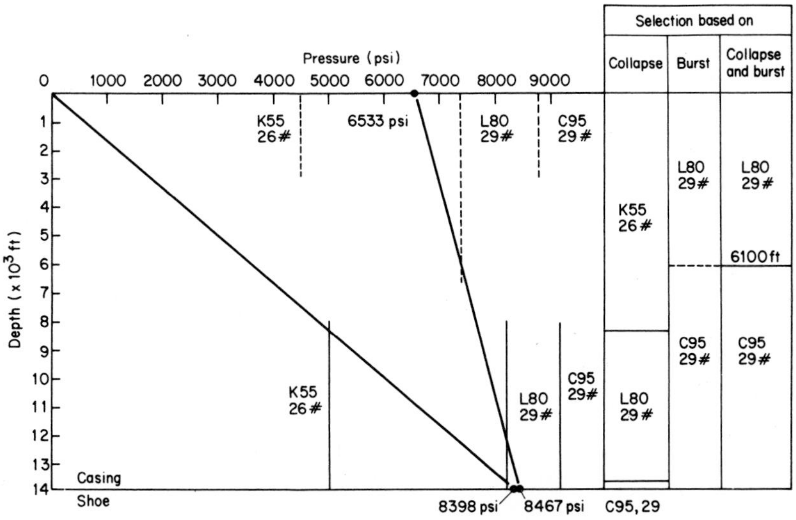

Fig. 10.15. 7-in casing design: (———) collapse resistances; (-------) burst resistances.

TABLE 10.10 $9\frac{5}{8}$ in casing design

Depth (ft)	Selected grades from collapse and burst criteria	Weight in air carried by top joint (× 1000 lb)	Buoyant weight[1] carried by top joint (× 1000 lb)	Bending force[2] $= 63\theta \times d \times W_N$ (× 1000 lb)	Shock load[2] $= 3200\ W_N$ (× 1000 lb)	Total tensile load at each joint = buoyant weight + bending load + shock load (× 1000 lb)	$SF = \frac{\text{yield strength}}{\text{total tensile load}}$ [3]
0	C95, 53.5#	474.75	474.75 × 0.851 = 404.012	63 × 9 × 9.625 × 53.5 = 97.32	3200 × 53.5 = 171.2	672.53	$\frac{1458}{672.53} = 2.14$
	weight = 42.8 × 1000 lb			97.32	171.2	629.73	$\frac{1458}{629.73} = 2.3$
800							
	L80, 47#	431.95	404.012 − 42.8 = 361.21	63 × 3 × 9.625 × 47 = 85.498	3200 × 47 = 150.4	597.11	$\frac{1086}{597.11} = 1.82$
	weight = 122.8 × 1000 lb			85.498	150.4	484.31	$\frac{1086}{484.31} = 2.24$
3200							
	C75, 43.5#	319.15	361.21 − 112.8 = 248.41	63 × 3 × 9.625 × 43.5 = 79.13	3200 × 43.5 = 139.2	466.74	$\frac{942}{466.74} = 2.02$
	weight = 239.25 × 1000 lb			79.13	139.2	227.49	$\frac{942}{227.49} = 4.14$
8700							
	L80, 47#	79.9	248.41 − 239.25 = 9.16	63 × 3 × 9.625 × 47 = 85.498	3200 × 47 = 150.4	245.06	$\frac{1086}{245.06} = 4.43$
10 400	weight = 79.9 × 1000 lb	0	9.16 − 79.9 = −70.74	85.498	150.4	165.16	$\frac{1086}{165.16} = 6.58$

(1) Buoyancy factor = 0.851. Buoyant weight recorded by weight indicator is valid for the total weight of string. Buoyant weight of sections below top joint is obtained by subtracting the weight in air of each section from the buoyant weight, as shown in the above table. Although the lowest joint is not carrying any tensile load, it is under compression due to fluid pressure acting from below. This explains the negative sign of buoyant loads at depths 10 400 ft and 9700 ft. (Tension is assumed positive and compression negative.)

(2) Bending and shock forces are calculated for each grade and at each change-over point for the grades above and below this point.

(3) The yield strength values are determined from Table 10.6 for both body and joint, and the lower of the two is used.

Collapse resistances can now be drawn as vertical lines in Figure 10.15.

Burst

$$\text{Burst pressure} = \text{internal pressure} - \text{external pressure}$$

(a) Burst at shoe

$$\text{Internal pressure} = P_f = 0.57 \times 13\,900 = 7923 \text{ psi (546.5 bar)}$$

$$\text{External pressure} = G_{mud} \times \text{CSD}$$

For added safety, the external pressure resisting internal pressure is assumed to be that of a mud column outside the casing, even though the casing is cemented. Also, the mud is assumed to deteriorate so that its gradient decreases to that of salt water, largely because of settlement of solids. Hence,

$$G = 0.465 \text{ psi/ft (0.1052 bar/m)}$$

$$\text{Burst at shoe} = 7923 - 13\,900 \times 0.465 = 1460 \text{ psi (100 bars)}$$

(b) Burst pressure at surface

$= \text{internal pressure} - \text{external pressure}$

$= 7923 - 13\,900$

$\times$ gradient of invading fluid (assumed gas)

$= 7923 - 13\,900 \times 0.1 = 6533$ psi (450.4 bars)

Worst conditions In practice, hydrocarbon production is carried out through a tubing (single or dual) sealed in a packer, as shown in Figure 10.16. Thus, under ideal conditions only the casing shoe will be subjected to burst effects. However, a situation may arise in practice when the production tubing leaks gas to the 7 in casing. In this case, the surface pressure (6533 psi) is now acting on the column of packer fluid between the casing and the tubing; see Figure 10.16. Hence, burst calculations for production casing should be modified as follows.

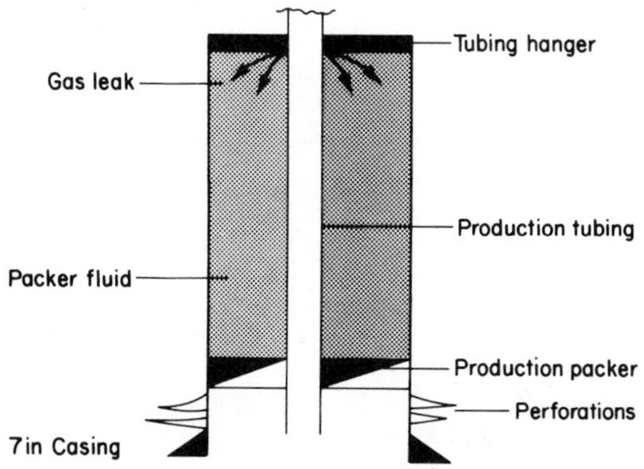

Fig. 10.16. 7-in casing design.

(a) Burst pressure at shoe = surface pressure + hydrostatic pressure of packer fluid − external pressure. Normally

$$\rho_{\text{packer fluid}} = \rho_{\text{drilling mud}} = 87 \text{ pcf (1.394 kg/l)}$$

$$\text{Burst at shoe} = 6533 + \frac{87 \times 13\,900}{144} - 13\,900 \times 0.465 = 8467 \text{ psi (584 bar)}$$

(b) Burst at surface = 6533 psi (450 bar)

(*Note:* All these calculations assume that there is no cement outside the casing.)

The burst line is drawn between 6533 psi at the surface and 8467 psi at 13 900 ft, as shown in Figure 10.15. From Table 10.7 burst resistances as adjusted for an SF of 1.1 for the available grades are:

Grade	*Weight* (lbm/ft)	*Burst resistance* *SF = 1*	*SF = 1.1*
K55	26.0	4980	$\frac{4980}{1.1} = 4527$
L80	29.0	8160	$\frac{8160}{1.1} = 7418$
C95	29.0	9690	$\frac{9690}{1.1} = 8809$

Adjusted burst lines (for SF = 1.1) can now be drawn as vertical lines in Figure 10.15.

Selection based on burst and collapse From Figure 10.15 selection based on burst and collapse is as follows:

Depth (ft)	*Grade and weight*
0–6100	L80, 29#
6100–13 900	C95, 29#

Tension The suitability of selected grades in tension can be checked by considering the cumulative weight

Depth (ft)	*Grade*	*Weight of each section in air* (× 1000 lb)	*Cumulative weight* (× 1000 lb)	*SF = yield strength ÷ weight carried*
13 900–6100	C95, 29#	(13 900–6100) × 29 = 226.2	226.2	$\frac{803}{226.2} = 3.55$
6100–0	L80, 29#	6100 × 29 = 176.9	403.1	$\frac{676}{403.1} = 1.68$

carried by each section. Hence, starting from the bottom, see tabulation above.

The above table considered the weight of casing sections in air only and a marginal safety factor of 1.68 was obtained. This value is below the required SF of 1.8 and it is instructive to check the suitability of this grade by adding the effects of shock loading (bending effects are assumed to be negligible) as shown in Table 10.11.

From Table 10.11 it is evident that grade L80 29# is not suitable as a top joint. Further refinement can be made when pressure testing is considered.

Pressure testing Normally, casing is tested to 60% of its mill burst pressure. Taking the lowest grade (L80), therefore,

$$\text{test pressure} = 0.6 \times 8160 = 4896 \text{ psi (338 bar)}$$

The weakest joint is the top joint of the lower grade. Therefore, load at top joint = buoyant weight of casing + tensile force, resulting from extra pressure on inside of casing

$$= W_{\text{air}} \times \text{BF} + \frac{\pi}{4}(d_i)^2 \times \text{test pressure}$$

where d_i is the internal diameter of the casing = 6.184 in and BF = 1 − (87/489.5) = 0.822. Therefore,

$$\text{total load at surface} = 0.822 \times 403.1 \times 1000$$

$$\times \frac{\pi}{4}(6.184)^2 \times 4896$$

$$= 478\,400 \text{ lb}$$

Therefore,

$$\text{SF in tension during pressure test} = \frac{676\,000}{478\,400}$$

$$= 1.41$$

Thus, the top joint of L80 must be replaced by a higher grade casing if the SF in tension of 1.8 is to be maintained throughout the running and testing of the casing. Hence, the maximum load, W, that L80 can carry and still produce an SF = 1.8 is given by

$$1.8 = \frac{676\,000}{W}$$

Therefore,

$$W = 375\,556 \text{ lb}$$

Hence,

weight of usable L80 section

= total weight carried (W) − air weight of C95, 29#

= 375 556 − air weight of C95, 29#

= 375 556 − 226 200

= 149 356 lb

and

$$\text{usable length of L80, 29\#} = \frac{149\,356 \text{ lb}}{29 \text{ lb/ft}} = 5150 \text{ ft}$$

(*Note:* The length of L80 as selected above (page 236) was 6100 ft. Hence,

length to be replaced = 6100 − 5150

= 950 ft (or 24 joints)

The length to be replaced must be of a grade higher than L80, 29#; in our example use grade C95, 29#.)

From Table 10.11 the total tensile load at top joint is still 424.15 × 1000 lb, since the two different grades have the same weight per foot of 29. If the weights were different, then another table should be constructed.

If grade C95 is used as the top section (950 ft long), then

$$\text{SF at top joint} = \frac{803\,000}{424\,150} = 1.89$$

(*Note:* Yield strength of C95, 29# = 803 000 lb.)

During pressure tests,

$$\text{SF} = \frac{803\,000}{478\,400} = 1.68$$

TABLE 10.11 7 in casing design

Depth (ft)	*Selected grades from collapse and burst criteria*	*Weight in air carried by top joint (× 1000 lb)*	*Buoyant weight*[1] *carried by top joint (× 1000 lb)*	*Shock load* $= 3200\ W_N$ *(× 1000 lb) (× 1000 lb)*	*Total tensile load = buoyant load + shock load (× 1000 lb)*	$SF = \frac{\text{yield strength}}{\text{total tensile load}}$
	L80, 29 lb weight = 176 900 lb	403.1	331.35	3200 × 29 = 92.8	424.15	$\frac{676}{424.15} = 1.59$
6100				92.8	267.55	$\frac{676}{267.55} = 2.53$
	C95, 29 lb weight = 226 200 lb	226.2	331.35 − 176.9 = 154.45	92.8	247.25	$\frac{803}{267.55} = 3.00$
13 900		0	154.45 − 226.2 = −71.75	92.8	21.05	$\frac{803}{21.05} = 38.15$

[1] Buoyancy factor $= 1 - \frac{87}{489} = 0.822$.

Thus, even with the higher grade, the SF during pressure testing is still below 1.8. To maintain an SF of 1.8, decrease the test pressure below 4896 psi. Hence,

$$1.8 = \frac{803\,000}{\text{buoyant weight} + \text{tensile force due to pressure test}}$$

$$= \frac{803\,000}{331\,348 + \frac{\pi}{4}(6.184)^2 \times P}$$

where P is the required pressure test. Therefore, P = 3821 psi (263 bar).

Hence, the new selection is:

Depth (ft)	*Grade*
0–950	C95, 29#
950–6100	L80, 29#
6100–13 400	C95, 29#

Biaxial effects Biaxial loading reduces collapse resistance of casing and is most critical at the joints of the weakest grade. Two positions will be investigated.

(a) At 950 ft

tensile ratio

$$= \frac{\text{buoyant weight carried by top joint of L80}}{\text{yield strength}}$$

tensile ratio (TR)

$$= \frac{(13\,900 - 950) \times 29 \times \text{BF}}{676\,000} = 0.46$$

Table 10.8 shows that for a TR = 0.46, collapse resistance reduces to 69% of its original collapse value (i.e. under zero load). Therefore, actual collapse resistance of L80 = 0.69 × 7020 = 4844 psi at 950 ft

$$\text{SF in collapse at 950 ft} = \frac{\text{collapse resistance of casing}}{\text{collapse pressure of mud}}$$

$$= \frac{4844}{\left(\frac{87 \times 950}{144}\right)} = 8.4$$

(b) At 6100 ft

$$\text{tensile ratio} = \frac{154\,450}{676\,000}\text{(from Table 10.4)} = 0.23$$

Table 10.8 shows that for a TR = 0.23, collapse resistance reduces to 86% of its original value. Therefore, adjusted collapse resistance of L80 = 0.86 × 7020 = 6037 psi at 6100 ft

$$\text{SF in collapse at 6100 ft} = \frac{6037}{\left(\frac{87 \times 6100}{144}\right)} = 1.6$$

Final selection

Depth (ft)	*Grade and weight*
0–950	C95, 29#
950–6100	L80, 29#
6100–13 900	C95, 29#

COMPRESSION IN CONDUCTOR PIPE

The conductor pipe must be checked for compression loading, since it carries the weight of other strings. The procedure is to determine the total buoyant weight of strings carried and then compare this with the yield strength of the conductor pipe (i.e. pipe body or coupling yield strength, whichever is the lowest). A minimum safety factor of 1.1 should be obtained.

In this analysis it is assumed that the tensile strength of casing is equal to its compressive strength.

For our example, we shall consider the submerged weights of $13\frac{3}{8}$ in, $9\frac{5}{8}$ in and 7 in strings in a mud weight of 0.465 psi/ft, so that the worst case is taken into account.

Hence,

$$\text{BF} = \left(1 - \frac{0.465\ \text{psi/ft}}{3.39\ \text{psi/ft}}\right) = 0.863$$

where 3.39 psi/ft is the pressure gradient of steel.

Casing	*Air weight* × 1000 lb
$13\frac{3}{8}$ in	439.2
$9\frac{5}{8}$ in	481.25
7 in	403.1

Total air weight carried by conductor pipe

= 1 317 050 lb

Total buoyant weight carried by conductor pipe

= 1 317 050 × 0.863 = 1 136 614 lb

Yield strength of coupling of top joint of K55, 94#

= 1 479 000

Hence,

$$\text{SF in compression} = \frac{1\,479\,000}{1\,136\,614}\ \text{lb} = 1.3$$

MAXIMUM LOAD CASING DESIGN FOR INTERMEDIATE CASING

The method presented in Example 10.4 should be used for designing surface and production casing used in artificial lift operations. For intermediate casing, the design can be refined further by critically examining the two assumptions used in solving Example 10.4. The two assumptions used were: (1) in collapse the casing was 100% evacuated; (2) in burst a gas kick was encountered such that it filled the entire hole.

Assumption (1) can only be satisfied in shallow casing during lost circulation when the mud level drops below the casing shoe. In intermediate casing lost circulation will cause the mud to drop to a height such that the remaining hydrostatic pressure of the mud column inside the casing is equal to the pressure of a full column of salt water outside the casing[9]. Assuming a salt water gradient of 0.465 psi/ft and the setting depth of the casing to be CSD, then during lost circulation the following relationship should hold:

$$\frac{L \times \rho_m}{144} = \text{CSD} \times 0.465 \qquad (10.50)$$

where L = height of mud remaining inside casing after lost circulation; and ρ_m = weight of mud used to drill the next hole. Therefore,

$$L = \frac{\text{CSD} \times 0.465}{\rho_m} \times 144\ \text{ft} \qquad (10.51)$$

Thus, the external collapse pressure at the shoe is now resisted by a mud column of length L. Collapse at the surface is still zero, while the collapse pressure at depth L is equal to the external pressure only. This will be explained further in Example 10.5.

Assumption (2), which states that a gas kick will fill the entire casing, is never satisfied in most casing strings, except when gas is injected from the surface in artificial lift. Thus, in most kick situations the gas may fill 40–60% of the hole, at which time the well is shut in and the gas kick is circulated out. Exception to this rule is that some experienced drilling crews may shut in the well before the gas occupies 40–60% of the hole. Thus, the 40–60% range is chosen to represent the worst conditions. Also, gas being lighter than mud, it will percolate through the mud and will occupy the upper half of the well during the shut-in period. Burst calculations under these conditions are presented in detail in Example 10.5.

Example 10.5

Redesign the $9\frac{5}{8}$ in casing previously given in Example 10.4 by modifying the assumptions for collapse and burst.

Collapse

(a) External pressure at surface = 0

$$\text{External pressure at shoe} = \frac{73 \times 10\,400}{144}$$

$$= 5272\ \text{psi (364 bar)}$$

It should be noted that external pressure calculations are based on the weight of the mud in which the casing was run.

(b) Owing to lost circulation, the mud column remaining inside casing (L) is

$$L = \frac{\text{CSD} \times 0.465}{\rho_m} \times 144 \qquad (10.51)$$

$$= \frac{10\,400 \times 0.465}{87} \times 144$$

$$L = 8004\ \text{ft}$$

Depth to top of mud column = 10 600 − 8004 = 2396 ft

Therefore, the back-up load (or internal pressure) for the $9\frac{5}{8}$ in casing will be provided by a column of heavy mud from 2396 ft to 10 400 ft. In other words, the length of back-up fluid inside the $9\frac{5}{8}$ in casing is 10 400 − 8004 = 2396 ft.

(c) Collapse pressure at surface = 0

Collapse pressure at 2396 ft = external pressure − internal pressure

(*Note:* The top of the mud column inside the casing is at 2396 ft; therefore, the internal pressure at this depth is zero.)

$$\text{Collapse pressure at 2396 ft} = \frac{2396 \times 73}{144} - 0$$

$$= 1215\ \text{psi (84 bar)}$$

Collapse at casing shoe = external pressure − internal pressure

$$= \frac{10\,400 \times 73}{144} - \frac{8004 \times 87}{144}$$

$$= 437\ \text{psi (30 bar)}$$

(d) Draw a collapse line from 0 psi at surface to 1215 psi at 2396 ft and finally to 437 psi at 10 400 ft, the casing setting depth. The collapse line is shown in Figure 10.17.

Burst Assume that a gas kick originates from the next TD at 13 900 ft and that the gas occupies 50% of the hole. Gas is also assumed to occupy the upper half of the well. Therefore,

$$\text{top of mud column in hole} = 50\% \times \text{TD} = 0.5 \times 13\,900 = 6950 \text{ ft}$$

Internal pressure of gas at 6950 ft

$$= 6950 \times \text{formation fluid gradient of next TD}$$
$$= 6950 \text{ ft} \times 0.57 \text{ psi/ft}$$
$$= 3962 \text{ psi (273 bar)}$$

Internal pressure of gas at surface

$$= 3962 - \text{pressure loss due to hydrostatic head of gas column from 6950 ft to surface}$$
$$= 3962 - (6950 \times 0.1)$$

where 0.1 psi/ft is the gas gradient.

Internal pressure of gas at surface = 3267 psi (225 bar)

Using the equation

$$\text{burst pressure} = \text{internal pressure} - \text{external pressure}$$

the burst pressure at various parts of the hole will be calculated as follows:

(a) Burst pressure at surface

$$= 3267 - 0 = 3267 \text{ psi (225 bar)}$$

(b) Burst pressure at 6950 ft

$$= 3962 - \left(\frac{6950 \text{ ft} \times 73 \text{ pcf}}{144}\right)$$
$$= 438 \text{ psi (30 bar)}$$

(c) Burst pressure at casing shoe

$$= \text{internal pressure} - \text{external pressure}$$

Internal pressure at casing shoe

$$= 10\,400 \text{ ft} \times 0.57 \text{ psi/ft}$$
$$= 5928 \text{ psi (408 bar)}$$

External pressure at shoe due to mud in which casing was run

$$= \frac{10\,400 \times 73}{144} = 5272 \text{ psi (364 bar)}$$

Therefore,

$$\text{burst pressure at shoe} = 5928 - 5272 = 656 \text{ psi (45 bar)}$$

(d) Draw a burst pressure line from 3267 psi at surface to 438 psi at 6950 ft and finally to 656 psi at 10 400 ft. The burst line is shown in Figure 10.17.

Selection The collapse and burst resistances of the available $9\frac{5}{8}$ in casing grades are plotted as vertical lines in Figure 10.17. The collapse resistances are adjusted for an SF of 0.85, while the burst resistances are adjusted for an SF of 1.1, as shown in Example 10.4.

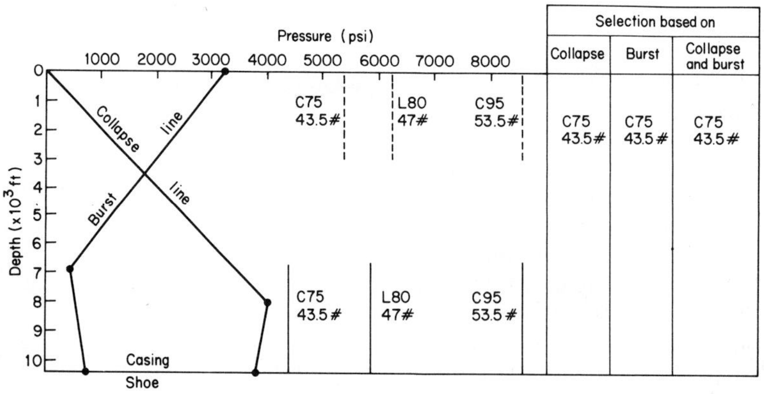

Fig. 10.17. Maximum load, $9\frac{5}{8}$ in casing design: (——) collapse lines; (-------) resistance lines.

From Figure 10.17 it can be seen that the lightest grade, C75, 43.5#, is in fact suitable for both burst and collapse. Refinement for the effects of biaxial loading, pressure testing, etc., are left as an exercise for the reader.

Comparison of the two design methods for the intermediate casing

The data of Example 10.4 were taken from an exploration well recently drilled in the Middle East. The selection of the grades was determined according to the method presented in Example 10.4. However, results of Example 10.5 clearly show that the casing has been overdesigned, since the worst possible conditions were considered.

The author believes that the assumptions of 100% empty casing and 100% gas-filled hole should be used for designing exploration wells only. For development wells where the formation pressures are well documented, the method presented in Example 10.5 should be used. This will lead to cost saving resulting from the use of lighter casing grades and less wear on the derrick, owing to handling lighter casing.

Another approach to burst design, presented by Prentice[8], uses the formation fracture pressure and maximum wellhead operating pressure as the end points of the burst line. The actual trend of the line is determined by solving two simultaneous equations. Finally, the back-up load is subtracted from the burst line to give the burst design load. This approach, however, still results in overdesign, since the surface pressure will never be allowed to approach the maximum working pressure of wellhead equipment.

References

1. API Specification 5A (1982). *Specification for Casing, Tubing, and Drillpipe.* American Petroleum Institute, Production Department.
2. API Bulletin 5C3 (1980). *Formulas and Calculations for Casing, Tubing, Drillpipe and Line Pipe Properties.* American Petroleum Institute, Production Department.
3. Mannesmann, Rohrenwerke (1980). *Tubular Goods for Oil and Gas Fields — Casing.* Mannesmann Publications.
4. API Rp 5B1 (1983). *Recommended Practice for Gauging and Inspection of Casing, Tubing and Line Pipe Threads.* American Petroleum Institute, Production Department.
5. Goins, W. C., Jr., Collings, B. J. and O'Brien, T. B. (1965, 1966). *A New Method to Tubular String Design. World Oil,* November and December 1965, January and February, 1966.
6. Vreeland, Thad., Jr. (1961). Dynamic stresses in long drillpipe strings. *Petroleum Engineer,* May, B58–B60.
7. Coates, D. F. (1970). *Rock Mechanics Principles.* Department of Energy, Mines and Resources, Canada, Mines Branch Monograph 874.
8. Prentice, C. M. (1970). Maximum load casing design. *Journal of Petroleum Technology,* 805–811.

Chapter 11

Cementing

In this chapter the following topics will be discussed:

Functions of cement
The manufacture and composition of cement
Classes and types of cement
Basic components of cement
Cement slurry
Hydration of cement
Sulphate resistance
Strength retrogression and the use of silica flour
Properties of cement slurry
Casing accessories
Cement contamination
Mechanics of cementing
Liner cementing
Practical calculations
Squeeze cementing
Plugging back operations

FUNCTIONS OF CEMENT

The functions of cement include:

(1) Restriction of fluid movement between permeable zones within the well.
(2) Provision of mechanical support for the casing string.
(3) Protection of casing from corrosion by sulphate-rich formation waters.
(4) Support for the well-bore walls (in conjunction with the casing) to prevent collapse of formations. In weak and unconsolidated formations the cement sheath must establish and maintain grain-to-grain stress loading to prevent the inflow of sand grains with the produced fluids.

THE MANUFACTURE AND COMPOSITION OF CEMENT

The basic raw materials[1,2] for cement are derived from calcareous and argillaceous rocks, such as limestone, clay and shale, and any other material containing a high percentage of calcium carbonate. The raw material may also include sand, iron ore, etc.

The dry materials are finely ground and mixed thoroughly in the correct proportions. The chemical composition of the dry mix is determined and adjusted if necessary. The dry mix is described as the 'kiln feed'.

The kiln feed is fed at a uniform rate into the upper end of a sloping rotary kiln. While the mixture travels slowly to the lower end of the kiln, powdered coal, fuel oil or gas is fired into the kiln; the temperature can reach 2600–2800 °F (1427–1538 °C). Under these conditions, chemical reactions between the raw materials take place, which result in a new material called clinker. The clinker varies in size from dust to particles of several inches in diameter. The clinker is then sent to air coolers, where it is quenched with air and put into storage[1].

After a specified period of storage time, the clinker is ground with a controlled amount of gypsum and other additives to form a new product, referred to as Portland cement.

Gypsum ($CaSO_4 \cdot 2H_2O$) is added, mainly to control the setting and hardening properties of the cement slurry. The amount of gypsum required varies between 1.5% and 3% by weight of the ground clinker.

Figure 11.1 gives a schematic drawing of the stages taken to develop a natural rock into cement and back to hard, rock-like set cement.

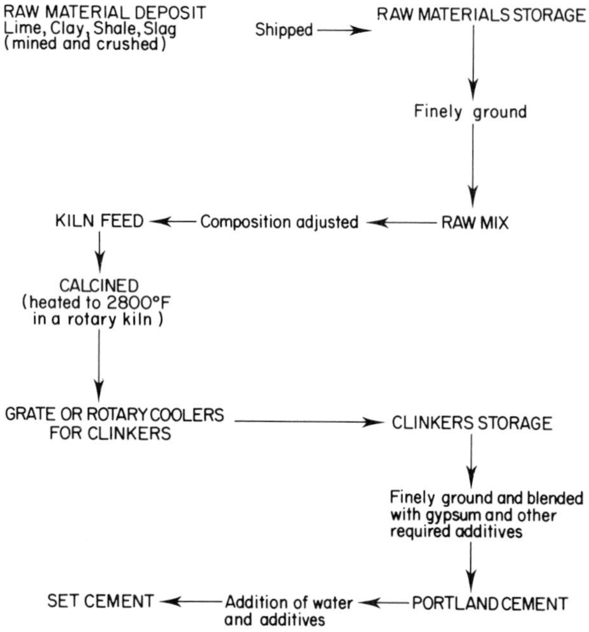

Fig. 11.1. Stages in the manufacture of cement[1].

CLASSES AND TYPES OF CEMENT

The API[3] has classified nine types of cement, depending on depth and conditions of hole to be cemented. These are as follows.

Class A. Intended for use from surface to 6000 ft (1830 m) depth when special properties are not required. Available only in ordinary type (similar to ASTM C 150, Type 1).

Class B. Intended for use from surface to 6000 ft (1830 m) depth when conditions require moderate to high sulphate resistance. Available in both moderately (similar to ASTM C 150 Type II) and highly sulphate-resistant-types.

Class C. Intended for use from surface to 6000 ft (1830 m) depth when conditions require high early strength. Available in moderately (similar to ASTM C 150, Type III) and highly sulphate-resistant types.

Class D. Intended for use from 6000 ft to 10 000 ft (1830–3050 m) depth, under conditions of moderately high temperatures and pressures. Available in both moderately and highly sulphate-resistant types.

Class E. Intended for use from 10 000 ft to 14 000 ft (3050–4270 m) depth, under conditions of high temperatures and pressures. Available in both moderately and highly sulphate-resistant types.

Class F. Intended for use from 10 000 ft to 16 000 ft (3050–4880 m) depth, under conditions of extremely high temperatures and pressures. Available in both moderately and highly sulphate-resistant types.

Class G. Intended for use as a basic cement from surface to 8000 ft (2440 m) depth as manufactured, or can be used with accelerators and retarders to cover a wide range of well depths and temperatures. No additions other than calcium sulphate or water, or both, shall be interground or blended with the clinker during manufacture of Class G cement. Available in moderately and highly sulphate-resistant types.

Class H. Intended for use as a basic cement from surface to 8000 ft (2440 m) depth as manufactured, and can be used with accelerators and retarders to cover a wide range of well depths and temperatures. No additions other than calcium sulphate or water, or both, shall be interground or blended with the clinker during manufacture of Class H cement. Available in moderately and highly (tentative) sulphate-resistant types.

Class J. Intended for use as manufactured, from 12 000 ft to 16 000 ft (3660–4880 m) depth, under conditions of extremely high temperatures and pressures, or can be used with accelerators and retarders to cover a range of well depths and temperatures. No additions of retarder, other than calcium sulphate, or water, or both, shall be interground or blended with the clinker during manufacture of Class J cement.

BASIC COMPONENTS OF CEMENT

The main components of Portland Cement are as shown in Table 11.1.

From Table 11.1 it can be seen that the main components of cement are C_3S and C_2S. The C_3S shows the fastest rate of hydration and is responsible for the overall strength characteristics and for early strength of the cement. The C_3S component is largely responsible for the protection of the set cement against sulphate attack. The C_2S is a slow-reacting component and is responsible for the gradual increase in the strength of cement. The C_3A is responsible for the initial set and early strength due to its rapid hydration.

The C_4AF contribution to the setting of cement is similar to that of C_3A, but is highly dependent on temperature and the percentage of additives used.

From Table 11.1 it is clear that the main components of cement are in an anhydrous form. The addition of water to cement converts these components to their hydrous form and the resulting liquid is described as 'cement slurry'.

HYDRATION OF CEMENT

When the cement slurry is placed around the casing it will be subjected to a high differential pressure in the direction of rock, and water is lost to the formation

TABLE 11.1 Chemical composition of cement components

Component	*Formula*	*Trade name*	*Amount*
Tricalcium silicate	$3CaO \cdot SiO_2$	C_3S	50%
Dicalcium silicate	$2CaO \cdot SiO_2$	C_2S	25%
Tricalcium aluminate	$3CaO \cdot Al_2O_3$	C_3A	10%
Tetracalcium aluminoferrite	$4CaO \cdot Al_2O_3 \cdot Fe_2O_3$	C_4AF	10%
Various other oxides such as gypsum, sulphates, magnesia, free slaked line (CaO) and any specialised additives			5%

mainly by dehydration and evaporation. While water is lost to the formation, the hydrous compounds form an interlocking crystalline structure which bonds to the casing and rock surfaces.

The actual mechanism of cement bonding is complex and depends on the type of surface to be cemented. For example, cement bonds to the rock by a process of crystal growth in which the pore space is interlocked to the main cement column by the growing crystals. Similarly, cement bonds to the outside of the casing by filling in the pit spaces in the casing body and casing collars.

SULPHATE RESISTANCE

Sulphate chemicals, including magnesium and sodium sulphates, have detrimental effects when they come into contact with the set cement. The sulphate minerals react with the lime in the cement to form magnesium and sodium hydroxide and calcium sulphate[1]. The calcium sulphate reacts with the C_3A components of cement to form sulphoaluminate, which causes expansion and disintegration of the cement. This is due largely to the fact that the sulphoaluminate particles are larger than the C_3A particles which they replace, which leads to disintegration of the set mass.

Sulphate minerals are abundant in some underground formation waters which may come into contact with the set cement in two ways: (a) through perforations; and (b) through micro-channels which may form in the set cement due to inefficient displacement of mud by cement (see below).

Resistance to sulphate attack can be increased by decreasing the amount of C_3A and free lime in the cement, or by adding a pozzolanic material. The pozzolanic material reacts with the lime, thereby reducing the percentage of lime available for reaction with magnesium sulphate.

STRENGTH RETROGRESSION AND THE USE OF SILICA FLOUR

At well-bore temperatures of less than 230 °F (110 °C) cement continues to hydrate and develop strength over a long period, which may extend from a few days to a few years, until ultimate strength is achieved. At a temperature greater than 230 °F (110 °C) the cement will develop its maximum strength during the first few weeks, and thereafter, the strength begins to decrease. This phenomenon is commonly known as 'strength retrogression'.

Strength retrogression can continue to the point of failure in some cases, and its severity increases with increasing temperature. The reason for this reduction in strength is attributed to the conversion of the main components of the set cement, known as tobermorite (calcium silicate hydrate), to dicalcium silicate hydrate at a temperature of approximately 250 °F (121 °C). The new compound has more porosity than tobermorite and is, consequently, more amenable to attack by corrosive fluids.

It has been established that the addition of 30–40% of silica flour to the dry cement prevents the formation of the undesirable dicalcium silicate[1] and will, instead, help to form monocalcium silicate hydrate (xonotlite), which is less porous and is stronger than the dicalcium silicate hydrate. Silica flour is normally used for cementing wells with a static temperature in excess of 230 °F (110 °C).

PROPERTIES OF CEMENT SLURRY

Thickening time

As soon as water is added to cement, a reaction begins between the various cement components and leads to an increase in the viscosity of the cement slurry. Thickening time tests are normally run to determine the length of time a given cement slurry remains in a fluid state under given laboratory conditions, and thus serve as a method of comparing various cements (3).

slurry is no longer pumpable. Thus, viscosity can be used as a base for defining the thickening time.

Thickening time is defined as the time required for the slurry to reach 100 Bearden units of consistency, each 100 Bearden units being equivalent to the spring deflection observed with 2080 g-cm of torque when using the weight-loaded type calibration device; see reference (3) for more detail. It is of vital importance that the thickening time be made longer than the actual cementing job, to avoid cementing surface pipes, cementing heads, drill pipe or the inside of the casing. The consequences of such a mishap are extremely costly, requiring the drilling out of cemented-up casing and remedial cementing of casing, which is usually less efficient than primary casing and is time-consuming. Therefore, an accurate estimate of the total job time should be made prior to cementing. Thickening time should then be equal to the total job time, plus a safety factor of, say, 1 h or 30 min.

Hence, thickening time is equivalent to the time required to keep cement pumpable and this can be expressed in equation form as

$$\begin{aligned}\text{thickening time} = {} & \text{mixing time} + \text{surface time}\\ & + \text{displacement time}\\ & + \text{plug release time}\\ & + \text{safety factor}\end{aligned}$$

The mixing time, T_m, is the time required to mix cement with water and additives:

$$T_m = \frac{\text{volume of dry cement}}{\text{mixing rate}} = \frac{V_c}{\text{sacks/min}}$$

where V_c = volume of dry cement (in sacks).

Surface time, T_s, is the time required for the cement slurry to be prepared at the surface and retained for testing. This time is normally small and can be included as part of the mixing time.

Displacement time, T_d. During mixing, the cement slurry is pumped inside the casing until the entire dry volume of cement is mixed. Displacement time is defined as the time required to displace cement (by mud) from inside the casing to the annulus. This time is dependent on casing capacity and the displacement rate, which is equivalent to the speed at which the pump is operated. Thus

$$T_d = \frac{\text{amount of fluid required to displace top plug}}{\text{displacement rate}}$$

Note that the volume of cement to be displaced excludes the volume inside the shoe track (as discussed later).

A safety factor of approximately 30 min to 1 h is normally used.

The thickening time of cement may be increased or decreased by the addition of special chemicals.

(1) *Cement accelerators.* Accelerators are added to decrease the thickening time of the cement slurry. In other words, an accelerator is used to speed up the gelling of cement so that it assumes the required compressive strength (usually 500 psi or 35 bars) in a short time.

The most commonly used accelerator in the oil industry is calcium chloride ($CaCl_2$). A less frequently used accelerator is sodium chloride (NaCl). These salts accelerate the setting time of cement by increasing the ionic character of the aqueous phase. This causes the main components of cement, C_3S, C_2S and C_3A, to hydrate and release calcium hydroxide more quickly. Calcium silicate hydrates can then rapidly form, which give rise to an early strength for the cement. Accelerated cement slurries are used to cement surface casing strings where waiting-on-cementing (WOC) must be kept to a minimum.

(2) *Retarders* are normally added to delay the setting of cement slurry in order to allow sufficient time for correct slurry emplacement[11]. Most retarders are organic in nature and are available in both powdered and liquid forms. Retarders include: (a) lignosulphonate; (b) cellulose derivatives; and (c) sugar derivatives and organics.

Slurry density

The specific gravity of cement is 3.14 and that of fresh water is 1. The specific gravity of a mixture of cement and water (sometimes referred to as neat cement) can be determined as follows:

slurry density

$$= \frac{\text{mass of dry cement} + \text{mass of mix water}}{\text{volume of cement} + \text{volume of water}}$$

If 100 kg of cement is mixed with 50 l of water, then the slurry density, ρ_s, is

$$\rho_s = \frac{100 + 50}{\frac{100}{(3140)} + 50 \times 10^{-3}} = 1833\ \text{kg/m}^3$$

where

$$\text{volume of 100 kg of cement} = \frac{100}{3140}\ \text{m}^3$$

$$\begin{aligned}\text{mass of 50 litres of water} &= \text{density} \times \text{volume}\\ &= 1000\ \text{kg/m}^3 \times 50\ \text{l}\\ &\quad \times \frac{10^{-3}\ \text{m}^3}{1} = 50\ \text{kg}\end{aligned}$$

$$\text{density of cement} = 3140\ \text{kg/m}^3$$

and

$$\text{density of water} = 1000\ \text{kg/m}^3$$

Hence, the resulting specific gravity of the slurry is 1.833 (1833/1000).

Higher slurry densities can be obtained by decreasing the volume of mix water or adding materials of high specific gravity. Low slurry densities are sometimes required to reduce the dangers of lost circulation, which can be caused by the excessive hydrostatic pressure of the cement column. Low slurry densities are prepared by using additives of low specific gravities. A cement slurry which has no additives is described as 'neat slurry'.

Various additives are available for either reducing or increasing slurry density.

Density-reducing additives

All density-reducing additives result in a set cement of lower strength than can be obtained from a neat slurry. It must be appreciated that water is the main density-reducing additive and that materials such as bentonite, etc., help carry the extra water needed to lighten the slurry.

Density-reducing additives include the following.

Bentonite. Bentonite, having a specific gravity of 2.65, is widely used as an additive for reducing slurry density, mainly because of its high requirement for water. With Class G-based slurries, the density can be reduced from 15.8 ppg (1.89 kg/l) to 12.6 ppg (1.51 kg/l) by the simple addition of 12% bentonite.

Diatomaceous earth. Addition of this material has the effect of reducing the slurry density by increasing the water requirement. The diatomaceous materials are characterised by their very high surface area.

Gilsonite. This is a light inert asphaltic mineral, having a low specific gravity of 1.07, which, when mixed with cement, results in a mix of low specific gravity. Owing to its bridging properties, this material is also used for combating lost circulation problems. Gilsonite has the characteristic property of adding volume and not weight to the cement slurry.

Pozzolan. This is a siliceous material of volcanic origin having a specific gravity of 2.5. When this material is mixed with Portland cement in a ratio of 50:50, and with 2% of bentonite, it will produce a slurry with a specific gravity of 1.6. Pozzolan uses more water than cement, thereby giving a lower final slurry density. This also results in a cost saving, since pozzolan is less expensive than cement. Pozzolan has the advantage that it can react with the calcium hydroxide, producing a set cement which is resistant to leaching. It should be borne in mind that calcium hydroxide is the main product of the reaction between water and cement compounds.

Density-raising additives

Density-raising additives include the following.

Barite has a specific gravity of 4.25 and increases the slurry density by virtue of its high density.

Ilmenite is an iron–titanium oxide with a specific gravity of 4.6 and is used when high slurry densities are required.

Hematite has a specific gravity of 5.02 and is normally preferred to ilmenite, because of its low water requirements and less adverse effects on other slurry properties. It also results in a set cement of high compressive strength.

Cement strength

Cement strength refers to the strength of the set cement and can be specified as tensile or compressive strength. Compressive strength is by far the most widely used index for quantifying the cement strength. Cement having a compressive strength value of 500 psi (35 bars) is normally considered adequate for most operations. The compressive strength is dependent upon the percentage of water and the curing time.

Prior to running the casing string, a sample of the proposed cement mixture is mixed with water under laboratory conditions equivalent to the expected downhole temperature and pressure. The slurry is poured in suitably sized cubes and left to dry for a specified period of time. The set cement cubes are then broken in a compression testing machine to determine

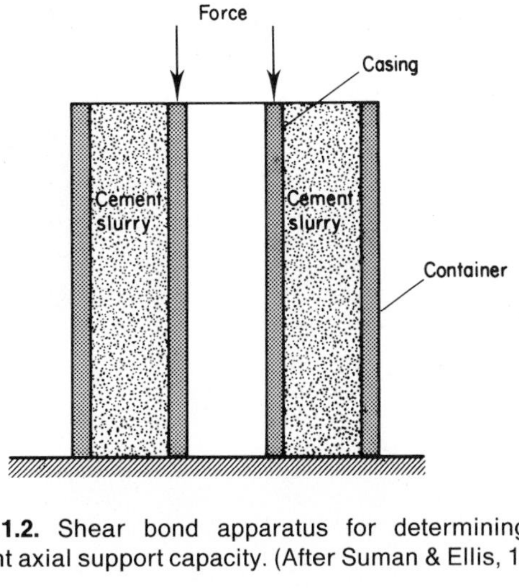

Fig. 11.2. Shear bond apparatus for determining the cement axial support capacity. (After Suman & Ellis, 1977[4])

the compressive strength of the samples. The average value of four or five samples is taken as the compressive strength of the set cement.

The strength of cement can also be determined from shear bond tests[4]. The cement column shown in the test apparatus in Figure 11.2 is loaded axially through the inner casing until cement fractures. The fracture load, divided by the surface area between cement and pipe, is taken as the shear bond strength[4]. The shear bond strength was found to increase with increasing tensile or compressive strength, but decreases significantly when the pipe surface is mud-wetted[4].

The support capability of a cement sheath can be determined from the following formula:

$$F = 0.969\ S_c\ d\ H \qquad (11.1)$$

where F = support capability or fracture load (lb);

Regular Pattern
Guide Shoe

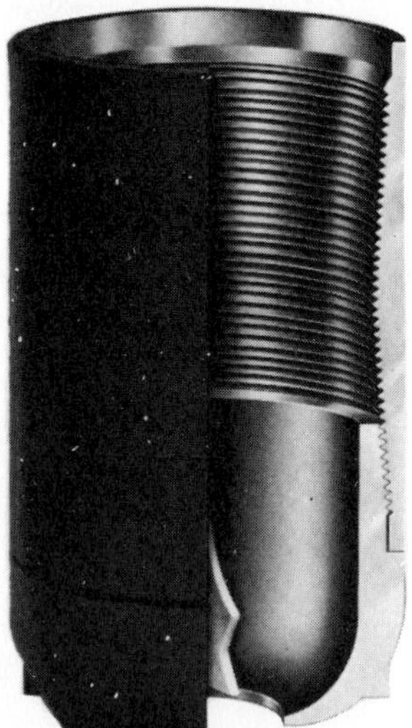

Swirl Guide Shoe —
Aluminum Type

Fig. 11.3. Guide shoes: (above) regular pattern guide shoe; (below) swirl guide shoe — aluminium type. (Courtesy of Dowell Schlumberger)

S_c = compressive strength of cement (psi); d = outside diameter of casing (in); and H = height of cement column.

Example 11.1

Determine the height of cement to be placed around 9000 ft of $13\frac{3}{8}$ in casing to support a total weight of 2 000 000 lb. Assume that the cement has a compressive strength of 500 psi.

Solution

From Equation (11.1),

$$H = \frac{F}{0.969\, S_c\, d} = \frac{2\,000\,000}{0.969 \times 500 \times 9.625} = 429 \text{ ft}$$

CASING ACCESSORIES

The following mechanical equipment is required during the cementing of a casing string: (a) guide shoes; (b) float collars; (c) stage cementer or D.V. tool; (d) centralisers and scratchers and; (e) cement plugs.

Guide shoes

A guide shoe is used to guide the casing through the hole. A guide shoe can be a simple sub, as shown in Figure 11.3, or may contain a ball valve or a flapper valve, as shown in Figure 11.4.

When a guide shoe contains a valve element, it is referred to as a 'float shoe'. A float shoe prevents cement from flowing back into the casing once it is displaced behind the casing.

Float collars

A float collar is basically a one-way valve placed at one or two joints above the shoe. It is used to provide the following functions:

(a) To prevent mud from entering the casing while it is being lowered inside the hole thereby allowing the casing to float during its descent and resulting in a reduction in loading of the derrick.

(b) To prevent cement from backflowing after it has been displaced behind the casing.

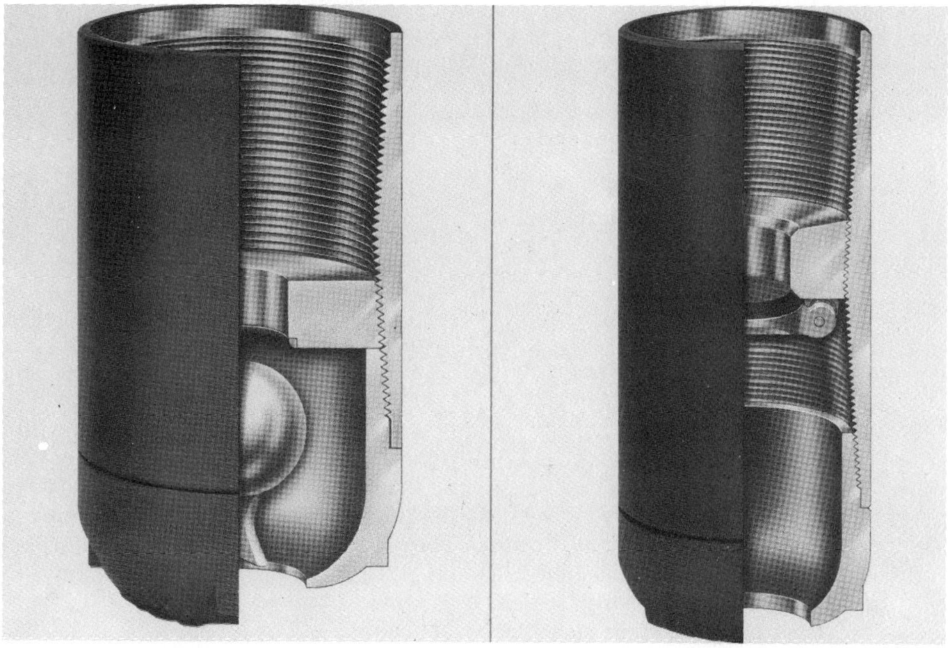

(a) (b)

Fig. 11.4. (a) Ball float shoe — aluminium type; (b) flapper float shoe — aluminium type. (Courtesy of Dowell Schlumberger)

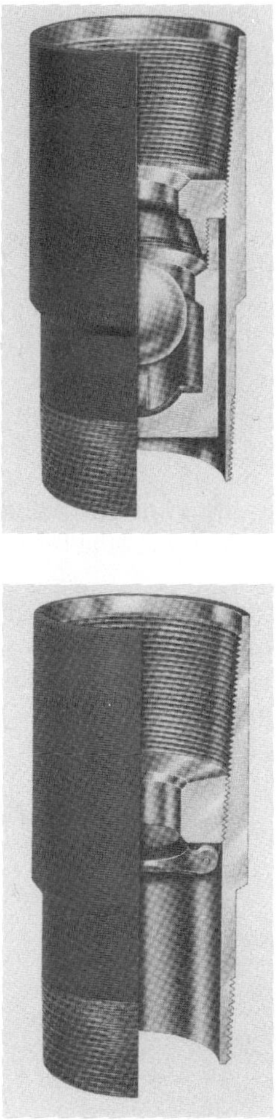

(a)

(b)

Fig. 11.5. (a) Ball float collar — aluminium type; (b) flapper float collar — aluminium type. (Courtesy of Dowell Schlumberger)

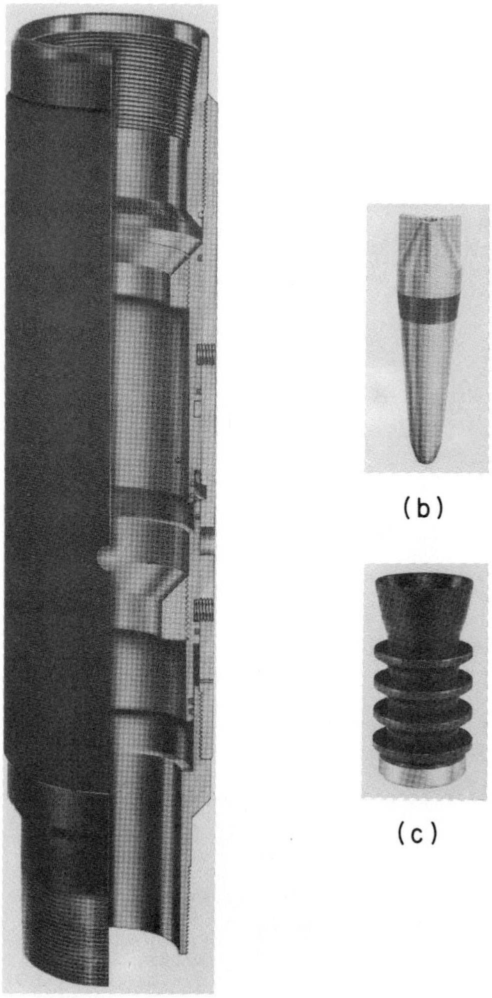

(a) (b) (c)

Fig. 11.6. Stage collar (a), with opening bomb (b) and closing plug (c). (Courtesy of Dowell Schlumberger)

(c) To act, if required, as a landing for the cementing wiper plugs.

Typical float collars are shown in Figure 11.5.

Stage cementer (or D.V. tool)

A stage cementer is used when cementing long strings of casing to prevent weak formations being subjected to excessive hydrostatic pressures of long cement columns. Figure 11.6 shows a typical stage cementer, together with opening and closing bombs. A stage cementer is manufactured from the same grade of steel as that used for the string of casing. It is equipped with two sleeves — an upper sleeve and a lower sleeve. Both are held on the inside of the collar by means of shear pins. The collar is also equipped with ports bored right through its body. The ports are covered initially by the lower sleeve and can be uncovered by pushing the lower sleeve downward, with an opening bomb. The ports can be covered again by dropping a closing plug which pushes the upper sleeve to its closing position.

Scratchers and centralisers

Drilling mud serves one of its main functions of preventing formation fluids from entering the well bore by forming a mud cake on the wall of the hole.

This mud cake, however, prevents cement from contacting the borehole surface, which results in an inferior cement job. Scratchers are used to help remove this mud cake by means of abrasion, brought about by rotation or reciprocation of the casing string. Figure 11.7 shows typical scratchers used in oil-well cementing. Scratchers are placed on the outside of the casing, the number used being dependent on hole conditions.

Practical experience has shown that mud displacement is greatly improved when the casing is centralised. If a casing string is not centralised, the cement slurry tends to by-pass the drilling mud, leaving uncemented channels around the casing. A centraliser

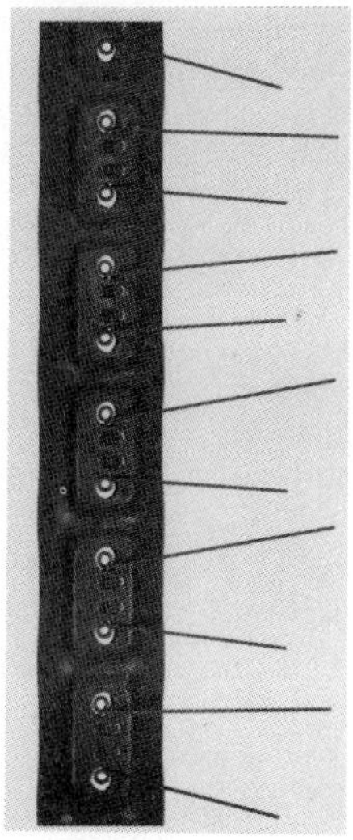

(a)

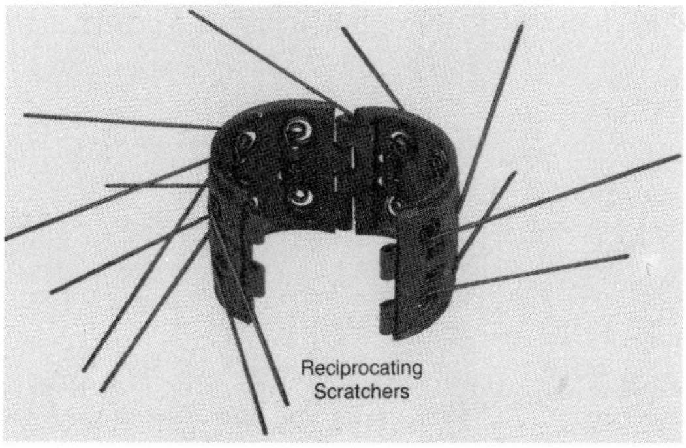

(b)

Fig. 11.7. (a) Rotating scratchers; (b) reciprocating scratchers. (Courtesy of Dowell Schlumberger)

Fig. 11.8. Centraliser. (Courtesy of Dowell Schlumberger)

(as illustrated in Figure 11.8) is a device used to hold the casing off the walls of the borehole, thus allowing a more uniform cement sheath to be placed around the casing. Centralisers are most effective in near-gauge zones of the hole and are normally placed against producing zones. Centralisers are produced in different sizes to match the various hole sizes and are designed to provide ample clearance for fluid passage.

The number of centralisers required varies according to the degree of hole deviation and length and number of zones, as detailed in API specification 10D[5]. Special arrangements of centralisers, scratchers and stop collars are often employed on casing joints to enhance cementing in regions where good isolation is paramount. These arrangements are variously known as cleavage barriers, post plugs etc, as shown in Figure 11.9.

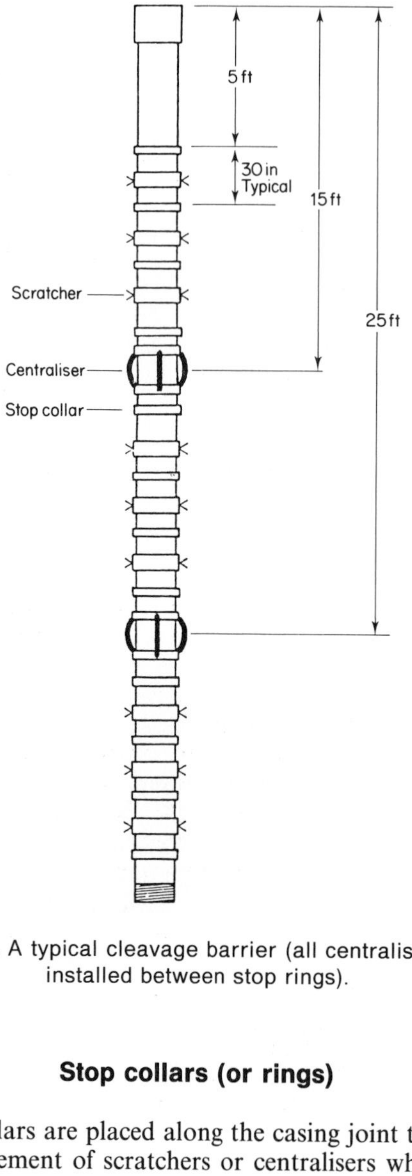

Fig. 11.9. A typical cleavage barrier (all centralisers are installed between stop rings).

Stop collars (or rings)

Stop collars are placed along the casing joint to limit the movement of scratchers or centralisers while the casing is run or reciprocated (Figure 11.10).

Wiper plugs

Wiper plugs are used to limit the contamination of cement by mud and are run inside the casing, ahead of and behind the cement slurry. A wiper plug is constructed of cast aluminium with a rubber moulded body[1] so that it can be easily drilled. The rubber is formed in the shape of fins in order to wipe the inside of the casing free of mud or cement. Wiper plugs are classified as bottom and top plugs.

The bottom wiper plug (Figure 11.11a) precedes the cement down the casing and wipes off the mud film from the inside of the casing. The bottom plug is designed to have a hollow central core covered with a diaphragm. The bottom plug is released from the surface prior to the arrival of the cement slurry. The advancing cement slurry will push down the bottom wiper plug until the latter is seated on the float collar, or in a special landing collar inside a casing coupling. The seating of the plug is indicated at the surface by a sudden increase in pumping pressure. Further increase

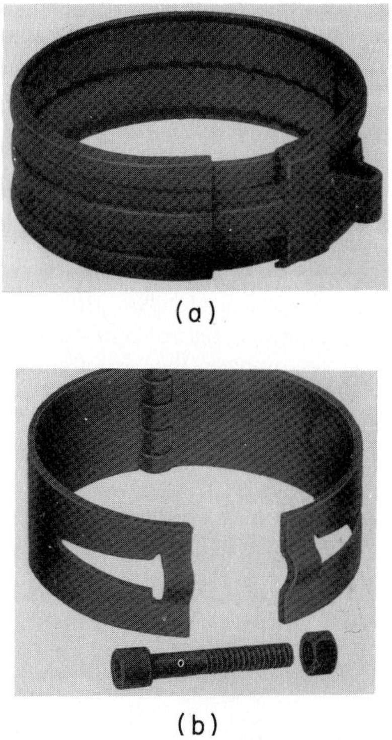

Fig. 11.10. (a) Hammer stop rings; (b) hinge-type stop rings. (Courtesy of Dowell Schlumberger)

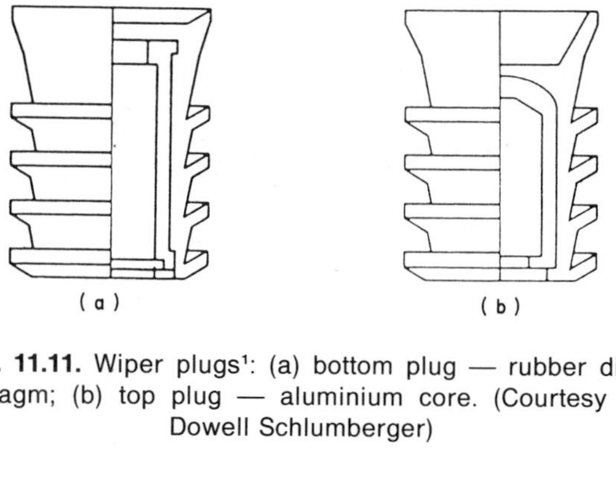

Fig. 11.11. Wiper plugs[1]: (a) bottom plug — rubber diaphragm; (b) top plug — aluminium core. (Courtesy of Dowell Schlumberger)

in surface pressure will rupture the diaphragm in the bottom plug, allowing cement to pass through and eventually be placed around the casing.

When the complete volume of cement slurry is pumped down the casing, a top plug (Figure 11.11b) is released from the surface by simply dropping it down the casing, or by unscrewing it from the cementing head. The top plug separates cement from mud and is used to wipe off cement from the inside of the casing.

When the entire volume of cement is displaced behind the casing, the top plug reaches the float collar, where it seats on top of the bottom plug. At this position a sudden increase in surface pressure is observed. A moderate back-pressure is normally kept on the top plug and is only released if there is no cement back-flow from the annulus. If there is a back-flow, the pressure is maintained on the inside of the casing until the cement assumes a hardness of 500 psi.

The seating positions of top and bottom plugs are shown in Figure 11.12.

CEMENT CONTAMINATION

The contamination of cement slurry by mud has adverse effects on the thickening time of cements. This, in turn, leads to unpredictable setting properties of cement. Contamination usually leads to loss of strength and, in extreme situations, to unpumpable mixtures.

Contamination usually takes place when cement contacts mud during emplacement of cement behind the casing. Owing to the difference in the physical properties of mud and cement, the interface between the two liquids is not stable and changes with time. As shown in Figure 11.13, small volumes of cement slurry separate from the leading edge of the main stream and diffuse into the drilling mud.

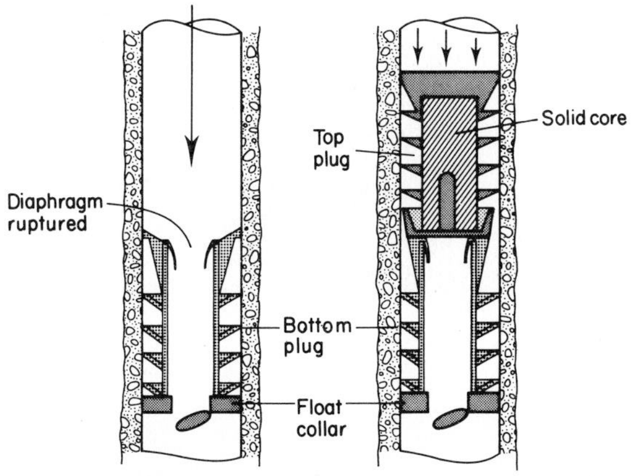

Fig. 11.12. Seating positions of top and bottom plugs[1]. (Courtesy of Dowell Schlumberger)

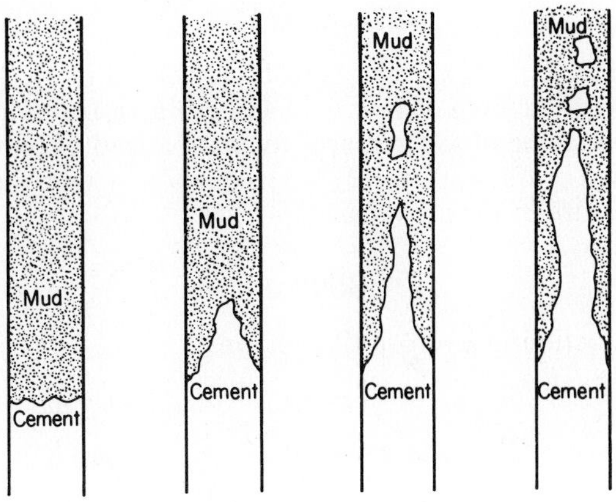

Fig. 11.13. Cement contamination by mud during displacement.

The degree of cement diffusion into mud is dependent largely on pipe centralisation within the hole, type of flow and on density differences. In an off-centred casing, cement by-passes mud on the wide side of the annulus, where resistance to flow is minimal. The mud channels left behind represent passages along which communication between the various zones can take place.

When the casing is severely off-centre, rotation appears to be more beneficial in reducing cement contamination, while reciprocation is more suited to a well-centred pipe. Reciprocation is also more efficient when a water spacer is used to displace mud ahead of cement.

Spacers and washers

Contamination of cement by mud can be greatly reduced by placing a spacer between mud and cement columns. Water is by far the most widely used spacer, because of its low viscosity which allows turbulent flow to take place at much lower pressures than would occur with a fluid of a higher viscosity. Water has also a low density, which allows it to channel through the mud and, together with turbulence, helps break up the gel structure of mud. A chemical wash is a mud-thinning agent dispersed in water. It is designed[1] to thin and disperse mud thereby allowing efficient removal of mud from the hole.

Example 11.2

A 10 000 ft, 7 in casing is to be cemented in an $8\frac{1}{2}$ in hole. If the mud density is 13.4 ppg (1.6 kg/l), calculate the volume of water spacer required ahead of cement which causes a 300 psi (21 bar) reduction in hydrostatic pressure in the annulus.

Solution

Reduction in hydrostatic pressure

$$=(\rho_{mud}-\rho_{water})\,gh$$

$$300 \text{ psi}=0.052\,(\rho_{mud}-\rho_{water})\,h$$

(where ρ_{mud} and ρ_{water} are in ppg)

$$=0.052\,(13.4-8.33)\times h$$

$$h=1138 \text{ ft}$$

(*Note*: the maximum reduction in hydrostatic pressure of the annulus occurs when the full volume of water fills the annulus)

$$\text{Annular area}=\frac{\pi}{4}(8.5^2-7^2)\text{ in}^2 \times\left(\frac{\text{ft}^2}{144 \text{ in}^2}\right)$$

$$\text{Annular capacity per foot}=0.1268 \text{ ft}^2\times 1 \text{ ft}$$

$$=0.1268 \text{ ft}^3/\text{ft}$$

Therefore,

$$\text{volume of water required}=0.1268\times 1138$$

$$=144.31 \text{ ft}^3$$

$$=\frac{144.31 \text{ ft}^3}{5.62\,\dfrac{\text{ft}^3}{\text{bbl}}}=25.7 \text{ bbl}$$

MECHANICS OF CEMENTING

Preparation and pumping of the cement slurry

Depending on hole depth and expected bottom hole temperature, chemical additives are added to control the setting properties of the slurry.

The cement slurry is prepared by vigorously mixing dry cement with a water jet. The resulting mixture is directed to a tank, where it is tested for density and viscosity. The cement slurry is then picked up by a powerful triplex pump and pumped at high pressure into the casing via the cementing head.

The cementing head shown in Figure 11.14 connects the top of the casing with the pumping unit. It contains two retainer valves for retaining the bottom and top wiper plugs. It is also equipped with a manifold that can be connected to the cement pumping unit or the rig pump. The cementing head also

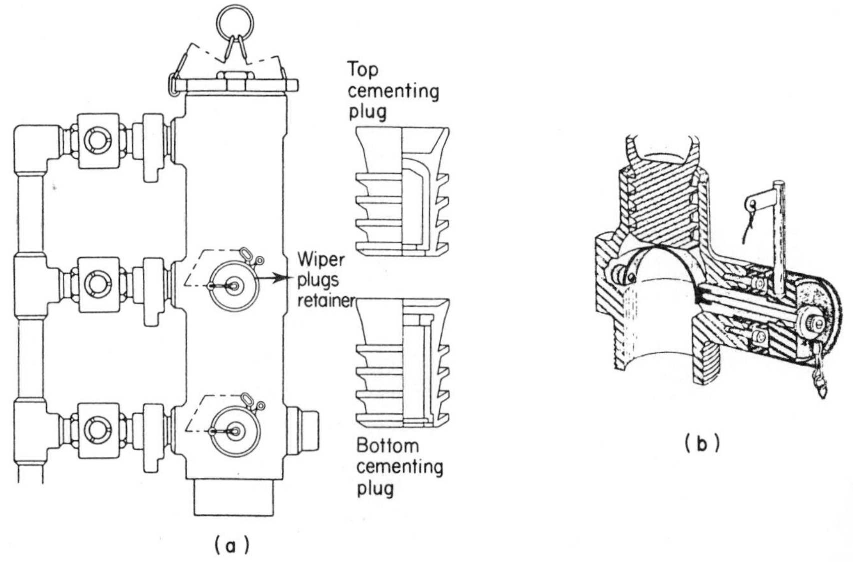

Fig. 11.14. (a) Double-plug cement head; (b) section through a retainer valve. (Courtesy of Dowell Schlumberger)

contains a quick-change cap that can be removed when it is desired to drop the wiper plugs by hand instead of by use of the retainer valves. Two types of cementing head are currently in use: (1) a single plug head containing one retainer valve; and (2) a double plug head containing two retainer valves (as shown in Figure 11.14). The retainer valve is simply a bow spring connected to a lever as shown in Figure 11.14. The plug rests on the convex side of the bow and a quarter-turn of the lever allows the plug to drop inside the casing.

The cementing operation proceeds by opening the bottom wiper plug retainer valve and directing the cement slurry through the top valve. The slurry will then push the bottom plug down the casing until the plug seats on the float collar. Continued pumping ruptures the central diaphragm in the plug which allows the cement to pass through and eventually be placed around the casing. When the complete volume of cement is mixed, pumping is stopped and the top wiper plug is placed in the cementing head. Drilling mud is then pumped through the top valve, which pushes the top wiper plug down the casing. When the top plug seats on the bottom plug, the well is shut in and the cement slurry is left to set.

Methods of cementing

Single-stage cementing

Single-stage cementing is normally used to cement conductor and surface pipes. A single batch of cement is prepared and pumped down the casing. The mechanics of single-stage cementing are shown in Figure 11.15 and are discussed above.

It should be noted that all the internal parts of the casing tools, including the float shoe, wiper plugs, etc., are easily drillable.

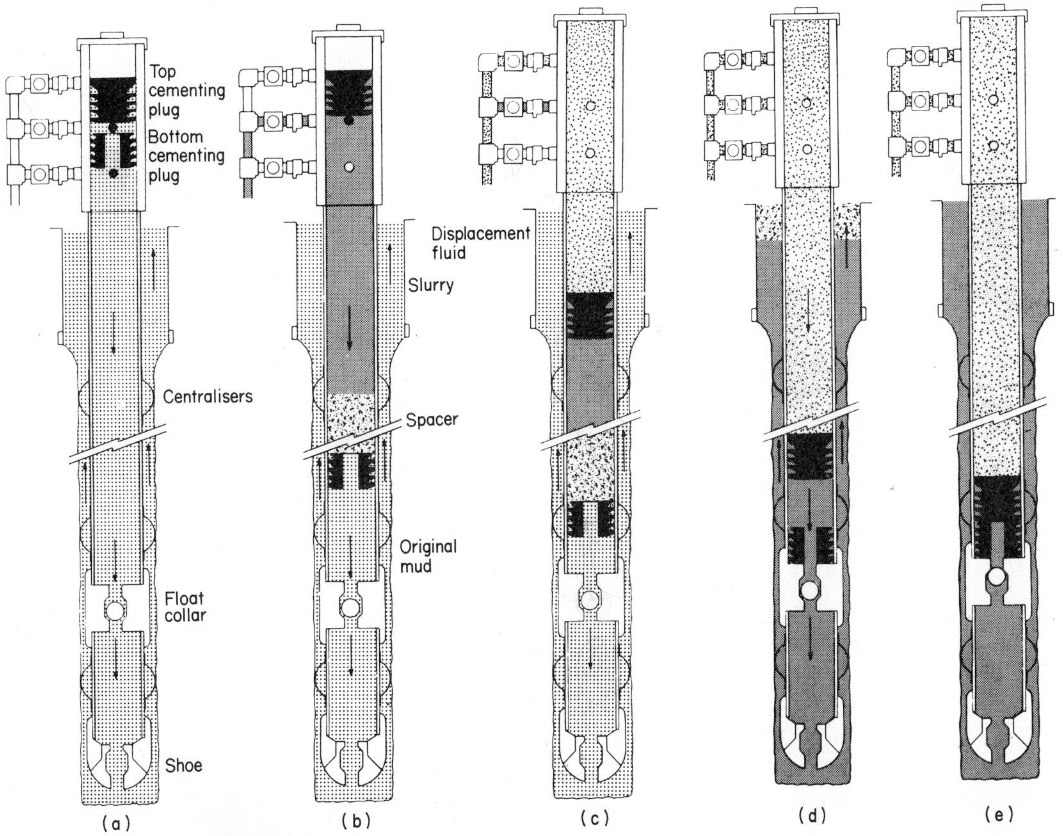

Fig. 11.15. Single-stage cementing[1]: (a) circulating mud; (b) pumping spacer and slurry; (c) and (d) displacing; (e) end of job. ●, plug-releasing pin in; ○, plug-releasing pin out. (Courtesy of Dowell Schlumberger)

Multi-stage cementing

Multi-stage cementing is employed in cementing long casing strings in order to: (a) reduce the total pumping pressure; (b) reduce the total hydrostatic pressure on weak formations, thereby preventing their fracture; (c) allow of selective cementing of formations; (d) allow the entire length of casing to be cemented; and (e) ensure effective cementation around the shoe of the previous casing string.

In multi-stage cementing a stage cementer is installed at a selected position in the casing string. The position of the stage cementer is dictated by the total length of the cement column and the strength of formations.

For a two-stage cementing job, a one-stage cementer is used in the casing string. The casing is run to the bottom of the hole. The casing is then circulated with a volume equal to twice the hole capacity. This is done primarily to ensure that the floating equipment, casing shoe and float collar are functioning properly and are free of foreign objects. This operation also allows circulation pressure to be established.

The first stage of cementing is as for a single-stage operation, but the top of the cement column ends just below the stage cementer.

The second stage is commenced by dropping an opening bomb from the surface and allowing it to gravitate to the opening seat in the stage collar. When the bomb is seated, a pumping pressure of 1200–1500 psi above the circulation pressure is applied to shear the retaining pins and allow a bottom sleeve to move downward. The movement of the sleeve uncovers ports, thereby establishing communication between the inside of the casing and the annulus. Mud is then circulated to condition the well prior to

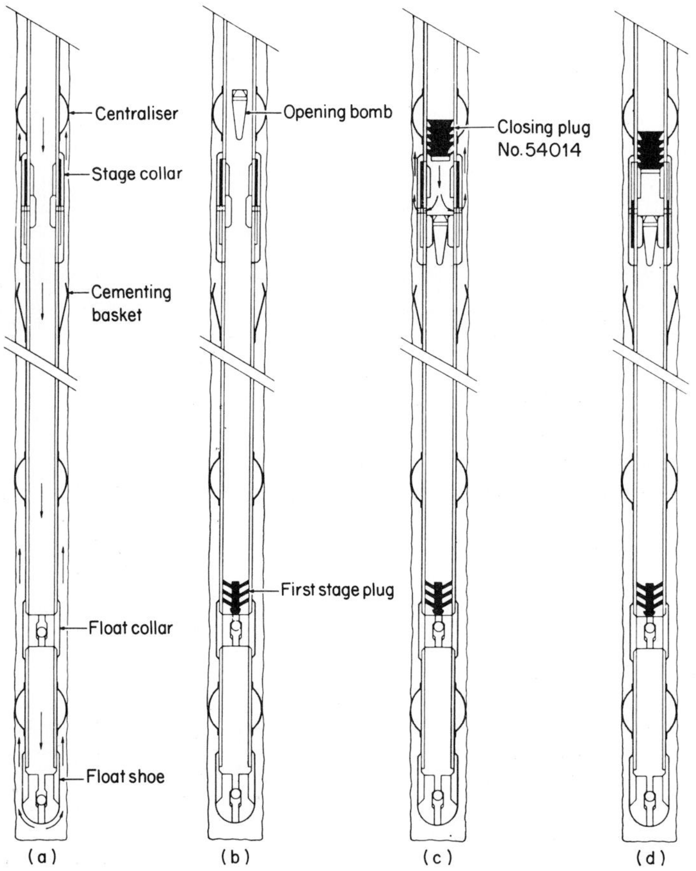

Fig. 11.16. Normal two-stage cementing[1]: (a) circulating well prior to cementing; (b) first-stage cement displaced and opening bomb approaching stage collar; (c) second-stage cement being displaced and closing plug approaching stage collar; (d) cementing operation is complete and stage collar closed. (Courtesy of Dowell Schlumberger)

commencement of the second stage. During the cementation period, the cement slurry from the first stage will have begun to set and assume an early strength.

The required cement volume for the second stage is then pumped in and followed by a closing plug. The cement slurry passes through the ports of the stage cementer and is eventually placed in the annular area above. When the plug reaches the stage cementer a pressure of 1500 psi above the pressure required to circulate the cement is then applied on the closing plug. Under this pressure the closing plug pushes the upper sleeve downward so that it covers the ports and prevents communication between the inside of the casing and the annulus. Thus, the entire casing string is now cemented.

The two-stage cementing and the operation of the stage cementer are schematically shown in Figure 11.16.

In three-stage cementing two stage cementers are used to allow the casing to be cemented in three stages. The first cementing stage is carried out through the casing shoe, the second stage through the first stage collar, while the third stage is carried out through the second stage cementer. A three-stage cementing operation is shown in Figure 11.17.

LINER CEMENTING

As outlined in Chapter 10, a liner is a short string of casing which does not reach the surface. It is hung from the bottom of the previous casing string, by use of a liner hanger. A liner is run on drillpipe and cemented by pumping the cement slurry through the drillpipe and liner and, finally, displacing it behind the liner to just above the liner hanger. In practice, liner cementing represents one of the most difficult cementing operations, as leaks are often found across the hanger or across the zones covered by the liner.

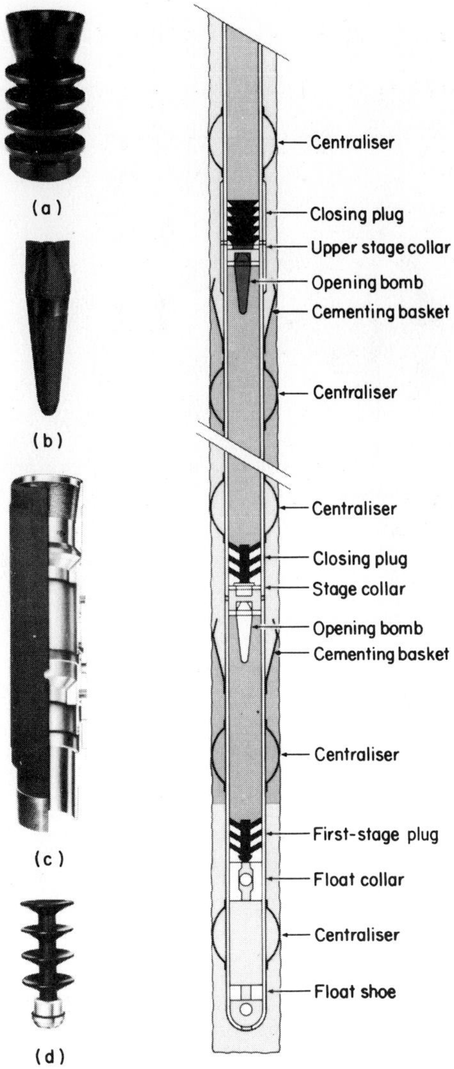

Fig. 11.17. Three-stage cementing[1]. (a) Closing plug; (b) opening bomb; (c) upper stage collar; (d) special closing plug (used to close bottom stage collar). (Courtesy of Dowell Schlumberger)

Liner equipment

Figure 11.18 gives a schematic drawing of liner equipment[4]. This equipment includes the following items.

Float shoe. As stated previously, a float shoe is required to guide the liner into the hole and also to act as a back-pressure valve.

Float collar. This is an optional piece of equipment and is used as a back-up for the float shoe.

Landing collar. This is a short sub placed at two joints above the casing shoe; it provides a seat for the liner wiper plug. The position of the loading collar is selected so that mud-contaminated cement is contained within the two casing joints above the shoe and only neat cement is placed around the liner shoe.

Liner. This is the string of casing used to case off the open hole. The liner strength properties are chosen by the graphical method of casing design described in Chapter 10. The length of overlap between the liner and the previous casing varies between 200 and 500 ft.

Liner hanger. The liner is hung on the previous casing string by means of a hanger installed at the top

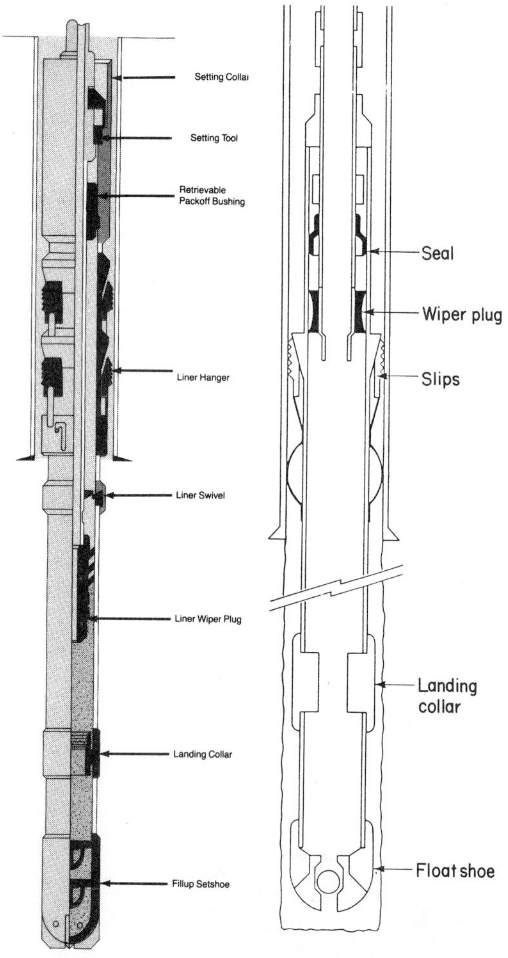

Fig. 11.18. Liner assembly (left, courtesy of Texas Ironworks) and setting tool and hanger assembly[1]. (Courtesy of Dowell Schlumberger)

of the liner. A liner hanger employs a set of slips to engage the previous casing. Liner hangers are classified, according to the method used to activate the slips, as hydraulic or mechanical hangers.

Hydraulic liner hangers are set by means of fluid pressure applied from the surface through the connecting drillpipe. The liner assembly is equipped with a ball-catcher sub placed immediately below the landing collar. To set the liner, the setting ball is dropped from the surface where it seats on the catcher sub, thereby closing one end of the drillpipe-liner assembly. Fluid pressure is then applied from the surface via the drillpipe to the liner hanger. The fluid pressure is directed to an annular piston inside the body of the hanger, which drives the slips and wedges them between the static cones (on the hanger body) and the casing body. Figure 11.19 shows a typical hydraulically set liner hanger.

Mechanically set liner hangers are provided with a J-latch which holds the slips in a retracted position, as shown in Figure 11.20. The J-latch is held by a pin welded to the body of the liner hanger. Upon reaching the hole bottom[6], the running string is raised a few inches to bring the pin (or key) to its top position (Figure 11.20) and then turned to the right to unlatch

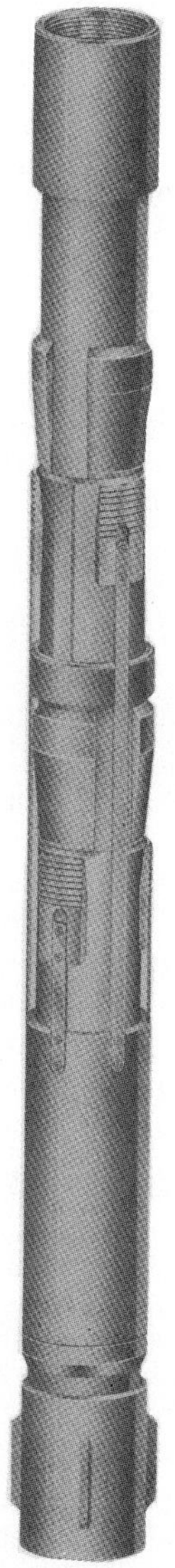

Fig. 11.19. Hydraulic set liner hanger[6]. (Courtesy of Brown-Hughes)

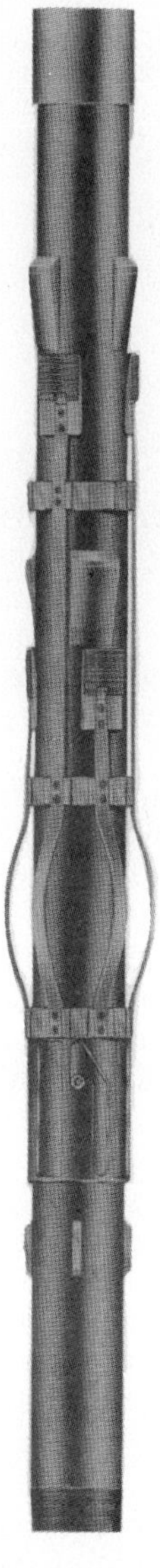

Fig. 11.20. Mechanical set liner hanger[6]. (Courtesy of Brown-Hughes)

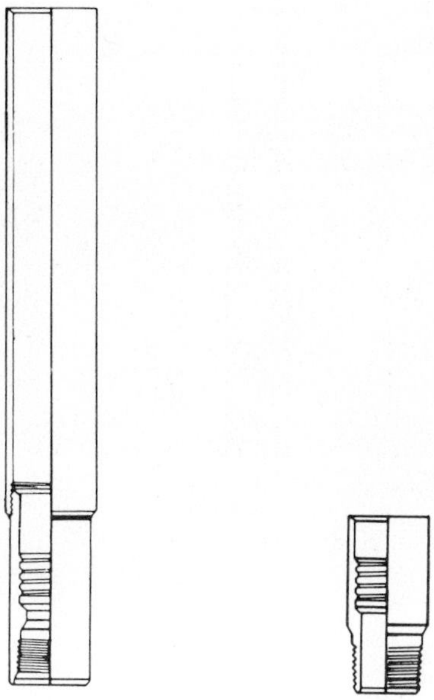

Fig. 11.21. (a) Liner setting sleeve with 6 ft extension[6], (b) standard liner setting sleeve-pin end[6]. (Courtesy of Brown-Hughes)

the J-slot and release the slips. The liner string is then lowered, to transfer the hanger weight to the slips. The weight is transferred via the key to the centraliser, which, in turn, wedges the slips between the cones and the casing body.

Liner setting sleeve and liner setting tool. The liner setting sleeve is used to carry the liner to depth and is provided with a thread to match the liner setting tool thread (Figure 11.21). The liner setting tool (Figure 11.22) connects the liner to the running-in string. It is used to run the liner and to set the liner hanger. It also provides, if required, attachment for the liner wiper plug. The setting tool can be released from the liner setting sleeve by right-rotation[6]. To prevent upward flow around the setting tool during cementing operations, the setting tool is provided with a sealing system such as a slick stinger and a pack-off bushing or inverted swab cups.

Swivel assembly and liner wiper plug. A swivel assembly is installed between the liner and the liner hanger to enable the hanger to be set even if the liner below the hanger is stuck. The liner wiper plug is used to separate the displacing mud from cement and also to clean the inside of the liner. It can be installed in the swivel assembly or can be shear-pinned to the bottom of the setting tool. Figure 11.23(a) shows a liner wiper plug.

Pump down plug. This is a solid plug (Figure 11.23b) used to clean the running-in string from cement. The plug is contained in a surface dropping head connected to the top of the drill pipe. The plug is run immediately behind the cement to latch to the liner wiper plug.

Cementing a liner

With the liner at its desired setting depth, the liner and hole are circulated to condition the mud and to ensure that circulation is possible before the liner is hung. The

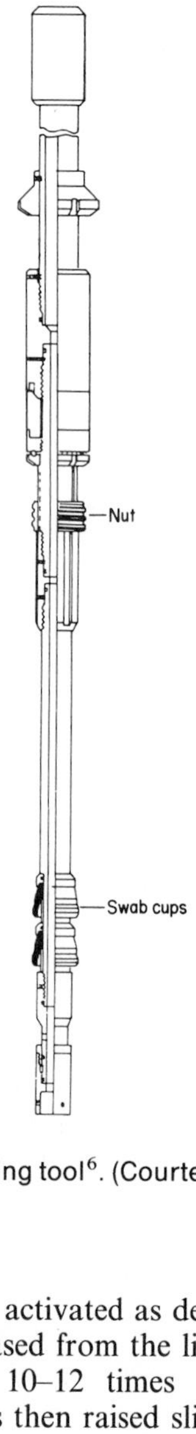

Fig. 11.22. Liner setting tool[6]. (Courtesy of Brown-Hughes)

liner slips are then activated as described above. The setting tool is released from the liner by rotating the running-in string 10–12 times to the right. The running-in string is then raised slightly to check that the setting tool is released from the liner. Disengagement is observed at the surface as a decrease in the hook load, due to weight loss of the liner. With the setting tool still inside the liner, the cementing operation can now begin.

Figure 11.24 shows a conventional liner cementing method. Cement is prepared and pumped through the running-in string, the setting tool, the liner hanger assembly and the liner, to come out of the liner shoe. When the complete volume of cement is pumped, the pump-down plug is released from the surface and is followed by the drilling mud. When the pump-down plug connects to the wiper plug, a sudden increase in pumping pressure is observed. Further increase in

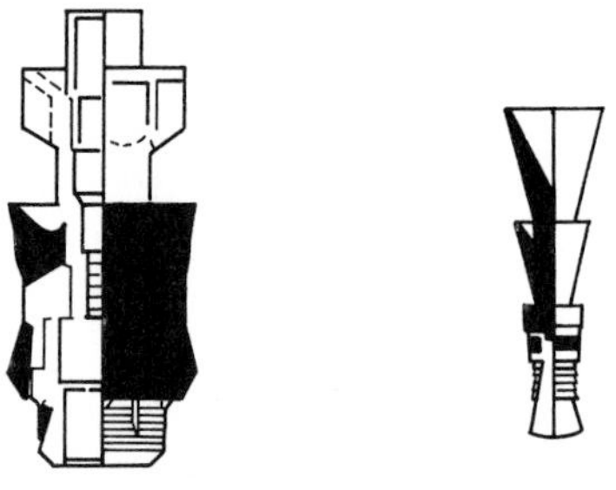

(a) (b)

Fig. 11.23. (a) Liner wiper plug; (b) Liner pump down plug[6]. (Courtesy of Brown-Hughes)

pressure shears the pins connecting the wiper plug to the setting tool, to allow them to pass down the liner as a single assembly.

The wiper plug then displaces cement around the liner until it seats on the landing collar. This action is shown at the surface by a sudden increase in pumping pressure and indicates that the total volume of cement has been placed around the liner.

The liner hanger slips are then activated as described above. The setting tool is released from the liner by rotating the running-in string 10–12 times to the right. The string is then picked up to completely disengage the setting tool from the liner assembly. Disengagement is observed at the surface by a decrease in the hanging weight, due to weight loss of the liner.

Excess cement inside the running string can be reversed out by raising the setting tool above the liner hanger and circulating through the annulus.

The success of liner cementing is directly related to the effectiveness of mud removal by cement. Centralising the casing is an important factor in proper placement of cement. Where multiple zone production or injection is practised, it has been found that running cleavage barriers opposite the producing /injection zones particularly provides good cement seal between the zones.

Despite all precautions, liner cementing rarely results in a satisfactory job, and leaks are found, particularly near the liner top and between the producing/ injection zones.

Problems in liner cementing can be attributed to[7]: (a) lack of adequate annular clearance; (b) temperature differential between top and bottom of liner; (c) inability to move liner while cementing; (d) inability to pump cement at high rates (thus establishing turbulent flow) due to high circulation pressures or to protect formations from breaking down; and (e) formation gas

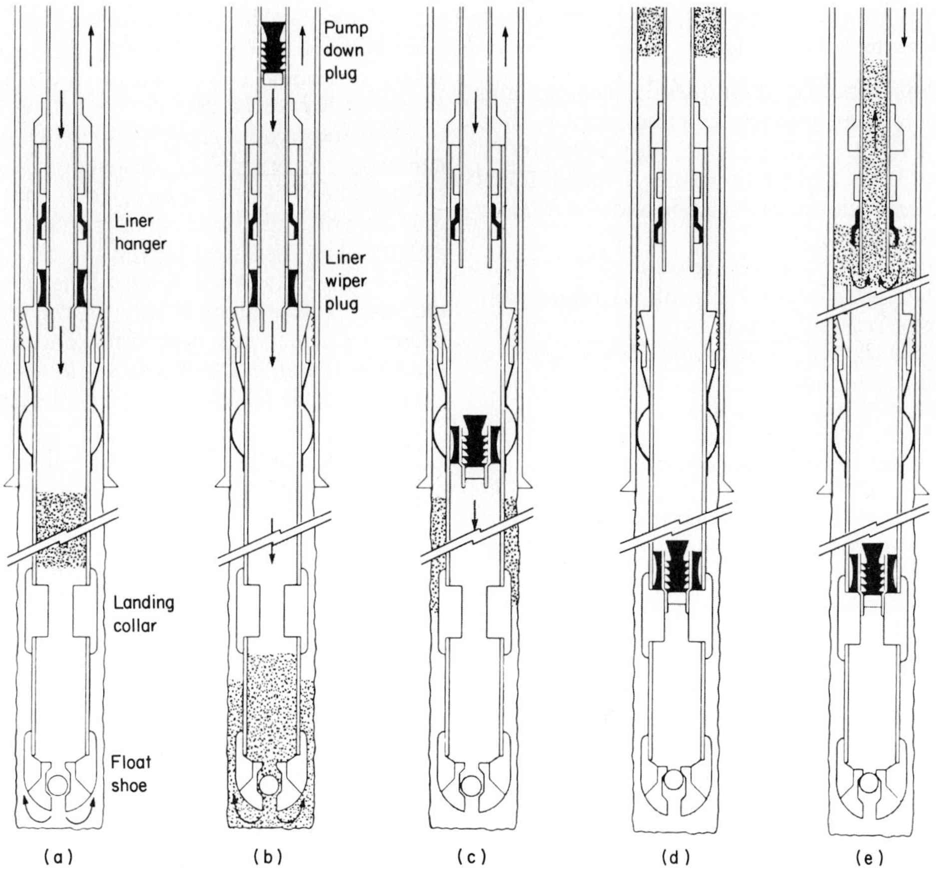

Fig. 11.24. Liner cementing[1]: (a) mixing; (b) and (c) displacing; (d) end of job; (e) reversing. (Courtesy of Dowell Schlumberger)

migrating up the cement slurry and creating channels in the set cement.

PRACTICAL CALCULATIONS

The following example calculations, together with sketches, provide a guide to the steps involved in determining the quantities of cement and additives, thickening time, displacement volume and differential forces:

Example 11.3: Conductor pipe

The following data are given.

Casing dimensions:	OD 20 in (508 mm), ID 18.73 in (475.7 mm), 133 lbm/ft (198 kg/m)
Hole size:	26 in (660.4 mm)
Casing setting depth:	350 ft (107 m)
Mud weight:	65 pcf (8.7 ppg or 1.041 kg/l)

Cement properties:
Cement API Class G with 4% bentonite

slurry weight:	106 pcf (1.7 kg/l)
slurry yield:	1.5 ft^3/sack (43.03 l/sack)
water requirement:	7.6 gal/sack (28.8 l/sack)

(*Note:* Cement data are obtained from cementing companies' handbooks.)

Pumping rate:	through drillpipe	100 gal/min (1700 l/min)
	through casing	300 gal/min (2385 l/min)
Drillpipe:	OD/ID 5 in/4.276 in (127 mm) 19.5 lb/ft (29.02 kg/m)	

Allow 15 min for the release of plugs and assume casing to be cemented to surface.

(1) Calculate required quantities of cement and bentonite for a conventional cementing job. A shoe track of 80 ft (24 m) is to be used. Also allow 100% excess cement in the open hole. (*Note:* A shoe track is the distance between the casing shoe and the float or landing collar.)

(2) Calculate volume of mixing water.

(3) Calculate total time for the job, assuming that the mixing rate is 10 sacks/min.

(4) Calculate the forces developed when using a conventional cementing operation and the safety factor in tension. Will the casing float?

Solution

Cementing of large casing can be carried out by use of either a conventional technique or an inner string method. In the conventional technique cement is pumped and displaced through the casing, as previously described. In the inner string method a drillpipe is used for displacing the cement behind the casing. The drillpipe is equipped with a stinger which fits inside a sealing sleeve, within the casing shoe, as shown in Figure 11.25.

Prior to cementing, the drillpipe with the stinger is run to the casing shoe and the assembly is stung into the casing shoe. The hole is circulated, and a spacer and the cement slurry are then pumped through the drillpipe and displaced behind the casing.

The choice between conventional and inner sting methods is dependent largely on the forces resulting from the cementing operation. The method which produces the highest safety factor in tension (i.e., yield strength/tensile force) is normally chosen.

(1) Refer to Figure 11.26 for slurry volume calculations.

Annular area between 26 in hole and 20 in casing

$$= \frac{\pi}{4}(26^2 - 20^2)\ \text{in}^2$$

$$= 216.77\ \text{in}^2$$

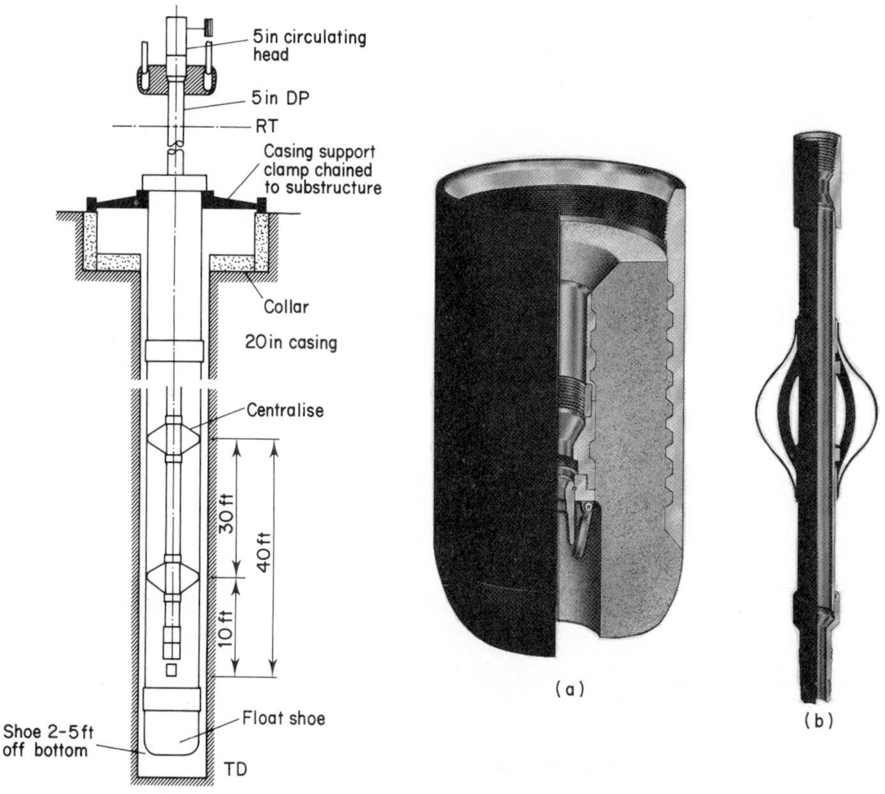

Fig. 11.25. Cementing of conductor casing, with, on right: (a) stab-in cementing shoe; (b) stab-in unit.

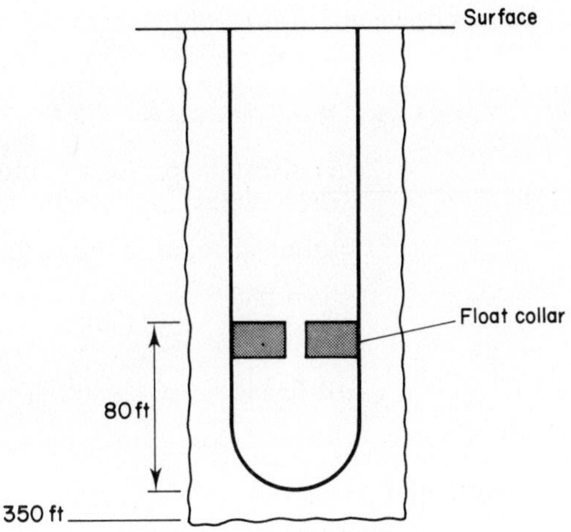

Fig. 11.26. Example 11.3.

$$\text{Annular capacity} = \left(\frac{216.77}{144}\right) \times 1\ \text{ft}$$

$$= 1.5053\ \text{ft}^3/\text{ft}$$

$$\text{Annular slurry volume} = 1.5053\ \frac{\text{ft}^3}{\text{ft}} \times 350\ \text{ft}$$

$$= 527\ \text{ft}^3$$

(a) Total annular volume = calculated volume

$$+ \text{excess}$$

$$= 527 + 100\%\ (527)$$

$$= 527 + 527 = 1054\ \text{ft}^3$$

$$\text{Capacity of 20 in casing} = \frac{\pi}{4}\,\frac{(\text{ID}^2)}{144} \times 1\ \text{ft}$$

$$= \frac{\pi(18.73)^2}{4 \times 144}$$

$$= 1.9134\ \text{ft}^3/\text{ft}$$

(b) Cement volume in shoe track $= 1.9134\ \frac{\text{ft}^3}{\text{ft}} \times 80\ \text{ft}$

$$= 153\ \text{ft}^3$$

Total required slurry volume

$$= \text{a} + \text{b} = 1054 + 153$$

$$= 1207\ \text{ft}^3\ (34\,172\ \text{l})$$

$$\text{Number of sacks of cement} = \frac{\text{slurry volume (ft}^3)}{\text{slurry yield (ft}^3/\text{sack)}}$$

$$= \frac{1207}{1.5}$$

$$= 805$$

Mass of bentonite = 4% × total weight of cement

Weight of cement = volume × density

$$= 1207\ \text{ft}^3 \times 106\ \text{lbm/ft}^3$$

$$= 127\,942\ \text{lbm}$$

Therefore,

$$\text{quantity of bentonite} = 0.04 \times 127\,942$$

$$= 5117.7\ \text{lbm}$$

Number of sacks of bentonite

$$= \frac{5117.7}{94\ \frac{\text{lb}}{\text{sack}}} = 54$$

(2) Volume of mix water = number of sacks × water requirements per sack

$$= 805\ \text{sacks} \times 7.6\ \frac{\text{gal}}{\text{sack}}$$

$$= 6118\ \text{gal}$$

$$= \frac{6118}{42} = 145.7\ \text{bbl}\ (23\ \text{m}^3)$$

(3) Total job time = mixing time

+ time for release of plug

+ displacement time

(a) Job time for a conventional job

$$= \frac{805\ \text{sacks}}{10\ \text{sacks/min}} + 15\ \text{min}$$

$$+ \frac{\text{internal capacity of casing excluding shoe track}}{\text{pumping rate}}$$

$$= 80.5 + 15 + \frac{1.9134\ \frac{\text{ft}^3}{\text{ft}} \times (350 - 80)\ \text{ft}}{300\ \frac{\text{gal}}{\text{min}} \times \frac{\text{ft}^3}{7.48\ \text{gal}}}$$

$$= 80.5 + 15 + 12.9$$

$$= 108.4\ \text{min (or 1 h 48 min)}$$

(b) Job time for an inner string job

$$= 80.15 + \text{displacement time}$$

The capacity of drillpipe is 0.0997 ft^3/ft and total internal volume in our example is $0.0977 \times 350 \simeq 35\ ft^3$.

The total volume of cement slurry is 1207 ft^3, which is much larger than the capacity of drill pipe of 35 ft^3. Thus, the displacement time is negligible and can be incorporated as part of the mixing time. In other words, as soon as the cement is mixed, it is displaced behind the casing. The actual value of displacement time

$$\frac{35\ (ft^3)}{100\ (gal/min) \times \dfrac{ft^3}{7.48\ gal}}$$

Thus,

$$\text{total job time} = 80.15 + 2.6 = 83\ \text{min}$$

(4) For a conventional cementing job, the worst conditions occur when the casing is entirely full of the 106 lb ft^3 cement slurry. Also, the calculations will include a pumping pressure of 500 psi as a displacement force.

$$\text{Differential force} = (\text{forces down}) - (\text{forces up})$$

$$\text{Forces down} = \text{weight of casing in air} + \text{weight of cement} + \text{force due to pumping pressure.}$$

(i) Weight of casing in air $= 133 \dfrac{lbm}{ft} \times 350\ ft$

$$= 46\,550\ lb$$

(ii) Weight of cement inside casing

$$= \text{casing inside area} \times \text{hydrostatic pressure due to cement column}$$

$$= \frac{\pi}{4}(18.73)^2 \times \frac{106 \times 350}{144}$$

$$= 70\,987\ lb$$

(iii) Force due to pumping pressure

$$= 500\ (psi) \times \frac{\pi}{4}(18.73)^2\ (in^2)$$

$$= 137\,764\ lb$$

$$\text{Forces down} = (i) + (ii) + (iii)$$

$$= 255\,301$$

$$\text{Forces up} = \text{Bouyancy force}$$

$$= \text{Hydrostatic pressure of mud in the annulus} \times \text{outside area of the casing}$$

$$= \frac{65 \times 350}{144} \times \frac{\pi}{4} \times (20)^2$$

$$= 49\,633\ lb$$

$$\text{Differential forces} = \text{forces down} - \text{forces up}$$

$$\text{scan at top joint} = 255\,301 - 49\,633$$

$$= 205\,668\ lb$$

Hence, casing will stay in the hole during cementing, i.e. it will not float.

Safety factor in tension

$$= \frac{\text{yield strength of body pipe or coupling}}{\text{Tensile force at top joint}}$$

Pipe body yield strength for casing is 2012000 lb. Hence,

$$SF = \frac{2012000}{205668} = 9.8$$

The calculation of forces developed using inner sting method is left as a problem for the reader (see Problem 11.1).

Example 11.4: Primary cementing of 7 in (215.9 mm) production casing

Hole depth:	13 900 ft (4237 m)
Hole size:	$8\frac{1}{2}$ in (215.9 mm)
Casing shoe:	13 891 ft (4234 m)
Mud weight:	87 pcf (1.394 kg/l)
Casing dimensions:	OD/ID = 7 in/6.184 Grade C95 29 #
Cement:	cement column should be 6562 ft (2000 m) long, as follows:

from shoe to 656 ft (200 m), use API Class G cement from 656 ft to 6562 ft (200–2000 m), use API Class H cement with 2% bentonite and 0.3% HR-4 (*Note:* HR-4 is a type of cement retarder)

To prevent contamination of cement by mud, 30 bbl (4770 l) of fresh water should be pumped ahead of the cement.

Allow 15 min for release of plugs

Shoe track: 80 ft (24 m)

Calculate:

(1) quantity of cement from each class;
(2) volume of mix water;
(3) total time for the job
(*Note:* Mix cement at the rate of 25 sacks/min and displace cement at 300 gpm (1136 l/min.));
(4) pressure differential prior to bumping the plug;
(5) annular velocity during chase;
(6) total mud returns during the whole cementing operation.

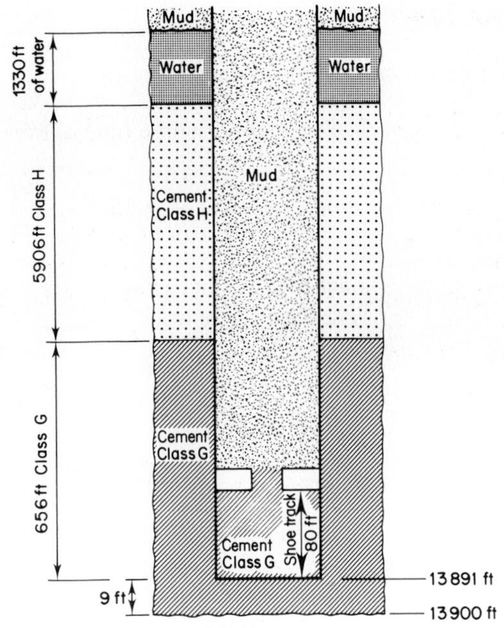

Fig. 11.27.

Solution

From cementing tables (Halliburton or Dowell Schlumberger), the properties of the two classes of cement including the additives are as follows:

	Class G cement	*Class H cement*
Slurry weight	118 pcf or 15.8 ppg	115 pcf or 15.5 ppg
Slurry volume	1.15 ft^3/sack	1.22 ft^3/sack
Mix water	5 gal/sack	5.49 gal/sack

(1) *Sacks of cement required (refer to Figure 11.27)*

Class G

Volume of Class G slurry

= volume of shoe track
+ volume of pocket
+ volume of 656 ft of annular space

$$= \frac{\pi}{4} \times (6.184)^2 \times \frac{1}{144} \times (80 \text{ ft})$$

$$+ \frac{\pi}{4} \times (8.5)^2 \times \frac{1}{144} \times (9 \text{ ft})$$

$$+ \frac{\pi}{4} (8.5^2 - 7^2) \times \frac{1}{144} \times (656)$$

$$= 16.7 + 3.5 + 83.2$$

$$= 103.4 \text{ ft}^3$$

Number of sacks of Class G cement

$$= \frac{103.5 \text{ ft}^3}{1.14 \text{ ft}^3/\text{sack}} = 90$$

Class H

Volume of slurry

$$= (6562 - 656) \times \text{annular capacity}$$

$$= 5906 \times \frac{\pi}{4} (8.5^2 - 7^2)$$

$$\times \frac{1}{144} = 748.9 \text{ ft}^3$$

Number of sacks of Class H cement

$$= \frac{748.9 \text{ ft}^3}{1.22 \text{ ft}^3/\text{sack}}$$

$$= 614$$

(2) *Volume of mix water*

Volume of mix water

= water required for Class G and Class H cement

= (90 sacks × 5 gal/sack) Class G
+ (614 sacks × 5.49 gal/sack) Class H

= 3820.9 gal

= 91 bbl

(3) *Total job time*

$$\text{Job time} = \text{mix time} + (\text{time for release}) + \text{displacement or chase time of plugs}$$

$$= \frac{\text{total number of sacks}}{\text{mixing rate}} + 15 + \frac{\text{inner capacity of casing excluding shoe track}}{\text{pumping rate}}$$

$$= \frac{(614+90)\text{ sacks}}{25\text{ sacks/min}} + 15 + \frac{\frac{\pi}{4}\times(6.184)^2\times\frac{1}{144}(13\,891-80)\text{ ft}^3}{\left(\frac{300\text{ gal}}{[\text{min}]}\times\frac{\text{ft}^3}{7.48\text{ gal}}\right)}$$

$$= 28.2 + 15 + 71.8$$

$$= 115\text{ min}$$

(4) *Differential pressure* The 30 bbl of water pumped ahead of the cement will occupy in the annulus a height, h, given by

$$h = \frac{30\text{ bbl}\times\left(\frac{5.62\text{ ft}^3}{\text{bbl}}\right)}{0.1268\,\frac{\text{ft}^3}{\text{ft}}} = 1330\text{ ft}$$

(annular capacity $= 0.1268\text{ ft}^3/\text{ft}$).

A pressure differential exists during the cementing operation due to density differences between mud, cement and the water spacer. Referring to Figure 11.29, the total pressure differential, Δp, is given by

Δp = pressure differential due to density difference between:
(i) mud in casing and cement (Grade G) in annulus for a height of $(656-80)=576$ ft
+ (ii) mud in casing and cement (Grade H) in annulus for a height of 5906 ft
+ (iii) mud in casing and water spacer in annulus for a height of 1330 ft

Assuming the density of fresh water is 62 pcf, then

$$\Delta p = \frac{576\times(118-87)}{144} + \frac{5906\times(115-87)}{144} + \frac{1330\times(62-87)}{144}$$

$$= 124 + 1148.4 + (-230.9)$$

$$= 1042\text{ psi}$$

(5) *Annular velocity*

Using $Q = VA$ (where V = velocity; Q = volume flow rate; A = annular area)

$$V = \frac{Q}{A} = \frac{300\text{ gal/min}}{\frac{\pi}{4}(8.5^2-7^2)}\text{in}^2\;\frac{\left(\frac{\text{ft}^3}{7.48\text{ gal}}\right)}{\left(\frac{\text{ft}^2}{144\text{ in}^2}\right)}$$

$$= 316\text{ ft/min}$$

(6) *Mud returns*

Mud returns = steel volume + volume of water ahead + total slurry volume

$$= \frac{\pi}{4}(7^2-6.184^2)\times\frac{1}{144}\left(\frac{\text{ft}^3}{\text{ft}}\right)\times 13\,891\text{ ft} + (30\text{ bbl}) + (748.9+103.5)\text{ ft}^3$$

$$= 815.1\text{ ft}^3 + 30\text{ bbl}\times\frac{5.62\text{ ft}^3}{\text{bbl}} + 852.4\text{ ft}^3$$

$$= 1836.1\text{ ft}^3$$

$$= 326.7\text{ bbl}$$

SQUEEZE CEMENTING

Squeeze cementing may be defined as the process of injecting cement into a confined zone behind the casing.

Squeeze cementing is used in the following situations.

(1) To repair a faulty primary cementing job. Most primary jobs fail because of inefficient removal of mud cake or inefficient displacement of mud, which leads to mud channels being left behind the casing. Squeeze cementing here is a 'remedial' operation.
(2) To eliminate water intrusion from perforations above, within or below the producing zone by squeezing off the unwanted perforations.
(3) To adjust the gas–oil ratio of a producing zone by plugging off the perforations adjacent to the gas zone.
(4) To plug off oil-producing zones opened up primarily for evaluation purposes when they are no longer required for production.

The success of any squeeze cementing job is dependent on the correct placement of the cement slurry into the perforations or zones to be squeezed off. The success of squeeze cementing can be greatly enhanced by conditioning[8] the surfaces of the perforation to be plugged off. The rock surfaces can be conditioned by means of an acid wash or matrix acidising using hydrochloric acid (HCl) or other acceptable acids. The acid reacts with the mud filter cake and other deposits, which can then be removed by circulating the well with a clean fluid.

Three types of squeeze cementing are in current use.

The Bradenhead method

In the Bradenhead method the drillpipe is run to just above the perforations (or zone) to be squeezed off. Cement is then displaced from the drillpipe to cover the entire zone. The pipe rams are then closed (Figure 11.28) and a precalculated pressure is then applied from the surface to squeeze off the open perforations.

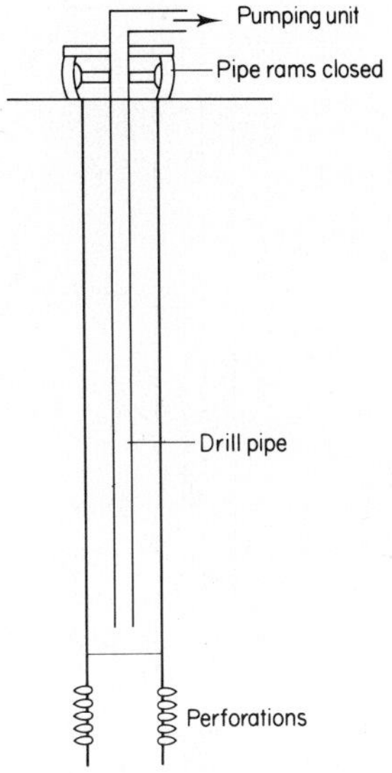

Fig. 11.28. Bradenhead squeeze method (no squeeze tool is required).

Packer squeeze method

In the packer squeeze method either a retrievable packer or a retainer packer is run to just above the zone to be squeezed off. The retrievable packer, shown in Figure 11.29, is run on drillpipe and can be easily set and unset mechanically. The retainer packer is run on wire line and set by a special setting kit, explosive charge being used to activate the packer slips.

When a retrievable packer is used, the cement slurry is first spotted against the perforations to be squeezed off. The packer is then set and the cement is squeezed into the open zone, as shown in Figure 11.29.

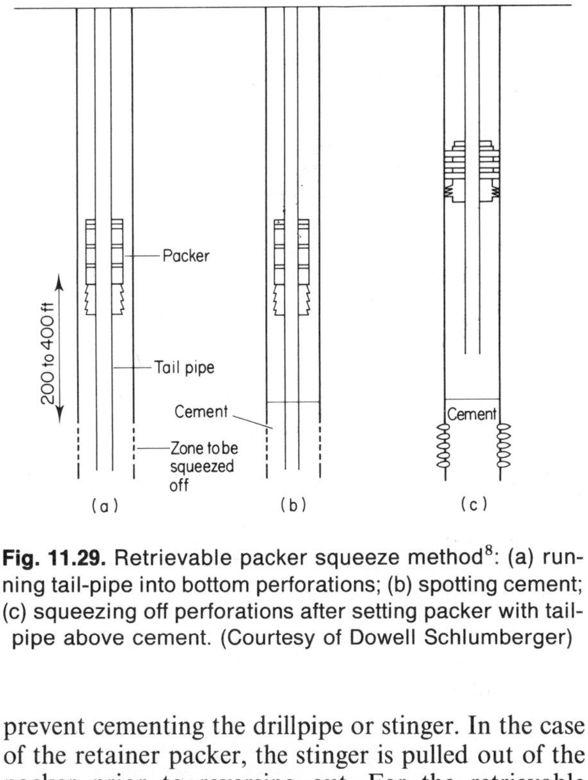

Fig. 11.29. Retrievable packer squeeze method[8]: (a) running tail-pipe into bottom perforations; (b) spotting cement; (c) squeezing off perforations after setting packer with tail-pipe above cement. (Courtesy of Dowell Schlumberger)

When a retainer packer is used (Figure 11.30) cement is first circulated to the shoe of the stinger. The stinger is then forced through (stung in) the packer bore and squeeze pressure is applied from the surface.

When the complete volume of cement is squeezed off, the excess cement should be reversed out to prevent cementing the drillpipe or stinger. In the case of the retainer packer, the stinger is pulled out of the packer prior to reversing out. For the retrievable packer method, the packer is first unset and excess cement is then reversed out.

The packer method has the advantage that it

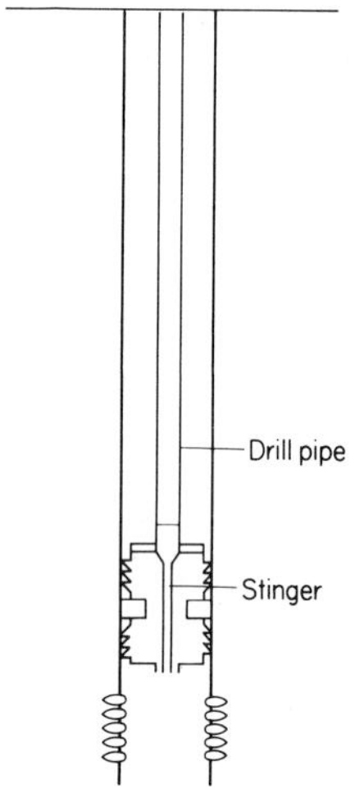

Fig. 11.30. Squeezing cement with a drillable cement retainer and a stinger.

isolates the casing string and wellhead equipment from the squeeze pressures.

Hesitation squeeze

The hesitation squeeze method is particularly advantageous for squeezing off low permeability zones. The equipment required is similar to that shown in Figure 11.29. A drillpipe is used to place the cement across the zone of interest and the slurry is alternately pumped and 'hesitated' (i.e. no pumping)[4]. The purpose of hesitation is to encourage a filter cake to deposit on the exposed surfaces of the open zone.

Plugging-back operations

Plugging-back operations involve the placement of a cement plug across the zone(s) to be plugged off or abandoned. Cement plugs are usually used to[4]: (a) abandon lower depleted zones; (b) plug off or abandon an entire well or part of an open hole; (c) provide a 'kick-off point' for sidetrack drilling operations; and (d) seal off lost circulation zones in open hole. Two types of plugging-back operation are in use: the balanced plug and the dump bailer.

The balanced-plug method

The balanced-plug method is performed with drillpipe only. A pre-flush is pumped ahead of the cement and a spacer fluid is pumped behind the cement. The principle involved is to displace the preflush and part of the cement out of the drillpipe until the cement columns inside the drillpipe and in the annulus are equal. This is done basically to maintain equal hydrostatic pressures in the drillpipe and annulus. It should be remembered that a greater volume of cement is required to occupy a given height inside the annulus compared with the volume required to occupy the same height inside the drillpipe.

The drillpipe is then slowly pulled out, so that the remaining cement falls through and occupies the volume previously held by the drillpipe steel. This will then leave a balanced plug in place (see Figure 11.31). The plug can be cut off at the desired height by reverse circulation.

A spacer is used behind the cement to provide a column of fluid of the same length and density as the preflush column in order to keep the columns balanced (see Figure 11.31).

The height of a balanced plug can be calculated from the following formula:

Volume of cement = Height of plug × (annular area between casing and drillpipe or tubing + internal capacity of drillpipe or tubing)

$$V = H(A + C)$$

$$H = \frac{V}{A + C} \tag{11.2}$$

where V = volume of slurry to be used (ft^3); A = annular volume between tubing or drillpipe and open

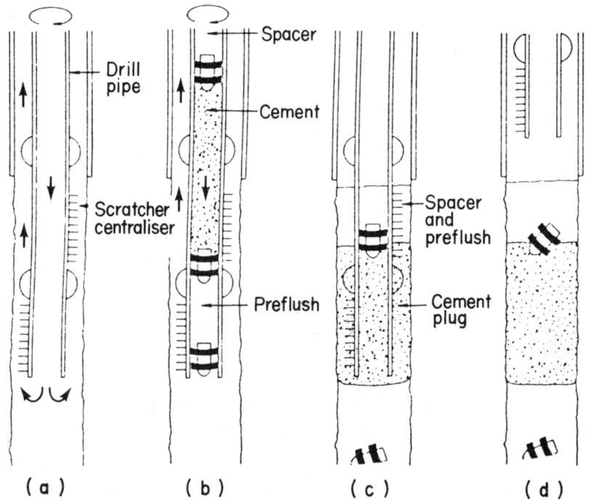

Fig. 11.31. Schematic drawing of balanced plug method[4]: (a) condition mud, rotate pipe; (b) displace cement and fluids; (c) spot balanced plug; (d) pull pipe slowly. All fluid volumes are carefully calculated so that hydrostatic pressure on plug in final location is identical in drillpipe and annulus. (After Suman & Ellis, 1977[4])

hole or casing (ft^3/ft); and C = capacity of tubing or drillpipe (ft^3/ft).

Example 11.5

It is required to balance 100 sacks of Class G neat cement in an $8\frac{1}{2}$ in open hole by use of a $3\frac{1}{2}$ in OD/3.068 in ID, 8.9 lb/ft tubing. The hole depth is 6000 ft and 10 bbl of water is to be used as preflush ahead of the cement slurry.

Calculate:

(1) total slurry volume, annular volume and tubing volume;
(2) height of the balanced plug;
(3) volume of water to be used as a spacer behind the cement;
(4) volume of mud chase (or displacement volume);
(5) number of strokes required to displace cement to just below the drillpipe shoe, assuming that the pump capacity is 0.1 bbl per stroke;
(6) volume of cement and number of sacks required if the height of the plug is 500 ft.

Solution

Using the ID of the tubing, the reader can verify that the capacity of the tubing is 0.049 68 ft^3/ft. The annular capacity between 3.5 in tubing and 8.5 in hole is 0.3068 ft^3/ft, as determined from manufacturers' handbooks. The slurry yield of Class G neat cement is determined from service companies' handbooks, of which Table 11.2 is an example. From Table 11.2 the slurry yield of Class G neat cement is 1.15 ft^3/sack and the slurry weight is 118.2 lbm/ft^3.

(1) Total slurry volume

$$= \text{number of sacks} \times \text{yield per sack} = 100 \times 1.15 = 115 \text{ ft}^3$$

Annular volume, C

$$= 0.3068 \text{ ft}^3/\text{ft}$$

(*Note:* By calculation the annular capacity is 0.3272 ft^3/ft, as no account is taken of the tubing collars.)

Tubing volume, A

$$= 0.049\,68 \text{ ft}^3/\text{ft}$$

(2) Height of balanced plug is

$$H = \frac{V}{A+C} = \frac{115 \text{ ft}^3}{(0.049\,68 + 0.3068) \text{ ft}^3/\text{ft}} = 322.6 \text{ ft} \simeq 323 \text{ ft}$$

(3) For a cement plug to be balanced, the height of the spacer inside the tubing must be equal to the height of the preflush in the annulus. Thus,

$$\frac{\text{volume of preflush}}{\text{annular capacity}} = \frac{\text{volume of spacer inside tubing}}{\text{capacity of tubing}}$$

$$\frac{10}{0.3068} = \frac{V}{0.049\,68}$$

$$V = 3.1 \text{ bbl}$$

In general, the volume of spacer required is calculated from

$$V = \frac{\text{capacity of tubing}}{\text{capacity of annulus}} \times \text{volume of preflush} \quad (11.3)$$

(4) Height of water inside the tubing

$$= \frac{3.1 \text{ bbl} \times \frac{5.62}{\text{bbl}} \text{ ft}^3}{0.049\,68 \text{ ft}^3/\text{ft}} = 356 \text{ ft}$$

TABLE 11.2 API neat cement slurries

Class	*Slurry weight* (lb/gal)	*Mixing water/sack* (ft^3/*sack*)	*Mixing water water/sack* (*gals/sack*)	*Slurry/sack cement* (ft^3/*sack*)	*Slurry weight* (lbm/ft^3)	*Percentage mixing water*
A	15.60	0.696	5.20	1.18	116.70	46
B	15.60	0.696	5.20	1.18	116.70	46
C	14.80	0.844	6.32	1.32	111.10	56
D	16.46	0.573	4.29	1.05	123.12	38
G	15.80	0.664	4.97	1.15	118.12	44
H	16.46	0.573	4.29	1.05	123.12	38

Height of cement and water inside tubing

$$= 323 + 356$$

$$= 679 \text{ ft}$$

Height of mud to fill remaining length of tubing

$$= 6000 - 679$$

$$= 5321 \text{ ft}$$

(4) Displacement volume

$$= 5321 \times \text{capacity of tubing}$$

$$= 5321 \times 0.049\,68$$

$$= 264.4 \text{ ft}^3$$

$$= 47 \text{ bbl}$$

(5) Number of strokes to displace cement

$$= \frac{47 \text{ bbl}}{0.1 \text{ bbl/stroke}} = 470$$

(6) Volume of cement

$$= \text{height of plug} \times \text{hole capacity}$$

$$= 500 \text{ ft} \times \frac{\pi}{4} \frac{(8.5)^2}{144}$$

$$= 197 \text{ ft}^3$$

Number of sacks

$$= \frac{\text{volume of slurry}}{\text{slurry yield}} = \frac{197}{1.15}$$

$$= 171 \text{ sacks}$$

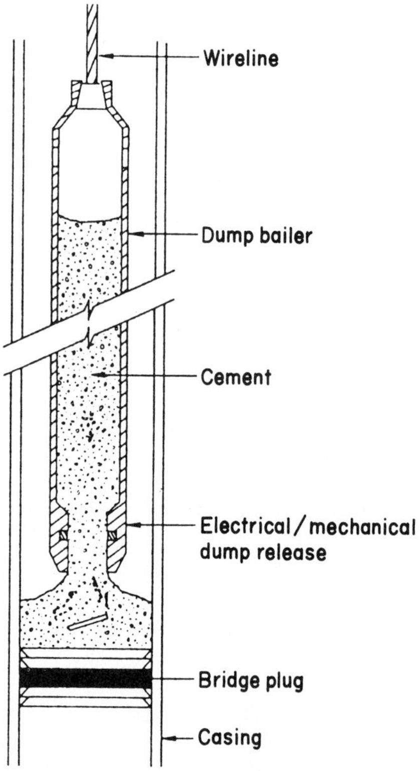

Fig. 11.32. Dump bailer method of cement placement[4]. A bailer of variable length is lowered on wire line. A device in the bottom releases the plate or opens ports to dump cement on the bridge plug or sand/gravel fill. (After Suman & Ellis, 1977)

The dump bailer method

The dump bailer method is shown schematically in Figure 11.32. A packer similar to that shown in Figure 11.30 is run on wire line and set just above the zone to be plugged. The packer bore is closed with a plug which gives the packer the characteristic name 'bridge plug'. A bailer containing a precalculated volume of cement is then run on wire line to the packer, where the cement slurry is released mechanically by breaking a glass plate at the end of the bailer.

The dump bailer method is normally used in cased holes to plug off low-pressured formations[4].

References

1. Dowell Schlumberger. *Cementing Technology*. Dowell Schlumberger Publications.
2. Halliburton (1972). *Outline of Oilwell Cementing*. Halliburton Publications.
3. API RP10B (1979). *API Recommended Practice for Testing Oilwell Cements and Cement Additives*. American Petroleum Institute, Production Department.
4. Suman, G. and Ellis, R. (1977). *Cementing Handbook*. Including *Casing Handling Procedures*. Reprinted from *World Oil*, Gulf Publishing Co.
5. API Spec. 10D (1973). *API Specification for Casing Centralisers*. American Petroleum Institute, Production Department.
6. Brown-Hughes (1984). *Liner Equipment*. Brown-Hughes Catalogue.
7. Halliburton, 1972. *Liner Cementing*. Halliburton Publications.
8. Dowell Schlumberger. *Squeeze Cementing Handbook*. Dowell Schlumberger Technical Literature.

Problem 11.1

Determine the magnitude of the differential force developed from Example 11.3 when using the inner string method of cementing. Also, calculate the safety factor in tension.

Answer:

47 870 lb

SF = 42

Problem 11.2

In an exploration well 6200 ft of 20 in casing 133 lb/ft with buttress connection is to be cemented to the surface in a 26 in hole with API Grade G cement. The cement column is to consist of 200 ft of Class G neat cement, having a 118 pcf density, and 6000 ft of Class G slurry, having a 95 pcf density.

Given

Casing ID:	18.73 in
Collapse resistance:	1500 psi
Burst resistance:	3060 psi
Joint strength:	2 012 000 lb

Compare the forces developed during the cementation of this string using both the conventional and inner string cementing methods. Which method would you propose?

Answer:

(a) Inner string method SF (in tension) = 5.3

(b) Conventional method SF (in tension) = 2.5

Chapter 12

Hole problems

Problems associated with the drilling of oilwells are largely due to the disturbance of earth stresses around the borehole caused by the creation of the hole itself and by drilling mud/formation interaction. Earth stresses, together with formation pore pressure, attempt to re-establish previous equilibrium by forcing strata to move towards the borehole.

A hole is kept open (or stable) by maintaining a balance between earth stresses and pore pressure on one side and well-bore mud pressure and mud chemical composition on the other side. Any time this balance is disturbed, hole problems may be encountered.

This chapter will highlight the most common hole problems encountered during the drilling of an oil well. Most of the material used in this chapter is based on recent articles published in the industrial literature.

Hole problems can be classified under three major headings: (1) pipe sticking; (2) sloughing shale; and (3) lost circulation.

PIPE STICKING

This is the condition when part of the drillpipe or collars are stuck in the hole. When this occurs, pipe movement and, in turn, further drilling progress, are inhibited.

In practice, pipe sticking problems are conveniently classified as: (a) differential sticking; (b) mechanical sticking; and (c) key-seating.

Differential pipe sticking

Differential pipe sticking arises when the differential pressure (the difference between hydrostatic pressure of mud and formation pore pressure) becomes excessively large across a porous and permeable formation such as sandstone or limestone (Figure 12.1). Other conditions most conducive to differential sticking include a thick filter cake and when a drill string is left motionless for some time inside the open hole. Differential pipe sticking can normally be recognised when pipe movement in the upward or downward direction is impossible but free circulation is easily established, since obstruction exists on only one side of the pipe. In a complete stuck pipe situation, neither circulation nor pipe movement are possible. Figure 12.1 gives a schematic drawing of a differentially stuck pipe.

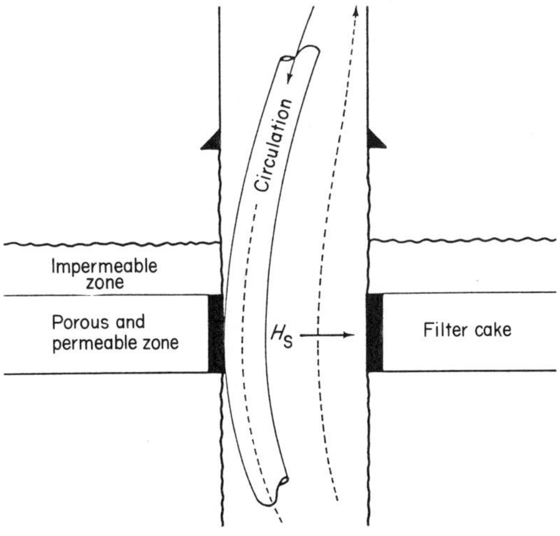

Fig. 12.1. Differential pipe sticking.

From Figure 12.1,

$$\text{differential force} = (H_S - P_f) \times \text{area of contact} \times \text{friction factor} \quad (12.1)$$

where H_S = hydrostatic pressure of mud; and P_f = formation pressure.

$$\text{Area of contact} = \text{thickness of permeable zone} \times \text{thickness of filter cake}$$
$$= h \times t \quad (12.2)$$

The friction factor (which will be denoted by f) is used to allow for variation in the magnitude of contact between steel and filter cakes of different composition.

Substituting Equation (12.2) in Equation (12.1) yields

$$\text{differential force (DF)} = (H_S - P_f) \times (h \times t) \times f \quad (12.3)$$

In Imperial units, Equation (12.3) becomes

$$\text{DF} = (H_S - P_f)\ \text{psi} \times h\left(\text{ft} \times \frac{12\ \text{in}}{\text{ft}}\right) \times t\ (\text{in}) \times f$$

$$\text{DF} = 12\,(H_S - P_f) \times h \times t \times f \quad (12.4)$$

The magnitude of the differential force is very sensitive to changes in the values of the contact area and the friction factor, which are both time-dependent. As the time in which the pipe is left motionless increases, the thickness of the filter cake increases. Also, the friction factor increases by virtue of more water being filtered out of the filter cake[1].

The differential force is also extremely sensitive to changes in differential pressure ($H_S - P_f$). In normal drilling operations an overbalance of between 100 and 200 psi (6.8–13.6 bar) is maintained. Excessive overbalance may arise as a result of the following situations: (a) sudden increases in mud density resulting in an increase in the hydrostatic pressure of the mud and, in turn, in the value of the overbalance; (b) drilling through depleted reservoirs and pressure regressions.

Pressure regression[1] is encountered in deep drilling when the formation pressure gradient is receding while the mud gradient remains constant to counteract the pore pressure of upper formations.

Figure 12.2 gives a possible picture of the situation at the start of differential sticking and after several hours.

Example 12.1

Determine the magnitude of the differential sticking force across a permeable zone of 30 ft (9.1 m) in thickness using the following data: differential pressure = 1000 psi (68 bar); thickness of filter cake = $\frac{1}{2}$ in (12.7 mm); friction factor = 0.1.

Solution

From Equation (12.4),

$$\text{DF} = 12\,(H_S - P_f) \times h \times t \times f$$
$$= 12 \times 1000\ \text{psi} \times 30\ \text{ft} \times \tfrac{1}{2}\ \text{in} \times 0.1$$
$$= 18\,000\ \text{lb}$$

Example 12.2

A drill string consists of 15 000 ft of 19.5 lbm/ft (29.02 kg/m) drillpipe and 500 ft of 150 lbm/ft drill collars. It is established that the drill string is differentially stuck at the first drill collar below the drillpipe.

Given that:

mud weight = 13 ppg (1.6 kg/l)

differential force = 108 000 lb (480 kN)

grades of drillpipe available are E, X, G and S.

Determine: (a) the buoyant weight of the drillpipe; (b) the total hook load when pulling on the differentially stuck pipe; (c) the magnitude of the margin of overpull (MOP) for the four grades, assuming the conditions of pipes to be premium.

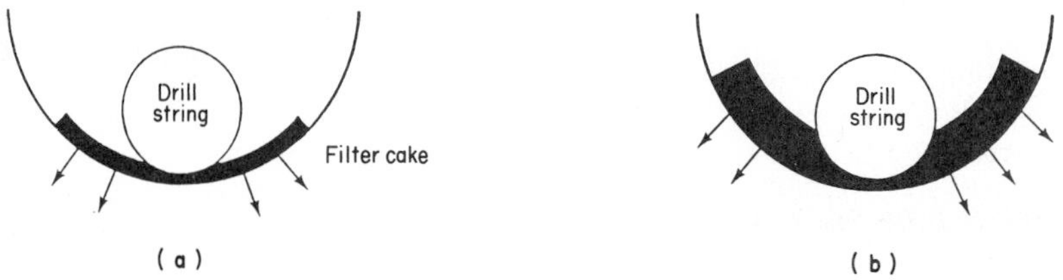

Fig. 12.2. Development of differential sticking with time: (a) initial; (b) after several hours.

Solution

(a) Note that only the weight of drillpipe will be considered, as the weight of drill collars is taken up by the stuck point.

Buoyant weight of drillpipe

$$= 15\,000 \times 19.5 \times \text{BF}$$

$$= 15\,000 \times 19.5 \left(1 - \frac{13 \text{ ppg} \times \frac{7.48 \text{ gal}}{\text{ft}^3}}{489.5 \frac{\text{lb}}{\text{ft}^3}}\right)$$

$$= 234\,394 \text{ lb}$$

(b) Hook load including differential force

$$= 234\,394 + 108\,000$$

$$= 342\,394 \text{ lb}$$

(c) This part can best be illustrated in tabular form, as follows:

Grade	*Weight* (lbm/ft)	*Pipe body yield** (lb)	*MOP = yield strength − hook load from part (b)* (lb)
E	19.5	311 540	311 540 − 342 394 = −30 854
X	19.5	394 600	394 600 − 342 394 = 52 206
G	19.5	436 150	93 756
S	19.5	560 760	218 366

*All drillpipe grades are assumed to be premium class and yield values are obtained from Tables 2.1–2.4.

Thus, for the existing conditions, Grade E gives a negative MOP, implying that the pipe will part if the required force of 342 394 lb is applied to free the pipe. Only Grades X, G and S can be used in this type of well, where the magnitude of the differential force is 108 000 lb.

Prevention of differential sticking

Observation of Equation (12.4) shows that the differential sticking force can be reduced by:

(1) Reducing the differential pressure, $H_S - P_f$. This means drilling with a minimum overbalance necessary to contain formation pressure and to allow for surging and swabbing effects. Mud density increase can be monitored by controlling the rate of penetration, especially in large holes where large quantities of drill cuttings are produced, resulting in an excessive increase in mud density and, in turn, an increase in the value of the differential pressure, $H_S - P_f$.

(2) Reducing the contact area, $h \times t$. Since the thickness, h, of the porous formation cannot be physically changed, the contact area can only be decreased by reducing the thickness of the filter cake, t.

This, in effect, means reducing the solids in mud to a minimum and using a mud of low water loss.

The friction factor, f, is directly related to the rate of water loss, and its value should be kept to a minimum by use of a mud of low water loss. Hence, oil-base muds appear to be ideal for drilling formations susceptible to differential sticking, if conditions allow.

The contact area is also related to the area of pipe steel in contact with the permeable formation. Most pipe sticking problems are associated with drill collars, and the ideal solution is to use drill collars with minimum surface area. A spirally grooved drill collar has 50% less area than a smooth drill collar[1] and, consequently, produces half as much differential sticking force. The reduction in surface area of drill collars reduces the weight by only 4–7% and, if extra weight is required, additional drill collars can be used.

The contact area can also be reduced by using stabilisers which centralise the drill collars within the hole.

(3) Since both contact area and friction factor increase with time, a reduction in the time during which the drill string is kept stationary directly results in less chance of severe differential sticking.

(4) Oil and walnut hulls can be used to reduce the friction factor, f, when drilling formations with potential differential sticking problems.

Freeing differentially stuck pipe

If, despite the above precautions, the pipe does become stuck, a number of methods can be used to free the stuck pipe.

The most commonly used methods include: (a) hydrostatic reduction; (b) spotting fluids; (c) back-off operations; (d) DST (for recovering the fish); and (e) fishing. Only methods a–c will be discussed here.

Hydrostatic reduction

The normal method used to reduce the hydrostatic pressure of mud is the U-tube method. The drill string and annulus can be thought of as a U-tube, with the

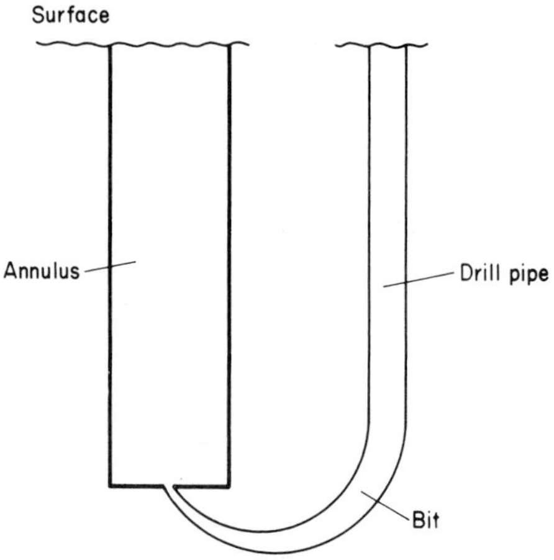

Fig. 12.3. U-tube configuration of a well.

drill bit connecting the two limbs, as shown in Figure 12.3.

Two situations of differential sticking can be recognised: (1) when the formation pressure is known (e.g. in development wells); and (2) when the formation pressure is unknown, as in exploration drilling.

When the formation pressure is known, the overbalance, $H_S - P_f$, can be gradually reduced to a safe level such that the hydrostatic pressure is always greater than the formation pressure.

Hydrostatic pressure is reduced either by pumping a new mud of lower density, or by pumping a small volume of a fluid of low specific gravity. Diesel oil is normally used because of its low specific gravity, but fresh or saline water can also be used for hydrostatic pressure reduction. The volume of fluid of low specific gravity is determined by calculating the required reduction in hydrostatic pressure and then converting this value to height and volume of diesel oil (or water) as shown in Example 12.3. Diesel oil is then pumped down the drill string until the complete volume is used. Since diesel oil has a lower pressure gradient than mud, the total pressure in the drillpipe will be less than that in the annulus and a back-pressure will be exerted on the drillpipe. The excess pressure is contained by closing a kelly cock on top of the drillpipe. A safe tension which is equal to the original hook load plus an extra overpull is then applied to the drill string.

The drillpipe is then allowed to back-flow at equal intervals until the entire volume of diesel oil is reversed out. At this point, the annulus level has dropped such that the hydrostatic pressure is equal to or slightly higher than the formation pressure.

During the back-flow operation, the drill string should be worked continuously until the pipe is free. If a drilling jar is used in the drill string, it should be activated to provide additional force for freeing the pipe. The jar is only useful if it happens to be above the stuck zone.

Also, during back-flow operations, the drillpipe and annulus pressures should be monitored continuously. When the well is static (or dead), the drillpipe pressure declines slowly with back-flow and there is no movement of fluid up the annulus. If the well is kicking, a gradual rise in the annulus level will be observed, and the drillpipe pressure increases slowly with back-flow. When this situation is encountered, the operation of freeing of the drill string should be stopped, and the well killing operation should commence.

Example 12.3

Calculate the volume of oil required to reduce the hydrostatic pressure in a well by 500 psi (34 bars), using the following data:

mud weight	= 10 ppg (1.2 kg/l)
hole depth	= 9843 ft (3000 m)
drillpipe	= OD/ID = 5 in/4.276 in (127 mm/109 mm)
hole size	= 12.25 in (311.2 mm)
specific gravity of oil	= 0.8

Solution

$$\text{Initial hydrostatic pressure} = \frac{(10\text{ ppg} \times 7.48) \times 9843}{144}$$

$$= 5113 \text{ psi}$$

$$\text{Required hydrostatic pressure} = 5113 - 500$$

$$= 4613 \text{ psi}$$

Thus,

New hydrostatic pressure

= pressure due to (mud and oil) in drillpipe

$$4613 = \left(\frac{10 \times 7.48 \times Y}{144}\right)_{\text{mud}} + \left(\frac{0.8 \times 62.3\,(9843 - Y)}{144}\right)_{\text{oil}}$$

where Y = height of mud in drillpipe. Therefore, $[Y = 6959 \text{ ft}]$

Hence,

$$\text{height of oil} = 9843 - 6959 = 2884\ \text{ft}$$

$$\text{volume of oil} = \text{capacity of drillpipe} \times \text{height}$$

$$= \frac{\pi}{4}(4.276)^2 \times \frac{1}{144} \times 2884$$

$$= 287.61\ \text{ft}^3$$

$$= 51.2\ \text{bbl}$$

Note that when the required volume of diesel oil is pumped inside the drillpipe, the hydrostatic pressure at the drillpipe shoe becomes 4613 psi, while the hydrostatic pressure in the annulus is still 5113 psi. This difference in the pressure of the two limbs of the well causes a back-pressure on the drillpipe which is the driving force for removing the diesel oil from the drillpipe and reducing the level of mud in the annulus. It is only when the annulus level decreases that the hydrostatic pressure against the formation is reduced.

When formation pressure is unknown, it is customary to reduce the hydrostatic pressure of mud in small increments by the U-tube technique until the pipe is free.

A variation of the U-tube method is to pump water into both the annulus and the drillpipe to reduce hydrostatic pressure to a value equal to or just greater than the formation pressure. This method is best illustrated by an example.

Example 12.4

The following data refer to a differentially stuck pipe at 11 400 ft:

formation pressure	= 5840 psi (403 bar)
intermediate casing	= $9\frac{5}{8}$ in, 40 # at 10 600 ft (3231 m)
drillpipe	= OD 5 in/ID 4.276 in (127 mm/109 mm)
mud density	= 92 pcf (11 kg/l)

It is required to reduce the hydrostatic pressures in the drillpipe and the annulus so that both are equal to the formation pressure.

Calculate the volumes of water required in both the annulus and the drillpipe, assuming that the density of saltwater = 65 pcf.

Solution

Assume the height of water in the annulus to be Y.

Required hydrostatic pressure at stuck point = 5840 psi or

$$5840 = \left(\frac{Y \times 65}{144}\right) + \frac{92\,(11\,400 - Y)}{144}$$

$$Y = 7696\ \text{ft}$$

Required volume of water in annulus

= annular capacity between drillpipe and $9\frac{5}{8}$ in casing × height of water

$$= 0.0515\ \frac{\text{bbl}}{\text{ft}} \times 7696$$

$$= 396.3\ \text{bbl}$$

Hence, pump 396.3 bbl of water in the annulus to reduce the hydrostatic pressure in the annulus to 5840 psi at the stuck point. When 396.3 bbl of water is pumped into the annulus, the drillpipe is still filled with the original mud of 92 pcf having a hydrostatic pressure at the stuck point of $(92 \times 11\,400)/144 = 7283$ psi. Thus, a back-pressure equivalent to $7283 - 5840 = 1443$ psi will be acting on the annulus and will be attempting to equalise pressures by back-flowing water from the annulus.

In order to contain the 396.3 bbl of water in the annulus, the drillpipe must contain a column of water equal in height to that in the annulus.

Thus,

volume of water required in drillpipe to prevent back-flow from annulus

= capacity of drillpipe × height of water

$$= 0.0178\ \frac{\text{bbl}}{\text{ft}} \times 7696\ \text{ft}$$

$$= 137\ \text{bbl}$$

Balancing of the columns of water in the drillpipe and the annulus can be achieved as follows: (a) circulate 396.3 bbl of water down the annulus; (b) displace 137 bbl of water down the annulus; (c) circulate 137 bbl of water in the drillpipe to remove 137 bbl of water from the annulus and to reduce the hydrostatic pressure in the drillpipe to 5840 psi.

If the well should kick during the operation, reverse-circulate down the annulus using the 92 pcf (i.e. original density) mud to recover all the water from the drillpipe. Then circulate in the normal way through the drillpipe, using 92 pcf mud until all the water is removed from the annulus.

Spotting organic fluids

Organic fluids are normally spotted across the stuck zone to reduce the filter cake thickness and the friction factor. A mixture of surfactant and diesel oil is by far the most widely used fluid, owing to its ability to wet the circumference of the pipe, thereby creating a thin layer between pipe and mud cake.[1,2] This action decreases the value of the coefficient of friction, thereby increasing the effectiveness of mechanical attempts to pull free. The normal procedure is to pump the organic fluid into the drillpipe and gradually pump small volumes into the annulus until the entire stuck zone is covered. The pipe should be worked continuously during the spotting of organic fluid. The success of this operation is dependent on the volume of organic fluid used, the characteristics of the mud cake, the magnitude of the differential force and spotting the fluid against the correct zone. For effective freeing of stuck pipe a minimum volume of 150 bbl of organic fluid is suggested[1]. The fluid should be left for a minimum of 8 h to work through the filter cake properly. An organic solution may also be added to the mud used to drill formations which are amenable to differential sticking. The use of oil will produce a reduction in the hydrostatic pressure of mud, and weighting materials can be used to compensate for the loss of pressure gradient. This is most important in wells with potential kick problems.

Back-off operations

If none of the above-mentioned methods are successful in freeing the pipe, back-off operations are a final solution.

Back-off operations involve the removal of the free portion of drill string from the hole. This effectively means parting the drill string at or above the stuck zone and removing the free portion from the hole. The remaining portion of the drill string (the fish) can then be removed by using either DST tools or washover tools. Alternatively, the hole may be plugged backed and side-tracked.

Before a back-off operation can be attempted, the position of the stuck pipe should be determined as accurately as possible. Two methods are normally used: (1) the pipe stretch method using surface observations; and (2) the pipe stretch method using specialised strain gauge tools, commonly known as 'free point indicators'.

Surface measurements of pipe stretch Brouse[2] details the method for estimating the position of the stuck zone from surface measurements as follows:

(1) Pull to normal hook load and mark the position above the rotary table using a stick, say X_1.
(2) Pull an additional 20 000 lb and release slowly until the weight indicator reads the hook load again. Mark the new position as X_2.
(3) Note the average distance between X_1 and X_2 as

$$Y_1 = \frac{X_1 \times X_2}{2}$$

(4) Increase tension load to 40 000 lb and mark position X_3 above the rotary table.
(5) Increase tension to 60 000 lb above hook load and release until the weight indicator reads HL + 40 000 lb. Mark the new position as X_4.
(6) Note the average distance between X_3 and X_4 as

$$Y_2 = \frac{X_3 \times X_4}{2}$$

(7) The pipe stretch is then measured as the difference between Y_2 and Y_1.

Using Hooke's law,

$$E = \frac{F/A}{e/L}$$

where L is the free length of drill string. Therefore,

$$L = \frac{AE\,e}{F}$$

where $e = Y_1 - Y_1$ and $F = (\text{HL} + 40\,000) - \text{HL}$ or $F = 40\,000$ lb. Therefore,

$$L = \frac{AE\,(Y_2 - Y_1)}{F} \quad (12.5)$$

Equation (12.5) can be simplified by replacing the cross-sectional area by the weight per unit length, using the relation

$$A = \frac{W_{dp}}{3.4}$$

where W_{dp} is the weight of drillpipe (lbm/ft). Thus,

$$L = \frac{\dfrac{W_{dp}}{3.4} \times e\left(\text{in} \times \dfrac{1\text{ ft}}{12\text{ in}}\right) \times E}{F}$$

$$L = \frac{W_{dp} \times e \times E}{40.8F} \quad (12.6)$$

Using $E = 30 \times 10^6$ psi, Equation (12.6) becomes

$$L = \frac{735\,294 \times e \times W_{dp}}{F}\text{ ft} \quad (12.7)$$

The drillpipe stretch measurements do not account for drill collars or heavy-wall drillpipe stretch. Pipe stretch will also be influenced by hole conditions such as dog-legs, hole angle, drag force, etc.

The above procedure can be normally applied in the field in the following simplified version:

(1) Pull drill string to normal hook load and mark position X_1.
(2) Pull additional 40 000–60 000 lb and mark a new position X_2.
(3) The difference between X_2 and X_1 is the stretch due to additional pull.

Hence,

$$L = \frac{AE\,e}{F} = \frac{AE\,(X_2 - X_1)}{F}$$

where F is the additional pull, or

$$L = \frac{735\,294 \times (X_2 - X_1) \times W_{dp}}{F}$$

Free point indicators Two types of free point indicator are in use: (1) strain gauge; and (2) sub-surface probe.

Strain gauge method Strain gauge methods rely on measurement of axial strain and angular deformation of the drill string at selected positions. The strain gauge tool (shown in Figure 12.4) measures the pipe stretch or angular deflection between the length of two prespaced belly-type springs. The strain gauge tool is run on a wire line containing electrical connections to a surface display unit which translates extension into a percentage of free pipe.

The tool is run to hole bottom and the driller applies a pull equal to the buoyant weight of the entire string in the hole[3]. The strain gauge tool is then positioned against the drill string and an additional pull Δp above the buoyant weight is applied. The strain measured by the tool is compared with the predicted strain due to the differential pull, to determine whether the pipe is completely free or partially or completely stuck[3]. The theoretical or predicted strain is calculated by direct application of Hooke's law. The tool is slowly moved up the hole and the tensioning procedure is repeated until the first 100% free point is located.

The free point indicator is also designed to measure the angle of twist between the two measuring springs for a given amount of torque as applied at the surface. The angle of twist, θ, for a given torque, T, can be determined[7] from

$$\theta = \frac{TL}{E_s J} \tag{12.8}$$

where L = length (ft); J = polar moment of inertia (in^4) $= (\pi/32)\ (\text{OD}^4 - \text{ID}^4)$; E_s = modulus of elasticity in shear (psi).

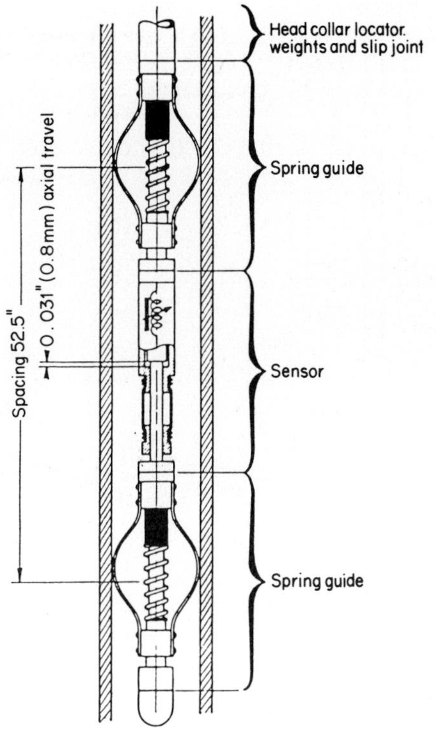

Fig. 12.4. Stuck point indicator tool[3]. (Courtesy of Schlumberger)

The free point indicator is designed to measure the angular strain, θ/L, in revolutions/1000 ft. Using Equation (12.8), it can be shown[3] that the angular strain at any section of drill string can be calculated from

$$\frac{\theta}{L} = \frac{\theta_t}{\dfrac{L_1}{E_{s1} J_1} + \dfrac{L_2}{E_{s2} J_2}} \times \frac{1}{E_{sx} \cdot J_x} \times 10^3 \text{ rev/1000 ft} \tag{12.9}$$

where subscript 1 refers to first section of drill string; subscript 2 refers to second section of drill string; E_{sx} = modulus of elasticity in shear for section under consideration; J_x = polar moment of inertia for section under consideration; and θ_t = total number of revolutions applied at the surface.

Once again, by comparing the measured angular strain with the calculated angular strain from Equation (12.9), the percentage of free pipe in torsion can be determined (Figure 12.5).

The pipe stretch and pipe torsion data are used to construct a graph of percentage of free pipe (in both tension and torsion) against depth. Figure 12.5 is an

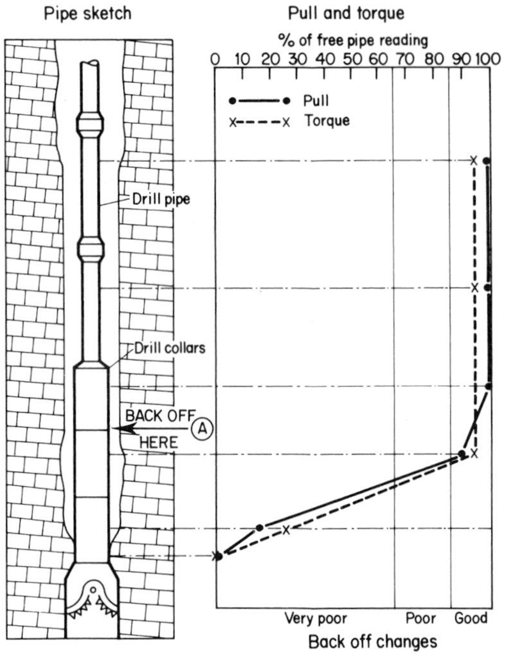

Fig. 12.5. A graphical presentation of back-off data for a straight hole, straight pipe stuck in drill collars[3]. (Courtesy of Schlumberger)

example of such a graph in a straight hole stuck at the drill collars.

Sub-surface probe A sub-surface probe is, like the strain gauge tool, run on wire line and positioned against the drill string while tension is being applied. The instrument consists of an oscillator which sends a high-frequency current, and a receiver.

The principle of operation of this tool is that, during tensioning, the molecular structure of the pipe changes, which alters the high-frequency signal. The change in the signal is proportional to the degree of pipe distortion. The frequency change of the signal is picked up by the receiver and transmitted to a surface display unit. The frequency change is then converted to strain reading by use of calibration charts.

This instrument is not capable of producing readings unless it is positioned against a free portion of pipe which can stretch under tension. The tool is normally run to hole bottom and gradually pulled up until a reading is obtained.

Back-off procedure A back-off shot is positioned against a drillpipe tool joint that is found to be free in both tension and torsion (point A in Figure 12.5). Point A is described as the back-off point. A left-hand torque and a slight positive tension above the back-off weight (pre-stuck hook load minus stuck pipe weight) are applied at the back-off point, and the back-off shot is detonated. The pipe should come free, which will be indicated by a sudden decrease in hook load. The pipe is rotated to the left and picked up to confirm back-off.

The portion of stuck drillpipe, drill collars and bit that are left in the hole are described as 'fish'. Fishing operations attempt to remove the equipment from the open hole.

Mechanically stuck pipe

A pipe can become mechanically stuck when: (a) drill cuttings or sloughing formations pack off the annular space around the drill string; (b) a drill string is run too fast, such that it hits a bridge or a tight spot or the bottom of the hole[2]; or (c) pulling into a keyseat.

Tight spots can result from drilling undersized (undergauged) holes due to the use of worn drillbits or undersized diamond coring bits. Tight spots can normally be recognised during tripping out as extra overpull (i.e. load in excess of the buoyant weight of the string). To prevent mechanical sticking, tight spots should be reamed prior to drilling new sections of hole.

The usual method used to free a mechanically stuck pipe is to work the drill string either by rotating and pulling it or by activating a drilling jar, if the latter is used. If this method is unsuccessful, an organic fluid should be spotted and the above procedure repeated.

If everything else fails, then drill string should be freed by use of back-off operations as previously discussed.

Key-seating

In a dog-legged hole containing soft formations, a drillpipe tool joint can drill an extra hole or a key-seat in addition to the major hole created by the bit, as shown in Figure 12.6.

During drilling, the drillpipe is always kept in tension (see Chapter 2) and as it passes through a dog-leg, it tries to straighten, thereby creating a lateral force as depicted in Figure 12.6. This lateral force causes the drillpipe joint to dig into the formation at the dog-leg bow, creating a new hole as the drill string is rotated. The new hole is described as a 'key-seat'.

A key-seat can only be formed if the formation drilled is soft and the hanging weight below the dog-leg is large enough to create a substantial lateral force.

The problem of key-seating can be diagnosed when the drill-string can be moved downwards but not upwards. Other symptoms include increased drag, increased noise at the rotary table and the ability to have full circulation.

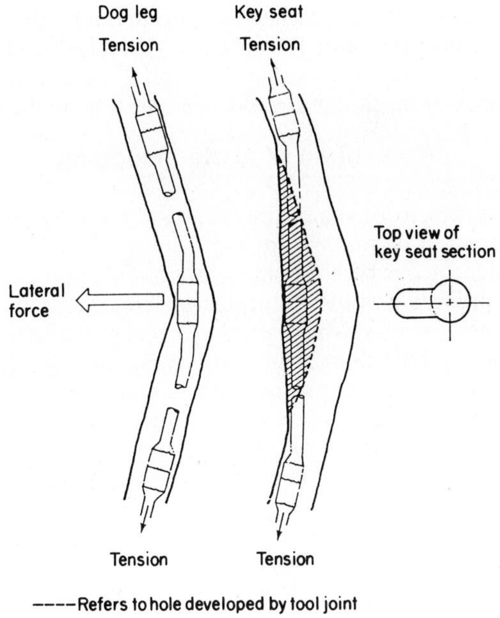

Fig. 12.6. Development of a key-seat.[4] (After Wilson, 1976)

To remove a key-seat, the hole should be reamed, and if a jar is used, an upward jarring action should be applied. Organic fluids can be spotted to reduce friction round the key-seat, thus facilitating the working of the pipe.

Key-seating can be prevented by drilling straight holes or avoiding sudden changes in hole inclination and/or direction in deviated wells.

SLOUGHING SHALE

Shale is a sedimentary rock formed by the deposition and compaction of sediments over periods of geological time. It is primarily composed of clays, silt, water and small quantities of quartz and feldspar. Depending on water content, shale may be a highly compacted rock or a soft, unconsolidated rock, normally described as mud or clay shale. Shale may also exist in a metamorphic form such as slate, phyllite and mica schist.

In oil well drilling, two types of sedimentary shale are normally encountered: unconsolidated shale (or clay) and compacted shale. Drilling of both types results in sloughing or caving of the shale section. Drillers normally refer to the type of hole instability resulting from drilling shale sections as 'sloughing shale'.

Research has shown that the severity of shale sloughing is related to the percentage content of montmorillonite (or active clay content) and the age of the rock.

Darley[5] used a scale for characterising the degree of shale dispersion based on the percentage of 50 μm (1 μm = 1 micron = 10^{-6} m) particles produced when shale is brought into contact with water. It was found that the degree of dispersion is 100% when a sample of shale is composed of 100% sodium montmorillonite. A 60% dispersion is produced with pure calcium montmorillonite. The degree of dispersion was also found to be dependent upon the age of shale; older shales containing a high percentage of montmorillonite dispersed less than younger shales with a lower content of montmorillonite.

Factors influencing shale sloughing can be conveniently divided into three groups: (1) mechanical factors; (2) hydration factors; and (3) miscellaneous factors.

Mechanical factors

Mechanical factors affecting shale sloughing are attributed largely to the erosion effects caused by the annular flow of mud. Erosion of shale is directly related to the degree of turbulence in the annulus, and mud viscosity. Most hydraulic programmes are designed with the object of providing laminar annular flow.

Other mechanical effects include the breakage of shale due to impaction by the drill string and caving due to horizontal movement of the shale section. The latter effect is due to the fact that creating a hole in the earth disturbs the local stress system, which leads to dynamic movement within the shale section. This movement leads to breakage of the shale bed adjacent to the well into small fragments which fall into the hole.

Hydration factors

A number of factors are involved in the hydration of shale. For practical purposes, shale hydration force and osmotic hydration are recognized and are quantifiable. Shale hydration force is related to the relief of compaction on the shale section. Osmotic hydration is related to the difference in salinity between the drilling mud and formation water of the shale.

During sedimentation, the shale section is progressively compacted by the weight of the overburden. The force of compaction squeezes out a large percentage of the adsorbed water and water from the pores of the shale. The compaction force is equal to the matrix stress (= overburden pressure − pore pressure). The drilling of a shale section relieves the compaction force on the borehole face and, as a result, a shale hydration force is developed. The shale hydration force is approximately equal to the matrix stress.

Osmotic hydration occurs when the salinity of the formation water of shale is greater than that of the drilling mud. In water-base muds, the shale surface acts as a semi-permeable membrane across which osmotic hydration takes place. In oil-base muds, the semi-permeable membrane is the oil film and the layer of emulsifier around the water droplet.

Since osmotic hydration is dependent on the difference in salinity between the formation water in shale and drilling mud, the process can result in either an adsorption or desorption force. An adsorption force is developed when the formation water of shale is more saline than the drilling mud. A desorption force is produced when the salinity of drilling mud is greater than that of the formation water of the shale.

Adsorption of water by shale usually leads to dispersion and swelling. Dispersion occurs when the shale subdivides into small particles and enters the drilling mud as drill-solids[10]. Swelling occurs due to increase in size of the silicate minerals making up the clay structure, and if the developed swelling pressure increases the hoop stress around the borehole above the yield strength of the shale, hole destabilization takes place[11]. Hole destabilization manifests itself by means of caving or sloughing shale.

For a more complete discussion of this subject, the reader is advised to consult References 6, 10 and 11 at the end of this chapter.

Miscellaneous factors

Shale sloughing has been correlated with a number of factors which were found to accelerate the rate of shale heaving into the well bore.

Dipping shales were found to slough more than horizontally laid shales. This is because during the adsorption of water, shale expansion takes place in a direction perpendicular to its bedding planes, which results in a greater shale heaving when the section is highly dipping.

The process of heaving in brittle shales containing no active clays is explained by the penetration of water between the bedding planes and microfissures of the shale. This results in high swelling pressures which break the cohesive forces between the fracture surfaces causing the shale to fall apart.

In abnormally- or geo-pressured shales the water content of the rock is much higher compared with that of normally pressured shales. In addition, the plasticity of the shale is abnormally high relative to the overburden load. Hence, when a hole is drilled through a shale section containing abnormal pressure, the shale will be squeezed into the hole owing to the difference between formation pore pressure and mud hydrostatic pressure. It follows that if such abnormally high pressures were catered for prior to drilling the shale section, sloughing of shale could be reduced.

Prevention of shale sloughing

The problem of shale sloughing is directly related to adsorption of water from the drilling mud. Hence, a change in the type or chemical composition of mud will provide possible solutions to shale sloughing.

The use of oil-base mud has been successful in reducing shale sloughing. The success is related to the fact that the oil phase provides a membrane around the hole which prevents water contacting the shale.

The water phase of an oil-base mud may also be prepared in such a way that its salt concentration matches that of the shale section. In this case, the osmotic (or dehydration) force is equal to the shale hydration force and the pressure causing water to flow between mud and shale is zero (10).

Potassium chloride polymer muds have also been successful in preventing shale sloughing. These muds reduce the swelling of shale due to the replacement of sodium ions, Na^+ (by cationic exchange) by potassium ions (K^+) which allow the clay sheets to be strongly bonded. Dispersion is also reduced due to encapsulation of the broken edges of shale by the polymer. Other types of mud which have been successful in reducing sloughing problems include lime mud, gyp mud, calcium chloride and silicate muds, surfactant mud, polymer muds, lignosulphonate mud, etc.

Other preventive measures include minimisation of the time for which an open hole containing a shale section is left uncased.

Hole deviation should be kept to a minimum, and swabbing and surging effects should be reduced, to avoid fracturing of the open hole sections.

High annular velocities are also be be avoided, to limit hole erosion and shale sloughing by mechanical action.

LOST CIRCULATION

Lost circulation is defined as the partial or complete loss of drilling fluid during drilling, circulating or running casing, or loss of cement during cementing. Lost circulation occurs when the hydrostatic pressure of mud exceeds the breaking strength of the formation, which creates cracks along which the fluid will flow. For lost circulation to occur, the size of the pore openings of the induced fractures must be larger than the size of the mud particles. In practice, the size of opening which can cause lost circulation is in the range 0.1–1.00 mm.

All rock types are susceptible to lost circulation,

but weak and cavernous formations are particularly vulnerable. In soft formations, such as sandstone, lost circulation is primarily attributed to their high permeability and the ease with which fractures can be induced. In hard rocks, such as limestone, dolomite and hard shale, lost circulation occurs due to the presence of vughs, caverns, natural fractures and fissures, and induced fractures.

The mechanics of production of fractures is controlled by several factors, including *in situ* earth stresses, hydrostatic pressure of mud, tensile strength of the rock, natural rock permeability, etc. Thus, the prediction of the length and direction of fracture is quite complex and is beyond the scope of this book.

As previously mentioned, lost circulation occurs as a result of the sudden increase in hydrostatic pressure of mud which can arise from a sudden density increase or pipe movement. The fast running of pipe in the hole causes fluid to surge up the pipe, creating additional pressures in the annulus. The total pressure due to surging effects and the hydrostatic pressure of mud can in certain cases be high enough to fracture uncased formations.

In surface holes lost circulation can cause a large washout with a possible loss of the drilling rig[7]. High penetration rates produce large quantities of drill cuttings which, if not immediately removed, will cause excessive increase in mud density and, in turn, an increase in hydrostatic pressure. Most operating companies limit the rate of penetration in surface holes to reduce the equivalent circulating density in the annulus and, in turn, limit the dynamic pressure to which the formation is subjected. Hence, a close monitoring of mud properties is required to detect sudden increases in mud density.

In intermediate holes, most lost circulation problems are caused when encountering a depleted zone whose formation pressure is markedly lower than that of upper formations. Sudden increases in the hydrostatic pressure of mud due to surging effects can fracture weak formations and lead to lost circulation.

Location of the lost circulation zone

Usually when losses occur during drilling, lost circulation material is spotted across the suspect zone to combat fluid losses. However, in severe lost circulation cases the location of the 'thief' (or lost circulation) zone must be determined prior to combating fluid losses. There are a number of established methods used for this purpose, including temperature survey, radioactive trace survey and spinner survey.

Temperature survey

A temperature recording device is run in hole on a wire line to provide a record of temperature against depth. Under normal conditions, a constant increase in temperature with increasing depth is observed. This trend (as shown in Figure 12.7) is recorded under static conditions to provide a base log. A quantity of cool mud is then pumped in the hole and another survey is made. The cool mud will cause the device to record a lower temperature than previously recorded, down to the 'thief' zone where mud is lost. Below the thief zone, the mud level is static and its temperature is higher than the mud flowing in the thief zone.

It follows that the new temperature log will show an anomaly across the thief zone, and the location of this zone can be determined by reading the depth at which the temperature line changes its gradient. Figure 12.7 shows two logs, one under static conditions and the other with flowing mud. Location of the lost circulation zone is clearly seen.

Radioactive tracer survey

A gamma-ray log is first run to establish the normal radioactivity of formation in hole and act as a basis for comparison. A small quantity of radioactive material is then displaced into the hole around the area where the thief zone is expected. A second gamma-ray log is run and compared with the base log. The point of loss of circulation is shown by a decrease in radioactivity of the new log, where the radioactive material is lost to the formation.

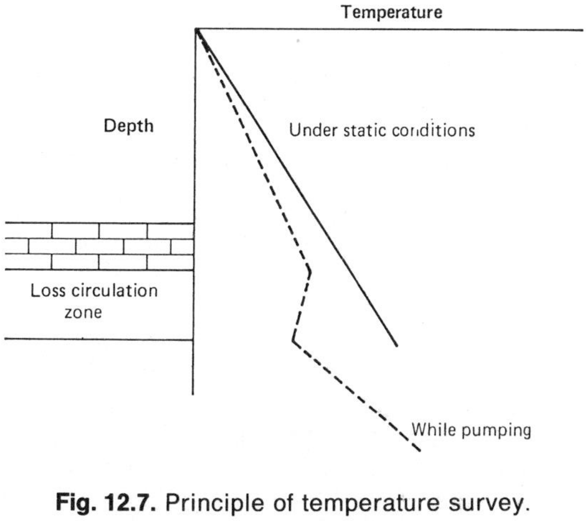

Fig. 12.7. Principle of temperature survey.

Spinner survey

A spinner attached to the end of a cable is run in hole to the place where loss of circulation is suspected. The spinner will rotate in the presence of any vertical-motion of mud such as encountered near a 'thief' zone. The speed of the rotor is recorded on a film as a series of dashes and spaces.

The spinner survey method was found to be ineffective when large quantities of sealing (lost circulation) material is used in the mud.

Combating lost circulation

Loss of circulation has a number of detrimental effects, which can be summarised as follows[8]: (a) loss of drilling mud and costly constituents; (b) loss of drilling time; (c) plugging of potentially productive zones; (d) blowouts resulting from the decrease in hydrostatic pressure subjected to formations other than the thief zone(s); (e) excessive inflow of water; and (f) excessive caving of formation.

Loss of circulation can be reduced or cured by one of the following methods:

(1) Reducing mud weight until the hydrostatic pressure of mud is equal to the formation pressure.

(2) Spotting of a pill of mud containing a high concentration of bridging materials against the thief zone. Bridging or lost circulation materials may also be used as additives in the circulating mud during drilling of formations susceptible to loss of circulation[8].

Lost circulation materials can be classified as fibres, flakes, granular material, and a mixture of all three. The fibres include plant fibres (such as hay or wood shavings), glass fibres, mineral fibres and leather. The flakes include cellophane, mica, cotton seed hulls, nut hulls, etc. Granular material includes ground rubber tyres, crushed rock, ground asphalt, asbestos, etc.

The fibres and flakes were found to be effective with low mud weight, while granular materials are best suited to weighted muds. Blends of mica and cellophane, and fibrous, flaky or ground materials are particularly effective in reducing lost circulation, as the mixture provides a gradation of sizes which can build up an effective seal.

Lost circulation material is normally mixed with a sufficient quantity of mud to prepare a pill which can be pumped to the lost circulation zone. The pill is spotted against the thief zone and gradually squeezed into the formation while the mud level in the annulus is continuously observed. If the mud level is still falling when the complete pill is squeezed into the zone, another pill is prepared and the procedure is repeated until the loss of circulation is stopped.

For severe lost circulation, the bridging materials form part of the mud additives and must, therefore, be capable of being pumped through the hole without causing severe pressure losses.

The severity of any lost circulation problem is related to the width and length of fractures created by the excessive hydrostatic pressure. Thus, for lost circulation to be cured, the openings or the fractures must be tightly packed with lost circulation materials. This can be achieved only if the lost circulation material contains a gradation of sizes such that large particles form bridges through the pores or fractures and small particles pack off the spaces between the large particles. Such a size distribution will produce an effective seal.

The performance of lost circulation materials is pressure-dependent; a good seal at 1000 psi differential pressure may fail at a higher pressure of, say, 2000 psi[7].

When lost circulation materials are used as part of the circulating mud additives, the shale shakers should be by-passed, to prevent the loss of these materials through the shakers.

(3) Spotting of bentonite–diesel oil or cement–diesel oil plugs across the thief zones. Several plugs may be required in the case of bentonite plugs before loss of circulation is stopped.

When spotting cement plugs, a wait-on-cement (WOC) time must be allowed for prior to resuming drilling, to permit the cement to set and seal the pore and fracture spaces of the thief zone. Cement plugs are normally spotted as a last resort when everything else fails. Calculating the quantity of cement and height of plug is treated in Chapter 11.

Bentonite or cement plugs are spotted with an open-ended drillpipe (OEDP) and the balanced plug method described in Chapter 11.

(4) Adoption of special drilling methods such as blind drilling, drilling with underbalance or drilling with air. Blind drilling refers to drilling without returns at the surface, so that the generated cuttings are used to seal off the fractures of the thief zone. The hydraulic programme must be adjusted so that the cuttings have sufficient annular velocity to reach the thief zones. For this technique to be effective, a plentiful supply of water is required to replace the mud lost to the formation.

Example 12.5

During drilling of an $8\frac{1}{2}$ in hole at 8000 ft, a complete loss of circulation was observed. Drilling was stopped and the mud level in the annulus was observed to fall rapidly. The well was filled with water of 62 pcf density until the annular level remained stationary. If the volume of water used was 65.7 bbl and mud density 75 pcf, determine the formation pressure and the new mud weight required to balance the formation pressure. Assume the intermediate casing to be $9\frac{5}{8}$ in, 40 # set at 6000 ft. Drillpipe is Grade E, 5 in OD.

Solution

Capacity of annulus between 5 in drillpipe and $9\frac{5}{8}$ in casing

$$=0.0515 \text{ bbl ft}$$

Height of water column

$$=\frac{65.7 \text{ bbl}}{0.0515 \dfrac{\text{bbl}}{\text{ft}}}=1276 \text{ ft}$$

When the well is balanced, we have

Formation pressure

= pressure due to mud column

+ pressure due to water column

$$=\frac{75\times(8000-1276)}{144}+\frac{62\times 1276}{144}$$

$$=4051 \text{ psi}$$

(*Note:* Hole depth = 8000 ft, of which 1276 ft is filled with water and 6724 ft with mud.)

$$\text{Required mud weight}=\frac{144\times 4051}{8000}$$

$$=72.9\approx 73 \text{ pcf}$$

References

1. Adams, N. (1977). How to control differential pipe sticking. *Petroleum Engineer*, October, November, December.
2. Brouse, M. (1982, 1983). How to handle stuck pipe and fishing problems. *World Oil*, November, December 1982, January 1983.
3. Schlumberger (1977). *Sit-back Off.* Schlumberger Publications.
4. Wilson, G. (1976). How to drill a usable hole. *World Oil.*
5. Darley, H. C. H. (1973). *A Laboratory Investigation of Borehole Stability.* Society of Petroleum Engineers Reprint Series No. 6a, SPE, Houston, TX.
6. Chenevert, M. E. (1973). *Shale Control with Balanced Activity Oil-continuous muds.* Society of Petroleum Engineers Reprint Series No. 6a, SPE, Houston, TX.
7. Moore, P. L. (1974). *Drilling Practices Manual.* Penwell Publishing Co, SPE, Houston, TX.
8. Howard, G. and Scott, P. (1951). An analysis and the control of lost circulation. Annual meeting of AIME, St. Louis, Mo., February.
9. White, R. J. (1956). Lost circulation materials and their evaluation. Spring Meeting of Pacific Coast District. Division of Production, Los Angeles, May.
10. Baroid/NL Industries Inc. (1979). *Manual of Drilling Fluids Technology.* Baroid/NL Industries Publishers.
11. Gray, G. and Darley, H. (1980). *Composition and Properties of Oil Well Drilling Fluids.* Gulf Publishing Company.

Chapter 13

Blowout Control

A blowout can be defined as an uncontrolled influx of formation fluid which has sufficient pressure to cause damage to rig equipment and injury to rig personnel. A blowout does not take place suddenly but develops gradually as the hydrostatic pressure of the drilling mud drops below the level required to contain formation fluids. Formation fluids will then enter the well bore and the well is said to be 'kicking'. If the kick is not detected and the well begins to flow at an uncontrolled rate, the well is said to be 'blowing out'.

Every well is capable of kicking any time the mud density decreases below the level required to contain formation pressure, or when an abnormally pressured formation is encountered, having a pore pressure greater than the hydrostatic pressure of mud. Because of this possibility, every well is equipped with specially designed equipment to contain possible well kicks. Such equipment is described as blowout preventers (BOPs). Blowout preventers are basically valves that can be hydraulically or manually operated to close the well any time a kick situation is encountered. Once the well is shut in, the drilling engineer can prepare a plan to kill the well by circulating the influx fluids and replacing the original light-weight mud by a heavier one.

This chapter will discuss blowout control under two major headings:

(1) Blowout preventers
(2) Kick control

BLOWOUT PREVENTERS

Blowout preventers (BOPs) are devices placed on top of the well to provide a line of defence against possible well kicks which may produce dangerously high pressures within the annulus of the well. The number, size and rating of BOPs used will depend on the depth of hole in question and the maximum anticipated formation pressures. In general, there are two types of BOP: annular and ram types.

Annular preventers

Annular BOPs are designed to shut off around any size and shape of equipment run through the hole. Thus, annular preventers can close around drillpipe, drill collars and casing, and can also pack off an open hole. An annular preventer is, therefore, the well's master valve and is normally closed first in the event of a well kick, owing to flexibility of the closing rubbers. Figure 13.1 shows a typical annular blowout preventer.

Referring to Figures 13.1 and 13.2, the main components of an annular preventer include (a) steel body; (b) operating piston; (c) opening and closing chambers; and (d) a ring of reinforced synthetic rubber having a high tensile strength. The rubber is moulded around a series of flanged steel ribs, as shown in Figure 13.2. The rubber rings can be squeezed to engage drillpipe, tubing or casing, and to provide an annular seal around the pipe.

Annular preventers can only be closed hydraulically by directing fluid under pressure to the operating cylinder through the closing chamber; see Figure 13.2. In the particular design shown in Figure 13.2, the hydraulic pressure raises the piston, which, in turn, squeezes the steel-reinforced packing unit inward to a sealing engagement with the drill string. The applied force is designed to compress the rubber element sufficiently to provide the required seal.

Fig. 13.1. Annular blowout preventer. (Courtesy of Hydril)

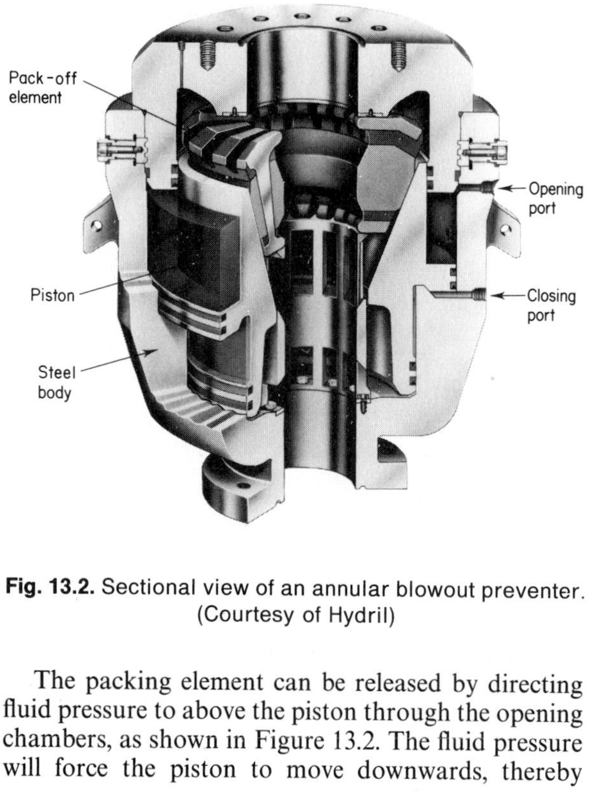

Fig. 13.2. Sectional view of an annular blowout preventer. (Courtesy of Hydril)

The packing element can be released by directing fluid pressure to above the piston through the opening chambers, as shown in Figure 13.2. The fluid pressure will force the piston to move downwards, thereby allowing the packing elements to expand and retain their original retracted position.

A variation of the Hydril annular preventer, shown in Figure 13.1, is the Regan Type K preventer used on conductor and surface casings. This type does not employ an operating piston and simply uses fluid pressure to compress a packing element.

The normal hydraulic closing pressure of an annular preventer is 1500 psi, which is applied when a kick is first detected, to provide an initial seal. The pressure is then adjusted to the manufacturer's recommended value to allow pipe to be removed, if necessary, through the preventer.

Ram preventers

Ram preventers can be fitted with four types of ram: (1) pipe rams; (2) variable-bore rams (VBRs); (3) blind rams; and (4) shear rams.

Pipe rams

Pipe rams (Fig. 13.3) are designed to close around a particular size of drillpipe, tubing or casing. The pack-off is provided by two steel ram blocks containing semi-circular openings with each ram being fitted with a two-piece rubber seal, as shown in Figure 13.4(a). The semi-circular openings can seal around the outside diameter of the drillpipe, tubing, drill collar, kelly or casing, depending on the size of the rams chosen. Pipe rams are designed to centre the pipe in the hole before providing a complete pack off.

Pipe rams can be closed manually or hydraulically to seal off the annular space below them. Hydraulically-operated pipe rams can be closed from the driller's console on the rig floor or remotely from a position on the ground some distance away from the rig floor.

Figure 13.4(b) shows a section through a pipe ram preventer together with a connecting piston rod which controls the movement of the rams. Under hydraulic pressure the piston rod pushes the two rams towards the centre of the hole, where the pipe is gripped and centred. The increased pressure energises the packer rubbers, allowing them to flow around the pipe and provide a complete pack-off. Figure 13.5 shows the hydraulic control system which closes and opens the rams through the hydraulic connections.

Most pipe rams can be closed by a hydraulic pressure in the range 500–3000 psi.

When the drill string contains two different pipe sizes, it is the practice to employ two pipe rams — one for the smaller pipe and the other for the large pipe. Also, whenever a string of casing is run, pipe rams

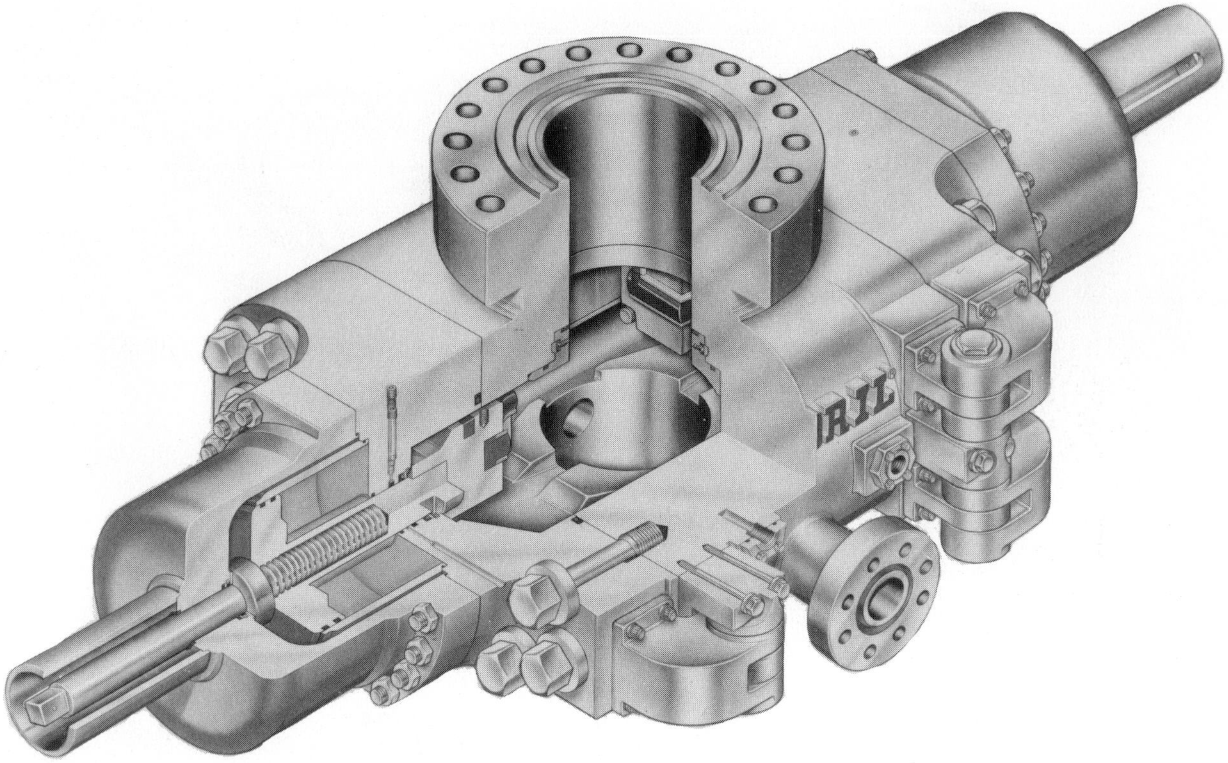

Fig. 13.3. Ram-type blowout preventer, $13\frac{5}{8}$ in, 10 000 psi manual lock. (Courtesy of Hydril)

have to be replaced by new rams to fit the casing outside diameter, unless variable-bore rams are used.

Pipe rams are designed to be changed easily and have the ability to direct hole pressures to the back of the preventer, to maintain the seal in the event of hydraulic pressure loss.

Variable-bore rams (VBRs)

In normal drilling operations a BOP ram has to be changed every time a new drillpipe or casing size is run. In offshore operations changing a BOP is both time-consuming and expensive.

Variable-bore rams have been developed to close and seal on a range of pipe sizes. Their use eliminates the need to disconnect a BOP stack to accommodate a new pipe size.

A popular design of VBR has a variable front packer with interlocking I-beam inserts moulded into the rubber. These inserts confine the rubber within the packer and against the pipe, thus preventing extrusion of the sealing elements. VBRs can be actuated in the same manner as pipe rams.

Blind rams

Blind rams (Figure 13.6) are similar to pipe rams, shown in Figures 13.4 and 13.5, except that packers are replaced by ones that have no cutouts in the rubber. They are designed to seal off the bore when no drill string or casing is present.

Shear rams

Shear rams are a type of blind ram that can cut the pipe and seal off the open hole. Most shear rams require 3000 psi to cut pipe.

Blowout preventers stack arrangements

Depending on expected pressures, a combination of one annular preventer, one or more ram-type preventers and a drilling spool can be used as wellhead control equipment. API[1] recommends the use of a single designation to distinguish various BOPs' stack

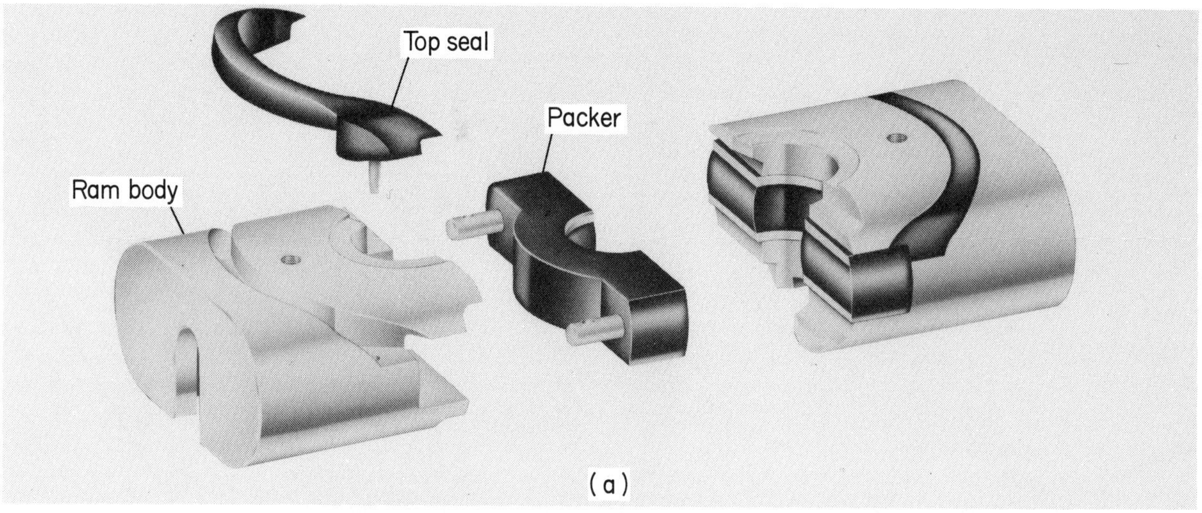

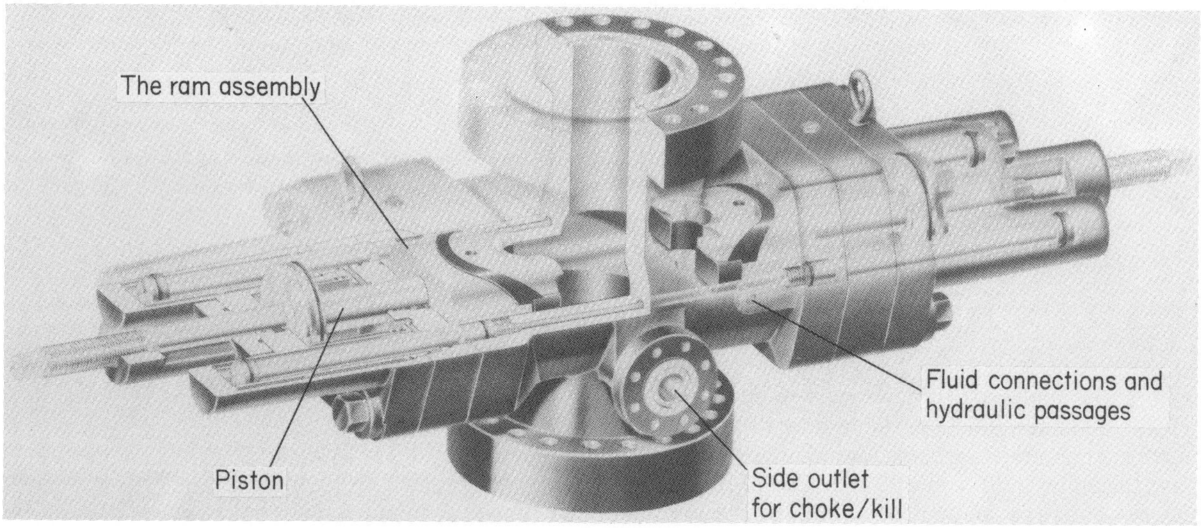

(b)

Fig. 13.4. Ram-type blowout preventer: (a) pipe rams; (b) sectional view. (Courtesy of Cameron Iron Works)

arrangements. The designation uses the working pressure of the stack, the through-bore of the preventers and the type of arrangement. Thus, a stack designated as 3M-$13\frac{5}{8}$-SRA means that the rated pressure is 3000 psi (1M = 1000 psi), the through-bore is $13\frac{5}{8}$ in and an arrangement of one drilling spool (S), one ram type preventer (R) and one annular preventer (A) is used. A drilling spool is normally used as a cross-over spool between the BOP and the casing housing.

Typical blowout preventer arrangements for 2M (2000 psi) to 13M (13 000 psi) working pressures are given in Figure 13.7.

The BOPs control system

The BOPs are designed to be closed remotely using hydraulic pressure supplied by an operating or control unit. The control unit is designed to close and open each individual BOP through a system of pipings and remotely controlled valves. The control unit is normally built on a skid-mounted assembly and placed at a safe distance from the rig floor.

As shown in Figure 13.8, the main components of a control system include: (a) an accumulator bank; (b) charging pumps; (c) a fluid reservoir; and (d) a mani-

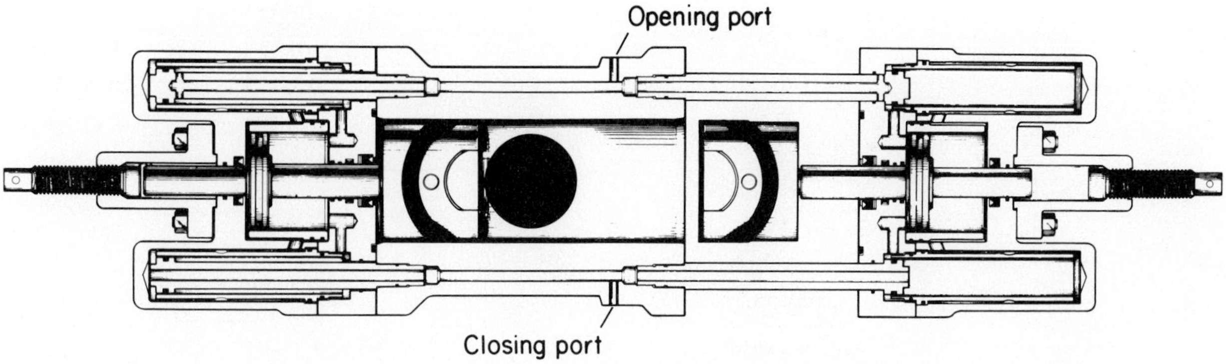

Fig. 13.5. Hydraulic control system for ram-type blowout preventer.

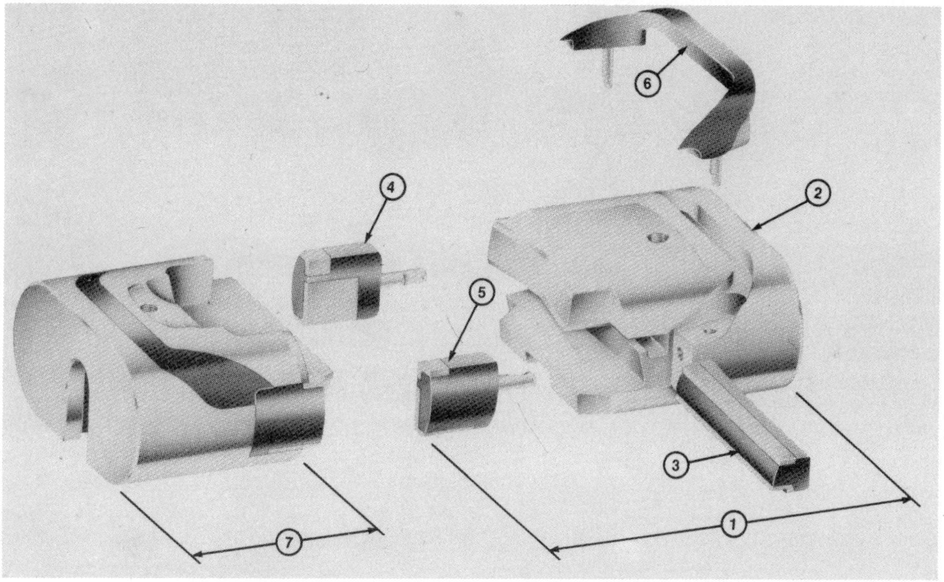

Fig. 13.6. Blind rams: (1) upper ram subassembly; (2) ram; (3) blade packer; (4, 5) packers; (6) top seal; (7) lower ram assembly. (Courtesy of Cameron Iron Works)

fold and pipings for directing the fluid to the appropriate preventer. The pumps can be powered by either the rig generators or a separate power source. A separate power source is normally used, to allow the unit to be used even when the rig engines are shut down.

The prime function of the control system is to store energy (in the accumulators) which can be released within 30 s or less. This energy is used to close the BOPs. The rig's air-operated pumps or any manually-operated pump can be used to effect the closing of the preventers, but these devices are very slow-acting and used only as a back-up to the main accumulator pumps.

Accumulators

The heart of the operating system is the bank of accumulators. An accumulator is a high-pressure cylinder containing precharged nitrogen gas and hydraulic fluid. The gas is separated from the fluid by a rubber diaphragm or a float, as shown in Figure 13.9. The hydraulic fluid can be just water or hydraulic oil with anticorrosion additives.

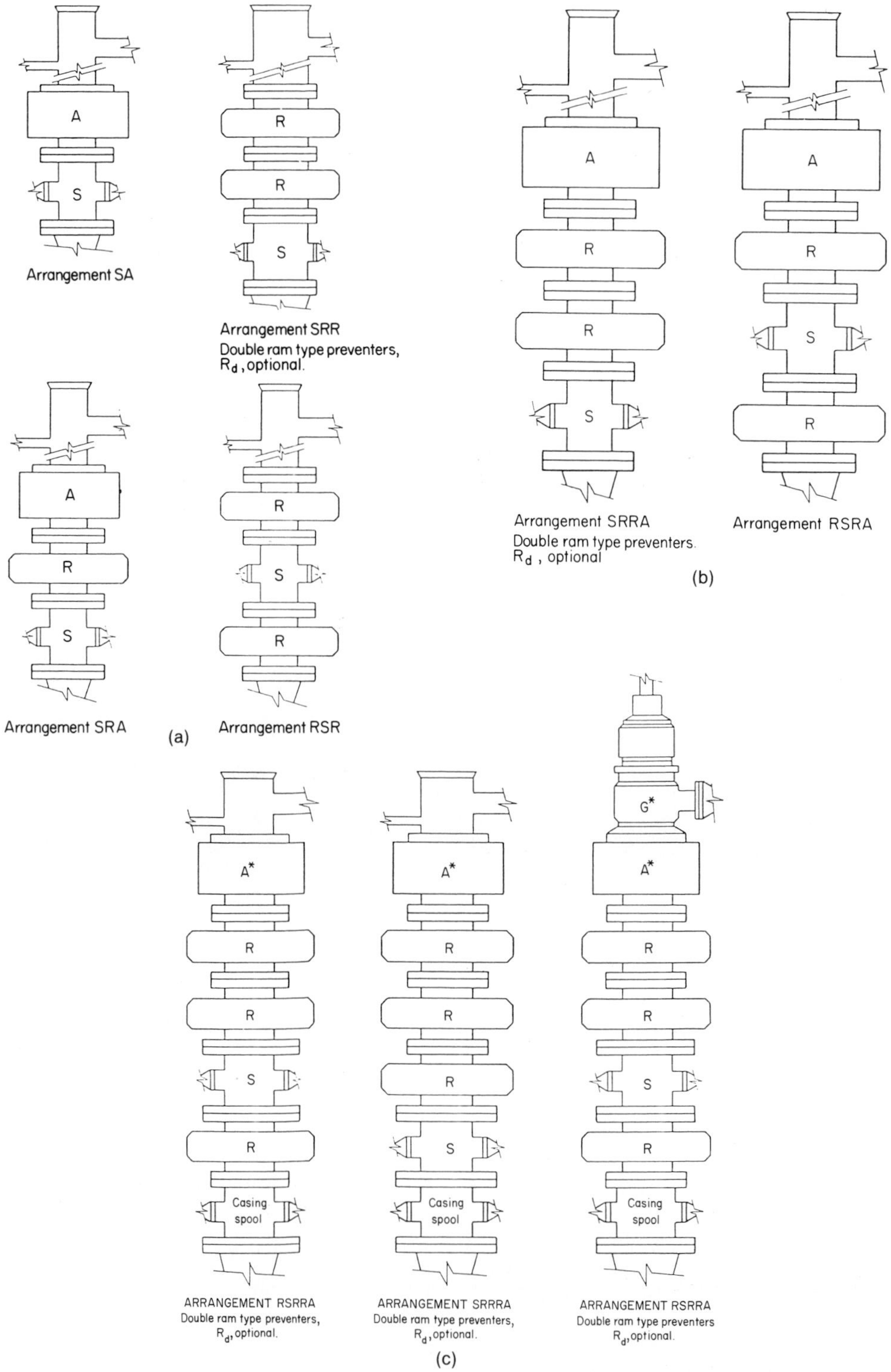

Fig. 13.7. Typical blowout preventer arrangements[1]: (a) for 2M rated working pressure service, surface installation; (b) for 3M and 5M rated working pressure service, surface installation; (c) for 10M and 15M working pressure service, surface installation. (Courtesy of API)

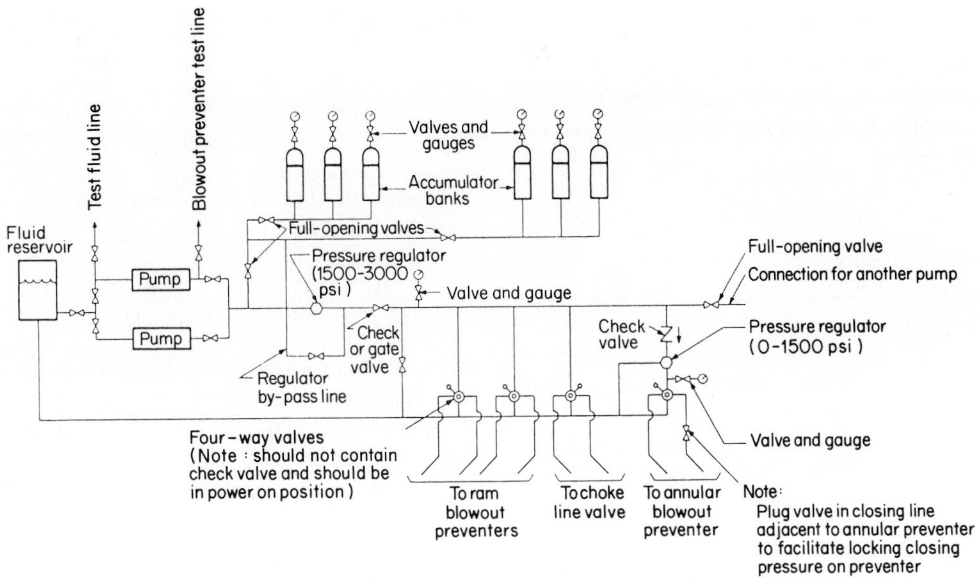

Fig. 13.8. Typical blowout preventer closing unit arrangement[1]. (Courtesy of API)

Air-operated or electrical pumps are used to force hydraulic fluid from a reservoir into the bank of accumulators until the working pressure of the system is achieved. For example, the working pressure required to close an annular preventer is 1500 psi.

Entry of the hydraulic fluid into the accumulator causes the nitrogen gas to occupy a much smaller volume and, in turn, have a much higher pressure; see Figure 13.9.

The increased gas pressure will help in releasing the fluid at a much faster rate than can be achieved by most pumps, thereby allowing the preventers to be closed quickly. The accumulator is provided with a valve on the outlet connection which closes when the usable fluid charge is exhausted. This is required to preserve the precharged nitrogen gas. The usable fluid charge is normally $\frac{2}{3}$ of the total fluid charge, and is defined as the amount of fluid which can be recovered as the pressure in the accumulator drops from the working pressure to 1200 psi. The 1200 psi represents the pressure required to hold an annular preventer closed.

The procedure for determining the total volume of hydraulic fluid required and, in turn, the number of accumulators for closing a set of BOPs is best illustrated by an example.

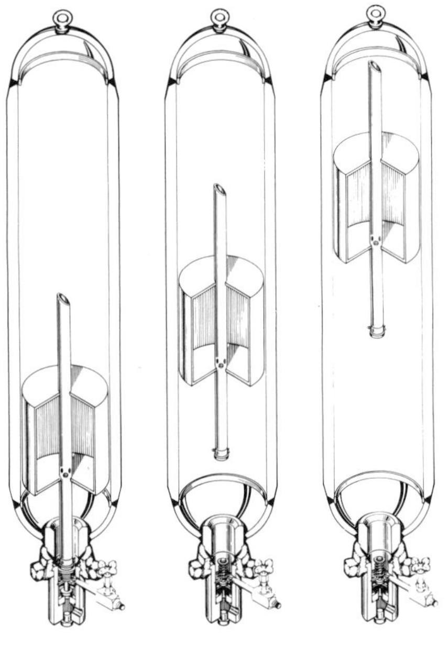

Fig. 13.9. Float separating nitrogen and fluid: (a) 1000 psi, precharge; (b) 2000 psi, charge; (c) 3000 psi, full charge. (Courtesy of Cameron Iron Works)

Example 13.1

Determine the volume of fluid and number of accumulators required to close, open and close a stack of BOPs consisting of an annular preventer and two ram-type preventers.

Solution

Construct the following table:

Preventer type	*Volume of fluid* (gal)			
	Close	*Open*	*Close*	*Total*
Annular	x	y	x	$2x+y$
Ram	z	L	z	$2z+L$
Ram	z	L	z	$2z+L$
Grand total				$2x+y+2L+4z$

The required volume to close, open and close the preventers is the usable volume of fluid, the latter being $\frac{2}{3}$ of the total fluid charge of an accumulator.

Thus,

$$\text{total volume required, } V_T = \frac{\text{usable volume}}{\frac{2}{3}}$$

$$= 1.5 \times \text{usable volume}$$

$$V_T = 1.5 \times (2x + y + 2L + 4z)$$

Number of accumulators

$$= \frac{\text{total volume}}{\text{capacity of each accumulator}}$$

$$= \frac{V_T}{\text{capacity of each accumulator}}$$

If Cameron-type preventers are used, the usable volume of fluid required to close, open and close a $13\frac{5}{8}$ in, 5000 psi working pressure BOP stack is as shown below:

	Close (gal)	*Open* (gal)	*Close* (gal)	*Total* (gal)
Annular BOP	12.1	10.3	12.1	34.5
Ram (Type U) preventer	5.8	5.5	5.8	17.2
Ram (Type U) preventer	5.8	5.5	5.8	17.1
Grand total:				68.7

Allowing a 25% reserve, the total usable volume

$$= 1.25 \times 68.7 = 85.9 \text{ gal}$$

$$\text{Total volume required} = 1.5 \times \text{usable volume}$$

$$= 1.5 \times 85.9 = 128.9 \text{ gal}$$

Using Boyle's Law, determine the total accumulator volume (V_3) of Nitrogen and fluid from:

$$V_3 = \left(\frac{V_R}{(P_3/P_2) - (P_3/P_1)}\right)$$

where V_R = total usable fluid including a safety factor; P_1 = maximum pressure of accumulator when completely charged (3000 psi); P_2 = minimum operating pressure of accumulator (1200 psi); P_3 = Nitrogen precharged pressure (1000 psi)

$$\therefore\ V_3 = \frac{128.9}{\frac{1000}{1200} - \frac{1000}{3000}} = 257.8 \text{ gal}$$

$$V_3 \simeq 258 \text{ gal}$$

Assuming an accumulator size of 10 gal, the required number of 10 gal bottles is given by

$$\frac{258}{10} = 25.8 \simeq 26$$

Using the total accumulator volume (V_3) the required size and type of pump may be selected by reference to the manufacturer's catalogue.

Charging pumps

Charging pumps can be air- or electrically-driven and are normally powered by two independent sources. The air power can be supplied by an air compressor or an air storage tank. A separate generator is required to provide the electric power.

Figure 13.10 shows a dual air/electric and dual electric system for powering the charging pumps.

Fluid reservoir

The fluid reservoir contains the hydraulic fluid used to charge the accumulators and acts to receive the hydraulic fluid upon opening the preventers. The total capacity of the reservoir should at least be equal to twice the usable fluid capacity of the accumulator system[1].

Hydraulic oil or fresh water-soluble oils are normally used with glycol as an additive when working at temperatures below 32 °F (0 °C).

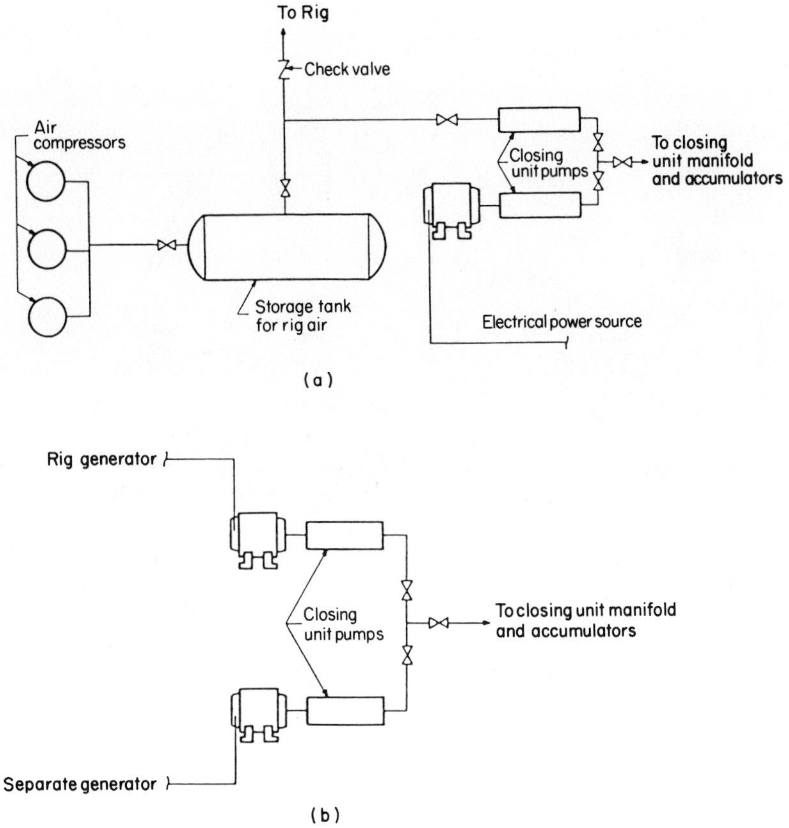

Fig. 13.10. (a) Typical dual air/electric systems for operating closing unit pumps; (b) typical dual electric systems for operating closing unit pumps[1]. (Courtesy of API)

Manifold and pipings

Each preventer is provided with two lines, an opening and a closing line, and a four-way valve (see Figure 13.8). The lines are made of seamless steel with a working pressure equal to or greater than the working pressure rating of the BOPs stack, up to 5000 psi[1].

Choke manifold

When a well kick is detected, the fluid influxes are circulated out through a pipe connected to the ram preventer, known as a 'choke line' (see Figure 13.11).

The choke line is connected to a choke manifold comprising adjustable choke valves, fittings and lines to direct the fluids either to the mud pit or to a flare line. The choke manifold is normally positioned at a short distance from the rig floor, as shown in Figure 13.11. The manifold usually employs, as a minimum, two adjustable chokes, one manually operated and one hydraulically operated (Figure 13.12).

To adjust the flow, the choke employs a cylindrical gate, or needle, moving in and out of a cylindrical seat, the range of adjustment being from fully closed up to the maximum effective bore of the seat.

It should be noted that chokes are not always capable of being closed off pressure-tight; indeed some types may actually suffer damage if an attempt is made to do so. Gate valves upstream of the choke are used for this purpose. The chokes have a device built in to indicate the degree of opening to which the choke is set. On drilling chokes this is usually fairly coarsely calibrated (i.e. $\frac{1}{8}$, $\frac{1}{4}$, $\frac{1}{2}$, etc.) and relates to the linear position of the gate relative to the seat and not to effective flow area.

Occasionally, manifolds may be equipped with production type chokes where the degree of opening is indicated in $\frac{1}{64}$ in of effective seat diameter, i.e. a setting of 32/64 indicates an opening equivalent to a $\frac{1}{2}$ in diameter orifice (or bean, as choke restrictions are commonly referred to) and a 2 in maximum orifice choke would have a maximum setting of 128/64.

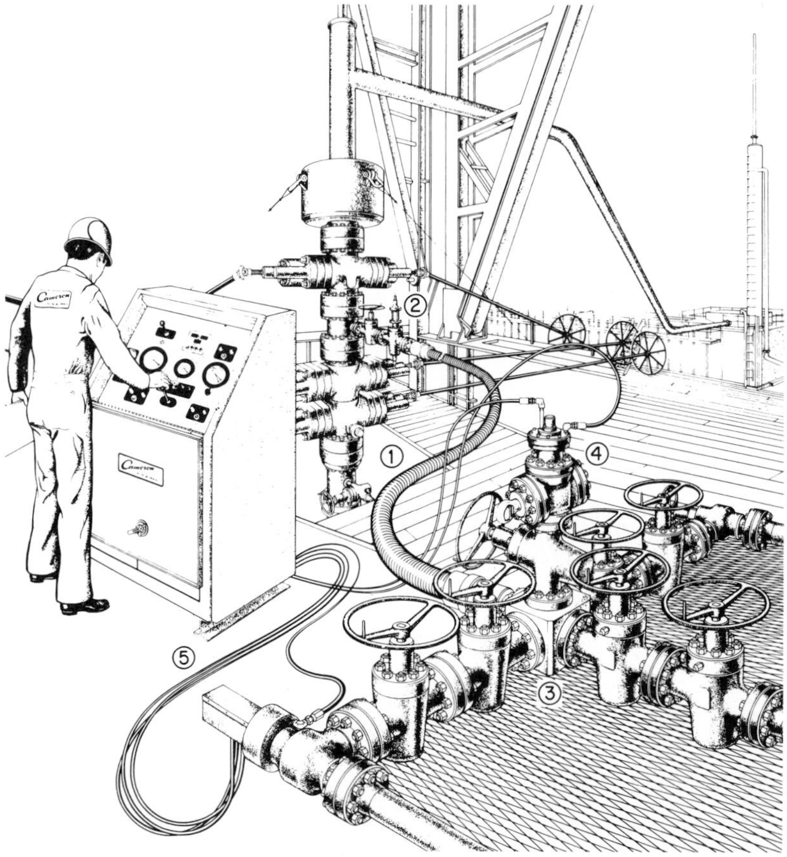

Fig. 13.11. Drilling choke system: (1) choke line; (2) hydraulically actuated valve; (3) choke manifold; (4) remotely-actuated drilling choke; (5) pressure hoses. (Courtesy of Cameron Iron Works)

The manual chokes are usually operated with a simple hand-wheel.

The hydraulically operated chokes do not usually have a means of local operation but must be manipulated from a control panel mounted adjacent to the drill floor. The control panel contains gauges for indicating drill pipe pressure, manifold pressure and choke position. It may also be equipped with a means of automatic choke control which will open and close the choke as required to maintain a pre-selected value of manifold pressure.

A section through a hydraulically actuated choke is shown in Figure 13.12(b).

The API recommended manifolds for pressures of 2000 and 3000 psi, 5000 psi and 10 000 psi are shown in Figure 13.13. For high-pressure operations (5000 psi or more), the choke line is fitted with a remotely actuated valve, usually fitted on the BOP outlet.

The kill line

In the event that heavy mud cannot be circulated down the drillpipe (or kelly), an auxiliary line must be used to allow pumping of heavy mud down the annulus. This line is known as a 'kill line'. One end of the kill line is connected to a side outlet below the ram preventer that is likely to be closed, while the other end is connected to the rig mud pumps. API recommends[1] that the kill line should be equipped with valves according to the pressures expected, as shown in Figure 13.14.

KICK CONTROL

As mentioned previously, a well kick is an unwanted flow of formation fluids into the well bore which may (if not controlled) develop into a blowout. Serious

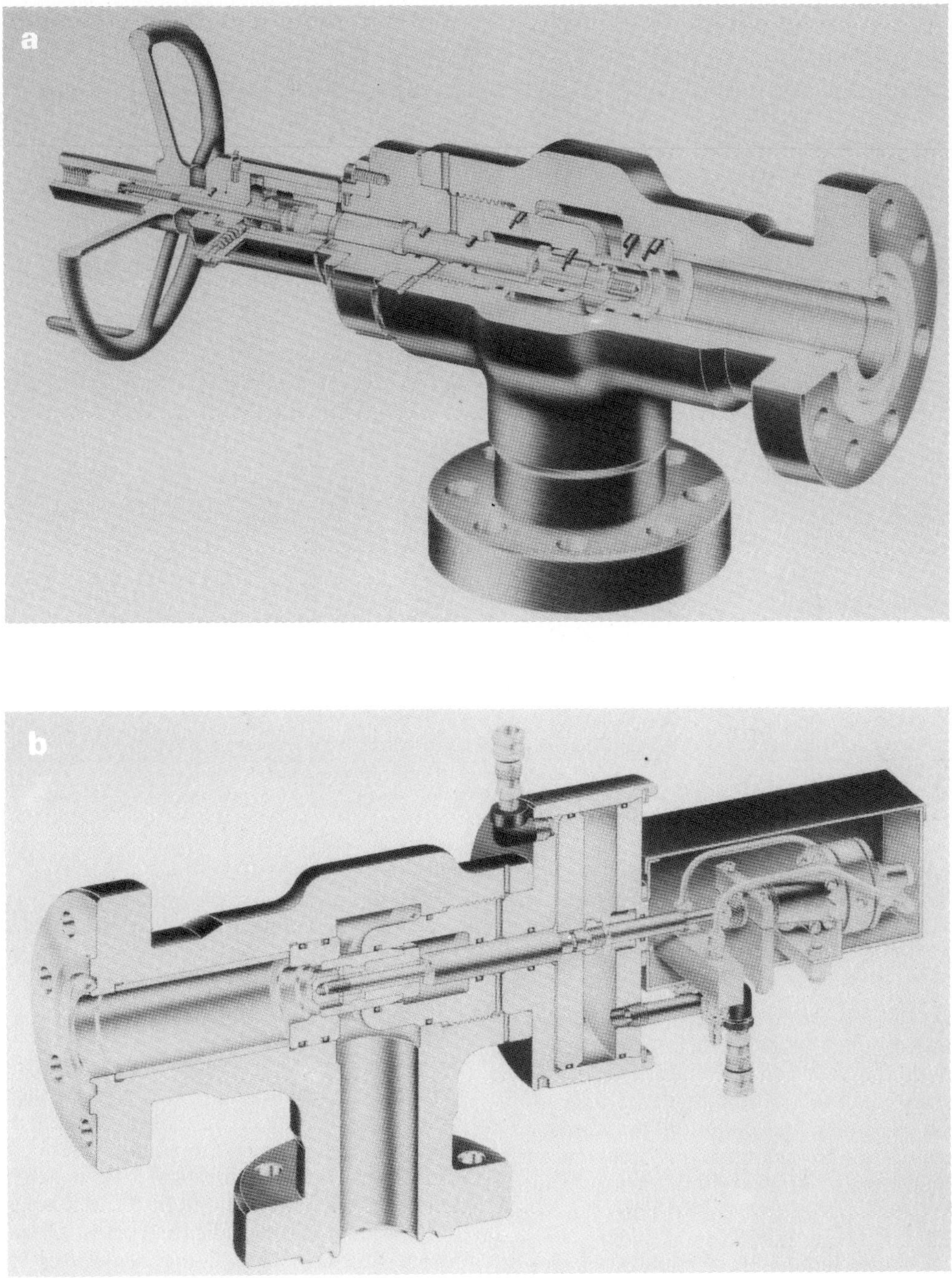

Fig. 13.12. (a) Manually-actuated drilling choke; (b) hydraulically-actuated drilling choke. (Courtesy of Cameron Iron Works)

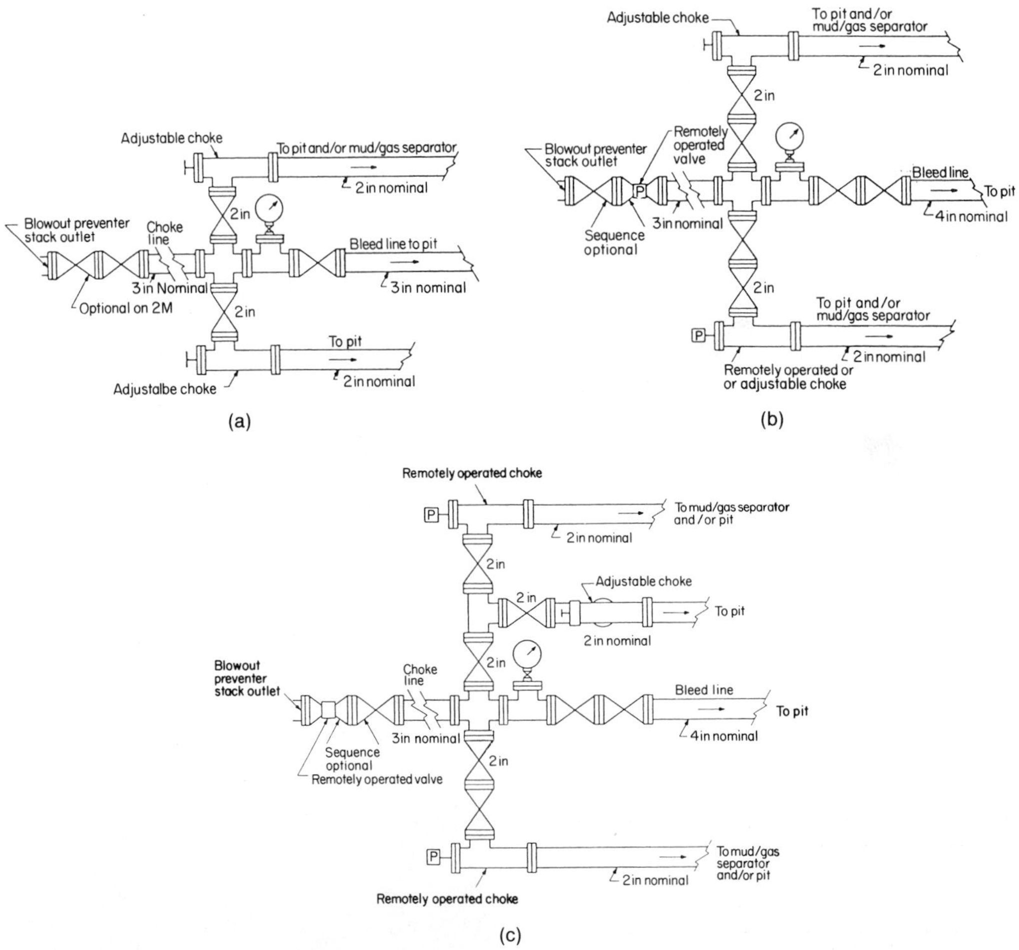

Fig. 13.13. (a) Typical choke manifold assembly for 2M and 3M rated working pressure service, surface installation; (b) typical choke manifold assembly for 5M rated working pressure service, surface installation; (c) typical choke manifold assembly for 10M and 15M rated working pressure service, surface installation[1]. (Courtesy of API)

consequences of a blowout include: (a) damage to rig equipment; (b) injury to personnel; (c) wastage of a great deal of the reservoir's natural energy to the extent that the reservoir can be rendered uneconomical; and (d) environmental pollution and damage.

A blowout is not an instantaneous occurrence that is suddenly seen at the surface without prior warnings. Several surface signs are normally observed which indicate that a well kick is developing. A large volume of literature has accumulated on this subject and the interested reader is advised to consult Reference 2.

Surface indications of a well kick

The following is a summary of surface indications of a developing well kick.

Increase in pit level

In normal drilling operations, drilling mud is continuously circulated down the hole such that the volume pumped in is equal to the volume of returns at the surface (ignoring the small volume loss due to filtration). Two situations can arise which disturb this balance. First, the volume of returns is less than the volume pumped in, indicating a lost circulation situation. Second, the volume of returns exceeds the volume pumped in, indicating that formation fluids have invaded the well. In the second case the well is said to be 'kicking'.

During a kick, formations unload their fluids into the well bore, thereby increasing the volume of liquid inside the well. The excess fluid seen at the surface is

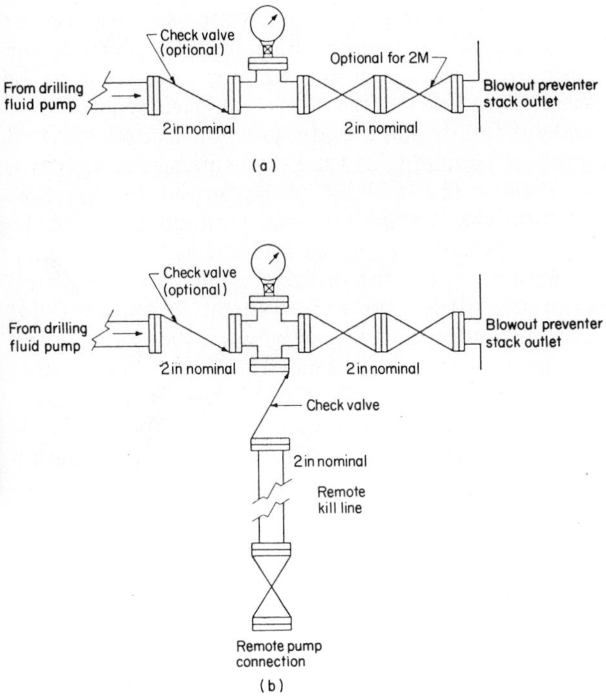

Fig. 13.14. Typical kill line assemblies[1]; (a) for 2M and 3M rated working pressure service, surface installation; (b) for 5M, 10M and 15M rated working pressure service, surface installation. Threaded connections optional for 2M rated working pressure service. (Courtesy of API)

the volume of the influx fluid. Normally the excess fluid manifests itself as an increase or gain in the pit volume. Mud engineers constantly monitor the volume of mud in the pits to watch for signs of a kick.

Increase in penetration rate

A sudden increase in penetration rate can take place when soft or abnormally pressured formations are encountered. Frequent changes in penetration rate occur due to changes in formation type being drilled. Hence, this indication is not always conclusive and must be used with other surface indications to determine whether a well is kicking or not.

In general, penetration rate is dependent on rotary speed, weight-on-bit, type of bit, formation hardness, type of drilling mud and hydraulics. For efficient drilling, these variables are kept at optimum levels and a normal drilling trend in a given area can be established. Any deviation from this trend is an indication of penetrating a soft formation or an abnormally pressured zone. Increased penetration rate can also be detected at the surface by the increased volume of cuttings on the shale shaker and by the large size of the individual chips.

Decrease in circulation pressure

During a well kick, fluids such as salt water, oil or gas enter the well bore and intermix with the existing drilling mud. This mixing often results in a new fluid of reduced viscosity and density, which, in turn, results in lower annular pressure losses. Thus, the total pressure losses seen at the surface are now smaller than before a kick is encountered. The decrease in circulation pressure is very pronounced when gas invades the well bore, since gas has a much lighter density than that of mud.

When lighter fluids enter the well bore, the circulation pressure is further reduced because of the U-tubing effect resulting from density differences between the heavy mud inside the drillpipe and lighter fluids in the annulus. The decrease in pump pressure is also accompanied by an increase in pump speed, as the same input power is now available to circulate the same volume of fluid against the reduced pressure losses. An obvious and definitive surface indication of a kick is “WELL FLOW” with pumps off.

Gas-, oil- or salt water-cut mud

Shows of gas, oil or salt water in drilling mud are clear indications that formation fluids have invaded the well.

Gas shows are particularly misleading, since, during tripping operations, the well can be swabbed (or sucked in), allowing small quantities of gas to enter the well bore. This gas is normally termed ‘trip gas’. It must not be confused with gas from a kicking formation. Swabbing occurs when there is a marginal overbalance or when the bit is balled up.

For each area, a normal trend for the quantity of trip gas can be established, and deviation from this trend can be taken as a sign of an impending gas kick.

Another form of gas feed that must not be confused with a gas kick is termed ‘drilled gas’. Drilled gas enters the mud as drill cuttings of gas-bearing formations travel up the annulus and release their gas content because of the reduced pressures up-hole. Drilled gas causes a decrease in mud density near the surface and does not cause a major reduction in the equivalent annular mud weight.

Shows of invading fluids are seen at the surface as slugs in the case of oil and salt water and as frothy bubbles in the case of gas. Gas-cut mud is particularly characterised by a lower value of surface mud density compared with the original density. A feed of formation water causes a much smaller change in density but can be detected by increase in pit volume and increased salinity.

Chloride increase

The quantity of chloride in mud is a good indicator for monitoring well kicks. The mud engineer routinely measures the quantity of chloride (Cl^-) in mud to establish a datum level for Cl^-.

Most formation fluids contain high levels of salinity compared with those of the drilling mud. An indication that formation fluids have invaded the well is obtained when the measured Cl^- content of mud exceeds the datum level.

Killing a well

Once a kick is adjudged to have taken place, the well is shut in at the preventers and at the drillpipe, as shown in Figure 13.15. The stabilised casing shut-in pressure (CSIP) and stabilised drillpipe shut-in pressure (DPSIP) are then recorded. It should be noted that the primary functions of shutting the well are to prevent further fluid entry and to permit the calculation of formation pressure, the DPSIP being used as a guide.

The killing operation is based on the premise that the drillpipe and annulus are two limbs of a U-tube configuration, such that any change in one limb directly produces an equivalent change in the other limb. Thus, when the well is dead, the pressures at the top of the U-tube limbs (i.e. at the surface) are zero. Any bottom hole pressure, such as arising from a well kick, will exert pressure on both limbs of the U-tube, which results in surface pressures at the drillpipe and the annulus.

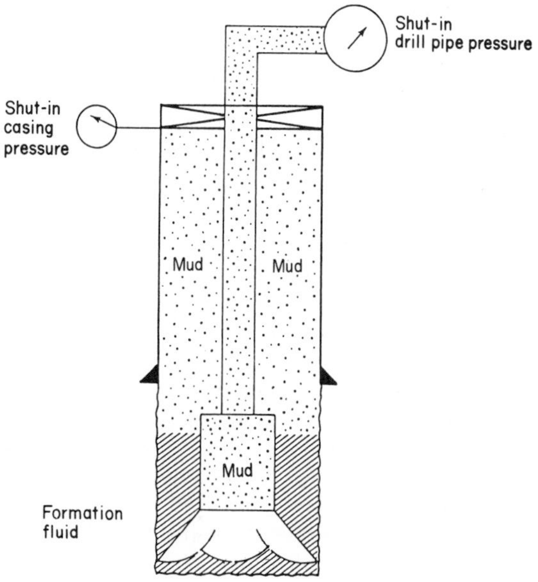

Fig. 13.15. Schematic drawing of a shut-in well.

Prior to shutting the well, it is necessary to raise the kelly above the rotary table and to position the drillpipe tool joint above the pipe rams. It must be remembered that pipe rams are designed to close around the drillpipe body and not around the tool joint (or coupling), or the kelly. Raising the kelly will also allow the drill bit to be moved off the hole bottom. The annular preventer is closed first, then the pipe rams and, finally, the drillpipe at the surface.

Assuming that the drillpipe is full of drilling mud prior to shutting the well, then the formation fluids can only enter the annulus between the drillpipe and the hole, as shown in Figure 13.15.

Thus, the fluid inside the drillpipe is the original drilling mud having a gradient G_m psi/ft (bar/m), while the annulus contains a mixture of mud and formation fluid(s) whose gradient is unknown. Formation fluids will continue to enter the well bore until the hydrostatic pressures exerted by mud inside the drillpipe and DPSIP, and by the fluid mixture in the annulus and CSIP, are equal to the formation pressure. When this happens, further entry of formation fluid(s) will cease and the shut-in pressures on drillpipe and casing assume constant values. The initial shut-in pressures are invariably less than the final stabilised shut-in pressures, and a finite time must elapse before the stabilised pressures are obtained.

On the drillpipe side, the stabilised shut-in pressure represents the amount by which the formation pressure exceeds the hydrostatic pressure of mud inside the drillpipe. Thus, the drillpipe acts as a bottom hole pressure gauge indicating the difference between the formation pressure and the hydrostatic pressure of mud. On the annulus side, the individual pressures are not precisely known, as part of the annulus is occupied by formation fluid and the other part by drilling mud. The density(ies) and height(s) of formation fluid(s) are not known, which makes the determination of formation pressure from annulus pressures difficult. This can be seen from the following equation which considers pressures on the annulus side:

formation pressure

= CSIP + hydrostatic pressure of mud

+ hydrostatic pressure of formation fluid(s)

However, on the drillpipe side, the following equation holds:

formation pressure

= shut-in drillpipe pressure

+ hydrostatic pressure of mud inside drillpipe

$$P_f = \text{DPSIP} + H_s \tag{13.1}$$

where P_f = formation pressure; DPSIP = drillpipe shut-in pressure; and H_s = hydrostatic pressure of mud inside drillpipe.

Thus, information on the drillpipe can be used directly to calculate the formation pressure. This is the reason why all calculations involving a well kick use drillpipe pressures only.

In summary, when a kick is detected, the well is shut in to record the DPSIP and CSIP. A remotely operated valve on the choke line is normally opened first prior to closing the preventers, to allow the annular mud to flow through the choke line via the choke manifold and finally to the mud pit.

The annular preventer is closed first, as this can close around any shape and size and can be used immediately a kick is observed. The ram preventer can be closed once the position of the pipe tool joint is determined. If the tool joint happens to be in the ram preventer, the kelly is raised to position the drillpipe body inside the ram preventer.

Once the preventers are closed, the choke is closed slowly, to avoid surging pressures which can be caused by the sudden stop of the fluid movement in the annulus.

The drillpipe is normally shut in by closing the lower kelly cock after the DPSIP is measured on the standpipe or on a remote choke panel.

Kill mud

Using Equation (13.1) the new mud density (P_2) required to balance the formation pressure is given by:

$$P_f = \rho_2 g h$$

or

$$\rho_2 = \frac{P_f}{gh}$$

In Imperial units

$$\rho_2 = \frac{144 \times P_f\,(\text{psi})}{h\,(\text{ft})}\ \text{pcf} \qquad (13.2)$$

or

$$\rho_2 = \frac{P_f\,(\text{psi})}{0.052 \times h\,(\text{ft})}\ \text{ppg} \qquad (13.2a)$$

In metric units

$$\rho_2 = \frac{P_f\,(\text{bar})}{0.0981 h\,(\text{m})}\ \text{kg/l} \qquad (13.3)$$

Thus, the well can now be killed by replacing the original mud (ρ_1) by a heavy mud of density ρ_2. After the well is killed, a safety factor of 100–200 psi is added to the drilling mud before normal operations resume.

The casing shut-in pressure plays a major role in the process of well killing. It is used as a monitor to prevent bursting of the casing and fracturing of the formation below the casing shoe. The CSIP must not be allowed to exceed the burst resistance of the casing, to prevent rupturing of the casing. In practice, the maximum allowable CSIP is taken equal to 70% of the burst resistance of casing.

Also, for a given casing setting depth, the formation just below the casing shoe has an initial strength (normally calculated as the product of casing setting depth and overburden stress of 1 or 0.8 psi/ft; see Chapter 9) which must not be exceeded. Thus, the maximum allowable CSIP is also dictated by the formation strength, as shown by the following equations;

maximum bottom hole pressure < formation strength

and

maximum bottom hole pressure

= hydrostatic pressure of mud in the annulus

+ surface pressure

In our case, surface pressure is the maximum allowable CSIP.

In practice, the maximum allowable CSIP to prevent fracturing of the formation below the last casing shoe and the 70% value of the burst resistance of the casing are recorded every time a new string of casing is run. The lowest of these two values is used during well-killing operations.

Pumping speed

An important consideration in well killing is the speed at which the heavy mud is circulated. The reader may recall from Chapter 7 that circulation pressure is proportional to volume flow rate raised to some power, *n*, normally between 1.8 and 2. The pump speed is normally quoted in strokes (of pump) per minute, each stroke being capable of delivering, for example, 0.1 bbl (0.016 m^3) of mud per minute. Thus, circulation pressure can be reduced by simply reducing the volume flow rate or number of strokes per minute. In well killing, excessive annular circulation pressures may cause lost circulation, thereby complicating the well-killing procedure even further.

Reduced pump pressure will give the rig personnel time to prepare the heavy mud and reduce the total pressure requirement from the mud pumps. In addition, pressure changes on the drillpipe and casing can be easily monitored with reduced pump pressure, allowing modifications to the killing procedure to be made, as necessary. The reduced pump speed is also described as the kill rate or kill speed.

Pressure variation during the kill operation

As discussed previously, only the drillpipe pressures are used for determining the required kill mud weight. Also required is the initial circulation pressure, P_{c1}, necessary to circulate the light mud at reduced pump speed.

Essential to all killing operations is the maintenance of constant bottom hole pressure (= formation pressure) to prevent further entry of formation fluid. This is achieved by maintaining a constant pumping speed while circulating the kill mud.

When the shut-in well is stable, Equation (13.1),

$$P_f = \text{DPSIP} + H_s$$

holds and the standpipe pressure is equal to DPSIP. When the pump is brought up to speed, the standpipe pressure reads P_{c1} plus DPSIP. However, the bottom hole pressure is still equal to DPSIP + H_s, since all the circulation pressure, P_{c1}, is lost in overcoming friction inside the drillpipe and drill collars and across the bit.

As the kill mud gradually fills the drillpipe, the DPSIP decreases and finally becomes zero when the entire drillpipe is full of the kill mud. At this point, the hydrostatic pressure of mud inside the drillpipe is equal to the formation pressure, P_f. The pressure on the standpipe is now equal to the pumping pressure required to circulate the kill mud, P_{c2}. Notice that the value of this circulation pressure is higher than P_{c1}, since the pump is now circulating a heavier mud. The new circulation pressure P_{c2} is calculated from Equation (13.4):

$$P_{c2} = \frac{\rho_2}{\rho_1} \times P_{c1} \tag{13.4}$$

where ρ_2 = kill mud weight; and ρ_1 = initial mud weight.

The drillpipe circulation pressure, P_{c2}, is maintained constant as the heavy mud fills the annulus until the CSIP pressure drops to zero. At this stage of the killing operation, the drillpipe and the annulus are full of the heavy mud, and the shut-in pressures on the drillpipe and the annulus are zero when the pump is shut off.

The pressure variation on the drillpipe side is plotted graphically in Figure 13.16 against the number of bbl (or time). Initial DPSIP is marked as point 1. When pumping of kill mud is started, the total surface pressure becomes DPSIP + P_{c1} and this is marked as point 2 on Figure 13.16. Point 3 represents the circulation pressure, P_{c2}, when the heavy mud fills the drillpipe. The straight line 3–4 represents the period during which the standpipe pressure is kept constant at P_{c2} as the annulus is gradually filled with the kill mud.

The construction of Figure 13.16 using actual well data will be presented in Example 13.2.

Using the pumping speed, the times required to fill the drillpipe and the annulus with the heavy mud can be calculated as follows:

time to fill drillpipe with heavy mud

$$= \frac{\text{capacity of drillpipe (bbl)}}{\text{pumping speed (bbl/min)}} \tag{13.5}$$

time to fill annulus with heavy mud

$$= \frac{\text{capacity of annulus (bbl)}}{\text{pumping speed (bbl/min)}} \tag{13.6}$$

Thus, points 5 and 6 in Figure 13.16 represent the times required to fill the drillpipe and the annulus, respectively. Line 0–5 can be divided into minutes and can be used to predict the pressures on the standpipe

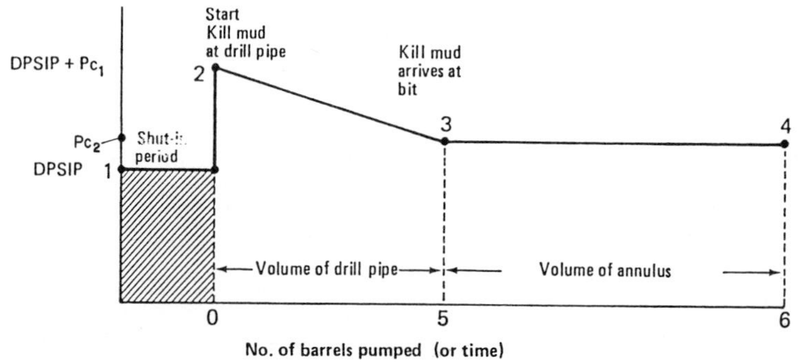

Fig. 13.16. Pressure variation across drillpipe.

while the killing operation is in progress. If there is any difference between the observed and the predicted values of standpipe pressure, the choke is adjusted until the predicted value is shown on the standpipe pressure gauge.

Pressure variation along the casing will depend on the nature of the influx fluid. In the case of oil or water kicks, the casing shut-in pressure increases slightly as the kill mud fills the drillpipe, mainly because of the U-tubing effect caused by the presence of heavy mud in one limb of the U-tube. Annulus pressure starts decreasing as the kill mud begins to fill the annulus and finally drops to zero when the kill mud arrives at the surface. This behaviour is shown in Figure 13.17. Pressure variation during a gas kick is discussed in the next section.

Effects of gas expansion on well-killing operations

Gas is a highly compressible fluid whose volume is dependent on both pressure and temperature. Gas is also much lighter than mud and, therefore, percolates to the top of a column containing mud and gas.

Figure 13.18 shows two hypothetical situations in which equal quantities of gas at the same pressure are injected to the bottom of an open tube and a closed tube. Both tubes are filled with the same type of mud having a density of 75 lbm/ft^3.

Since gas is lighter than mud, it rises up the tube and occupies the upper part of the tube. In the open tube the rise of the gas is accompanied by a gradual decrease in pressure and gradual increase in volume. Assume that 1 bbl of gas is injected to the bottom of tube at 5000 psi; then the volume of gas at the surface, V_S, can be determined by use of the gas law:

$$PV = ZRT$$

or

$$\frac{P_B V_B}{Z_B T_B} = \frac{P_S V_S}{Z_S T_S} \tag{13.7}$$

where P = absolute pressure; V = volume; B denotes bottom of tube; S denotes top of tube; T = temperature; and Z = compressibility factor, which is a function of temperature and pressure.

Since the pressure of gas at the top of the tube is equal to atmospheric pressure, P_S, and assuming a bottom hole temperature of 240 °F, then Equation (13.7) becomes:

$$\frac{5000 \times 1}{(240 + 460)} = \frac{14.7 \times V_S}{60 + 460}$$

$T_S = 60°F$ (surface temperature); $P_S = 14.7$ psi; and $Z_S = Z_B$ (assumed). Therefore,

$$V_S = 253 \text{ bbl}$$

Thus, when the gas arrives at the top of the open tube, its volume will increase from 1 to 253 bbl while its pressure reduces from 5000 psi to an atmospheric pressure of 14.7 psi.

When the tube is closed at the top, the gas rises up the hole without expansion, thereby increasing the pressure all the way along the tube. The surface pressure recorded when the gas reaches the top of the tube will be the gas injection pressure of 5000 psi, as shown in Figure 13.18b. Calculations of pressure at intermediate stages will depend on the position of the gas bubble within the tube and the density of the mud inside the tube. Assuming that gas has migrated to a distance of 5000 ft below the surface and the tube capacity is 0.05 bbl/ft, the gas bubble will occupy a height of

$$\frac{1 \text{ bbl}}{0.05 \text{ bbl/ft}} = 20 \text{ ft}$$

The gas pressure from 5000 to 5020 ft is practically constant at 5000 psi. At 5000 ft the hydrostatic pressure of mud is $(5000 \times 75)/144 = 2604$ psi. Thus, at the 5000 ft interface there will be a difference in pressure of $5000 - 2604 = 2396$ psi. This excess pressure is seen at the surface as a deflection in the pressure gauge.

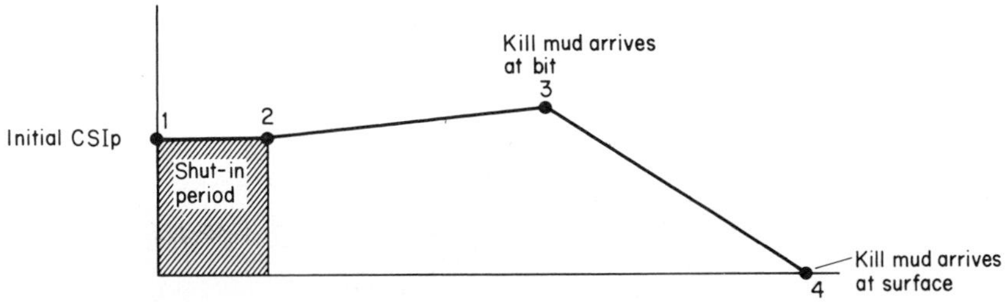

Fig. 13.17. Pressure variation across the annulus due to oil or water kick.

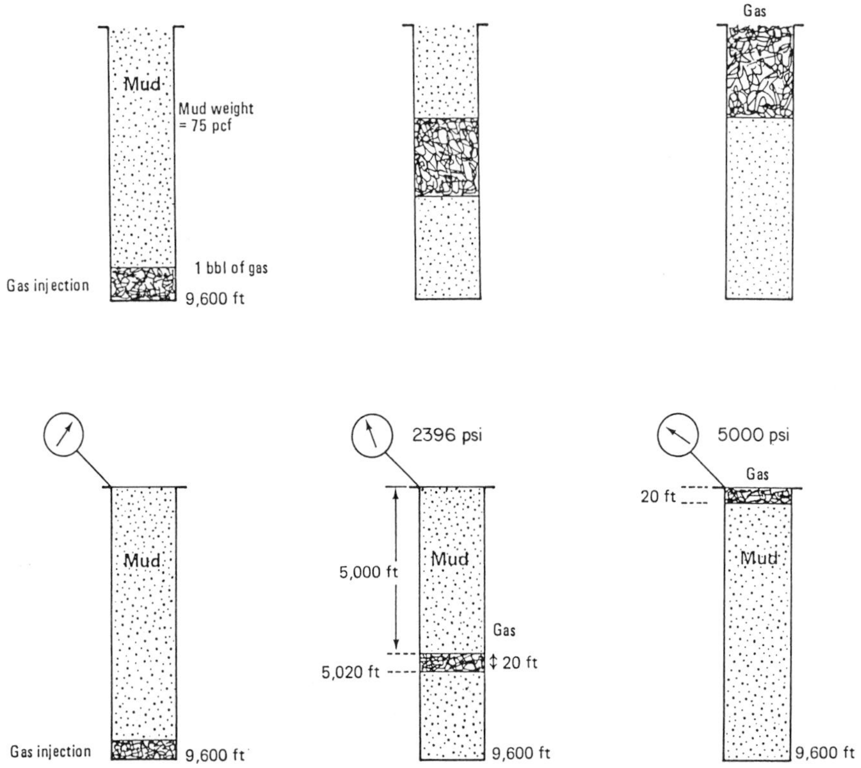

Fig. 13.18. (a) Tubes are open to the atmosphere; (b) tubes are closed at both top and bottom.

While the gas is rising up the tube, a continuous increase in the bottom hole pressure is produced. When the gas bubble reaches the surface, the bottom hole pressure (BHP) becomes

$$\text{BHP} = 5000 \text{ psi} + \text{hydrostatic pressure of mud}$$

$$= 5000 + \frac{75 \times (9600 - 20)}{144}$$

$$= 9990 \text{ psi}$$

(*Note:* The gas height is 20 ft.)

In our example, if the tube was a real well in which the formation fracture gradient is 0.8 psi/ft, then

formation fracturing pressure

$$= 0.8 \times 9600 = 7680 \text{ psi}$$

and

bottom hole pressure if gas expansion is prevented

$$= 9990 \text{ psi}$$

If the above situation is allowed to take place, the formation will fracture, causing a lost circulation situation together with the existing kick. If the formation should fracture near the surface, the existing kick could cause the gas to broach to the surface, causing a cratering of the surface and a possible loss of rig and injury to rig personnel.

If the formation should fracture near the hole bottom, the formation fluids must first be circulated out and the fracture zone must then be plugged off. This type of situation is difficult to control.

Gas expansion must, therefore, be allowed to take place, to reduce the surface casing pressure (a) to prevent bursting of casing and (b) to keep the total bottom hole pressure below the formation fracturing pressure.

Variation of casing pressure as a gas bubble is circulated out of a real well is graphically shown in Figure 13.19. Point (1) depicts the shut-in well as the kill mud is ready to be pumped in. Point (2) depicts the well position after the kill mud occupies the entire drill string. Between points (1) and (2) the casing shut-in pressure rises because of gas expansion which results in less hydrostatic pressure (P_f = CSIP + hydrostatic pressure of fluids in annulus) and, in turn, higher surface pressure.

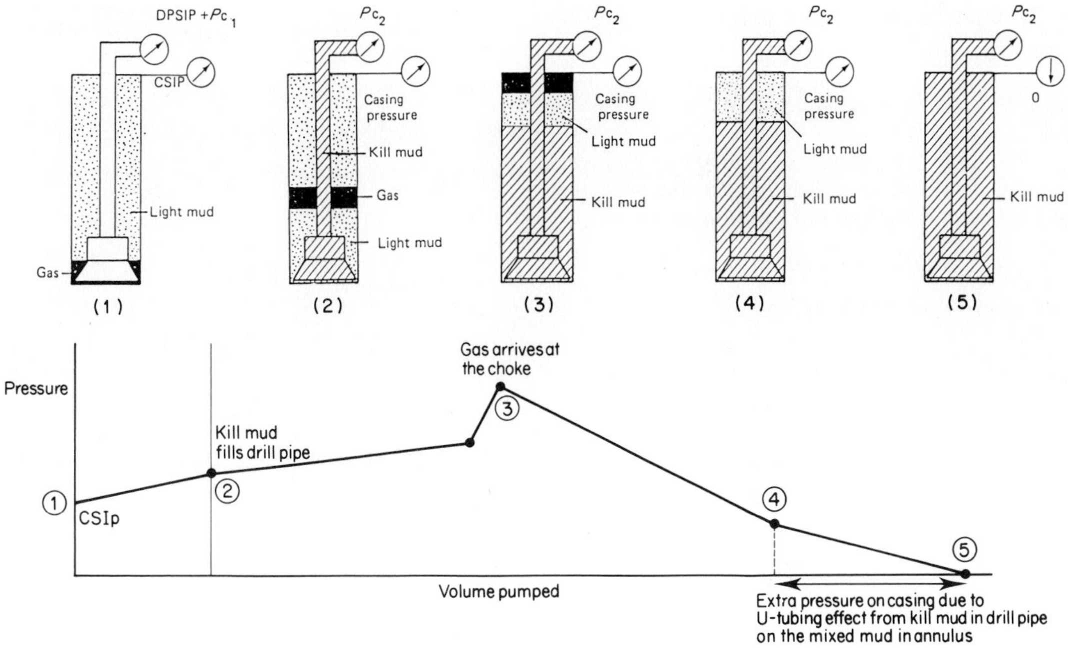

Fig. 13.19. Variation in annulus pressure as a gas bubble is circulated out of a well: DPSIP, drillpipe shut-in pressure; CSIP, casing shut-in pressure; P_{c1}, initial circulation pressure with light mud; P_{c2}, final circulation pressure with kill mud.

Beyond point (2), the kill mud begins to fill the annulus and the gas bubble approaches the surface. When the gas arrives at the surface, a sudden increase in casing pressure is observed, as depicted at point (3). Gradually the casing pressure decreases as the gas is bled off and the annulus is filled with the light and kill muds. At point (4) all the gas is bled-off and the light mud arrives at the surface, producing a small surface casing pressure due to the U-tubing effect. The U-tubing effect results from the kill mud in the drillpipe exerting a higher pressure on the hole bottom than the pressure of the mixed columns of light and kill mud in the annulus.

At point (5) the kill mud arrives at the surface and the casing pressure is zero.

Kill methods

Depending on whether the required kill mud can be prepared quickly in one or in two stages, two methods have been used extensively — the wait-and-weight method and the driller's method.

The wait-and-weight method The name 'wait-and-weight' arises from the fact that a given time (waiting time) must be allowed for the kill mud (or weighted mud) to be prepared. The method uses one circulation to remove the influx fluids and kill the well at the same time.

The heavy mud is pumped at a selected reduced speed and the total standpipe pressure, consisting of DPSIP and P_{c1}, is noted. The pumping speed is held constant until the kill mud fills the entire drillpipe. During this period, the drillpipe pressure is walked down from P_{c1} + DPSIP at the start of circulation to P_{c2} when the kill mud arrives at the bit by slowly opening the choke at the annulus. The manual reduction in drillpipe pressure is achieved in a convenient number of increments; see Example 13.2 and Figure 13.21.

The pressure P_{c2} is then held constant (by adjusting the choke size) until the kill mud fills the annulus and arrives at the surface.

The variation in casing pressure with this method will be dependent on the type of formation fluid circulated out. In the case of oil or water, fluctuations in casing pressure are much smaller than with gas. Circulation of gas causes increases in casing pressure until all the gas is removed from the well. During this period the casing pressure must not be allowed to exceed the 70% casing burst resistance value or the maximum allowable CSIP which can cause formation fracture.

The wait-and-weight method is best illustrated by an example.

Example 13.2

During drilling of an $8\frac{1}{2}$ in hole at 10 000 ft, a kick was encountered. The well was shut in and the pressures recorded on both drillpipe and annulus were:

$$\text{DPSIP} = 200 \text{ psi}$$

$$\text{CSIP} = 400 \text{ psi}$$

Other relevant data include:

last casing = $9\frac{5}{8}$ in, N80, 43.5 lbm/ft, ID = 8.755 in
casing setting depth = 8600 ft
drill collars = OD/ID:8 in/3 in, 500 ft
drillpipe = OD/ID:5 in/4.276, 19.5 lb/ft
circulation pressure at normal speed = 2000 psi at 60 strokes per minute
circulation pressure at 30 strokes/min = 500 psi
present mud weight = 75 pcf (10 ppg)

Calculate the following:

(1) capacity of drillpipe and drill collars;
(2) capacity of annulus;
(3) maximum allowable casing pressure;
(4) formation pressure;
(5) kill mud weight;
(6) standpipe pressure at start of circulation of heavy mud;
(7) final circulation pressure and standpipe pressure;
(8) time required to replace the contents of the drillpipe with the kill mud (this is also referred to as the 'travel time');
(9) total time to replace the contents of the well with the kill mud;
(10) total number of strokes required assuming that the pump capacity is 0.1 bbls per stroke;
(11) graphically show the variation of drillpipe pressure with time and number of strokes using the information calculated from steps 6 to 10.

Solution

The well can be schematically represented, as shown in Figure 13.20.

Length of drillpipe = 10 000 − 500 = 9500 ft

$$\text{Capacity of drillpipe} = \frac{\pi}{4}(4.276)^2 \times \frac{1}{144} \times 9500$$

$$= 947.4 \text{ ft}^3$$

$$= 168.6 \text{ bbl (using 5.62 conversion factor)}$$

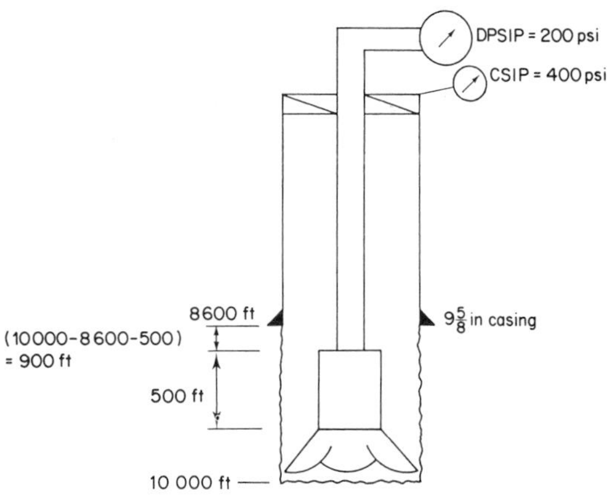

Fig. 13.20. Schematic representation of Example 2.

$$\text{Capacity of drill collars} = \frac{\pi}{4}(3)^2 \times \frac{1}{144} \times 500$$

$$= 24.54 \text{ ft}^3$$

$$= 4.4 \text{ bbl}$$

$$\text{Total capacity of drill string} = 168.6 + 4.4$$

$$= 173.0 \text{ bbl}$$

(2) Capacity of annulus = (a) annular capacity of cased hole around drillpipe + (b) annular capacity of open hole around drillpipe + (c) annular capacity of open hole around drill collars.

The annular capacities can be determined from the manufacturer's tables or by simply calculating the annular areas between the casing (or hole) and drillpipe (or drill collars) and multiplying by the appropriate length.

Thus, (a) annular capacity between $9\frac{5}{8}$ in casing of ID = 8.755 in and 5 in drillpipe = 0.0505 bbl/ft (or 0.2836 ft^3/ft). (b) annular capacity between open hole ($8\frac{1}{2}$ in diameter) and 5 in drillpipe = 0.0459 bbl/ft. (c) annular capacity between open hole ($8\frac{1}{2}$ in diameter) and 8 in drill collars

$$= \frac{\pi}{4}(8.5^2 - 8^2) \times \frac{1}{144}$$

$$= 0.0450 \text{ ft}^3/\text{ft}$$

$$= 0.008 \text{ bbl/ft}$$

Referring to Figure 13.20 for lengths of drillpipe in cased and open hole, we obtain

total annular capacity

$$= 0.0505 \times 8600 + 0.0459 \times 900 + 0.008 \times 500$$

$$= 479.6 \text{ bbl}$$

(3) From casing tables, the burst resistance of $9\frac{5}{8}$ in, N-80, 43.5 lb/ft is 5930 psi.

For added safety, the maximum casing shut-in pressure must not exceed 70% of the burst resistance.

Thus, the maximum allowable casing pressure (MACP)

$$= 0.7 \times 5930$$

$$= 4151 \text{ psi}$$

Since the CSIP is only 400 psi (much smaller than 4151 psi), it is therefore safe to shut in the well without bursting the casing. Note that in these calculations it is assumed that the casing is not supported, i.e. there is no back-up pressure behind the casing (see casing design example in Chapter 10).

(4) Formation pressure

$$= \text{hydrostatic pressure of mud inside drillpipe} + \text{DPSIP}$$

$$= \frac{75 \times 10\,000}{144} + 200$$

$$= 5408 \text{ psi}$$

(5) Mud weight required to just balance the formation pressure

$$= \frac{\text{formation pressure}}{\text{hole depth}} \times 144$$

$$= \frac{5408 \times 144}{10\,000} = 77.88 \text{ pcf}$$

Kill mud weight

$$= 77.87 + \text{safety factor (or overbalance)}$$

$$= 77.87 \text{ pcf} + 200 \text{ psi}$$

$$= 77.87 \text{ pcf} + \frac{200 \dfrac{\text{lb}}{\text{in}^2} \times \left(\dfrac{144 \text{ in}^2}{\text{ft}^2}\right)}{10\,000 \text{ ft}}$$

$$= 80.75 \approx 81 \text{ pcf}$$

Alternatively, the kill mud can be determined as follows.

Total hydrostatic pressure of kill mud

$$= \text{formation pressure} + 200 \text{ psi}$$

$$= 5408 + 200 = 5608 \text{ psi}$$

$$\text{Kill mud weight} = \frac{5608 \times 144}{10\,000} = 81 \text{ pcf}$$

(6) Standpipe pressure at start of circulation

$$= \text{DPSIP} + P_{c1}$$

$$= 200 + 500 = 700 \text{ psi}$$

(7) Initial circulation pressure with 75 pcf mud at reduced pump speed = 500 psi

Final circulation pressure

$$= P_{c1} \times \frac{\text{density of kill mud}}{\text{density of initial mud}}$$

$$= 500 \times \frac{81}{75}$$

$$= 540 \text{ psi}$$

(8) Time to fill drillpipe with heavy mud

$$= \frac{\text{drill string capacity}}{\text{pumping speed}}$$

$$= \frac{173 \text{ bbl}}{30 \text{ strokes/min}}$$

Assuming that the pump can deliver 0.1 bbl for every stroke, the travel time from surface to bit is

$$\frac{173 \text{ bbl}}{30 \times 0.1 \dfrac{\text{bbl}}{\text{min}}} = 57.7 \text{ min}$$

(9)

$$\text{Total time} = \frac{\text{total hole capacity}}{\text{pumping speed}}$$

$$= \frac{(173 + 479.6) \text{ bbl}}{30 \times 0.1 \text{ bbl/min}} = 217.5 \text{ min}$$

$$\simeq 218 \text{ min}$$

Thus, after 218 min, the kill mud should fill the complete well, and the DPSIP and CSIP will disappear when the pumps are shut off.

(10) Sometimes a stroke counter is used to monitor the volume of kill mud pumped in. Thus,

total number of strokes required to replace the entire well contents with the kill mud

$$= \frac{(173 + 479.6) \text{ bbl}}{0.1 \dfrac{\text{bbl}}{\text{stroke}}}$$

$$\text{Total number of strokes} = 6526$$

(11) A graph of standpipe pressure against time and number of strokes is shown in Figure 13.21, from

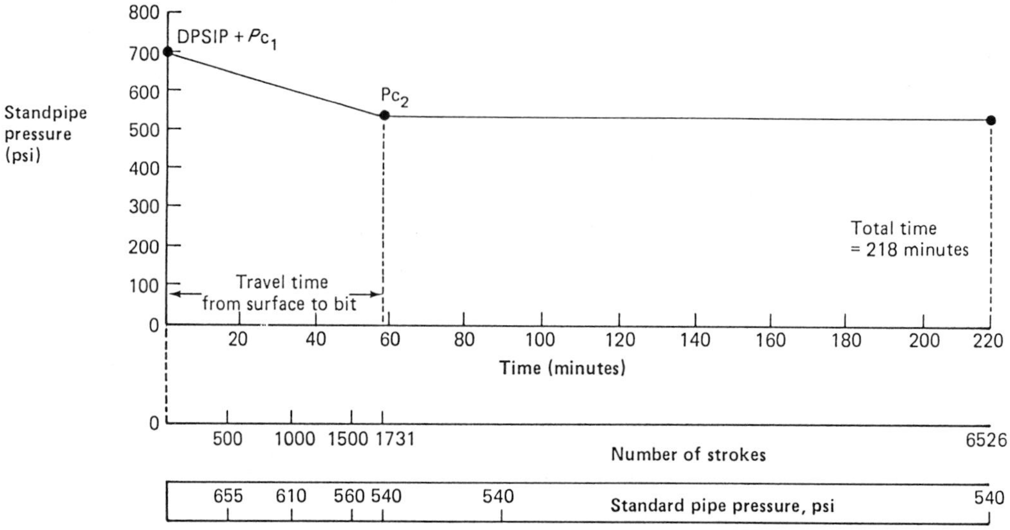

Fig. 13.21. Graphical representation of the wait-and-weight method.

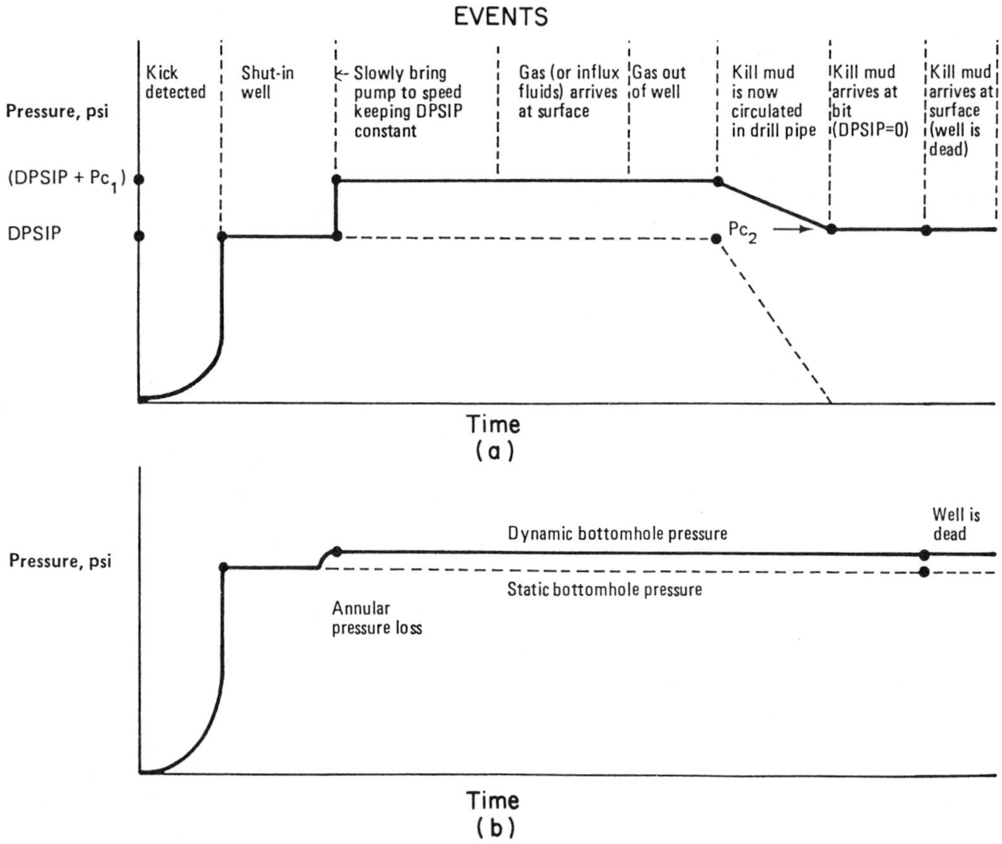

Fig. 13.22. Summary of the driller's method: (a) drillpipe pressure; (b) casing pressure.

which the standpipe pressure can be predicted as the heavy mud starts filling the well. Thus, at 500 strokes, the standpipe pressure should be 655 psi.

A table can be constructed of standpipe pressure against number of strokes, as shown in Figure 13.21.

The driller's method The driller's method uses two circulations to kill the well employing a slow pumping speed (kill speed).

In the first circulation the original mud is circulated, the drillpipe pressure being kept constant by continuously adjusting the choke setting. The initial standpipe pressure is the DPSIP plus the circulation pressure at reduced pump speed. Thus, during the first circulation, the casing pressure varies continuously and the aim is to remove all the fluid influxes. At the end of the first circulation, the CSIP should be equal to the DPSIP, since both limbs of the U-tube will have equal columns of the same mud.

In the second circulation the required heavy mud is circulated, the casing pressure being kept constant by adjusting the choke setting at the choke manifold until the drillpipe is filled with the kill mud. Thereafter, the standpipe pressure is maintained constant by varying the choke size until the kill mud arrives at the surface.

As the heavy mud starts rising up the annulus, the casing pressure is allowed to decrease gradually, the drill pipe pressure being kept constant until the heavy mud arrives at the surface. The pump is now stopped and the pressures on both drillpipe and casing should be zero. If this is not the case, the well has not been killed and shut-in pressures are noted as before and the killing procedure is repeated with a new mud weight. Figure 13.22 gives a graphical summary of the driller's method.

Example 13.3

Given the same information as in Example 13.2, determine, using the driller's method, (1) shut-in drillpipe and shut-in casing pressures at the end of the first circulation; and (2) standpipe pressure when the original mud fills the drillpipe.

Solution

Refer to Example 13.2 for the calculations of capacities, kill mud weight and formation pressure. The first circulation is performed using the initial mud weight of 75 pcf.

(1) Hydrostatic pressure of mud inside the drillpipe when circulating the original mud

$$= \frac{75 \times 10\,000}{144} = 5208 \text{ psi}$$

DPSIP = formation pressure − hydrostatic pressure of mud inside drill pipe

$$= 5408 - 5208$$

$$= 200 \text{ psi}$$

Also,

$$\text{CSIP} = 200 \text{ psi}$$

(2) Standpipe pressure after first circulation

= circulation pressure at a slow rate + DPSIP

= 500 + 200 = 700 psi

A graph similar to Figure 13.21 can be established to predict the standpipe pressures during the first circulation.

For the second circulation, the required kill mud is pumped, and the standpipe pressure will vary from an initial value of 700 psi to 540 psi when the kill mud fills the drill pipe, the 540 psi being the circulation pressure required to pump the kill mud, as calculated in Example 13.2.

From the above example, it follows that the driller's method requires two working sheets — one for the light mud and one for the kill mud.

The reader is encouraged to repeat Example 13.2 in full, using the driller's method.

References

1. API RP 53 (1978). *Recommended Practices for Blowout Prevention Equipment System.* API Production Department.
2. Fertl, W. H. (1976). *Abnormal Formation Pressures.* Elsevier, Amsterdam.
3. Kendal, H. A. (1977–78). Fundamentals of pressure control. *Petroleum Engineer International*, 6 part series, October, November and December 1977, and January, February and March 1978.

Appendix

Tables of Nomenclature and Conversion

TABLE A.1 Nomenclature

Quantity	Symbol	Units			
		Oilfield	Abbreviation	Metric	Abbreviation
Mass	m	pound-mass	lbm	kilogram	kg
Length	L (or l)	feet	ft	metre	m
Diameter	D	inches	in	millimetre	mm
Area	A	square inch	in^2	square metre	m^2
Density	ρ	pound-mass per cubic foot	lbm/ft^3	kilogram per cubic metre	kg/m^3
		or		or	
		pound-mass per gallon	ppg	kilogram per litre	kg/l
Time	t	hours, minutes, seconds	h, min, s	hours, minutes, seconds	h, min, s
Force	F	pound-force	lb	Newton	N
Pressure	P	pound-force per square inch	psi	Newton per square metre (or kilo N/m^2	N/m^2 kN/m^2)
Stress	σ	pound-force per square inch	psi	Newton per square metre	N/m^2
Strain	ε	dimensionless		dimensionless	
Velocity	V	feet per second	ft/s	metre per second	m/s
		or			
		feet per minute	ft/min		
Rotary speed	N	revolutions per minute	rpm		
Weight-on-bit	W	pound-force	lb	kilogram	kg
Poisson's ratio	ν	dimensionless		dimensionless	
Circulation rate	Q	gallons per minute	gpm	litre per minute	l/min
Yield point	YP	pound-force per 100 feet squared	$lb/100\ ft^2$	Newton per square metre	N/m^2
Plastic viscosity	PV	centipoise	CP	centipoise	cP
Reynold's number	R_e	dimensionless		dimensionless	
Temperature	T	degrees Fahrenheit	F°	degrees Celcius	°C

TABLE A2 Conversions

Metric to Imperial		*Imperial to Metric*	
		Length	
Inches	= 0.3937 × centimetres	Centimetres	= 2.54 × inches
Feet	= 3.280 84 × metres	Metres	= 0.3048 × feet
Yards	= 1.093 61 × metres	Metres	= 0.9144 × yards
Miles	= 0.621 373 × kilometres	Kilometres	= 1.609 34 × miles
		Area	
$(\text{Inches})^2$	= 0.155 × $(\text{centimetres})^2$	Centimetres^2	= 6.4516 × $(\text{inches})^2$
$(\text{Feet})^2$	= 10.7639 × $(\text{metres})^2$	Metres^2	= 0.0929 × $(\text{feet})^2$
		m^2	= acre × 4.046 873
Acres	= 2.471 05 × hectares	Hectares	= 0.404 686 × acres
		Volume	
$(\text{Feet})^3$	= 35.3147 × $(\text{metres})^3$	m^3	= 0.028 32 × $(\text{feet})^3$
Barrels (US)	= 6.289 44 × $(\text{metres})^3$	m^3	= 0.159 × barrels
Barrels per day	= 150.959 × $(\text{metres})^3$ per hour	$(\text{Metres})^3/\text{h}$	= 0:006 624 33 × bbl/day
		m^3	= gal × 3.785 × 10^{-3}
Gallons	= 0.264 178 × litres	Litres	= 3.7854 × gallons (US)
Gallon per foot	= 0.080 521 4 × litres per metre	Litres per metre	= 12.4191 × gallons per foot
		Mass	
lbm	= 2.2046 kg	kg	= 0.4536 × lbm
		Force	
lb	= 0.224 809 × newtons	N	= 4.448 22 × lb
Ton-force	= 0.907 185 × short tonnes-force (USA)	Short tonnes-force (USA)	= 1.102 31 × ton-force
Ton-force	= 1.016 05 × long tons-force (UK)	Long tons-force	= 0.984 204 × ton-force
		Density	
lbm/ft^3 (pcf)	= 0.0623 × kg/m^3	kg/m^3	= 16.04 lbm/ft^3
lbm/gal (ppg)	= 0.008 35 × kg/m^3	kg/m^3	= 119 lbm/gal
pcf	= 7.48 × ppg	kg/m^3	= 1000 kg/l
		Pressure	
lb/100 ft^2	= 2.0877 × N/m^2 (or Pa)	Pa	= N/m^2 = 0.479 lbf/100 ft^2
psi	= 0.145 × kPa	kPa	= 6.8947 × psi
psi/ft	= 0.0442 kPa/m	KPa/m	= 22.62 × psi/ft
psi	= 14.5038 × bars	Bars	= 0.068 947 6 × psi
psi/ft	= 4.4209 bar/m	Bars/m	= 0.2262 psi/ft
	1 bar = 1000 kPa; 1 kPa = 1000 Pa = 1000 N/m^2		
		Energy	
Foot-pounds-force	= 0.737 561 × joules	Joules	= 1.355 82 × foot-pounds-force
Short tonnes-force mile	= 0.684 944 × ton-force kilometre	Ton-force kilometre	= 1.459 97 × short tonnes-force mile
		Power	
Horsepower	= 0.001 341 02 × watts	Watts	= 745.7 × horsepower
		Viscosity	
$\frac{\text{lb s}}{\text{ft}^2}$	= 2.0886 × 10^{-5} × cP	1 poise	= 0.1 Pa s
		1 poise	= 100 cP
		1 cP	= 10^{-3} kg/m s = 10^{-3} Pa s
		1 cP	= $478.7896 \times 10^{2} \frac{\text{lb s}}{\text{ft}^2}$
		Temperature	
Degrees Fahrenheit	= $\frac{9}{5}$ °C + 32	Degrees Celcius	= (°F − 32) × $\frac{5}{9}$
		K	= R × $\frac{5}{9}$

Index

*Numbers in italic indicate figures (*109*).
Numbers in bold indicate tables (**135**).